AF789774

JACK PRICE
DEPT OCEAN ENGINEERING
MASS INSTITUTE / TECHNOLOGY
CAMBRIDGE MASS

11 SUSAN DRIVE
TEWKSBURY MASS 01876
PH 617-851-6921

Semiconductor Electronics

PRENTICE-HALL SOILD STATE PHYSICAL ELECTRONICS SERIES

Nick Holonyak, editor

ANKRUM
Semiconductor Electronics
BURGER & DONOVAN, eds.
Fundamentals of Silicon Integrated Device Technology:
Vol. I: Oxidation, Diffusion and Epitaxy
Vol. II: Bipolar and Unipolar Transistors
GENTRY, GUTZWILLER, HOLONYAK & VAN ZASTROW
Semiconductor Controlled Rectifiers: Principles and Applications of p-n-p-n Devices
LAUDISE
The Growth of Single Crystals
NUSSBAUM
Applied Group Theory for Chemists, Physicists and Engineers
NUSSBAUM
Electromagnetic and Quantum Properties of Materials
NUSSBAUM
Semiconductor Device Physics
PANKOVE
Optical Processes in Semiconductors
ROBERTS & VANDERSLICE
Ultrahigh Vacuum and Its Applications
VAN DER ZIEL
Solid State Physical Electronics, 2nd edition
WALLMARK & JOHNSON, eds.
Field Effect Transistors: Physics, Technology and Application
WESTINGHOUSE ELECTRIC CORPORATION
Integrated Electronics Systems

Semiconductor Electronics

PAUL D. ANKRUM

School of Electrical Engineering
Cornell University

Prentice-Hall, Inc., Englewood Cliffs, N.J.

Current printing (last digit):

10 9 8 7 6 5 4 3 2 1

SBN 13-806257-9

Library of Congress Catalog Card Number: 76-135025
Printed in the United States of America

PRENTICE-HALL INTERNATIONAL, INC., *London*
PRENTICE-HALL OF AUSTRALIA, PTY., LTD., *Sydney*
PRENTICE-HALL OF CANADA, LTD., *Toronto*
PRENTICE-HALL OF INDIA PRIVATE LIMITED, *New Delhi*
PRENTICE-HALL OF JAPAN, INC., *Tokyo*

Preface

For many years, the consideration of electronic circuits has provided an excellent opportunity for students to develop facility in the analysis and synthesis of electrical systems that include both linear and nonlinear circuit elements. Traditionally, electronics courses have introduced the student to electronic devices and have emphasized basic device principles and properties while considering their applications. However, as system consideration has become more extensive, the pressures of time have tended to reduce the attention received by electronic devices. As a result, the trend in electrical engineering curricula appears to be in the direction of courses oriented toward either electrophysics or electrical systems with limited mixing of the two. Evidence of this trend is to be found in recent textbooks that tend either to limit the consideration of electronic devices to little more than their terminal characteristics and circuit models or to emphasize devices almost to the exclusion of their applications. As a result of this trend at Cornell University, the author has found himself involved in undergraduate electronics instruction that is device-oriented with limited consideration of device applications. In general, applications are those deemed necessary to illustrate the usefulness of devices and motivate interest in their principles and properties. This type of course is believed to have an important place in present-day curricula, since an understanding of device principles is essential for judicious use of electronic systems, and a complete disregard of applications in a device course may render it somewhat sterile. Because of the difficulty encountered in finding a text having the desired relative emphasis on devices and their applications, the author began writing the supplementary notes that developed into this book.

Because the properties of semiconductor materials, junctions, and contacts are basic to those of semiconductor devices, these topics are emphasized in Chapters 1 to 3. Both valence-bond and energy-band models of materials are presented, but the latter may be omitted along with energy-level representations of junctions and contacts in Chapters 2 and 3. Such an omission does not detract appreciably from the material following Chapter 3, since the energy-band model is presented as another point of view, albeit an important one. The order of presentation of material following the consideration of diode principles and applications is somewhat arbitrary, and the author has often departed from it in using the text. For example, bipolar transistors are little more than alluded to in the field-effect transistor consideration, permitting use of Chapter 10 immediately after Chapter 4. Similarly, Chapter 12 could follow Chapter 5, since the ground work for thyristor consideration is provided by work on bipolar transistors. Also, the material on integrated circuits (Chapter 11) can be interspersed throughout the subject matter that follows the early part of Chapter 5. The author would like to believe the work on integrated circuits to be a strong feature of this book and has placed it in a separate chapter to emphasize its existence and facilitate its use.

An effort is made to show representative applications of almost every device under consideration. In general, applications emphasize the use of diodes and thyristors in rectifier and power-control circuits and the use of bipolar and field-effect transistors as amplifiers and switches. Applications selected are believed to be those that the student will find interesting, useful, and relatively easy to set up in the laboratory.

The level of this book is believed suitable for juniors who have had an introduction to electrical science or electrical engineering. It may also be used for senior-year courses in curricula too crowded at the junior level for more than an introduction to electronic devices. It has been used in mimeograph form for introductory electronics instruction in the junior year at Cornell University and in a senior-year elective course. The author is grateful for the helpful comments and suggestions of his colleagues in the School of Electrical Engineering who have had occasion to use the text and for the administrative support of Professor Joseph L. Rosson. The author is also indebted to students in the School of Electrical Engineering for their helpful criticism and gracious acceptance of the mimeographed edition.

PAUL D. ANKRUM

Contents

1 PROPERTIES OF SEMICONDUCTORS 1

1-1 Valence-Bond Models of Semiconductors 2
1-2 Conduction Processes in Semiconductors 12
1-3 Additional Electronic Processes in Semiconductors 26
1-4 Energy-Band Models of Semiconductors 36
References 60

2 PROPERTIES OF *pn* JUNCTIONS 64

2-1 The Nature of *pn* Junctions 64
2-2 Energy-Level Representation of *pn* Junctions 77
2-3 The Diode Equation 83
2-4 Junction Reverse-Saturation Current 84
2-5 Electron and Hole Currents in the *p*- and *n*-Regions 88
2-6 Junction-Region Width and Capacitance 94
2-7 Diffusion Capacitance 103
2-8 Junction Resistance 106

3 METAL-SEMICONDUCTOR CONTACTS 110

3-1 Rectifying *n*-type Semiconductor Contacts 110
3-2 Rectifying *p*-type Semiconductor Contacts 115
3-3 Ohmic Contacts 117

4 SEMICONDUCTOR DIODES 121

4-1 Junction Diodes 121
4-2 The Point-Contact Diode 129
4-3 The Hot-Carrier Diode 131
4-4 Metallic Rectifiers 133
4-5 Conductivity Modulation 136
4-6 Diode Switching Transients 137
4-7 Light-Sensitive and Light-Emitting Diodes 145
4-8 Tunnel Diodes 149
4-9 Backward Diodes 156
4-10 Diode Rectifier Circuits 158
4-11 Polyphase Rectifiers 165
4-12 The Shunt-Capacitor Filter 171
4-13 Regulated Power Supplies 177
4-14 The Diode Limiter 179
4-15 Diode Clamping Circuits and a-c Electronic Voltmeters 181

5 THE BIPOLAR TRANSISTOR 188

5-1 Types of Bipolar Junction Transistors (BJTs) 189
5-2 BJT Characteristics 192
5-3 Alpha of BJTs 220
5-4 The Drift-Field BJT 233
5-5 The Double-Diffused BJT 235
5-6 Comparison of Alloy and Diffused BJTs 236

6 BJT EQUIVALENT-CIRCUIT MODELS AND PARAMETERS 241

6-1 Two-Port Network Representations 241
6-2 BJT h-Parameters Representations 245
6-3 Two-Port Network-Parameter Conversion Equations 250
6-4 High-Frequency h-Parameter Representation 255
6-5 The Hybrid-π Equivalent Circuit 263
6-6 Equivalent-Circuit Parameters 274

7 BASIC BJT AMPLIFIER STAGES 293

7-1 The Three Configurations 293
7-2 Comparison of the Basic Configurations 297
7-3 Stages in Cascade; Iterative Gain 305
7-4 Power Gain 306

8 **BJT BIASING NETWORKS** 308

8-1 Bias-Stabilizer Analysis and Design 308
8-2 Bias-Stabilized Emitter Follower 315
8-3 Additional Bias-Stabilization Methods 317
8-4 Junction Temperature and Thermal Stability 320
8-5 Operating-Point Selection 324
8-6 The Difference Amplifier 325
References 327

9 **THE BJT AS A SWITCH** 332

9-1 Switching Modes 332
9-2 Transient-Response Characteristics 338
9-3 Nonregenerative Switching Circuits 342
9-4 Regenerative Switching Circuits 348

10 **FIELD-EFFECT TRANSISTORS (FETs)** 362

10-1 The Junction Field-Effect Transistor (JFET) 363
10-2 The Insulated-Gate Field-Effect Transistor (MOSFET) 379
10-3 Small-Signal (Linear) Amplifier Stages 391
10-4 Bias Arrangements 398
10-5 The FET as a Switch 408
References 415

11 **INTEGRATED CIRCUITS** 420

11-1 Types of Integrated Circuits 421
11-2 Monolithic Integrated Circuits and Their Isolation 422
11-3 Fabrication of Single-Crystal Monolithic Circuits 426
11-4 Fabrication of Dielectric-Isolated Monolithic Circuits 435
11-5 Integrated Resistors 439
11-6 Integrated Capacitors 443
11-7 Integrated Inductors 447
11-8 Monolithic Diodes 447
11-9 Monolithic Transistors 450
11-10 Integrated Field-Effect Transistors 453
11-11 MOS Integrated Circuits 467
11-12 Linear Integrated Circuits 474
References 487

12 THYRISTORS AND RELATED DEVICES 490

12-1 The Unijunction Transistor (UJT) 490
12-2 Applications of UJTs 498
12-3 The *pnpn* Diode 502
12-4 Applications of *pnpn* Diodes 506
12-5 The Silicon Controlled Recifier (SCR) 509
12-6 Applications of SCRs 514
12-7 The Gate-Controlled Switch (GCS) 526
12-8 The Bidirectional Thyristor 530
12-9 Miscellaneous Thyristors 533
References 538

INDEX 541

Semiconductor Electronics

1

Properties of Semiconductors

The electrical properties of semiconductors are of interest because they are fundamental to the operation of transistors, integrated circuits, and related devices.

Semiconductors are solids that have resistivities intermediate to those of metallic conductors and insulators. To cite typical examples, copper has a resistivity of the order of 10^{-6} ohm-cm; the resistivity of mica, an excellent insulator, is of the order of 10^{16} ohm-cm; pure germanium, an extremely useful semiconductor, has a resistivity of approximately 50 ohm-cm at room temperature.

Semiconductor devices have an interesting history, in that they were used in radio communication prior to the development of the vacuum tube by which they were replaced. However, as operating frequencies became higher, the interelectrode capacitance and transit time of vacuum tubes began to present serious problems. Because of their superiority at high frequencies, semiconductor diodes thereupon began replacing vacuum diodes in precisely the same applications in which they had previously been superseded. This revival of interest in semiconductor electronics led to the development of a host of devices that have replaced vacuum tubes in most applications.

1-1 VALENCE-BOND MODELS OF SEMICONDUCTORS

There are two significantly different ways of considering the electrical properties of semiconductors. These are the *valence-bond model,* which presents a visual picture of the conduction mechanism, and the *energy-band model.* Each point of view has merit, the valence-bond picture being convenient for qualitative purposes and the energy-band viewpoint lending itself more readily to quantitative consideration.

1-1.1 Atomic Arrangements in Solids

Atoms of matter may be pictured as miniature solar systems with positive nuclei surrounded by electrons in sufficient numbers to make each complete

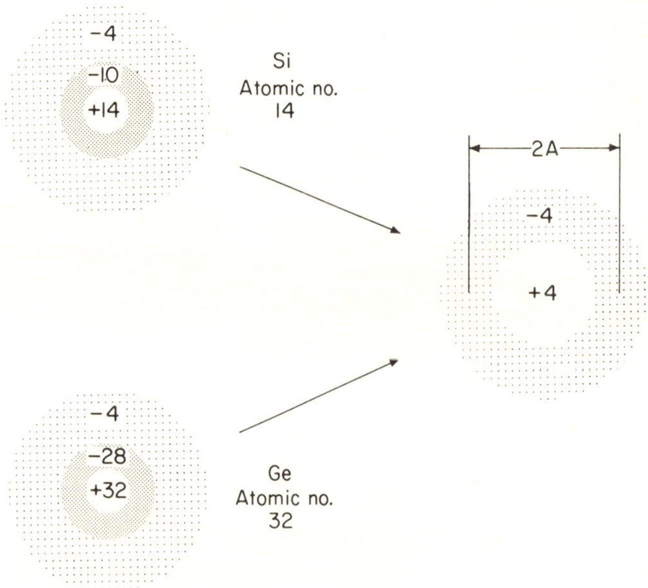

Fig. 1-1. Schematic pictures of atoms (nuclei shown greatly oversize). (a) Silicon and germanium, showing nuclei and electron orbit regions. (b) Simplified schematic in terms of an ionic core and valence electrons. The approximate diameter in angstroms of valence-electron orbits is indicated. (From Adler, Smith and Longini, *Introduction to Semiconductor Physics*, John Wiley and Sons, Inc., New York, 1964.)

atom electrically neutral, as shown in Fig. 1-1(a). The electrons of an atom are arranged in orbits about the nucleus, the electrons encircling the nucleus as planets encircle the sun. Substances differ in the charge of the nuclei and the number of neutralizing electrons. For example, nuclei of silicon and germanium, two widely used semiconductors, have charges of +14 and +32, respectively. The number of electrons in the outer orbit of each atom of a substance is the valence of the substance, both germanium and silicon having a valence of 4. Only the valence electrons are able to enter into chemical reactions, the remaining electrons being closely bound to the nucleus. These electrons, in conjunction with the nucleus, produce a stable ionic core having a net positive charge. Thus atoms of germanium and silicon may be represented by such cores, each having a charge of +4 units and four associated electrons, as shown in Fig. 1-1(b).

The phase (solid, liquid, or gaseous) of a particular substance depends upon the separation of the atoms of the substance. If the atoms are far apart, compared to the diameter of their valence-electron orbit, they have relatively little influence upon one another and form a gas. However, if the separation of the atoms is decreased, say by reducing the temperature, so that it becomes comparable to their outer-orbit diameter, the substance may become a liquid or a solid.

When the atoms of a substance form a solid, their outer orbits may overlap and be distorted by the proximity of the atoms. In extreme situations, as in metallic conductors, the atoms are so closely packed that the outer-orbit electrons of an atom are about as close to the nucleus of a neighboring atom as to their own nucleus. As a result, it is no longer possible to say which electrons are associated with which atoms, and the valence electrons are shared by all the atoms. Such electrons are referred to as *free*, or *unbound*, because they are free to move throughout the interior of the metal under the action of applied forces. The general movement of free electrons, when a voltage sets up an electric field in the metal, constitutes an electric-current flow. In the absence of an applied force, the electrons assume an erratic motion in every possible direction, moving at velocities dependent upon their thermal energy. The density of free electrons in a metal is of the order of the density of the atoms, 10^{23} per cubic centimeter, and is independent of temperature over a wide temperature range.

In contrast to metallic conductors, insulating materials have atomic separations such that almost all of the valence electrons remain bound to their parent atoms. As a result, few charges are available to move through the material in the presence of an applied electric field, and it is an extremely poor electrical conductor. This is true over a wide temperature range.

The atomic spacing in semiconductors is intermediate to that of metals and insulators, resulting in materials that are neither good conductors nor good insulators at room temperature. However, the conductivity of semiconductors is highly temperature-dependent, the materials being insulators at very low temperatures and rather good conductors at high temperatures.

The atoms of semiconductors such as germanium or silicon, and of metals, are arranged in a three-dimensional symmetrical periodic array known as a *crystal structure*. The structure is periodic in that a representative unit is repeated at regular intervals along any direction in the crystal. If this is true throughout a specimen of the material, it is said to be single-crystal material, If it occurs only in small randomly oriented grains that make up the specimen, the material is polycrystalline. Most semiconductor devices require the use of single-crystal material.

1-1.2 Intrinsic Semiconductors

Electrical conductivity in a semiconductor occurs either by means of charges that are released by atoms of the parent crystalline material or by means of those present as a result of foreign elements (impurities), or by a combination of the two effects. If impurities have negligible effect on conductivity, the material is an *intrinsic semiconductor*, since conductivity is an intrinsic property of the parent material. If the conductivity occurs largely because of impurity content, the material is an *extrinsic semiconductor*.

The atoms of a crystal of germanium or silicon are arranged in a *diamond structure*, as shown by Fig. 1-2. Each atom forms covalent or electron-pair

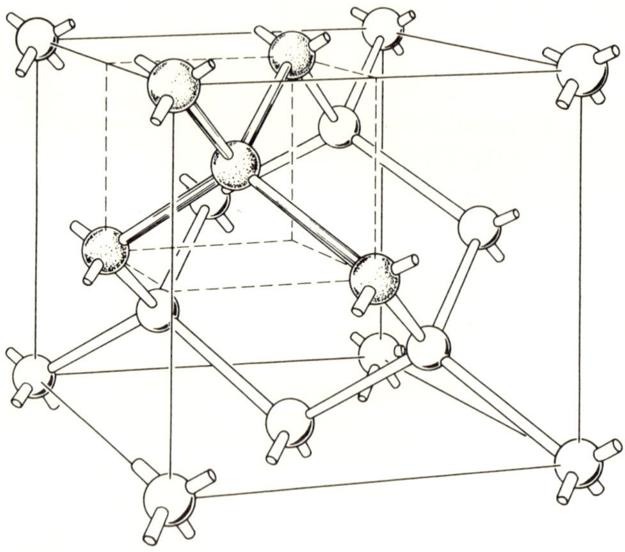

Fig. 1-2. Diamond crystal structure showing how each atom forms four bonds with neighboring atoms. (From W. Shockley, *Electrons and Holes in Semiconductors*, D. Van Nostrand Inc., Princeton, N.J. 1950.)

bonds with four neighboring atoms, as shown in the dotted cube. The covalent bond, commonly referred to as a *valence bond,* is an arrangement in which a pair of atoms share two valence electrons. The connecting tubes in Fig. 1-2 may be thought of as tracks along which two electrons travel back and forth between associated atoms. Valence bonds act as structural elements that hold together atoms of the diamond crystal structure. A simplified two-dimensional representation of the three-dimensional array is shown for germanium in Fig. 1-3, valence bonds being represented by pairs of adjacent electrons.

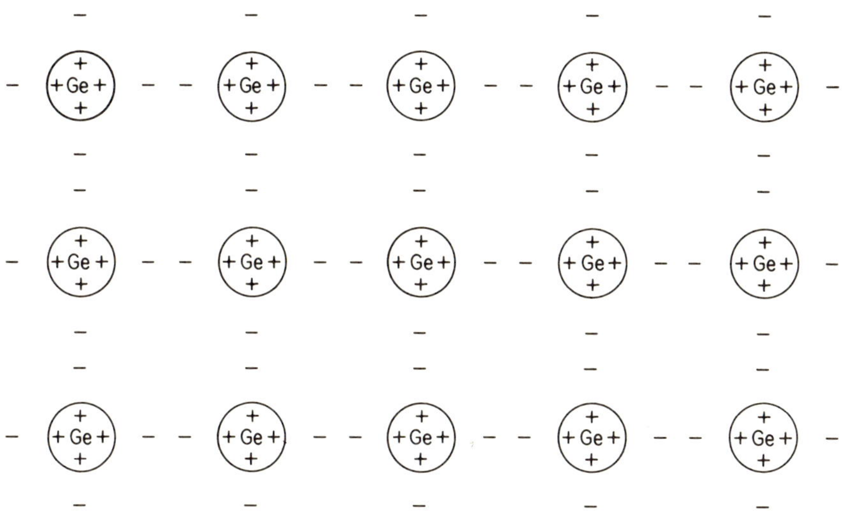

Fig. 1-3. Arrangement of atoms and valence bonds in a crystal of pure germanium at very low temperatures.

Figure 1-3 shows the arrangement of atoms and valence bonds for pure germanium at very low temperatures. At these temperatures all electrons are bound in the valence bonds, and the material is an insulator. At higher temperatures, however, thermal vibration of atoms in the crystal lattice results in some of the electrons being shaken loose. The phenomenon, known as valence-bond disruption, is an ionization process that produces free electrons which move about through the crystal in a random manner. The energy E_g required to break a valence bond is a property of the substance, being about 0.7 eV for germanium and 1.1 eV for silicon.* When a disruption occurs, a vacancy exists in the lattice structure at the spot vacated by the freed electron. This vacancy is referred to as a *hole* and is considered to have a positive charge of one unit, since its presence causes the atom to become a positively charged ion. This situation is shown in Fig. 1-4. Because of its

* An electron volt (eV) is the energy gained by an electron in falling through a potential difference of 1 volt, being 1.6×10^{-19} joule.

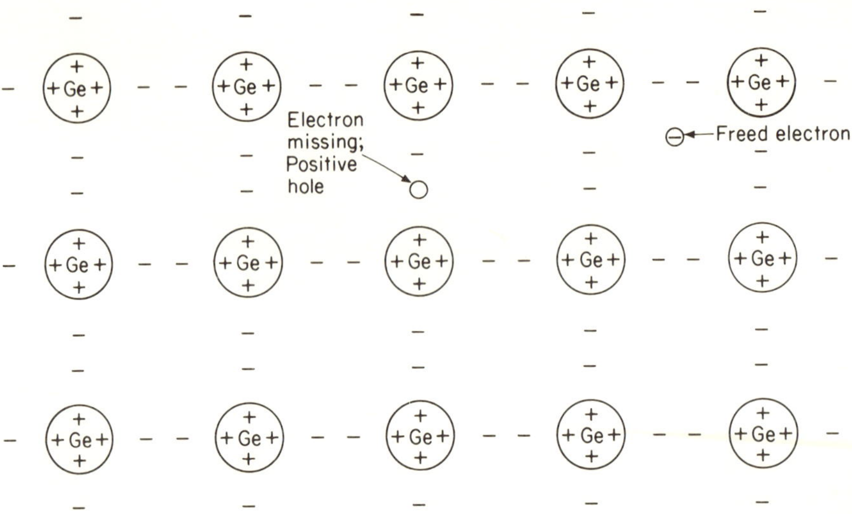

Fig. 1-4. Electron-hole pair caused by disruption of a valence bond.

positive charge, the ion may "rob" an adjacent atom of an electron and in so doing transfer the positive hole to a new position in the lattice structure. Thus the hole, like the freed electron, moves about the crystal in a random manner. It is important to recognize that *the ionized atom itself does not move*, but that the state of ionization is transferred from one atom to another by motion of bound electrons *between* their covalent bonds.

The hole and an electron may recombine (and disappear) with an accompanying release of energy in an amount (a photon) equal to that required to disrupt a valence bond. As electron-hole pairs are produced by increased thermal agitation, the increased densities of electrons and holes increase the likelihood of their recombination. When equilibrium conditions are established at a particular temperature, the densities of electrons and holes are such that they recombine at the rate at which they are produced by valence-bond disruption.

When an electric field is applied to an intrinsic semiconductor, the random motion of freed electrons acquires a systematic drift in the direction opposite that of the field (toward the positive terminal) in the same manner as free electrons in a metal. For a particular electric field, the electron-current density in a semiconductor is much smaller than that in a metallic conductor, owing to the smaller electron density. For example, in germanium at room temperature one electron is set free for approximately every 10^9 atoms, as compared to approximately one free electron per atom in a metal. In intrinsic silicon, the corresponding electron density is about three orders of magnitude lower than that of germanium, owing the larger energy required for disruption of a silicon valence bond.

It should not be concluded that the conduction process in a semiconductor is identical to that in a poorly conductive metal. A fundamental difference in the two types of material comes to light when we consider the effect of an applied electric field on the holes of a semiconductor. In the absence of such a field, an atom having a hole (that is, having become ionized by losing an electron) may fill the hole at the expense of one of its neighboring atoms. Therefore, the hole may be transferred in any direction and moves about in a random manner, as already indicated. In a moderate electric field, a force is developed on bound (valence) electrons that is insufficient to free them but sufficient to increase the likelihood of their filling holes in adjacent atoms in the direction of the field force. Therefore, the random motion of the holes acquires a systematic drift in the direction of the field. When a hole reaches the negative electrode, an electron moves from the electrode into the semiconductor, filling the hole and neutralizing the positive ion. This results in an electron shift throughout the circuit external to the semiconductor and an electron deficiency at the positive electrode. This deficiency (electric field) causes an electron to flow from the semiconductor to the electrode, creating a new positive hole to act as a current carrier.* Thus the hole behaves as though it were a new free particle with a positive electronic charge and a mass comparable to that of an electron. Electric current flow in an intrinsic semiconductor is a combination of electron current and hole current, the currents adding since the direction of electron flow in the circuit external to the semiconductor is the same whether caused by hole flow or electron flow within the semiconductor.

The explanation of the motion of a hole in an electric field by considering the field to push a particular valence electron in the direction opposite that of the field to move the hole in the direction of the field is an oversimplification that can lead to erroneous conclusions. In reality, hole motion is produced by a group af bound electrons, and it is often helpful to think of a hole moving through valence electrons as a bubble moves through a liquid.

Electrical neutrality. When ionization takes place in an intrinsic semiconductor, some individual atoms (those ionized) are no longer electrically neutral. Nevertheless, the crystal remains neutral as a whole, since free electrons and holes exist in equal numbers. Electrical or *space-charge neutrality* is one of the most important principles involved in the use of semiconductor devices. In brief, this principle states that the total number of positive charges in any region of a semiconductor must equal the total number of negative charges in the same region, provided that there are no large potential differences within the region.

* Those familiar with gaseous discharges will recognize a similarity between electric current due to hole flow in a semiconductor and that due to positive-ion flow in a gas.

1-1.3 Extrinsic Semiconductors

The most useful semiconductor materials are those produced by adding controlled amounts (of the order of a few parts per million) of appropriate impurities to pure material. In this process, known as *doping*, impurity atoms displace silicon or germanium atoms from their regular sites in the lattice structure. If the doping material is from the group of elements having a valence of 5—antimony, for example—four of the valence electrons form covalent bonds with adjacent silicon or germanium atoms. The fifth valence electron cannot fit into the bond structure, and, because of this, the ionization energy required to set it free is very small compared to that required to break a valence bond. As a result, at temperatures of 100°K or above the probability is high (0.95 or greater for germanium) that the electron is shaken free to become a conduction electron, as shown in Fig. 1-5. It is

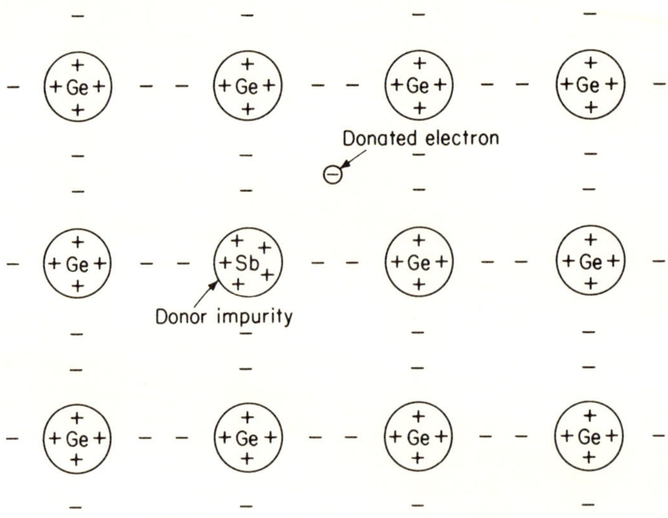

Fig. 1-5. Production of *n*-type germanium by the addition of impurity atoms having a valence of 5.

important to recognize that the positive ion left behind by the freed electron has the same number of electrons around it as do each of the host semiconductor atoms. Therefore, no vacancy exists in the bond structure; that is, the five-valence impurity atom has donated a free electron to the crystal *without producing a hole*. For this reason, such atoms are referred to as *donors*.

The positive donor ion is a charge that is fixed in the lattice and *cannot help carry current*. When an electric field is applied, current flow occurs as a result of the drift motion of donated electrons and of electrons and holes produced by valence-bond disruption. In practical semiconductor material, the donor-atom density is such that, over a wide temperature range, electrons are much more numerous than holes and are responsible for most of the current. Such material is said to be *n-type*, because the current carriers are negative charges. In n-type material, the electrons are referred to as *majority carriers* and the holes as *minority carriers*. The crystal as a whole is electrically neutral, since the electron density n_n is equal to the sum of the hole density p_n and the density of positive donor ions.

Minority carrier suppression. When the presence of donor atoms increases the electron density of a semiconductor, the likelihood of a particular hole's recombining with an electron is increased. Since, for a particular temperature, the recombination rate is unaffected by the presence of the donor atoms, *the increased electron density reduces the hole density below that of pure semiconductor material*. When energy-band models of semiconductors are considered, it will be shown that the product of the electron density and the hole density is a constant for a particular temperature and equal to the square of the electron density (or hole density) of pure material at the same temperature; that is,

$$n_n p_n = n_i^2 \tag{1-1}$$

where n_n and p_n are electron and hole densities, respectively, in n-type material and n_i is the intrinsic carrier density.

Acceptor impurities. If pure semiconductor material is doped with material from the group of elements having a valence of 3, indium for example, each impurity atom has only three valence electrons to form covalent bonds with adjacent atoms of the host material. As a result, a vacancy exists in the lattice structure. Because the bonding forces in the crystal are very large compared to the electrostatic forces on the electrons, the vacancy of the impurity atom is attractive to other bound electrons. Consequently, the ionization energy required to cause another bound electron to move into this vacancy is very small compared to that required to break a valence bond. As a result, at relatively low temperatures (say, 100°K) the probability is high that the impurity-atom valence bond is filled and a hole made available for conduction, as shown in Fig. 1-6. Since the impurity atom accepts an electron from a host atom, it is called an *acceptor*. When this occurs, the acceptor atom becomes negatively charged, but cannot contribute to conductivity, since it is fixed in the lattice. Current flows as a result of the drift motion (in an applied electric field) of holes created by acceptors and of holes and electrons produced by valence-bond disruption. When the doping level and temperature are such that positive holes are responsible for most of the current, the material is said to be *p-type*. Like n-type material, the

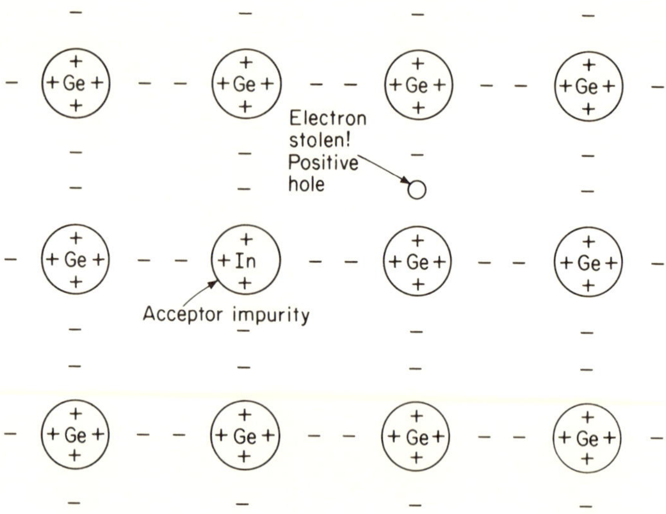

Fig. 1-6. Production of p-type germanium by addition of impurity atoms having a valence of 3.

crystal as a whole is electrically neutral, since the hole density p_p is equal to the sum of the electron density n_p and the density of negative acceptor ions.

Increasing hole density of a semiconductor by acceptor-atom doping results in suppression of the density of minority electrons. For p-type material, the relationship between the majority-carrier density p_p and the minority-carrier density n_p is

$$p_p n_p = n_i^2 \tag{1-2}$$

where n_i is the intrinsic-carrier density at the same temperature.

Intrinsic temperature. If the temperature of either n-type or p-type semiconductor material becomes large enough, carriers generated by valence-bond disruption outweigh those due to impurities, and conduction is essentially the same as if no impurities were present. When this occurs, the material is said to have become intrinsic or to have reached its intrinsic temperature. This occurrence places an upper limit on the operating temperature of semiconductor devices, since all are composed of extrinsic material. Clearly, this limit depends upon the impurity concentration or doping level of the material. Typically, this temperature is of the order of 100°C for germanium devices and 200°C for those of silicon.

Semiconductor-type determination. Because doped semiconductors are extremely pure materials by most purity standards, there is no difference in the physical appearance of intrinsic material and extrinsic materials, all being metallic-appearing crystalline solids *that are quite brittle.* It is often necessary to determine whether an extrinsic sample is p-type or n-type. There are several ways of doing so, but the most convenient test makes use of the thermoelectric properties of semiconductors. This is the *hot-probe method,* which we shall now briefly consider.

The hot-probe arrangement for determining semiconductor type is shown in Fig. 1-7. The hot probe, typically a pencil-type soldering iron, is brought

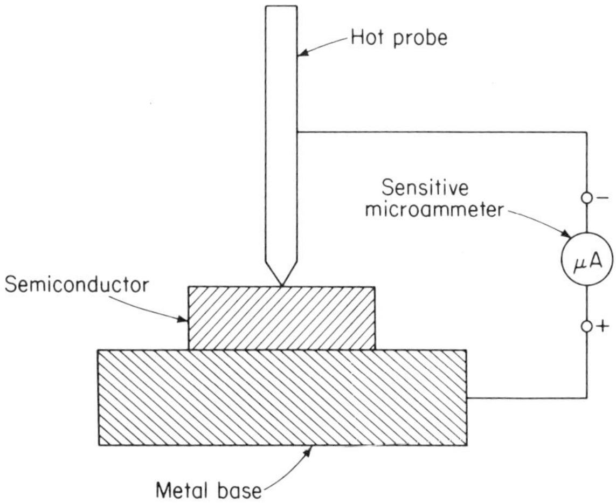

Fig. 1-7. Hot-probe method for determining semiconductor type. Upon momentary application of the probe, the microammeter deflects up-scale for p-type material and down-scale for n-type material.

in momentary contact with the semiconductor, and the direction of deflection of the microammeter pointer is observed. We shall not attempt a detailed analysis of the potential differences developed at the contacts and along conductors having temperature gradients, but shall merely observe that momentary heating of a point on the semiconductor surface increases the thermal velocity of majority carriers near the probe above that of carriers near the cold contact. This causes the random motion of carriers away from the probe to be greater than that toward it. Consequently, impurity ions near the probe are no longer completely neutralized by majority carriers, causing the probe to be positive (with respect to the metallic contact) for n-type material and negative for p-type material. A microammeter having a full-scale sensitivity of the order of 30 μA usually yields satisfactory results in this arrangement.

In summary, we see that the presence of three-valence (acceptor) impurity atoms in a semiconductor crystal at ordinary temperatures not only creates holes without producing electrons, resulting in p-type material, but also reduces the electron density of the material. Similarly, the presence of five-valence (donor) impurities produces electrons without creating holes, suppressing the hole density and resulting in *n*-type material. Commonly used acceptor materials are boron (B), aluminum (Al), gallium (Ga), and indium (In). Donor materials are phosphorus (P), arsenic (As), and antimony (Sb).

1-2 THE CONDUCTION PROCESS IN SEMICONDUCTORS

Perhaps the most important characteristic of semiconductors is their ability to carry current. Since electric current is the rate at which charge moves across a surface in the direction normal to it, the current depends on the number of charges free to move and the velocity with which they move.

There are two basic processes by which charges are transported through semiconductors. These are *drift* due to an electric field and *diffusion* due to variation in the charge density from point to point within the material. The familiar electric current resulting from charge drift in an electric field is known as *drift current* to distinguish it from the less familiar *diffusion current.*

1-2.1 Drift Current, Carrier Mobility, and Electrical Conductivity

The behavior of charges in a solid in an applied electric field is quite different from that in an evacuated medium. Since electrons and holes may be thought of as charged particles having effective mass, it might appear that they could move through a periodic array of crystal atoms without colliding with them, much as electrons move in a vacuum. This could occur in a perfect crystal if the atoms were stationary at their equilibrium positions in the lattice. However, at ordinary temperatures the thermal vibrations of atoms about their equilibrium positions are responsible for charges colliding with, or being *scattered* by, crystal atoms.

Thermal vibrations result in atoms being squeezed together or pulled apart over distances that may be comparable to the distance between their equilibrium positions in the lattice. These displacements alter the electric potential in the neighborhood of the atoms, resulting in the existence of electric fields that change the direction and magnitude of the momentum of moving charges. This phenomenon is known as *thermal* or *lattice scattering.*

A second important type of collision or scattering results from the presence of charged impurity atoms such as donor or acceptor ions. These fixed charges in the lattice deflect moving charge carriers, also changing the direction and magnitude of their momentum. This is known as *impurity scattering.*

Other types of scattering result from the presence of electrically neutral impurity atoms in the crystal and from the interactions of charge carriers. These are usually second-order effects *as far as majority carriers are concerned.*

In the absence of an applied electric field, the changes in momentum of randomly moving charge carriers by thermal and impurity scattering neutralize one another, and the net current is zero. In an applied field, however, carriers drift in the direction of the field force, experiencing momentum-changing collisions that limit their drift velocities and result in a steady current.* Therefore, as far as the drift velocity is concerned, we may think of the carriers as being accelerated from rest, in free flight, between inelastic collisions in which they lose the energy gained between collisions. This energy loss is the Joule heating associated with current flow through a resistance.

Carrier mobility. On the basis of the model just suggested, the effect of an electric field E on a charge e of effective mass m is to produce an acceleration a given by

$$a = \frac{eE}{m} \tag{1-3}$$

For moderate fields, the increase in total velocity of the charge is very slight and the time t between collisions is virtually unaffected by the field. Therefore, the velocity of the charge due to the field at the end of the free-flight period t is

$$v = at = \frac{eEt}{m} \tag{1-4}$$

where t is a property of the material and its operating temperature. Since, upon collision, the charge loses the velocity gained as a result of the field, the average velocity is

$$v = \frac{eEt}{2m} \tag{1-5}$$

and this is the drift velocity of the charge through homogeneous material.

Since, but for v and E, all terms in Eq. (1-5) are constants for a particular material and charge type, drift velocity v_d is seen to be proportional to the applied electric field. Thus Eq. (1-5) may be written as

$$v_d = \mu E \tag{1-6}$$

where μ is the *drift mobility* of the charges. From Eq. (1-6), we see that

* If the field produced a steady acceleration rather than a drift speed, the effect would be an ever-increasing current with an ever-increasing power input—an obvious impossibility.

drift mobility is drift velocity per unit electric field. Thus the dimension of μ, in the mks system, is meters per second per volt per meter or $(\text{meters})^2$ per volt-second. The dimension $(\text{cm})^2$ per volt-second is also in common use.

Experimental investigation discloses that carrier mobility is indeed constant over a broad electric-field range for a particular material operated at a fixed temperature. It has been found that the drift velocity of carriers in silicon varies directly with the electric field for field values up to about 1000 V per cm. For larger field values, the field begins to have significant effect on the time between collisions and drift velocity varies as the square root of the field. For still larger fields, the drift velocity has been found to "saturate" and be essentially independent of the field strength.

Drift-current density. In a semiconductor having an electron desntiy n in a volume of cross-section area A and length l, the total charge Q_t of the electrons in the volume is

$$Q_t = neAl \tag{1-7}$$

where e is the electronic charge. If t is the time required for Q_t to pass area A in the direction normal to it, as a result of an electric field E normal to A, the current I_n due to electron flow is given by

$$I_n = \frac{Q_t}{t} = \frac{neAl}{t} = neAv_d \tag{1-8}$$

where v_d is the drift velocity of the electrons. The current density J_n is seen to be

$$J_n = \frac{I}{A} = nev_d \tag{1-9}$$

which, in conjunction with Eq. (1-6), becomes

$$J_n = ne\mu_n E \tag{1-10}$$

where μ_n is the mobility of electrons in the semiconductor.

If the semiconductor has a hole density p, the electric-current density J_p due to hole flow is

$$J_p = pe\mu_p E \tag{1-11}$$

where μ_p is the mobility of holes in the material. Since the total current is the sum of the electron and hole currents, the *drift current* density J is seen to be

$$J = eE(n\mu_n + p\mu_p) \tag{1-12}$$

This expression is seen to be a form of Ohm's law, since current is proportional to J, applied voltage is proportional to E, and $(n\mu_n + p\mu_p)$ is a constant for the material at a particular temperature.

Electrical conductivity. The electrical conductivity σ of a material (the reciprocal of the resistivity ρ) is the conductance between the opposite surfaces of a unit cube of the material. Since the current through a unit cube is the current density J and the voltage between opposite surfaces is the voltage per unit length which (by definition) is the electric-field intensity E, we see that conductivity is the current density per unit electric field; that is,

$$\sigma = \frac{J}{E} \tag{1-13}$$

From Eqs. (1-12) and (1-13) we obtain, for semiconductors,

$$\sigma = e(n\mu_n + p\mu_p) \tag{1-14}$$

Equation (1-14) is a general expression that is applicable to any type of material. For *intrinsic material* it may be written

$$\sigma_i = en_i(\mu_n + \mu_p) \tag{1-15}$$

since the hole and electron densities are equal. For *n-type material* the conductivity expression becomes

$$\sigma_n = e(n_n\mu_n + p_n\mu_p) \cong n_n e\mu_n \tag{1-16}$$

since the hole contribution to the conductivity of n-type material is usually negligible. For *p-type* material, Eq. (1-16) becomes

$$\sigma_p = e(p_p\mu_p + n_p\mu_n) \cong p_p e\mu_p \tag{1-17}$$

In the preceding equations, σ is in mhos per meter, or (ohm-meters)$^{-1}$, for e in coulombs, n and p in carriers per cubic meter, and μ in (meters)^2 per volt-second.

Carrier mobilities in semiconductors.[1,2] From the preceding discussion, we see that electrical conductivity depends upon the carrier density-drift mobility product. For metals and extrinsic semiconductors, the carrier density is essentially constant over a broad temperature range and the conductivity variation with temperature is simply that of the drift mobility. If the carrier density of an extrinsic semiconductor is known from information about the impurity concentration, the drift mobility may be determined by measuring the conductivity. Experimental values of the drift mobility of holes and electrons in germanium and silicon are given in Table 1-1 for a temperature of 300°K. Holes are seen to have different mobilities than electrons in the same material, owing to their larger effective mass, and both holes and electrons are less mobile in silicon than in germanium.

[1] E. M. Conwell, "Properties of Silicon and Germanium," *Proc. IRE*, Vol. 40, No. 11 (November 1952).

[2] E. M. Conwell, "Properties of Silicon and Germanium II," *Proc. IRE*, Vol. 46, No. 6 (June 1958).

The temperature dependence of the drift mobility of carriers in germanium and silicon is shown by the empirical expressions of Table 1-2. These expressions and the mobility values of Table 1-1 were obtained by using samples pure enough that the effect of impurities on mobilities (that is, impurity scattering) was negligible. They therefore represent lattice mobilities and are the upper limits of carrier mobilities in practical semiconductors.

The mobility-temperature expressions of Table 1-2 indicate that lattice mobility decreases with increase in temperature, which is as expected, since thermal velocities of carriers and thermal vibrations of atoms increase with temperature. This decreases the time between collisions of carriers with lattice atoms, decreasing the velocity acquired by carriers between these collisions.

Carrier mobility in an extrinsic semiconductor may be markedly affected by impurities, depending on their concentration. Figure 1-8 shows a plot of

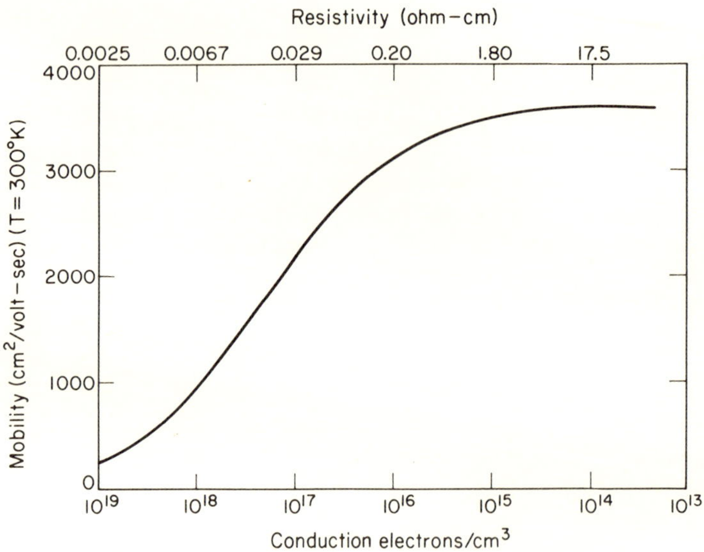

Fig. 1-8. Electron mobility vs. carrier density and resistivity for *n*-type germanium at 300°K. (From Reference 1, p. 1331.)

the mobility of electrons in *n*-type germanium at 300°K as a function of the electron density in, and the resistivity of, the material. Since the intrinsic carrier density of electrons in germanium at 300°K is of the order of 10^{13} cm^{-3}, and the resistivity is about 50 ohm-cm, the highest mobility shown in the plot is that for electrons in intrinsic material It is of interest to note that this mobility is somewhat lower than that of Table 1-1, suggesting that improvements in the quality of materials occurred between the years 1952 and 1958.

Table 1-1* Experimental values of drift mobility in cm^2/volt-sec for very pure material at 300°K

Carrier	*Germanium*	*Silicon*
Electron	3900 ± 100	1350 ± 100
Hole	1900 ± 50	480 ± 15

* Reference 2, p. 1284.

Table 1-2* Temperature dependence of drift mobility in cm^2/volt-sec for very-pure material

Carrier	*Germanium*		*Silicon*	
	MOBILITY	TEMPERATURE RANGE	MOBILITY	TEMPERATURE RANGE
Electron	$4.9 \times 10^7 T^{-1.66}$	100–300°K	$(2.1 \pm 0.2) \times 10^9 T^{-2.5 \pm 0.1}$	160–400°K
Hole	$1.05 \times 10^9 T^{-2.33}$	125–300°K	$(2.3 \pm 0.1) \times 10^9 T^{-2.7 \pm 0.1}$	150–400°K

* Reference 2, p. 1284.

For electron densities in Fig. 1-8 that are large compared to the intrinsic carrier density, the electron density is essentially equal to the impurity concentration, owing to 100 per cent donor ionization. It is seen that impurity scattering reduces mobility significantly for concentrations that result in resistivities below about 1 ohm-cm. The effect of donor-impurity concentra-

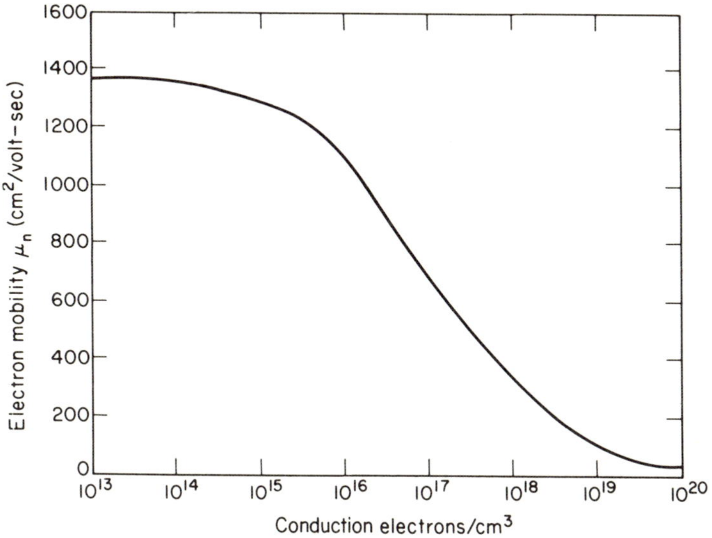

Fig. 1-9. Mobility vs. carrier concentration for electrons in silicon at 300°K. (Adapted from Reference 2, p. 1287.)

tion on the mobility of electrons in n-type silicon is shown in Fig. 1-9. Since the intrinsic carrier density in silicon is of the order of 10^{10} cm^{-3} at 300°K, the electron density may be considered to be the donor-ion density. Relationships similar to those of Figs. 1-8 and 1-9 exist between the mobility of holes in p-type material and the concentration of acceptor-impurity atoms.

The effect of temperature on impurity scattering is just the opposite of that previously described for lattice scattering, since the lower the speed of a charge carrier, the greater the deflecting effect of an impurity ion. Therefore, as temperature is decreased, the lower thermal velocities of the charge carriers result in increased impurity scattering. Since lattice and impurity scattering are affected oppositely by temperature change, a mobility versus temperature plot may have a maximum at intermediate temperatures, depending on the impurity concentration. Data showing the effect of temperature on the mobility of electrons in n-type material for several donor-atom concentrations are presented in Fig. 1-10. It is seen that the lower the temperature, the more prounced the effect of impurity concentration on mobility.

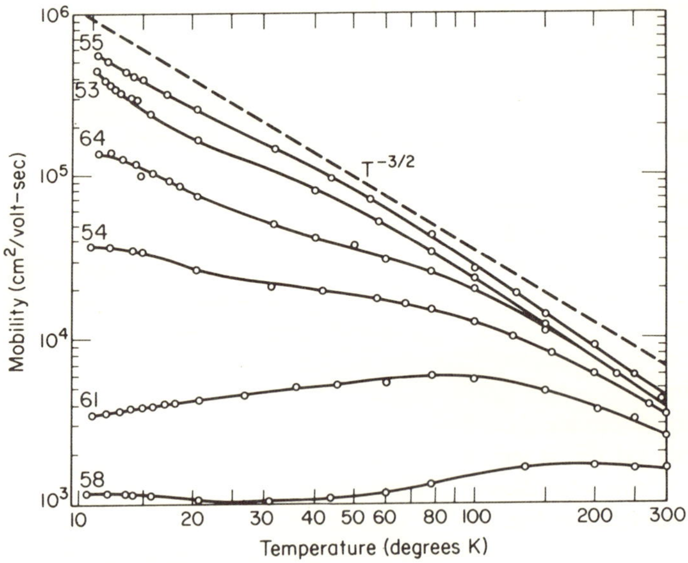

Fig. 1-10. Electron mobility vs. temperature for a set of arsenic-doped (n-type) samples. The numbers on the curves identify the samples as to donor-atom density N_d as follows: No. 55, $N_d = 1.0 \times 10^{13}$ cm^{-3}; No. 53, $N_d = 9.4 \times 10^{13}$ cm^{-3}; No. 64, $N_d = 1.7 \times 10^{15}$ cm^{-3}; No. 54, $N_d = 7.5 \times 10^{15}$ cm^{-3}; No. 61, $N_d = 5.5 \times 10^{16}$ cm^{-3}; No. 58, N_d 10^{18} cm^{-3}. The dashed line represents the theoretical slope for thermal scattering. (These data and those of Figs. 1-11 and 1-12 are from Reference 1, p. 1332.)

Carrier densities in semiconductors. At "ordinary temperatures" the majority carrier concentration in extrinsic material may be considered equal to the impurity-atom concentration, *providing only one type on impurity is present.* If both donor and acceptors are present, all will be ionized just as if they were alone in the material. However, if the donor density N_d and the acceptor density N_a are unequal, the carriers supplied by the less numerous atoms recombine with an equal number of carriers supplied by the more numerous. As a result, the density of carriers supplied by impurity atoms is equal to the difference between the donor and acceptor concentrations. This is responsible for the term *excess-donor concentration* N, where

$$N = N_d - N_a \tag{1-18}$$

If N is positive, it respresents the *effective concentration* of donors supplying free electrons. If N is negative, acceptors outnumber donors, and N represents the magnitude of the effective acceptor concentration that furnishes holes at "ordinary temperatures."

Thus we see that semiconductor material containing both donors and acceptors may be either n-type or p-type, depending on whether N is positive or negative, and the type may be changed by adding impurities such that N changes sign. Semiconductor-type change in this manner is common practice in the fabrication of semiconductor devices. If N is zero, the material is said to be *intrinsic* in that the hole and electron concentrations are those of pure material. A plot of (N_d-N_n) versus distance through a crystal is known as the *impurity profile* of the crystal. Using such a plot, one can identify the p-, n-, and intrinsic regions as well as junctions of these regions.

In the application of semiconductors, one needs to know the limits of the "ordinary"- or "extrinsic"- temperature range. The lower limit is the temperature below which impurity-atom ionization may not be considered to be complete, and the upper limit is the intrinsic temperature. These limits depend on the impurity concentration, as can be seen from the carrier concentration versus ($1/T$) plots of Fig. 1-11. The use of ($1/T$) as a coordinate serves to accentuate the phenomena at low temperatures where the number of ionized carriers is temperature dependent, but careful examination of the curves permits identification of the intrinsic range.

The samples of Fig. 1-11 are the same as those of Fig. 1-10 for which impurity-concentration information is provided. For the purest sample (No. 55), the "extrinsic" temperature range is seen to be between about 25°K and 300°K, the "intrinsic" temperature range is above about 300°K, and the "impurity-dionization" range is below about 25°K. For the somewhat less pure sample (No. 53) the "intrinsic" range begins at a somewhat higher temperature, since a larger intrinsic density n_i is required to compete with the larger number of donated electrons, and the "impurity-deionization" range is increased somewhat. These trends continue through successively

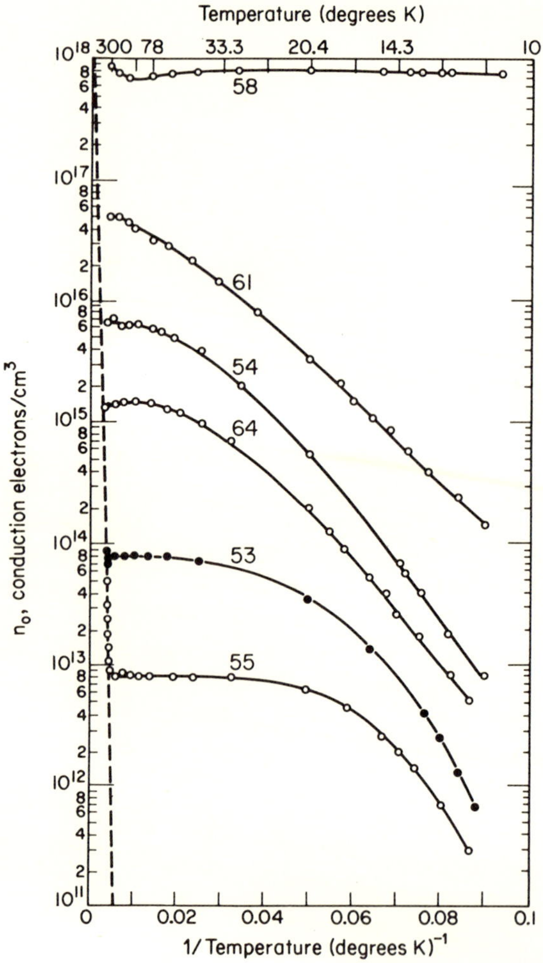

Fig. 1-11. Concentration of charge carriers vs. the reciprocal of absolute temperature for the arsenic-doped germanium samples of Fig. 1-10. The dashed line represents the concentration n_i of intrinsic carriers.

more heavily doped samples Nos. 64, 54, and 61. (The dip near the intrinsic line for sample No. 64 is a peculiarity of the measurement method rather than a characteristic of the electron concentration.) The extremely heavily doped sample (No. 58) behaves quite differently from the other samples for reasons to be discussed later. It is said to be doped until it has become *degenerate*, meaning that its properties are more like those of a metal than of a normal semiconductor. The hole concentrations p_n in these n-type samples are completely negligible in the extrinsic temperature range and below, as one can see by solving Eq. (1-1) for p_n, using the n_n and n_i values of Fig. 1-11.

Intrinsic-carrier density. The n_i versus $1/T$ characteristic, shown for germanium by the dashed line of Fig. 1-11, points out the enormous temperature dependence of the intrinsic-carrier concentration of semiconductors. It has been found that, for temperatures greater than about 50°K, n_i is given by

$$n_i = 1.76 \times 10^{16} T^{3/2} e^{-4550/T} \text{ cm}^{-3} \text{ for germanium} \qquad (1\text{-}19)$$

and

$$n_i = 3.88 \times 10^{16} T^{3/2} e^{-7000/T} \text{ cm}^{-3} \text{ for silicon} \qquad (1\text{-}20)$$

The change of n_i with T is largely determined by the exponential terms of Eqs. (1-19) and (1-20). For example, in the temperature range from 100°K to 300°K the term $T^{3/2}$ increases by the factor 5.2, which appears quite significant. However, the exponentials increase by approximately the factors $e^{30} \cong 1 \times 10^{13}$ for germanium and $e^{47} \cong 2.5 \times 10^{20}$ for silicon, indicating that changes with temperature of the $T^{3/2}$ term are relatively unimportant. Therefore, n_i is nearly an exponential function of $(1/T)$, as indicated by the dashed line of Fig. 1-11.

At a temperature of 300°K, which is approximately normal room temperature, intrinsic carrier densities are found to be

$$n_i = 2.4 \times 10^{13} \text{ cm}^{-3} \text{ for germanium}$$

and

$$n_i = 1.5 \times 10^{10} \text{ cm}^{-3} \text{ for silicon}$$

The room-temperature value of n_i for silicon is seen to be about three orders of magnitude lower than for germanium but to increase more strongly with temperature.

Conductivity-temperature characteristics. Since conductivity is proportional to the carrier density-carrier mobility product, information such as that of Figs. 1-10 and 1-11 regarding carrier density and mobility enables us to predict the temperature dependence of semiconductor conductivity. Figure 1-12 shows relationships between conductivity and temperature for the n-type germanium samples of Figs. 1-10 and 1-11. It is seen that, in the intrinsic range, the conductivity follows the intrinsic carrier density because n_i is so much more temperature dependent than are the carrier mobilities. In the extrinsic range, conductivity *rises* as the temperature *drops*, even though the carrier density is unaffected by temperature. This behavior reflects the inverse variation of mobility with temperature in this region. Eventually, as the temperature decreases, the carrier density decreases as the impurities deionize, and impurity scattering either prevents the mobility from increasing

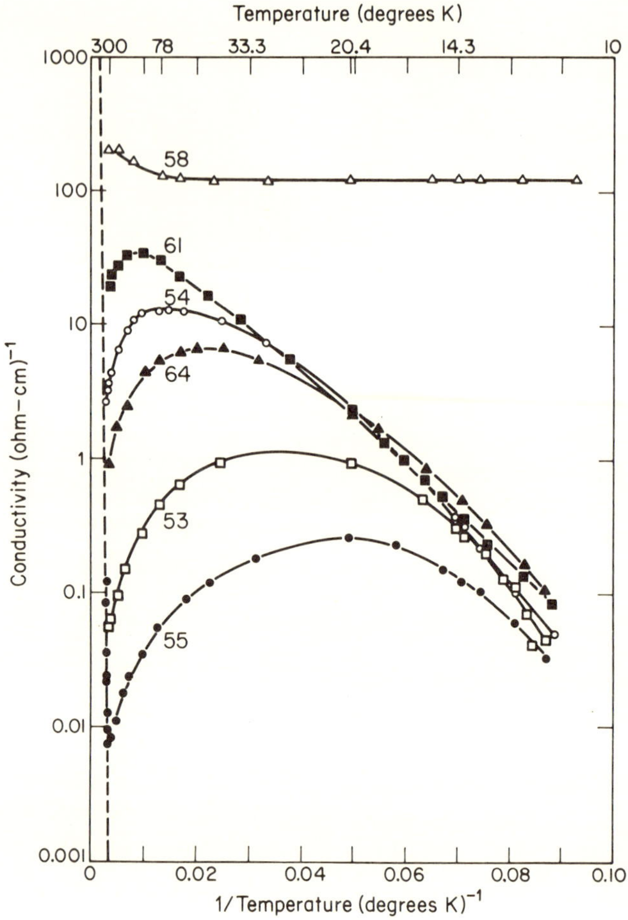

Fig. 1-12. Conductivity-temperature relationships for the arsenic-doped germanium samples of Figs. 1-10 and 1-11. The dashed line represents intrinsic conductivity.

so rapidly or may even cause it to decrease. As a result, a second point of inflection occurs in the conduction characteristics, and conductivity again decreases with temperature. *These points of inflection distinguish an extrinsic semiconductor from a metallic conductor.*

1-2.2. Hall Effect and Magnetoresistance

If a metal or *n*-type semiconductor specimen carrying a *conventional* d-c current I is placed in a transverse d-c magnetic field B as in Fig. 1-13, the electrons are acted upon by both a steady electric field and the steady

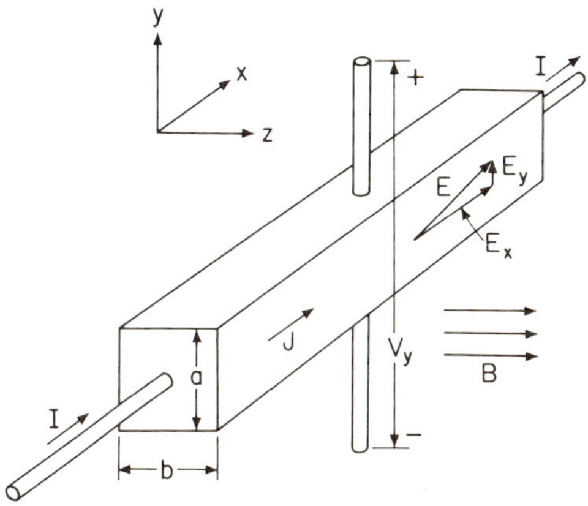

Fig. 1-13. Arrangement for the Hall-effect phenomenon.

magnetic field. The magnetic field results in the exertion of a force on the electrons in a direction at right angles to their direction of motion (along the x-axis) and to the direction of the magnetic field (along the z-axis). The direction of this force is, therefore, along the y-axis and downward (in the $-y$-direction) for the directions of I and B in Fig. 1-13. As a result, electrons are forced toward the lower surface, causing it to become negatively charged with respect to the upper surface. This results in an electric field E_y in the y-direction which exerts a *force* on the electrons in the upward or $+y$-direction. Under equilibrium conditions this electric-field force just balances the magnetic-field force; that is,

$$eE_y = Bev_d \tag{1-21}$$

where e is the electronic charge and v_d is the drift velocity of the electrons.

Recognizing that v_d is the product of electron mobility μ_n and the x-direction electric field E_x, we see that Eq. (1-21) becomes

$$E_y = B\mu_n E_x \tag{1-22}$$

and the voltage V_y in Fig. 1-13 is seen to be

$$V_y = aB\mu_n E_x \tag{1-23}$$

providing that V_y *is zero in the absence of the magnetic field.* Voltage V_y is known as the *Hall voltage* in deference to E. H. Hall, who discovered the effect in 1879.

If a p-type semiconductor specimen is used in the arrangement of Fig. 1-13, the polarity of V_y is found to be negative; that is, the polarity is opposite that for an n-type specimen. This occurs because, with p-type material, the magnetic field forces holes toward the lower surface, causing it to become positively charged. That is, the magnetic field pushes the current toward the same side of the specimen, whether it is carried by holes or electrons.

Thus we see that the type of an extrinsic semiconductor can be determined from the polarity of the Hall voltage.

From Eq. (1-22), we see that the Hall effect provides a means for determining carrier mobility; that is,

$$\mu = \frac{E_y}{BE_x} \tag{1-24}$$

μ being electron mobility μ_n or hole mobility μ_p, depending on the conductivity type of the specimen. If both types of carrier are present in significant concentrations, the effects tend to cancel and the measurement becomes meaningless. In Eq. (1-24), μ is in (meters)^2 per volt-sec for B in Webers/(meter)^2, providing the units of E_x and E_y are the same.

The arrangement of Fig. 1-13 lends itself readily to the measurement of the electrical conductivity of the specimen. If this is done simultaneously with the mobility determination, the carrier density may be determined, since, from Eqs. (1-16) and (1-17),

$$n_n \cong \frac{\sigma_n}{e\mu_n} \tag{1-25}$$

and

$$p_p \cong \frac{\sigma_p}{e\mu_p} \tag{1-26}$$

On the other hand, if the carrier density is known from information about the impurity concentration, the carrier mobility may be determined from the conductivity. When this is done, it is found that the mobility differs somewhat from that determined by the Hall-effect method. For this reason, mobilities measured by utilizing the Hall effect are knows as *Hall mobilities*, μ_H.

Hall mobility differs from drift mobility because the drift velocity of carriers is a long-time average of the instantaneous drift velocity values that determine the Hall effect. If this difference is taken into account for electrons by averaging the drift velocity over the ramdom thermal velocities, it may be shown that

$$\mu_n = \frac{8\mu_H}{3\pi} = \frac{\mu_H}{1.18} \tag{1-27}$$

This expression takes only lattice scattering into consideration. It has been found that impurity concentration also has an effect on μ_H/μ, causing it to be either increased or decreased, depending on the concentration. Values of μ_H/μ as low as 1.0 and as high as 1.7 have been observed.

The Hall mobility of holes may differ significantly from the drift mobility in the same specimen, *depending on the magnetic field.* The dependence on magnetic-field intensity occurs because there are both *heavy holes* and *light* holes, For large magnetic fields, the light holes are swept away and μ_H/μ becomes independent of field strength and is approximately unity.

Typical Hall mobilities of holes and electrons in very pure germanium and silicon are given in Table 1-3 for a temperature of 300°K. Since the minority-carrier density of lightly doped germanium is significant at 300°K, the values for germanium were extrapolated from values measured at lower temperatures. It is of interest to compare these mobility values with the drift-mobility values of Table 1-1.

Table 1-3* Hall mobility in cm²/volt-sec for very pure matrial at 300°K

Carrier	*Germanium*	*Silicon*
Electron	3950 ± 250	1900
Hole	3400 at 3000 gauss	425

* Reference 2, p. 1284.

Hall voltages developed in metals are very small, owing to the small electron-drift velocities that occur for practical currents. For this reason, it was believed for many years that the Hall effect had no practical value. This was changed by the advent of semiconductors with their much higher carrier-drift velocities that result from their lower carrier densities and larger drift mobilities. For example, Hall voltages of the order of millivolts can be obtained with moderate magnetic fields and moderate specimen currents.

In addition to its use in the investigation of semiconductor properties, the Hall effect is the basis of a number of practical devices. Perhaps the most common of these is the magnetic-field detector and intensity meter. Since the Hall effect is independent of frequency over a wide range, it can be used to measure magnetic fields from d-c to radio frequencies.

Since the Hall voltage is proportional to the product of the current I through the specimen and the magnetic field B, Hall devices can be used as electronic multipliers. For example, if the field is produced by a solenoid carrying a current proportional to the voltage of a circuit and the circuit current flows through the Hall element, the result is an electronic wattmeter. A microwave wattmeter can be made by orienting a semiconductor element in the electromagnetic field of wave guide in such a way that the electric field induces a longitudinal current and the magnetic field reacts with this current to generate the Hall voltage.

Magnetoresistance. Careful investigation using the Hall-effect arrangement of Fig. 1-13 discloses that the resistance of the specimen increases when the magnetic field is applied. This is known as the *magnetoresistance effect*.

In our discussion of the Hall effect, it was pointed out that carriers are deflected as a result of the application of a magnetic field at right angles to their direction of flow. If all carriers had the same drift velocity, the Hall-effect electric field would exert an equal and opposite force on the carriers and, in an equilibrium condition, no more carriers would be deflected. Therefore, there is no component of carrier motion in the direction of the Hall field, and no change would be expected in the conductivity of the material as a result of the magnetic field. However, in practice there is a distribution of drift velocities, and the only undeflected carriers are those whose velocity is precisely the value that satisfies the condition of equal and opposite force by the Hall electric field. Carriers having values of velocity larger or smaller than this value will be deflected downward or upward by the field of Fig. 1-13. This deflection results in the carriers following curved trajectories, and the average drift distance along the current-flow direction between successive collisions is reduced. This reduces the velocity of the carriers in the direction of the applied electric field, increasing the resistance of the specimen.

The change in specimen resistance due to this effect is usually so small that it is easily overlooked. It is found to be proportional to the square of the magnetic field strength and to the square of the carrier mobility. Its principal use is as an indicator of doping-atom density ($N_d + N_a$), owing to its dependence on mobility and the dependence of mobility on impurity concentration.

1-3 ADDITIONAL ELECTRONIC PROCESSES IN SEMICONDUCTORS

1-3.1 Minority-Carrier Injection

As pointed out in Sec. 1-1, free electron-hole pairs are produced in a semiconductor by valence-bond disruptions. The energy required to break a valence bond can be obtained thermally or it can be a photon of light, provided the frequency is high enough. The semiconductors in common use are quite sensitive to light in the near-infrared and the visible regions of the spectrum.

Suppose a small spot near the middle of a homogeneous bar of *p*-type material is illuminated by a pulsed beam of light of sufficient frequency to produce electron-hole pairs. As the light penetrates the material, electron-hole pairs are produced at a rate dependent upon the intensity of the light. Thus, immediately after the pulse, the electron density n_p and the hole density p_p are in *excess* of their thermal equilibrium values n_{po} and p_{po}. If the *excess* densities are denoted by n'_p and p'_p, we see that

$$n'_p = n_p - n_{po} \tag{1-28}$$

and

$$p'_p = p_p - p_{po} \tag{1-29}$$

If the bar is strongly p-type, $p_{po} \gg n_{po}$ and $n_{po} \ll n_i$ in equilibrium. Therefore, the electron-density increase n'_p produced by the light may be quite large on a percentage basis, but the hole density increases by a very small percentage. This process is referred to as *minority-carrier injection* in order to focus attention on the dominant percentage effect. In a similar manner, injection of minority holes occurs when illumination of n-type material increases the hole density p_n above the thermal equilibrium value p_{no}. Minority-carrier injection may be achieved by means other than illumination and is fundamental to almost all semiconductor-device operation.

It should be strongly emphasized that in minority-carrier injection by illumination, equal increments in hole and electron densities occur (that is, $p' = n'$) and *the material remains electrically neutral.* This will usually be true regardless of the method of injection.

1-3.2 Diffusion Current

It was pointed out in Sec. 1-2 that *diffusion* is an important process by which carriers are transported in semiconductors. Let us again consider an illuminated semiconductor bar, such as that of Sec. 1-3.1, having an excess carrier density in a small volume near the middle.

Carriers in a crystal have a random motion, owing to the temperature of the lattice, and move in a random manner from one region to another. Even though their motion is random, more carriers move away from the illuminated region of the semiconductor bar than toward it, owing to the higher carrier concentration in the region. This process is known as *diffusion* and is responsible for carrier flow in a semiconductor in a manner similar to the flow of a gas from a high-pressure region to regions of lower pressure. Thus the diffusion flow of carriers from the illuminated region of the semiconductor bar is proportional to the difference in the carrier density at adjacent points in the bar and may be expressed as

$$J_e = -D_n \frac{dn}{dx} \text{ electrons/sec/cm}^2 \tag{1-30}$$

and

$$J_h = -D_p \frac{dp}{dx} \text{ holes/sec/cm}^2 \tag{1-31}$$

where dn/dx and dp/dx are the carrier-density gradients in carriers/cm^4, and D_n and D_p are electron and hole proportionality factors having the dimension cm^2 per sec.

The proportionality factors D_n and D_p are the electron and hole *diffusion constants* and, like carrier drift mobility, are properties of the material and its temperature. Diffusion constant values for electrons and holes in germanium and silicon having mobilities approaching lattice values are given in Table 1-4. The negative signs in Eqs. (1-30) and (1-31) account for the fact the particles flow from the region of high density to that of low density.

Table 1-4 Diffusion constants of very pure material at 300°K in cm^2/sec

	Germanium	*Silicon*
D_n, electrons	100	35
D_p, holes	49	12

Equations (1-30) and (1-31) for particle-current density may be converted to electric-current density by including the electronic charge e, Thus the electron diffusion current is given by

$$J_n = -eD_n \frac{dn}{dx} \text{ amperes/cm}^2 \tag{1-32}$$

and the hole diffusion current by

$$J_p = -eD_p \frac{dp}{dx} \text{ amperes/cm}^2 \tag{1-33}$$

If the use of these expressions, consideration must be given to the algebraic sign of e and the relationship between the direction of particle flow and that of conventional electric current.

Since the diffusion flows of holes and electrons in our example are in the same directions (away from the illuminated region), their electric current flows tend to cancel. However, we should not conclude from this phenomenon that diffusion current is unimportant. Very often, in semiconductor devices, an electric-field region exists that accelerates one type of carrier but turns the other type away. Therefore, current flow by diffusion in such devices is extremely important, perhaps even more so than flow by the drift process.

The Einstein relationship. Since both drift and diffusion involve the collisions of carriers with host atoms and with impurity ions in the lattice of a crystal, one might expect the diffusion constant and drift mobility to be related. This is indeed true, and it may be shown that

$$D_p = \frac{kT}{e}\mu_p \tag{1-34}$$

and

$$D_n = \frac{kT}{e}\mu_n \tag{1-36}$$

These expressions are known as the *Einstein relationship* and will prove useful in future considerations.

1-3.3 Carrier Recombination

From the preceding discussion of an illuminated semiconductor bar, it might appear that continued illumination of a spot on the bar, and the diffusion of carriers therefrom, would continue to increase the carrier level in the bar. In reality, this is prevented by the recombination of excess carrier pairs. As a result of recombination, the carrier level some distance away from the illuminated spot, say at the ends of a long bar, will be unaffected by the illumination, all injected carriers having disappeared before having an opportunity to diffuse great distances from the spot. The average distance that an injected carrier travels by diffusion is known as the *diffusion length* L and the average time that an injected carrier exists is the *minority-carrier lifetime* τ.

In order for recombination to occur, a valence bond is refilled and an electron-hole pair must lose the energy acquired when the bond was disrupted. It might appear that a free electron and a hole could readily recombine, with the release of a photon of radiant energy, in the course of their random wanderings through a conducting medium. This does occur in some materials and is the phenomenon responsible for the electromagnetic radiation of *light-emitting semiconductor diodes* and *injection lasers*. In most materials, however, a momentum change as well as an energy change is required in order for recombination to occur, and the lack of appreciable inertia in light makes direct recombination unlikely. For this reason, most recombination occurs at *recombination centers*.

A recombination center is a third body that can either absorb or provide the momentum necessary to fulfill the requirement of the recombination process. Almost anything that disturbs the periodicity of the crystal lattice serves as a recombination center. Crystal imperfections such as impurity atoms (other than donors or acceptors), empty lattice sites, cracks, and surfaces promote recombination and reduce minority-carrier lifetime. It is for this reason that high-purity single-crystal material is necessary in the fabrication of semiconductor devices. Representative minority-carrier lifetimes range from 0.1 to 1000 microseconds, depending on the condition of the crystal.

Fortunately, the rather complicated recombination process can usually be represented by a simple relationship. Suppose our bar of p-type material is flooded with light, uniformly illuminating the entire bar rather than confining the illumination to a small spot near the middle. The electron density n_p throughout the bar increases above the equilibrium value n_{po} and stabilizes at a level $n_p(0)$ at which the carrier generation rate equals the recombination rate. If the light is then suddenly removed, the electron density gradually falls to n_{po} as a result of recombination. Since injected minority electrons cease to exist in an average time τ_n, where τ_n is the lifetime, we may write

$$\frac{dn_p}{dt} = -\frac{(n_p - n_{po})}{\tau_n} \tag{1-36}$$

Solving this expression for n_p yields

$$n_p = [n_p(0) - n_{po}]e^{-t/\tau_n} + n_{po} \tag{1-37}$$

or we may write

$$n'_p = n'_p(0)e^{-t/\tau_n} \tag{1-38}$$

where n'_p is the excess electron density and $n'_p(0)$ is the excess value of n_p at $t = 0$, the instant the light is removed.

1-3.4 Divergence of Carrier Flow

Suppose we return to our bar of *p*-type material having illumination confined to a small spot near the middle and consider the effect of suddenly removing the light. From our discussion of Secs. 1-3.2 and 1-3.3, we know that the excess electron density n'_p in the vicinity of the spot decreases with time as a result of both diffusion and recombination. The effect of diffusion on injected carrier density is known as *divergence*, and consideration of both recombination and divergence yields the *diffusion equation.*

An elementary volume in the vicinity of the illuminated spot is shown by Fig. 1-14. The diffusion electron-current density (in contrast to the electric

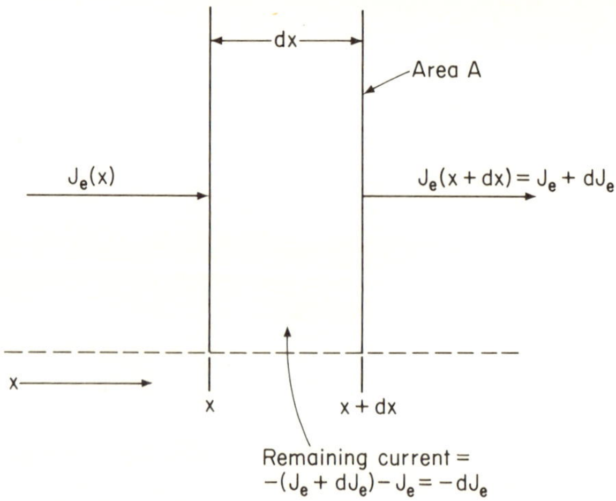

Fig. 1-14. One-dimensional flow of diffusion current.

current resulting from electron drift) into the volume is represented by J_e, and $(J_e + dJ_e)$ is the diffusion current out of the volume. The number of electrons entering the volume in time dt is seen to be $J_e A\, dt$ and those leaving number $(J_e + dJ_e)A\, dt$. The increase in the number of electrons in the volume in time dt is then

$$[J_e - (J_e + dJ_e)]A\, dt = -dJ_e A\, dt$$

and the electron-density increase dn_p is the electron increase divided by the volume. That is,

$$dn_p = \frac{-dJ_e A\, dt}{A\, dx} = -\frac{dJ_e\, dt}{dx}$$

or

$$\frac{dn_p}{dt} = -\frac{dJ_e}{dx} \tag{1-39}$$

The diffusion electron current J_e is given by Eq. (1-30) as,

$$J_e = -D_n \frac{dn_p}{dx}$$

from which

$$\frac{dJ_e}{dx} = -D_n \frac{d^2 n_p}{dx^2} \tag{1-40}$$

From Eqs. (1-39) and (1-40), we see that

$$\frac{dn_p}{dt} = D_n \frac{d^2 n_p}{dx^2} \tag{1-41}$$

In the interest of simplicity, we have restricted our divergence consideration to the one-dimensional case, even though we recognize that excess electrons diffuse in all direction from the illuminated region.

1-3.5 The Diffusion Equation

If the excess carrier density is small compared to the majority-carrier density, that is, if $n'_p \ll p_{po}$ and $p'_n \ll n_{no}$ the nonequilibrium forces of drift, recombination, and diffusion are small enough that each, acting alone, will produce only a small departure of the system from equilibrium. The resulting effect of all of them may then be considered to be a simple superposition of the individual effects. That is, the effects do not mutually interact, and the individual effects are linearly dependent upon their causes. This condition usually exists in ordinary device operation in which the excess minority-carrier density $(n_p - n_{po})$ may be thousands of times the equilibrium value n_{po} but small compared to the equilibrium majority-carrier density p_{po}.

If it is assumed that that "small-signal" (linear) conditions prevail, the effects on minority-carrier density of recombination, given by Eq. (1-36), and divergence, given by Eq. (1-41), may be combined, yielding

$$\frac{dn_p}{dt} = D_n \frac{d^2 n_p}{dx^2} - \frac{(n_p - n_{po})}{\tau_n} \tag{1-42}$$

For holes injected into *n*-type material, Eq. (1-42) becomes

$$\frac{dp_n}{dt} = D_p \frac{d^2 p_n}{dx^2} - \frac{(p_n - p_{no})}{\tau_p} \tag{1-43}$$

Equations (1-42) and (1-43) are known as *diffusion equations*, and their solutions give the variation of minority-carrier density as functions of both time and distance. In their derivation, it was assumed that no generation of carriers occurred in the region under consideration. If minority holes are generated by some process at a rate g, Eq. (1-43) becomes

$$\frac{dp_n}{dt} = D_p \frac{d^2 p_n}{dx^2} - \frac{(p_n - p_{no})}{\tau_p} + g \tag{1-44}$$

which may be written in terms of excess hole density as

$$\frac{dp'_n}{dt} = D_p \frac{d^2 p'_n}{dx^2} - \frac{p'_n}{\tau_p} + g \tag{1-45}$$

since changes in minority-carrier density are excess-density changes.

1-3.6 The Continuity Equation

Another factor affecting the minority-carrier concentration at a point in a semiconductor is the presence of an electric field, and adding this effect to that of recombination and diffusion results in the *continuity equation*.

Adapting drift-current Eq. (1-11) to the flow of particles and applying it to an excess of holes in an *n*-type semiconductor, we obtain

$$J_h = p'_n \mu_p E_x \tag{1-46}$$

where J_h is the density of hole flow across a surface normal to the flow. Since Eq. (1-39) for the effect of particle current on particle concentration applies for either drift current of diffusion current (or both), adapting it to hole flow results in

$$\frac{dp'_n}{dt} = -\frac{dJ_h}{dx} \tag{1-47}$$

Since p'_n is a function of x, we can differentiate Eq. (1-46) with respect to x to obtain

$$\frac{dJ_h}{dx} = \mu_p E_x \frac{dp'_n}{dx} \tag{1-48}$$

which, combined with Eq. (1-47) yields

$$\frac{dp'_n}{dt} = -\mu_p E_x \frac{dp'_n}{dx} \tag{1-49}$$

Equation (1-49) makes the not unexpected statement that an excess carrier concentration at a point is decreased by the presence of an electric field. Combining this effect linearly with the effects of recombination and diffusion, as given by the diffusion equation, results in

$$\frac{dp'_n}{dt} = D_p \frac{d^2 p'_n}{dx^2} - \mu_p E_x \frac{dp'_n}{dx} - \frac{p'_n}{\tau_p} + g \tag{1-50}$$

which is known as the *continuity equation.*

In summary, the term on the left of the equal sign of Eq. (1-50) is the time-rate-of-change of excess minority-carrier density. (The equation applies equally well to excess electron concentration in a p-type semiconductor.) The first term on the right is the rate of pileup of minority carriers per unit volume due to the difference between the infusion and effusion rates. The second term on the right is the rate of minority-carrier pileup per unit volume as a result of the presence of an electric field. The third term on the right is the rate of decay of excess minority-carrier density due to recombination. The final term in the equation is the rate of generation (or the loss rate in the presence of a sink) of carriers because of other sources.

1-3.7 Application of the Continuity Equation

We shall now consider a few examples of the application of the continuity equation to practical problems. Additional examples will be presented as device operation is considered, the continuity equation being fundamental to such consideration.

We should now recognize that the continuity equation was applied in the example considered in the recombination discussion of Sec. 1-3.3 and in arriving at Eq. (1-38). The conditions of this example were such that only the recombination term of the equation required consideration, the interior of the illuminated specimen being both field-free and free of minority-carrier-density gradient (owing to the uniform illumination), and there being no sources or sinks of minority carriers after the illumination was turned off. Applying Eq. (1-50) under these conditions to excess hole density in a bar of n-type semiconductor yields

$$\frac{dp'_n}{dt} = -\frac{p'_n}{\tau_p} \tag{1-51}$$

from which

$$p'_n = p'_n(0)e^{-t/\tau_p} \tag{1-52}$$

where p'_n is the excess hole density, τ_p is the minority-hole lifetime, and $p'_n(0)$ is the excess value of p_n at $t = 0$, the instant the illumination was removed.

As a second example, suppose our bar of n-type semiconductor material is continuously illuminated at a point from which excess holes spread out by diffusion in the $+x$ and $-x$ directions and we wish to determine the excess hole density p'_n as a function of distance x from the point of generation. As in our preceding example, the interior of the specimen is field-free, and there are no carrier sources or sinks except at the point of generation. With continuous illumination, the excess carrier density at each point in the bar is constant (that is, dp'_n/dt is zero), and the continuity equation becomes

$$D_p \frac{d^2 p'_n}{dx^2} - \frac{p'_n}{\tau_p} = 0 \tag{1-53}$$

which has a solution of the form

$$p'_n = A\mathrm{e}^{-x/\sqrt{D_p\tau_p}} + B\mathrm{e}^{x/\sqrt{D_p\tau_p}} \tag{1-54}$$

Coefficients A and B are evaluated from boundary conditions which are $p'_n = p'_n(0)$ at $x = 0$ and $p'_n = 0$ at $x = \infty$, owing to recombination. Applying these boundary conditions to Eq. (1-54) yields, for $x = 0$,

$$p'_n(0) = A + B \tag{1-55}$$

for $x = \infty$,

$$0 = B\mathrm{e}^{\infty/\sqrt{D_p\tau_p}} \tag{1-56}$$

From Eqs. (1-55) and (1-56), we see that $B = 0$ and $A = p'_n(0)$. Therefore, the solution of Eq. (1-53) is

$$p'_n = p'_n(0)\mathrm{e}^{-x/\sqrt{D_p\tau_p}} \tag{1-57}$$

The term

$$\sqrt{D_p\tau_p}$$

of Eq. (1-57) is known as the minority-hole *diffusion length* L_p. As defined earlier, L_p is the average distance traveled, by diffusion, by a minority hole before disappearing by recombination. Thus

$$L_p = \sqrt{D_p\tau_p} \tag{1-58}$$

and Eq. (1-57) may be written

$$p'_n = p'_n(0)\mathrm{e}^{-x/L_p} \tag{1-59}$$

Similarly, the minority-electron diffusion length L_n is given by

$$L_n = \sqrt{D_n\tau_n} \tag{1-60}$$

and the excess electron density n'_p in a bar of p-type semiconductor as a function of distance x from the point of illumination is given by

$$n'_p = n'_p(0)\mathrm{e}^{-x/L_n} \tag{1-61}$$

Equations (1-59) and (1-61) may be utilized in the experimental determination of minority-carrier lifetime.

As our third example, suppose that a short-duration pulse of holes is injected into a bar of *n*-type semiconductor material at a point from which excess holes spread out by diffusion in the $+x$- and $-x$-directions and we wish to determine the excess hole density p'_n as a function of time after the instant of injection. (Hole injection in this manner may be accomplished through the use of a pulsed light source or, as will be understood later, obtained from a pulsed emitter.) As in our preceding example, the specimen is field-free, and the quantity g is zero. Therefore, the continuity equation becomes

$$\frac{dp'_n}{dt} = D_p \frac{d^2p'_n}{dx^2} - \frac{p'_n}{\tau_p} \tag{1-62}$$

The solution to this equation may be shown to be

$$p'_n = \frac{N}{\sqrt{4\pi D_p t}} e^{-x^2/4D_pt - t/\tau_p} \tag{1-63}$$

where N is the total excess charge initially injected. Equation (1-63) indicates that the cloud of injected excess holes spreads out by diffusion and dies out by recombination as it does so, the two effects occurring simultaneously. Figure 1-15(a) shows a family of solutions to Eq. (1-63) for successive time intervals after the injection of the pulse.

The final example that we shall now consider is the same in every respect as the preceding one except that an electric field E_x is maintained along the semiconductor bar and the term $-\mu_p E_x \, dp'_n/dx$ of the continuity equation can no longer be neglected. It can be shown that if a solution such as Eq. (1-63) holds for the field-free condition, being of the form $p'_n = f(x, t)$, then the solution in the presence of an electric field E_x will be

$$p'_n = f(x - \mu_p E_x t, t) \tag{1-64}$$

This equation says that the solution will be the same as for the field-free example except that, in addition to spreading out by diffusion and decaying by recombination, the cloud of excess minority holes is also transported in the x-direction with the velocity $\mu_p E_x$ as a result of the electric field. Figure 1-15(b) shows a family of solutions for successive time intervals with this drift feature superimposed on the conditions illustrated by Fig. 1-15(a). The solution of the continuity equation for the conditions in this example is the basis of a practical method for the measurement of minority-carrier mobility and lifetime and the determination of minority-carrier diffusion constant.[3]

[3] J. R. Haynes, and W., Shockley, "Mobility and Life of Injected Holes and Electrons in Germanium," *Physical Review*, Vol. 81 (1951), p. 835.

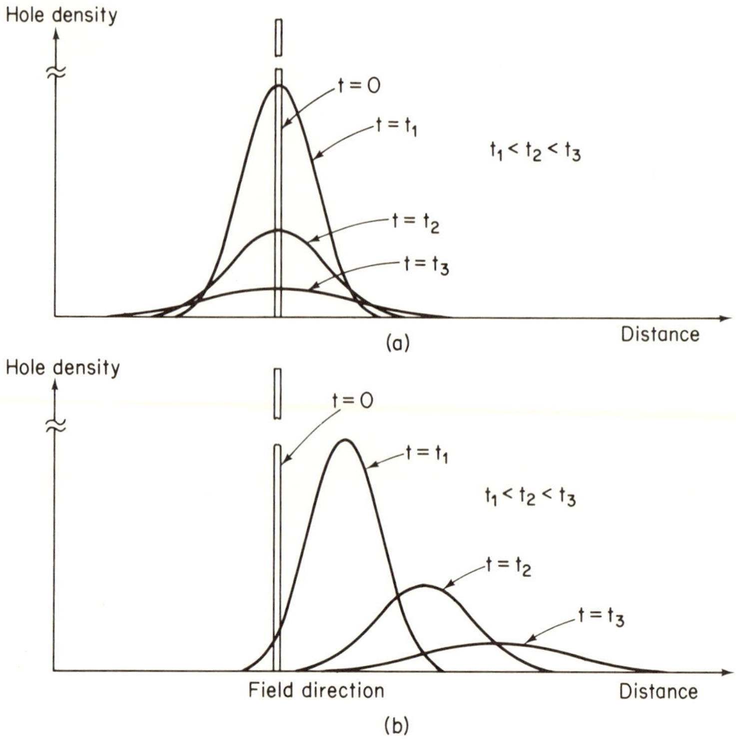

Fig. 1-15. Spread by diffusion and decay by recombination of a cloud of holes injected into an *n*-type semiconductor: (a) in the absence of an electric field; (b) in the presence of an electric field.

1-4 ENERGY-BAND MODELS OF SEMICONDUCTORS

So far, our consideration of the properties of solids has been primarily qualitative in nature, emphasizing the location *in space* of bound and free electrons, holes, donors, and acceptors, We shall now begin an approach that lends itself more readily to quantitative consideration. This is the *energy-band viewpoint*, which relates more to energy concepts than to space-location and velocity. In general, each point of view supplements the other. In a few instances, we shall be hard pressed to explain certain device characteristics without using the energy-band concept, but either point of view provides a basis for reasonable understanding of most of the properties of semiconductor materials and devices.

1-4.1 Energy Levels, Bands, and Gaps

The orbital arrangement of electrons about the nucleus of an atom, as presented in Sec. 1-1.1 and represented by the models of Fig. 1-1, is shown in greater detail by the model of Fig. 1-16(a). Figure 1-16(a) shows electrons in discrete orbits and includes an orbit that is free of electrons under normal circumstances. This is known as an *excitation orbit*, and an atom having one or more electrons in such orbits is said to be in an *excited state*.

In addition to their kinetic energy of rotation about the nucleus, the

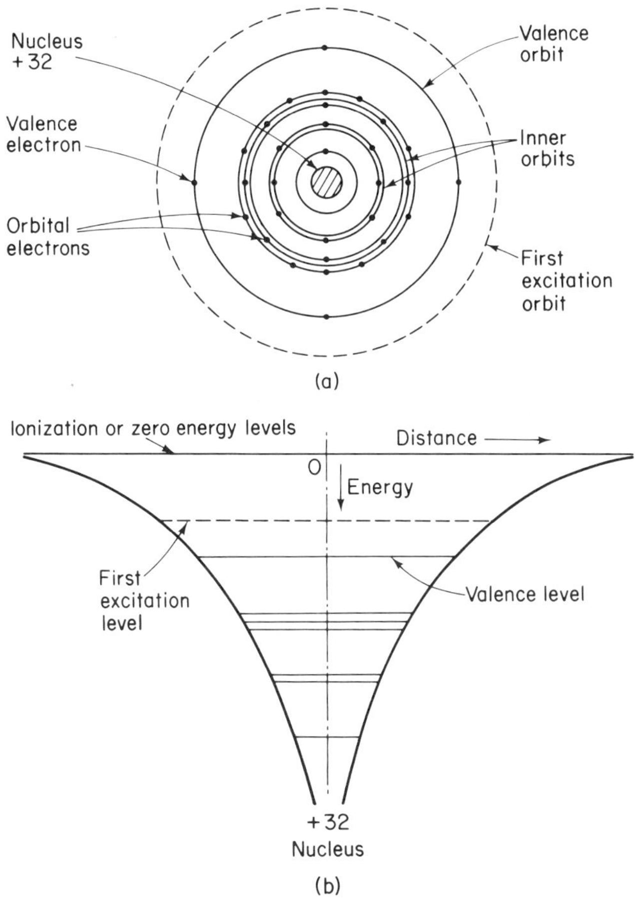

Fig. 1-16. (a) Simplified orbital model of an atom and (b) energy-level diagram of the same atom. (From L. P. Hunter, *Handbook of Semiconductor Electronics*, McGraw-Hill Inc., New York, 1962.)

electrons possess *potential energy* as a result of their positions with respect to the nucleus. An electron in an outer orbit possesses more energy than one in an inner orbit, since moving an electron from an inner orbit to an outer one involves motion against the attractive force of the positive nucleus. Therefore, work must be done on the electron to effect the change in orbit, and the potential energy of the electron is increased. A finite amount of energy is associated with each orbit of a particular type of atom, and the electrons are said to occupy *energy levels.* Many more possible energy levels, or orbits, exist than those normally occupied by the electrons of an atom. These are higher energy levels known as *excitation levels.*

An electron that normally occupies an energy level E_1 can move to a higher energy level E_2, provided an amount of energy $E_2 - E_1$, referred to as a *quantum of energy*, is received by the atom. After reaching energy level E_2, the electron may go to a still higher level E_3 if the atom receives an additional quantum of energy $E_3 - E_2$. Movement of the electron from energy level E_1 to level E_3 can be accomplished in a single operation instead of the two-stage process if the atom receives a quantum of energy $E_3 - E_1$. Thus, an electron can move in either a single- or a multiple-stage process from its normal energy level to any higher energy level, the higher level and the type of process being dependent upon the amount of energy received by the atom and the manner in which it is received.

As mentioned earlier, an electron that has one or more of its electrons raised to a higher than normal energy level is said to be in an excited state, and the energies required to produce the energy-level changes are called *excitation energies.* The least energy required to produce excitation of an atom is called the *first excitation energy.* If the atom receives sufficient energy, an electron may be moved from either a normal or an excited orbit to a point sufficiently far from the nucleus that it becomes free from the influence of the nucleus. In other words, the electron is removed from the atom, and the atom is said to be *ionized.* The electron becomes a free electron, and the atom becomes a single-charged positive ion. The minimum energy required to remove an electron from a normal atom is called the *ionization energy* of the atom. The electrons in an atom have a tendency to revert to the lowest possible energy levels. However, there is a limit to the number of electrons that can exist at any given energy level in a particular atom. When all electrons occupy the lowest permitted levels, the atom is said to be in a *normal state.*

An energy-level diagram for the atom model of Fig. 1-16(a) is shown in Fig. 1-16(b). In the energy-level diagram, the vertical position of each horizontal line represents the potential energy associated with each electronic orbit, with the ionization energy of the atom as the reference, and the length of each line corresponds to the diameter of the particular orbit. Thus, the curves bounding the horizontal lines represent a plot of electron energy as a function of distance from the nucleus. As mentioned earlier, only discrete values of energy and orbital diameter are possible. The allowed energy levels that normally contain electrons are illustrated in the energy-level diagram by the positions of solid horizontal lines. Additional allowed energy levels associated with normally unoccupied orbits exist between the valence level and the ionization level. These are the excitation levels, and the first excitation level is illustrated by the broken horizontal line in Fig. 1-16(b).

Let us now consider a volume consisting of atoms such as that of Fig. 1-16. If the atomic spacing is large enough, significant overlapping of the orbits *does not* occur, and the energy levels (or states of the multiatomic system are simply those of the individual atoms. This is essentially the situation in a gas, and the energy-level diagram of Fig. 1-16 can be the basis for the explanation of certain gaseous phenomena. For example, the light given off by an electric discharge in a gas is observed to consist of discrete frequencies that are associated with quanta of energy exactly equal to the differences between excitation levels and the valence level or those between individual excitation levels. Thus, the light is the result of electrons of excited atoms returning to their normal levels either directly or by jumps to intermediate excitation levels on the way. Excitation occurs as a result of the transfer of energy to atoms from colliding free electrons that have acquired energy from the electric field of the discharge. The excited state of an atom is temporary, usually lasting for a time of the order of 10^{-8} sec or less, after which the atom returns to its normal state by the process just described.

If the atoms of the volume are brought closer together, say in a crystalline structure, it is clear that the outermost electron of each atom will eventually come close enough to other atoms to feel the forces exerted upon it by these atoms. With closer spacings, the inner electrons are also affected. If two atoms identical to the atom of Fig. 1-16 are brought together to form a diatomic molecule, an energy-level diagram may be drawn for the combination as shown in Fig. 1-17. Two highly significant features of the diatomic-molecule energy-level diagram are the existence of two closely spaced energy levels near each of the levels of the single atom and the depressing of the energy plot in the regions between the nuclei. Energy-level splitting is a result of a coupling process similar to the production of two slightly different frequencies when two oscillators of identical frequency are coupled. The depressing of the energy versus distance plot between the nuclei is also a coupling effect. In addition to being attracted by its own nucleus, each electron is attracted by the nucleus of the other atom of its molecule. Thus, the effect of the second atom is to reduce the energy required for an electron to move toward the adjacent atom. As a result, the highest energy levels are shared by both atoms, as shown in Fig. 1-17. The sharing of the valence levels means that the valence electrons are not localized in either atom but are allowed to range throughout the molecule and serve to bind the atoms together.

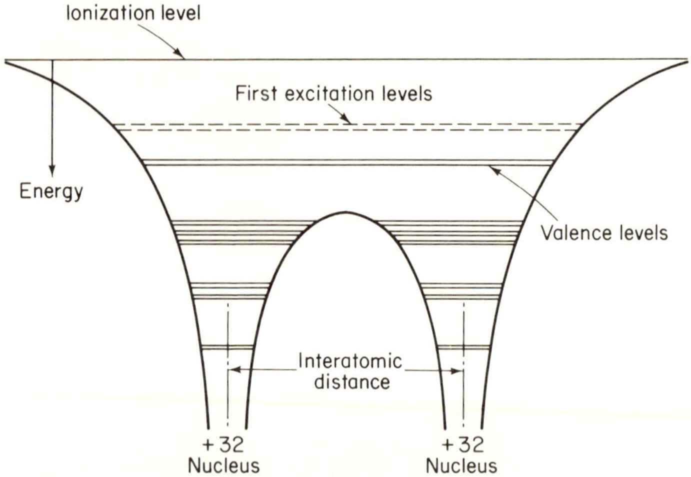

Fig. 1-17. One-dimensional energy-level diagram of a diatomic molecule. (From L. P. Hunter, *Handbook of Semiconductor Electronics*, McGraw-Hill, Inc., New York, 1962.)

As additional identical atoms are brought together to form a solid crystal, further energy-level splitting occurs until by the time a milligram of matter has been assembled there are approximately 10^{19} energy levels for each level of an isolated atom.[4] The levels are so close together that the spacing between them has no significance, and the group of levels is referred to as an *allowed band.* The energy band is treated as if a continuous distribution of energies were allowed for electrons within the band. Energy bands for such a crystal are shown in the energy-level diagram of Fig. 1-18. The band formed by excitation-level splitting is referred to as the *conduction band,* and valence-level splitting forms the *valence band.* The conduction band is significantly different from the other energy bands in that it is formed by the splitting of normally unoccupied energy levels, In the absence of excitation, the conduction band is an *empty band,* and the other bands are *filled bands.* As the band names suggest, each energy level of a filled band is occupied in that an electron having that particular energy exists, but each energy level of an empty band is unoccupied or vacant.

Energy gaps or *forbidden bands* are seen to exist between the energy bands shown in Fig. 1-18. An energy gap is an interval of energy in which there are no allowed energy levels. The gap between the valence band and the conduction band is particularly significant, being the energy E_g referred to in Sec. 1-1.2 as that required for the disruption of a valence bond and the freeing

[4] Sproull, Robert L., *Modern Physics, A Textbook for Engineers*, John Wiley & Sons, Inc. 2d ed. 1963, p. 266.

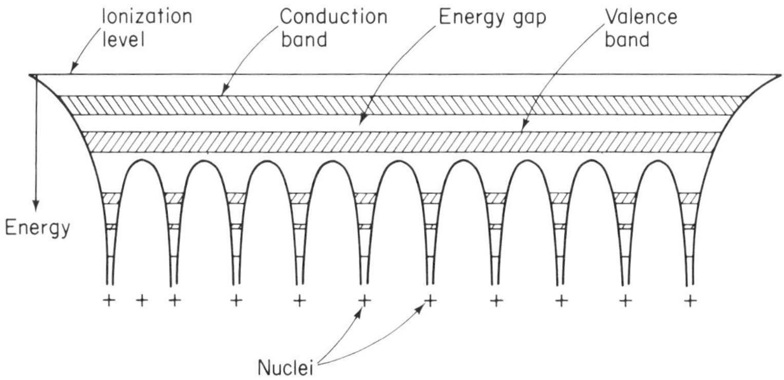

Fig. 1-18. One-dimensional energy-level diagram of a crystal. (Adapted from L. P. Hunter, *Handbook of Semiconductor Electronics*, McGraw-Hill, Inc., New York, 1962.)

of an electron. It is the gap referred to when one speaks of the energy gap of a material.

Energy levels are commonly referred to as *quantum states*, and an electron having the energy of a particular level is said to be in that quantum state. According to the exclusion principle discovered by Pauli, there can be at most one electron in each quantum state of an atom or closely spaced group of atoms. Therefore, an electron in a filled band can gain energy only at the expense of another electron in the band through an interchange of quantum states. However, at temperatures above absolute zero, certain valence-band electrons acquire sufficient energy to "jump the gap," filling quantum states in the conduction band and creating vacant states, or *holes*, in the valence band.

1-4.2 Band Structures of Crystalline Solids

The magnitude of the energy gap between the conduction band and the valence band of a crystal is a property of the material, being dependent upon the spacing between the atoms of the structure. The separation of the energy levels formed by the splitting of an atomic energy level depends upon the atomic spacing; the closer the spacing, the greater the separation. Therefore, energy-band widths depend upon the atomic spacing; if the spacing is close enough, the conduction and valence bands overlap, and there is no energy gap at the valence-electron level. In such materials, empty quantum states are closely adjacent to filled states, permitting small amounts of electron-energy increases. This condition is requisite for electrical conduction, since electrons accelerated in a solid by an applied electric field can acquire only small energy increases between collisions. Thus, we see that *electrical con-*

ductors are materials that have no energy gap at the valence-electron level. (Alkaline metals are believed to conduct as a result of a partially filled valence band rather than an overlapping of the valence and conduction bands.)

If the atomic spacing of a crystalline structure is large, a large energy gap (several electron volts) results, and very few valence-band electrons are excited into the conduction band at ordinary temperatures. For all practical purposes, the valence band is completely filled with electrons, there are no closely adjacent vacant quantum states, and it is not possible to produce small electron-energy increases by the application of an electric field. Therefore, high energy-gap materials are *nonconductors* or *insulators*. Since valence-band electrons are free to range throughout the crystal, we might expect them to contribute to electrical conduction. These electrons are affected by an externally applied electric field, but their motion is such that a balance of charge flow occurs and the net current is zero. This phenomenon may be explained by utilizing the *negative-mass* concept—a concept that will be useful in the consideration of hole flow in semiconductors. Higher-energy valence-band electrons are considered to have negative mass and are, therefore, accelerated in the direction of an applied electric field, the direction opposite that in which the lower-energy positive-mass valence-band electrons are accelerated. Thus, the charge flows neutralize each other and the net energy change of valence-band electrons is zero, since the gain in energy experienced by the accelerated positive-mass electrons equals the energy loss of those having negative mass.

If the energy gap is intermediate to the extremes of the gaps of conductors and insulators (say of the order of 1 eV), a few valence-band electrons will be thermally excited into the conduction band at ordinary temperatures, leaving a few vacancies in the valence band. Conduction-band electrons occupy quantum states that are closely adjacent to higher-energy empty states and, therefore, experience small energy increases in the presence of an applied electric field and contribute to electrical conduction. Since the density of these conduction-band electrons is much lower than the electron density in conductors, the resistivity of the material is several orders of magnitude higher than that of most metallic conductors, and the material is known as a *semiconductor*.

As we know from the discussion of Sec. 1-1.2, valence-band vacancies (holes) of a semiconductor also contribute to conduction. Since only high-energy electrons are excited to the conduction band, the vacancies exist near the top of the valence band. The electrons near the top of the band have negative mass and are, therefore, accelerated in the direction of the applied field, carrying the vacancies along with the same drift velocity. Let us say that each of these electrons has an effective mass $-m_v$, a charge $-e$, and an average drift velocity v_d. Suppose that we now consider two coincident particles to be placed in each vacancy, one having a mass $-m_v$, a charge

$-e$, and a drift velocity v_d in the direction of the field. This particle is seen to be identical with the surrounding electrons. Let the second particle in each vacancy have a mass $+m_v$, a charge $+e$, and a drift velocity v_d, also in the direction of the field. We see that the simultaneous addition of these two coincident particles to each vacancy has not altered the conditions of the system, since we have added zero net mass and zero net charge. However, the advantage of the addition is that the particles of the first type (electrons) together with all the other valence-band electrons completely fill the valence band. Since a completely filled band contributes no net current in an applied field, the only net current appearing in the valence band is that carried by the fictitious particle of mass $+m_v$ and charge $+e$ moving with velocity v_d in the direction of the field. Thus a valence-band vacancy or hole may be represented by a fictitious particle having a positive charge equal to the electronic charge and a positive mass equal to the effective mass of a valence-band electron. This representation of a hole is the basis for the hole contribution to the conductivity expression of Eq. (1-14). The lower mobility of holes, compared to that of conduction-band electrons, is the result of the greater effective mass of the more strongly bound valence-band electrons. Thus valence-band electrons, and the accompanying vacancies, experience a lower acceleration in an applied electric field and acquire a lower drift velocity per unit field.

Since both conductors and semiconductors conduct electrically, and low energy-gap semiconductors and certain metallic conductors may have comparable resistivities at elevated temperatures, it might appear difficult to distinguish between them. Practically, this may be done by the sign of their temperature coefficients of resistance, since semiconductors have a negative temperature coefficient over an appreciable temperature range and conductors have a positive coefficient at all temperatures. From Eq. (1-14), we see that the resistivity of a solid depends on carrier density and mobility, both of which may be temperature dependent. In a conductor with no energy gap at the valence-electron level, the number of electrons available for conduction is independent of temperature. The carrier mobility in both conductors and semiconductors decreases as temperature increases, as explained in Sec. 1-2.1. With electron density independent of temperature and electron mobility decreasing with increased temperature, a conductor has a resistivity that increases with temperature, showing a positive temperature coefficient. In a semiconductor, carrier density increases exponentially with temperature, overcoming the effect of temperature on carrier mobility, and resistivity decreases with increased temperature, showing a negative temperature coefficient.

Intrinsic semiconductors. The type of semiconductor considered so far in this section is the intrinsic semiconductor in which conduction-band electrons and valence-band vacancies (holes) exist in equal numbers and contribute to electrical conduction in proportion to their drift mobilities. An energy-level diagram for an intrinsic semiconductor (or an insulator) is shown in Fig. 1-19. As in Fig. 1-18, the ordinate of the plot is electron potential energy,

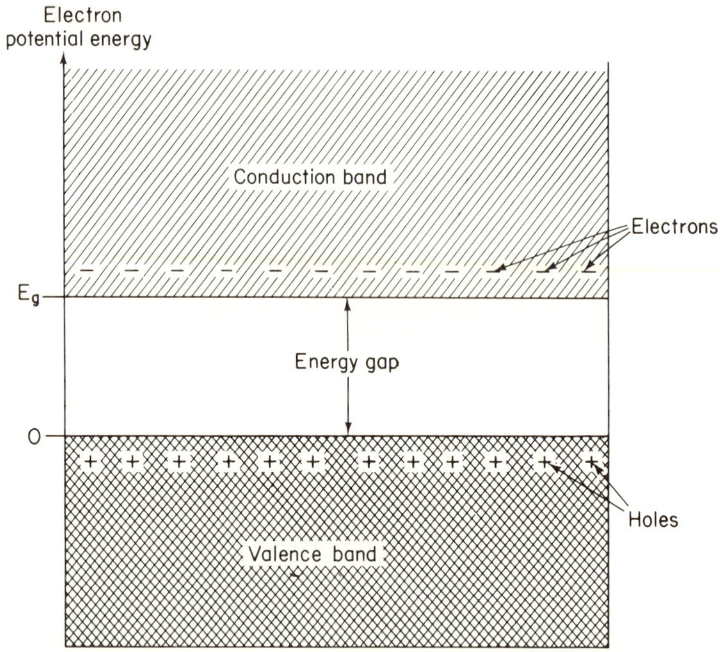

Fig. 1-19. Electron energy diagram for an intrinsic semiconductor or an insulator.

but the diagram differs from that of Fig. 1-18 in that the top of the valence band is considered to be the reference (zero energy level). This proves to be a convenient reference, since we are particularly interested in the region near the top of the valence band, where holes exist, and that near the botton of the conduction band, where free electrons are located. If this reference is used, energy level E_g is the width of the energy gap. The general locations of electrons and holes are shown in the diagram.

Extrinsic semiconductors. From the energy-level diagram point of view, the effect of five-valence impurity atoms in the lattice of a semiconductor is to cause the existence of an *occupied* (at a temperature of 0°K) energy level in the energy gap just below the conduction band, as shown in Fig. 1-20. A single energy level E_d exists, rather than an energy band, because the small impurity concentrations used result in large impurity-atom separations.

Energy E'_d, referred to as the *binding energy*, is the minimum energy that an impurity-atom valence electron must acquired in order to move into the conduction band. Since the binding energy is much less than the gap energy E_g, being 0.01 eV for germanium and 0.05 eV for silicon, the electrons of many impurity atoms are excited to the conduction band by thermal energies acquired at a temperature of a few degrees Kelvin, as shown by Fig. 1-11. For all practical purposes, donor exhaustion (100 percent ionization) occurs at temperatures well below room temperature, and the electron density may be considered equal to the donor-atom density.

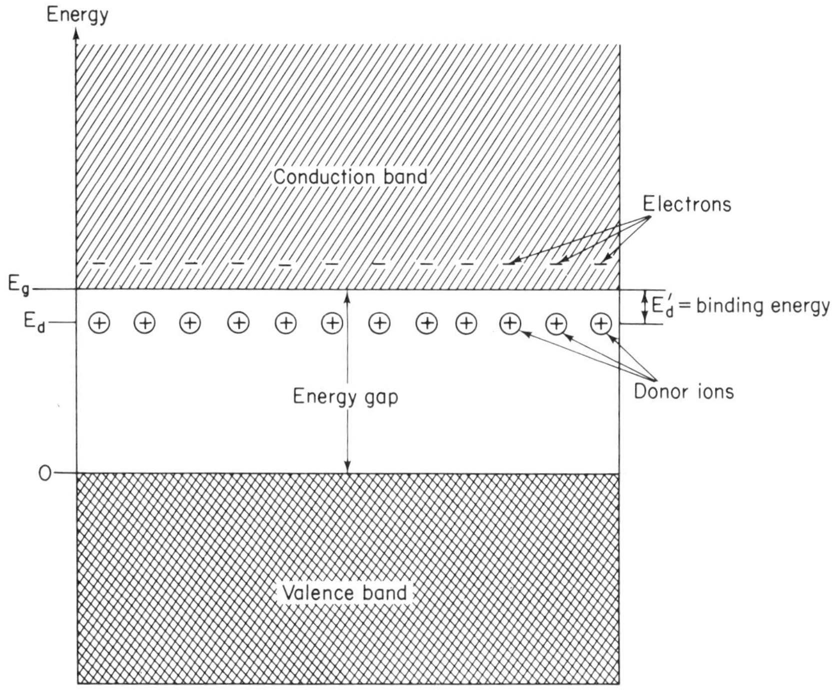

Fig. 1-20. Electron energy diagram for an *n*-type semiconductor.

Donated electrons and ionized donor atoms are shown in Fig. 1-20, but minority holes resulting from thermal disruption of valence bands are not shown. As pointed out in Sec. 1-1.3, *n*-type material is electrically neutral, since the densities of donated electrons and positive donor ions are equal.

The effect of three-valence impurity atoms in the lattice of a semiconductor is to cause the existence of an *unoccupied* (at a temperature of 0°K) energy level E_a in the energy gap just above the valence band, as shown in Fig. 1-21. Binding energy E'_a is the minimum energy that a valence-band electron must acquire in order to occupy energy level E_a. Like the binding energy E'_d of a donor atom, E'_a is small compared to E_g (being 0.01 eV for

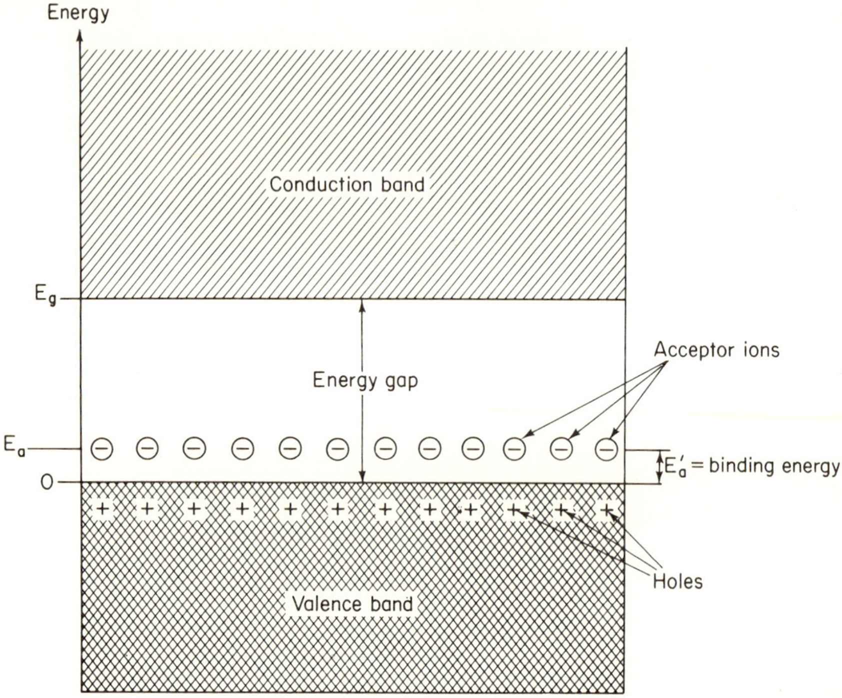

Fig. 1-21. Electron energy diagram for a *p*-type semiconductor.

germanium and 0.046 eV for silicon), and valence-band electrons are excited to energy level E_a by thermal energies at a temperature of a few degrees Kelvin. As a result, vacancies (holes) occur in the valence band, and the acceptor impurity atoms become negatively charged. At temperatures well below room temperature, 100 percent ionization occurs, and the hole density may be considered equal to the acceptor-atom density.

Ionized acceptor atoms and the resulting holes are shown in Fig. 1-21, but minority electrons resulting from thermal effects are not shown. Like *n*-type and intrinsic materials, *p*-type material is electrically neutral, electrical neutrality resulting from equal densities of positive holes and negative acceptor ions.

1-4.3 Distribution of Electron Energies and the Fermi Level

Except at a temperature of 0°K, at which the valence band is filled and the conduction band empty, our energy-level consideration so far has been a qualitative description of the energies of electrons in solids. Our ultimate objective is to determine the densities of conduction-band electrons and valence-band vacancies under realistic conditions. As a step in this direction, we shall now consider the distribution of energies of electrons in solids.

Let us first consider the meaning of an electron energy-distribution plot such as that of Fig. 1-22. Upon cursory consideration, this might appear to

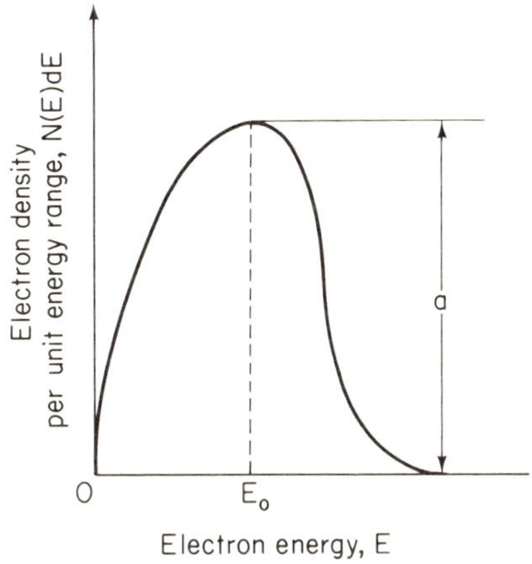

Fig. 1-22. An electron-energy distribution curve.

be simply a plot of electron density $N(E)$ versus electron energy E. However, the Pauli exclusion principle restricts to unity the number of electrons having a particular energy, such as E_0, and the density of electrons having a single energy value is meaningless. The density of electrons having energies within a particular range is meaningful, however, and the ordinate $N(E)\,dE$ of Fig. 1-22 is the number of electrons per unit volume with energies between the level E and the level $E + dE$. Thus, ordinate a in Fig. 1-22 is the density of electrons having energies within a band of differential width dE at the level E_0. The electron density n in the material represented by the plot of Fig. 1-22 may be obtained from the relationship

$$n = \int_0^\infty N(E)\,dE \tag{1-65}$$

The distribution of electron energies in a solid depends upon the density of the quantum states (energy levels) and the probability that the states are occupied; it is called the *Fermi distribution*.[5]

The Fermi distribution, which applies to all solids, is

$$N(E)\,dE = S(E)\,dE\,[f(E)] = S(E)\,dE\left[\frac{1}{e^{(E-E_f)/kT}+1}\right] \tag{1-66}$$

[5] For a derivation, see C. Kittel, *Introduction to Solid State Physics*, 2nd Ed. New York: Wiley & Sons, Inc., 1956, pp. 251–256.

where $N(E)\,dE$ is the number of electrons per unit volume with total energy between E and $E + dE$.

$S(E)\,dE$ is the number of quantum states per unit volume with total energy between E and $E + dE$.

$f(E)$ is the Fermi factor—the probability that a quantum state of energy E is occupied.

e is the Napierian base.

E_f is the *Fermi level* of energy (undefined so far, in our discussion).

k is Boltzmann's constant (1.38×10^{-23} joules/deg K or 8.61×10^{-5} eV/deg K).

T is temperature in degrees K.

The density of states $S(E)\,dE$ is different for different solids and is zero for values of E in the energy gap of a semiconductor, but the Fermi factor $f(E)$ applies to all solids.

The Fermi distribution function of Eq. (1-66) is similar to the *Maxwell-Boltzmann* function

$$f_{\mathrm{MB}}(E) = A\mathrm{e}^{-E/kT} \tag{1-67}$$

that describes gases and other rarefied systems in which there are many allowed energy states compared with the number of particles in the system. The Fermi function describes dense systems having comparable numbers of particles and states and is derived by application of the Pauli exclusion principle. For energies such that $E - E_f$ is large compared to kT (approximately 0.026 eV at room temperature), the Fermi factor expression is seen to be

$$f(E) = \mathrm{e}^{-(E-E_f)/kT} \tag{1-68}$$

and the two distribution functions are essentially the same.

Let us now examine the Fermi factor

$$f(E) = \frac{1}{\mathrm{e}^{(E-E_f)/kT} + 1} \tag{1-69}$$

for various values of temperature T, beginning with the limit $T \to 0$. For $T = 0°\mathrm{K}$ and $E < E_f$,

$$f(E) = \frac{1}{\mathrm{e}^{-\infty} + 1} = 1$$

For $T = 0°\mathrm{K}$ and $E > E_f$,

$$f(E) = \frac{1}{\mathrm{e}^{\infty} + 1} = 0$$

The relationship between Fermi factor $f(E)$ and energy E for $T = 0°\mathrm{K}$ is shown in Fig. 1-23. Every quantum state lying below the Fermi level E_f is filled, and every other quantum state is empty.

For a temperature of 300°K, $kT = 0.026$ eV, and the Fermi factor $f(E)$

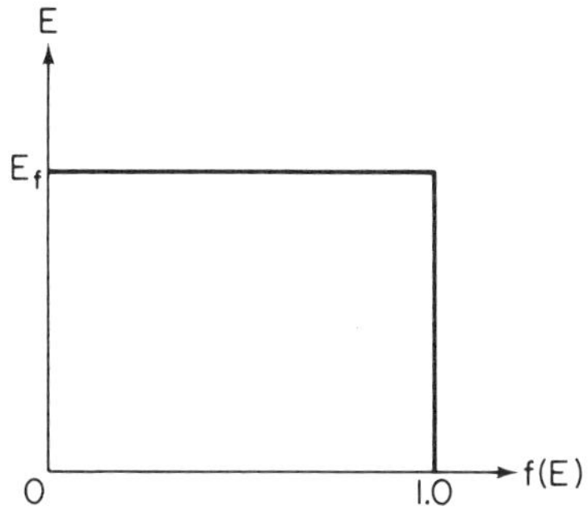

Fig. 1-23. Relationship between Fermi factor $f(E)$ and energy E at $T = 0°K$. All energy levels $E < E_f$ are filled; all other energy levels are vacant.

for $E = E_f + 0.026$ eV is

$$f(E) = \frac{1}{e + 1} = \frac{1}{3.718} = 0.27$$

For $T = 300°K$ and $E = E_f - 0.026$ eV, $f(E)$ is

$$f(E) = \frac{1}{e^{-1} + 1} = \frac{1}{1.37} = 0.73$$

For $E = E_f$,

$$f(E) = \frac{1}{e^0 + 1} = \frac{1}{2} = 0.5$$

Table 1-5 Fermi factor $f(E)$ for various values of energy E; $T = 300°K$ and $T = 1000°K$

E	$f(E)$, $T = 300°K$	$f(E)$, $T = 1000°K$
$E_f + 0.40$ eV	0.00	0.01
$E_f + 0.30$ eV	0.00	0.03
$E_f + 0.20$ eV	0.00	0.09
$E_f + 0.15$ eV	0.00	0.15
$E_f + 0.10$ eV	0.02	0.24
$E_f + 0.075$ eV	0.05	0.29
$E_f + 0.050$ eV	0.13	0.36
$E_f + 0.025$ eV	0.28	0.43
E_f	0.50	0.50
$E_f - 0.025$ eV	0.72	0.57
$E_f - 0.050$ eV	0.87	0.64
$E_f - 0.075$ eV	0.95	0.71
$E_f - 0.10$ eV	0.98	0.77
$E_f - 0.15$ eV	0.99	0.85
$E_f - 0.20$ eV	1.00	0.91
$E_f - 0.30$ eV	1.00	0.97
$E_f - 0.40$ eV	1.00	0.99

regardless of temperature. Values of $f(E)$ for additional values of E at $T = 300°K$ and for $T = 1000°K$ are given in Table 1-5 and plotted in Fig. 1-24. From Table 1-5, we see that $f(E)$ is nearly unity for E values a few kT less than E_f and nearly zero for values a few kT greater than E_f. Since kT is of the order of 0.1 eV or less at ordinary temperatures and the zero of E is arbitrary, a broken energy scale is used in the plot of Fig. 1-24 in order to show the details of the variation of $f(E)$ near $E = E_f$. But for the broken scale, the plot would look much like that of Fig. 1-23 for all ordinary temperatures.

The significance of the Fermi level E_f may now be seen. From the plots of Fig. 1-24 and from Eq. 1-69, we see that $f(E)$ is always 0.50 for $E = E_f$. *Therefore, the Fermi level E_f is the energy level having a probability of occupancy of 0.50.* At $T = 0°K$, the lowest energy levels are occupied by electrons until the electron supply is exhausted. *For a conductor*, the value of energy

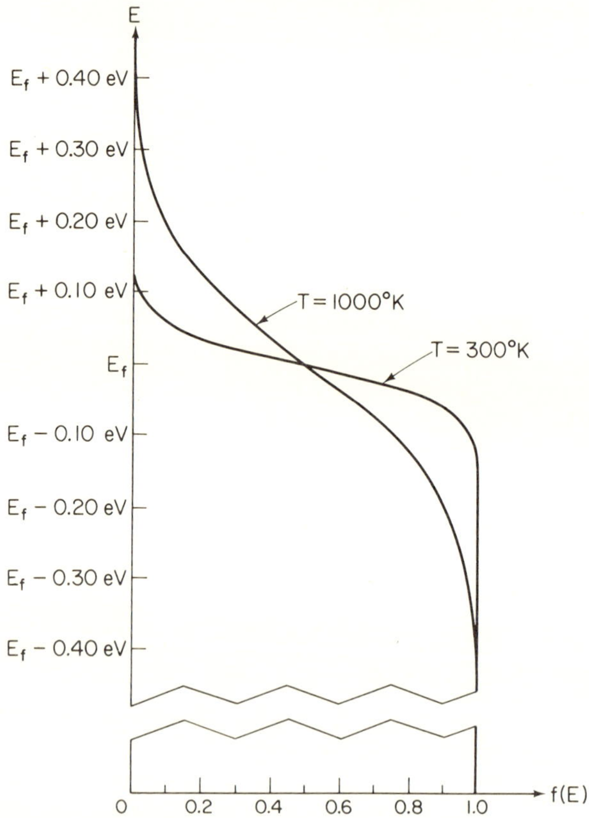

Fig. 1-24. Relationship between Fermi factor $f(E)$ and energy E at $T = 300°K$ and $T = 1000°K$.

E reached at this point is E_f. For this reason E_f is often called the *Fermi brim* and, *for a conductor, the Fermi level E_f is the highest energy level occupied at a temperature of 0°k.* This definition is not applicable to semiconductors because of the energy gap between the valence and conduction bands.

The Fermi level is of particular importance in the consideration of two solids in contact. When two solids are in contact and the system is in thermal equilibrium, the Fermi level E_f in one solid must be at exactly the same energy as E_f in the other solid. In other words, in an electron energy-level diagram for the system *the Fermi levels are aligned.* Suppose solid A is brought in contact with solid B, which has a lower E_f than that of A. When this contact occurs, electrons in A opposite vacant energy levels in B "spill over" into B, raising the energies of the bands and E_f in B and lowering them in A. This process continues until E_f in B is lined up with E_f in A. At the contact, a layer of negative charge (an excess of electrons) exists in B, and a layer of positive charge (a deficiency of electrons) occurs in A. This double layer of charge causes a step in electrostatic potential at the contact. Thus there is always a difference in electrostatic potential at the external surfaces of two dissimilar solids in contact. This is known as *contact potential.* If, instead of being in equilibrium, the two solids are connected to a voltage source, the Fermi levels are no longer in alignment. The separation in Fermi levels is the product of the electronic charge and the voltage applied to the contact (the source voltage, if negligible IR drop occurs in the solids). Thus, if the electron volt is the unit of energy in the energy-level diagram, the difference in Fermi levels in electron volts is equal to the applied emf in volts.

Conduction-band electrons in a metal. Application of the Fermi distribution to electrons in the conduction band of a metal requires knowledge of the density of states $S(E)$ in the band. It can be shown* that the density of states is zero at the band edges and varies as the one-half power of the energy level E in the lower part of the band; that is,

$$S(E) = kE^{1/2} \tag{1-70}$$

in the lower part of the band but $S(E)$ falls to zero at the top of the band.

The relationship between $S(E)$ and E is plotted in Fig. 1-25(a), and that between the Fermi factor $f(E)$ and E is shown by Fig. 1-25(b). Since $N(E)\,dE$, the electron density per unit energy range, is given by

$$N(E)\,dE = S(E)\,dE\,[f(E)]$$

the relationship between $N(E)$ and E is obtained from the point-by-point product of the plots of Figs. 1-25(a) and 1-25(b), as shown by the plot of Fig. 1-25(c). From Fig. 1-25(c), we see that the electron densities are greatest at energies near the Fermi level E_f. In fact, most (of the order of 85 percent)

* Reference 4, pp. 294–296, 608–612.

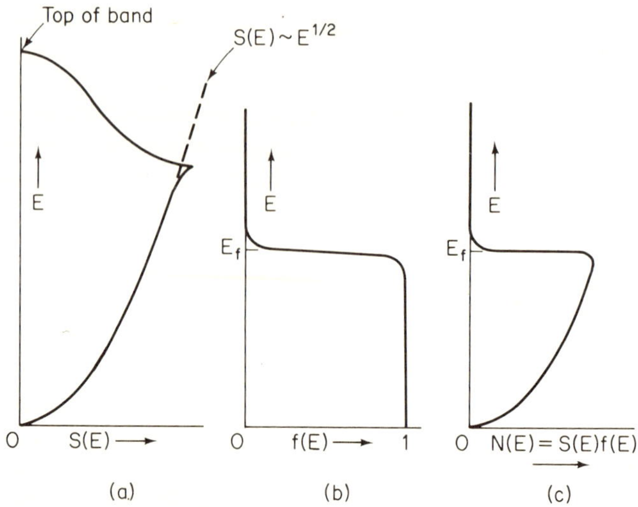

Fig. 1-25. Factors involved in the energy distribution of the conduction-band electrons of a metal. (a) Density of quantum states per unit energy range as a function of energy. (b) Probability that a state is occupied as a function of energy. (c) Density of electrons per unit energy range as a function of energy. (Adapted from Reference 4, p. 296.)

of the conduction-band electrons of a metal have energies that are essentially at the Fermi level.

1-4.4 Fermi Levels and Carrier Densities in Semiconductors

We shall now apply the concepts of energy bands and Fermi statistics to semiconductors. We are interested in the locations of the Fermi level in intrinsic and extrinsic materials and in the densities of carriers in these materials.

Intrinsic semiconductors. Since the location of the Fermi level E_f in a metal was determined in Sec. 1-4.3 by considering the special case $T = 0°K$, this approach *may be* helpful in locating E_f in an intrinsic semiconductor.

At $T = 0°K$, the valence band is filled and the conduction band is empty. Since the Fermi factor $f(E)$, or the probability of occupancy of an energy level, is unity at all energy levels in a filled band and zero at all energy levels in an empty band, $f(E)$ is seen to be unity at the top of the valence band and zero at the bottom of the conduction band. Thus, all we learn from conditions at $T = 0°K$ is that the Fermi level E_f of an intrinsic semiconductor is located somewhere in the energy gap. If it seems strange that the probability of occupancy is 0.5 somewhere in the energy gap, it should be remembered that $f(E)$ at an energy E is not the probability of the existence of an electron having energy E, but is the probability of the occupancy of an energy level E *if such a level exists*. Since there are no energy levels in the energy gap, the electron density $N(E)\,dE$ is zero in the gap regardless of the value of $f(E)$.

The location of E_f in an intrinsic semiconductor may be determined by considering a temperature above 0°K. At any such temperature, the number of conduction-band electrons must be the same as the number of holes in the valence band, since electron-hole pairs are produced by thermal disruptions of valence bonds. Since only the highest energy valence-band electrons are likely to be thermally excited to the conduction band where they are likely to occupy the lowest energy levels, the holes may be considered to exist at the top of the valence band at $E = 0$ (the top of the valence band being the reference) and the electrons at the bottom of the conduction band at $E = E_g$. The density of quantum states $S(E)\,dE$ near the bottom of the conduction band can be shown to be the same as that near the top of the valence band, as indicated in Fig. 1-26(a). With equal electron and hole densities and equal densities of states, the probability of occupancy of states at the bottom of the conduction band is equal to the probability of vacancy of states at the top of the valence band. Since the probability of vacancy is unity minus the probability of occupancy, $f(E)$ for $E = E_g$ is equal to $[1 - f(E)]$ for $E = 0$. That is,

$$f(E_g) = 1 - f(0) \tag{1-71}$$

Substituting in Eq. (1-71) the explicit expressions for $f(E)$ obtained from Eq. (1-69) gives

$$\frac{1}{e^{(E_g - E_f)/kT} + 1} = 1 - \frac{1}{e^{-E_f/kT} + 1} = \frac{e^{-E_f/kT}}{e^{-E_f/kT} + 1} \tag{1-72}$$

From Eq. (1-72),

$$e^{-E_f/kT} + 1 = e^{-E_f/kT}[e^{(E_g - E_f)/kT} + 1]$$

from which

$$e^{-E_f/kT} + 1 = e^{(E_g - 2E_f)/kT} + e^{-E_f/kT} \tag{1-73}$$

From Eq. (1-73),

$$1 = e^{(E_g - 2E_f)/kT}$$

indicating that

$$E_g - 2E_f = 0$$

or

$$E_f = \frac{E_g}{2} \tag{1-74}$$

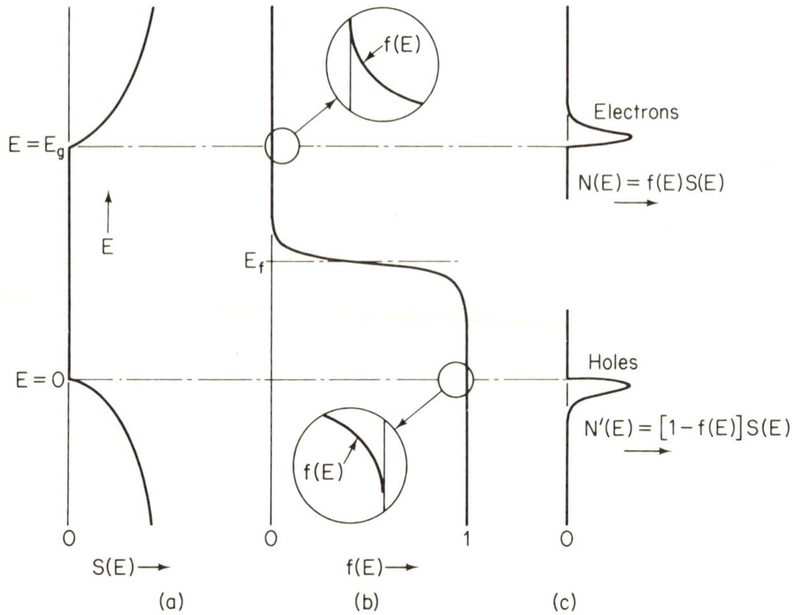

Fig. 1-26. (a) Density of states in a semiconductor. (b) Fermi factor of an intrinsic semiconductor; circled regions show $f(E)$ on a greatly enlarged scale. (c) Densities of electrons $N(E)$ and holes $N'(E)$ as functions of E in an intrinsic semiconductor; the horizontal scale is greatly enlarged compared to that of (a). (From Reference 4, p. 387.)

Thus we see that the Fermi level of an intrinsic semiconductor is in the middle of the energy gap midway between the top of the valence band and the botton of the conduction band.

A plot of the Fermi factor $f(E)$ of an intrinsic semiconductor is shown in Fig. 1-26(b), and the concentrations of conduction-band electrons, $N(E) = S(E)[f(E)]$, and valence-band holes, $N'(E) = S(E)[1 - f(E)]$, are plotted in Fig. 1-26(c). The electron density is seen to be zero at $E = E_g$ and the hole density is zero at $E = 0$, since $S(E)$ is zero at the edges of the conduction and valence bands. The electron density is also essentially zero for energies appreciably greater than E_g because the probability of occupancy $f(E)$ of such levels is very small, and the hole density is essentially zero for energies appreciably below the top of the valence band because of the small probability of vacancy $[1 - f(E)]$. Thus, as already suggested, most of the electrons have energies near the bottom of the conduction band, and most of the holes are at energy levels near the top of the valence band.

The density of conduction-band electrons n may be expressed as the product of an effective density of states in the conduction band, referred to as N_c, and the Fermi factor evaluated at the bottom of the conduction band. That is,

$$n = N_c f(E_g) \tag{1-75}$$

The expression for n given by Eq. (1-65) as

$$n = \int_0^\infty N(E)\,dE$$

may be modified for a semiconductor to

$$n = \int_{E_g}^\infty N(E)\,dE \tag{1-76}$$

since the energy gap is devoid of electrons. From Eqs. (1-75) and (1-76),

$$N_c f(E_g) = \int_{E_g}^\infty N(E)\,dE \tag{1-77}$$

which becomes

$$N_c f(E_g) = \int_{E_g}^\infty S(E)\,dE[f(E)] \tag{1-78}$$

by utilizing the expression for $N(E)\,dE$ given by Eq. (1-66). Since the density of states in the conduction band varies as the one-half power of E when the bottom of the band is the energy reference [see Eq. (1-70)], we may write

$$S(E) = K(E - E_g)^{1/2} \tag{1-79}$$

for a reference at the bottom of the energy gap. Substituting Eq. (1-79) into Eq. (1-78) and using the approximate $f(E)$ expression of Eq. (1-68), which is valid in the conduction band since $(E_g - E_f) \gg kT$, we obtain

$$N_c e^{-(E_g - E_f)/kT} = \int_{E_g}^\infty K(E - E_g)^{1/2} e^{-(E - E_f)/kT}\,dE \tag{1-80}$$

Solving Eq. (1-80) for N_c by multiplying both sides by $e^{(E_g - E_f)/kT}$ and evaluating the integral yields

$$N_c = K_1 T^{3/2} \tag{1-81}$$

Thus, from Eqs. (1-75) and (1-81),

$$n = K_1 T^{3/2} f(E_g) \tag{1-82}$$

Using the approximate $f(E)$ expression and recognizing that $E_f = E_g/2$, we find that Eq. (1-82) becomes

$$n = K_1 T^{3/2} e^{-(E_g - E_f)/kT} = K_1 T^{3/2} e^{-E_g/2kT} \tag{1-83}$$

Since one of the factors of constant K_1 of Eq. (1-83) is the effective electron mass, and this is unknown, the value of K_1 must be obtained experimentally. Using the conductivity of intrinsic germanium measured at a temperature of 300°K and an E_g value of 0.67 eV (the value measured optically at 300°K), K_1 is found to be 2.075×10^{21} for n in electrons per cubic meter. That is, the density of conduction-band electrons in intrinsic germanium is given by

$$n = 2.075 \times 10^{21} T^{3/2} e^{-E_g/2kT} \text{ m}^{-3} \tag{1-84}$$

In using this expression, it should be recognized that E_g is temperature dependent and must be chosen accordingly. The energy gap decreases with increase in temperature, owing to changes in the crystal lattice, and has been found to be given by

$$E_g = 0.785 - \left(\frac{0.785 - 0.67}{300}\right) T \text{ eV} \tag{1-85}$$

for germanium *at ordinary temperatures.* Equation (1-85) is in error at low temperatures, say below 50°K, since the value of E_g is found to be 0.75 eV at 0°K rather than 0.785 eV, as obtained by extrapolating the experimental $E_g(T)$ characteristic.

Combining Eqs. (1-84) and (1-85) yields

$$n = 1.76 \times 10^{22} T^{3/2} e^{-0.785/2kT} \text{ m}^{-3} \tag{1-86}$$

for intrinsic germanium. Substituting for Boltzmann's constant and recognizing that $n \equiv n_i$, we obtain

$$n_i = 1.76 \times 10^{22} T^{3/2} e^{-4550/T} \text{ m}^{-3} \text{ for germanium} \tag{1-87}$$

Similarly, for intrinsic silicon

$$n = 3.88 \times 10^{22} T^{3/2} e^{-1.21/2kT} \text{ m}^{-3} \tag{1-88}$$

and

$$n_i = 3.88 \times 10^{22} T^{3/2} e^{-7000/T} \text{ m}^{-3} \text{ for silicon} \tag{1-89}$$

Equations (1-87) and (1-89) have been found valid in the temperature range 300°K–500°K for germanium and 400°K–700°K for silicon.*

An expression for the conductivity of an intrinsic semiconductor may be obtained by combining Eqs. (1-15) and (1-83); by so doing we obtain

$$\sigma_i = K_1 T^{3/2} e(\mu_n + \mu_p) e^{-E_g/2kT} \tag{1-90}$$

Since the temperature has a much greater effect on the exponential factor of Eq. (1-90) than on the other factors, the logarithm of intrinsic conductivity is essentially a linear function of $1/T$, as shown by the intrinsic regions of Fig. 1-12. Therefore, the approximate value of the energy gap E_g can be experimentally determined by measuring intrinsic conductivity as a function of temperature.

Certain properties of intrinsic germanium and silicon are given in Table 1-6. The intrinsic carrier density of silicon is seen to be much smaller than that of germanium. From Eqs. (1-86) and (1-88), we see that this is due primarily to the difference in the energy gaps of the material and that E_g has an enormous effect on intrinsic-carrier density.

*Reference 2, p. 1291.

Table 1-6 Properties of intrinsic germanium and silicon

	Germanium	*Silicon*
Energy gap, eV		
0°K	0.75	1.15
300°K	0.67	1.106
Electron mobility (300°K) $cm^2/(volt)(sec)$	3900	1350
Hole mobility (300°K) $cm^2/(volt)(sec)$	1900	480
Intrinsic carrier density (300°K), m^{-3}	2.4×10^{19}	1.5×10^{16}
Resistivity (300°K), ohm-cm	46	2.3×10^5
Electron diffusion constant (300°K), cm^2/sec	100	35
Hole diffusion constant (300°K), cm^2/sec	49	12
Dielectric constant	16	12

Extrinsic semiconductors. The location of the Fermi level E_f in an extrinsic semiconductor may be determined by utilizing Eq. (1-82). Using the approximate $f(E)$ expression of Eq. (1-68) and the value of K_1 for germanium, we find that the general expression for the density of electrons in the conduction band of germanium is

$$n = 2.075 \times 10^{21} T^{3/2} e^{-(E_g - E_f)/kT} \tag{1-91}$$

electrons per cubic meter. In n-type germanium at temperatures for which n_n may be considered to be the donor-atom density, Eq. (1-91) may be solved for E_f. As an example, let us calculate the Fermi level at a temperature of 300°K, of germanium having a donor-atom density of 10^{22} atoms per cubic meter. If we assume $n_n = N_d$, an assumption whose validity will be checked, Eq. (1-91) becomes

$$10^{22} = 2.075 \times 10^{21} \times 5.2 \times 10^3 e^{-(E_g - E_f)/0.0258} \tag{1-92}$$

from which

$$e^{-(E_g - E_f)/0.0258} = 0.93 \times 10^{-3} \tag{1-93}$$

Solving Eq. (1-93), we obtain

$$E_g - E_f = 0.18 \text{ eV}$$

and

$$E_f = E_g - 0.18 = 0.67 - 0.18 = 0.49 \text{ eV}$$

indicating that the Fermi level is located in the energy gap 0.18 eV below the bottom of the conduction band.

The Fermi-level location may also be determined by writing Eq. (1-84) as

$$n_i = 2.075 \times 10^{21} T^{3/2} e^{-E_g/2kT} \tag{1-94}$$

since $n_i = n_n$ for intrinsic material. Dividing Eq. (1-91) by Eq. (1-94), we obtain

$$\frac{n}{n_i} = \frac{e^{-(E_g - E_f)/kT}}{e^{-E_g/2kT}} = e^{(E_f - E_g/2)/kT} \tag{1-95}$$

and solving this equation for E_f gives the location of the Fermi level with respect to the middle of the energy gap. From Eq. (1-95), we see that the Fermi level lies above the middle of the energy gap for material having an electron density greater than that of intrinsic material *at the same temperature*. Unlike Eq. (1-91), Eq. (1-95) is a general equation that is not limited to germanium. Perhaps the most important conclusion to be reached from the consideration of Eq. (1-95) is that *the Fermi level of an n-type semiconductor lies above the middle of the energy gap.*

The hole density p in the valence band can be expressed as the product of an effective density of states N_v in the valence band and the quantity unity minus the Fermi factor evaluated at the top of the band. That is,

$$p = N_v[1 - f(E_v)] \tag{1-96}$$

By the same type of development that led to Eq. (1-81) for N_c, the effective density of states in the conduction band, it can be shown that N_v and N_c are given by the same expression. That is,

$$N_v = K_1 T^{3/2} \tag{1-97}$$

and, since E_v is zero,

$$p = K_1 T^{3/2}[1 - f(0)] \tag{1-98}$$

which becomes, for germanium,

$$p = 2.075 \times 10^{21} T^{3/2}\left[1 - \frac{1}{e^{(0-E_f)/kT} + 1}\right] \text{m}^{-3} \tag{1-99}$$

Since $e^{-E_f/kT} \ll 1$, Eq. (1-99) has the approximate form

$$p = 2.075 \times 10^{21} T^{3/2} e^{-E_f/kT} \text{ m}^{-3} \tag{1-100}$$

Let us now use Eq. (1-100) to calculate the hole density in the n-type germanium of our example (material having an electron density of 10^{22} electrons per cubic meter and an E_f of 0.49 eV). In so doing we obtain

$$p = 2.075 \times 10^{21} \times 5.2 \times 10^3 e^{-0.49/0.0258} = 6 \times 10^{16}/\text{m}^3$$

This is the density of holes resulting from thermal disruption of valence bonds and equals the density of electrons produced by the same process. Thus we see the validity of our assumption that the density of electrons in the material of our example equals that of the donor impurities.

It is of interest to compare the hole density in the n-type material of our example with that of intrinsic material at the same temperature, given in Table 1-6 as 2.4×10^{19} holes per cubic meter. From this comparison, we see that the hole density is decreased by increasing the electron density through the addition of donor impurities. This phenomenon is known as *minority-carrier suppression* and, as explained earlier, occurs because the increased electron density increases the likelihood of the recombination of holes and electrons.

The hole density-electron density product pn proves to be an interesting and useful relationship. From Eqs. (1-91) and (1-100) this product is

$$pn = (2.075 \times 10^{21} T^{3/2})^2 e^{-E_g/kT} \tag{1-101}$$

which, from Eq. (1-94), is seen to be

$$pn = n_i^2 \tag{1-102}$$

Equation (1-102) is useful in the determination of the minority-carrier density for conditions such that the majority-carrier density may be considered to be equal either to the donor-atom density of n-type material or the acceptor-atom density of p-type material. In such determinations, n_i is the density of holes (or electrons) in intrinsic material *at the temperature of the n-type or p-type material under consideration.*

Equation (1-102) suggests yet another method for determining the Fermi-level location for n-type material when the donor-atom density is known. We may make this determination by solving the hole-density expression Eq. (1-100) for E_f, using the hole density obtained from Eq. (1-102).

The preceding analysis for n-type germanium is illustrated in Fig. 1-27. It is of interest to compare this illustration with that of Fig. 1-26 for intrinsic material. The horizontal scale of Fig. 1-27(c) is greatly enlarged compared to that of Fig. 1-27(a), but less so than that of Fig. 1-26(c) for the same temperature. That is, there are many more conduction-band electrons in Fig. 1-27(c) than in Fig. 1-26(c). There are practically no holes in the valence band of Fig. 1-27(c), because of the larger $f(E)$ values compared to those of Fig. 1-26(c), and practically no electrons are bound to donor atoms, owing to the small value of $f(E_d)$.

The location of the Fermi level of p-type material may be determined in the same manner as for n-type material. For temperatures such that the hole density p_p may be considered equal to the acceptor-atom density N_a, E_f may be obtained by solving the hole-density expression. As an example, let us calculate the Fermi level, at a temperature of 300°K, of germanium having an acceptor-atom density of 10^{22} atoms per cubic meter. Assuming $p_p = N_a$, we find that Eq. (1-100) becomes

$$10^{22} = 2.075 \times 10^{21} \times 5.2 \times 10^3 e^{-E_f/0.0258}$$

from which

$$e^{-E_f/0.0258} = 0.93 \times 10^{-3} \tag{1-103}$$

Solving Eq. (1-103), we obtain $E_f = 0.18$ eV, indicating that the Fermi level is located in the energy gap 0.18 eV above the top of the valence band.

The Fermi-level location for p-type material may also be determined from an expression relating hole density to intrinsic carrier density. This expression is obtained by dividing Eq. (1-100) by Eq. (1-94), resulting in

$$\frac{p}{n_i} = \frac{e^{-E_f/kT}}{e^{-E_g/2kT}} = e^{(E_g/2 - E_f)/kT} \tag{1-104}$$

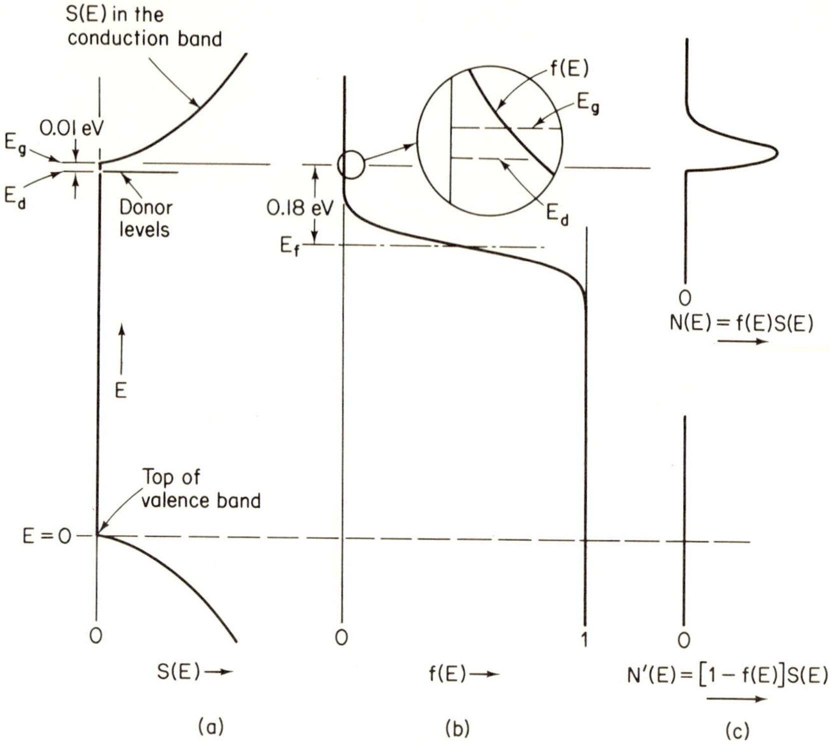

Fig. 1-27. Density of states $S(E)$, Fermi factor $f(E)$, and Fermi level E_f, and densities of electrons $N(E)$, and holes $N'(E)$ in n-type germanium at 300°K with 10^{22} donors per m^3. (Adapted from Reference 4, p. 394.)

From Eq. (1-104), we see that the Fermi level lies below the middle of the energy gap for material having a hole density greater than that of intrinsic material *at the same temperature*. Thus *the Fermi level of a p-type semiconductor lies below the middle of the energy gap*. As for n-type material, the Fermi-level location for p-type material may also be determined by using the minority-carrier density, obtained from Eq. (1-102), in conjunction with the appropriate carrier-density expression.

REFERENCES

1 E. M. Conwell, "Properties of Silicon and Germanium," *Proc. IRE*, Vol. 40, No. 11 (November 1952).

2 E. M. Conwell, "Properties of Silicon and Germanium II," *Proc. IRE*, Vol. 46, No. 6 (June 1958).

[3] J. R. Haynes and W. Shockley, "Mobility and Life of Injected Holes and Electrons in Germanium," *Physical Review*, Vol. 81 (1951), p. 835.

[4] Robert L. Sproull, *Modern Physics, A Textbook for Engineers*, 2nd ed. New York: John Wiley & Sons, Inc., 1963, p. 266.

[5] C. Kittel, *Introduction to Solid State Physics*, 2nd ed. New York: John Wiley & Sons, Inc., 1956, pp. 251–256.

PROBLEMS

1-1. The diameter of No. 18 copper wire is 1.01 mm, the resistance is 6.5 Ω per 1000 ft, and the electron concentration is 8.4×10^{22} cm^{-3}. Determine the mobility of the electrons.

1-2. Determine the drift velocity of electrons for a current of 1.0 A in the copper wire of Prob. 1-1.

1-3. A 1.0-ohm-cm germanium rod has the cross-sectional area of No. 18 copper wire (Prob. 1-1) and an electron mobility of 3600 cm^2/(volt) (sec). Determine the average drift velocity of electrons for a current of 0.10 A.

1-4. Using the intrinsic carrier-density and drift-mobility values of Table 1-6, determine the resistivity in ohm-cm of: (a) intrinsic germanium and (b) intrinsic silicon.

1-5. What value of intrinsic carrier density, carrier mobilities, and resistivity might be expected for intrinsic germanium at a temperature of 195°K (that of dry ice)?

1-6. A specimen of *n*-type germanium has a donor-atom density of 8.7×10^{14} atoms/cm^3. Neglecting the effect on drift mobilities of impurity-ion scattering, determine the minority-carrier density and resistivity at 300°K.

1-7. Repeat Prob. 1-6 for *p*-type germanium having an acceptor-atom density of 3.7×10^{14} atoms per cm^3.

1-8. If both donor atoms and acceptor atoms are present in germanium, determine the resistivity at 300°K for the impurity-atom densities of Probs. 1-6 and 1-7.

1-9. (a) A sample of *n*-type germanium has a resistivity of 5 ohm-cm at 300°K. What resistivity might be expected at 195°K (the temperature of dry ice)?
(b) Repeat (a) for *p*-type germanium.

1-10. In the Hall-effect arrangement of Fig. 1-13, $a = b = 0.1$ cm and $V_y = 0$ for $B = 0$. For $B = 3000$ gauss, $V_y = -20$ mV.
(a) What is the material type? Explain.
(b) What is the average drift velocity of the carriers?

1-11. If the Hall-effect specimen in Prob 1-10 is a 0.75-in. rod of 5-ohm-cm germanium having a carrier mobility of 3000 cm^2/(volt)(sec), determine the current through the specimen.

1-12. If the specimen in Probs. 1-10 and 1-11 is a copper rod having 8.4×10^{22} electrons per cm^3 with a mobility of 40 cm^2/(volt)(sec), determine the magnitude of the Hall voltage V_y for a specimen current of 1.0 A.

1-13. A 0.05-in. $\times$ 0.05-in. $\times$ 0.75-in. germanium rod arranged as in Fig. 1-13 has a Hall voltage of 27.5 mV for a current of 10 mA and a magnetic field of 3000 gauss. If the voltage applied to the rod is 5.0 V, determine the mobility and approximate density of carriers in the specimen.

1-14. The drift mobility of holes in a semiconductor sample is 400 cm^2/(volt)(sec) at a temperature of 300°K. What is the diffusion constant?

1-15. What diffusion constant might be expected for electrons in germanium at a temperature of 195°K (that of dry ice)?

1-16. As a result of minority-carrier injection, the hole density p_n in the n-region of a device has the distribution shown in Fig. P1-16. The n-region is a 5-ohm-

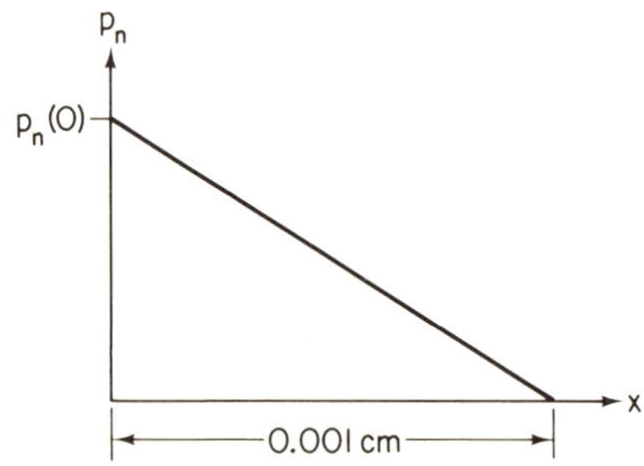

Figure P1-16

cm germanium section having a cross-sectional area of 0.01 cm^2 and a thickness of 0.001 cm. The hole density $p_n(0)$ at one edge of the region is 100 times the equilibrium value p_{no} and the hole density at the other edge is essentially zero. The temperature is 300°K. If the n-region is essentially field-free and holes are able to leave the region in the $+x$ direction but electrons are unable to do so, what electric current might be expected to flow through the region?

1-17. The diffusion length of elctrons in a p-type germanium specimen is found to be 0.010 in. at a temperature of 300°K. What is the average minority-carrier lifetime?

1-18. In measuring excess hole density p'_n in an n-type semiconductor as a function of distance from the point of illumination, p'_n is halved when the distance is increased by 0.01 in. Determine the hole diffusion length.

1-19. Determine the Fermi level at 300°K, considering the bottom of the energy gap as the reference, for n-type germanium having a donor-impurity density of 8.7×10^{14} atoms/cm^3.

1-20. Repeat Prob. 1-19 for p-type germanium having an acceptor impurity density of 3.7×10^{14} atoms/cm^3.

1-21. A plot of resistance R in ohms versus 1/T in $(\text{deg K})^{-1}$ for a semiconductor rod is shown in Fig. P1-21. Determine the approximate value of the energy gap of the semiconductor material.

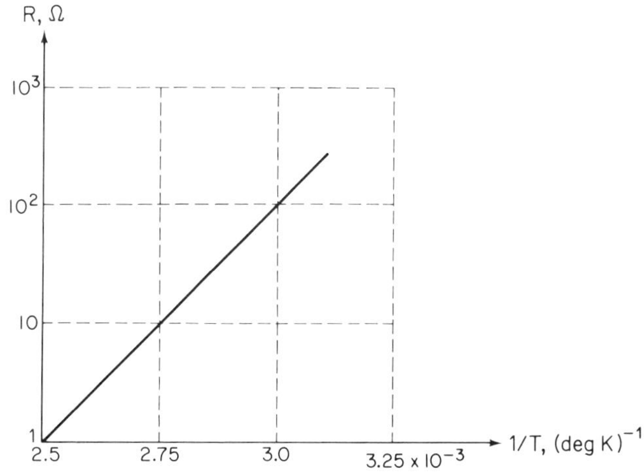

Figure P1-21

2

Properties of pn Junctions

Since almost all important phenomena that take place in semiconductor devices occur either at junctions of two different types of material or at metal-to-semiconductor junctions (most commonly referred to as metal-semiconductor contacts), the understanding of junction properties is basic to device considertion. Although, historically, the first semiconductor devices employed contacts rather than *pn* junctions, and connections to *pn* junctions require metal contacts, the *pn* junction is easier to understand, and the understanding of it is helpful in the consideration of metal-semiconductor contacts. Therefore, metal-semiconductor-contact consideration will be deferred to Chapter 3.

2-1 THE NATURE OF *pn* JUNCTIONS

A *pn* junction is the boundary region between *p*-type and *n*-type regions of a single crystal. Such a junction cannot be formed by merely pressing together two types of material, since intimate contact is prevented by surface roughness, surface oxide films, and other crystal imperfections; instead, it takes the form of a continuous single crystal having an impurity content (doping) that varies thorugh the crystal, causing one side to be *p*-type and the other *n*-type. In order to emphasize this point, and because we need to know some of the basic differences between junctions produced by different processes, we shall briefly discuss some methods for forming *pn* junctions.

Forming pn junctions. Junctions of the *pn* type are usually formed in one of three general ways: *grown-junction* processes, *alloy or fused-junction* processes, and *diffused-junction* processes. In the grown-junction method, a small piece of single-crystal material known as a seed crystal is dipped into molten germanium or silicon to which 5-valence donor impurities have been added to form *n*-type material. This adding of impurities is known as *doping*. As the seed is slowly withdrawn from the melt, the molten material adheres to it and solidifies with its atoms lining up with those of the seed. This is known as *crystal growing* or pulling and an entire ingot of single-crystal *n*-type material may be produced in this manner if desired.* It is also a purification method, since undesired impurity atoms have an affinity for the molten form of the material and most of them remain behind in the melt. Pulling or withdrawing rates are quite slow, ranging from less than 0.1 in. per hr to as high as 4 in. per hr, or higher. If a *pn* junction is desired, a portion of the crystal is formed, the melt is then doped with 3-valence acceptor impurities to form *p*-type material, and the remainder of the ingot is grown. In this way, a single crystal composed of both *n*-and *p*-type material is formed. In the conversion from *n*-type to *p*-type material, holes introduced by acceptor atoms recombine with electrons from donor atoms, as explained in Sec. 1-2.1, the process continuing until the acceptor atoms are in the majority. Thus acceptors and donors cancel each other, as far as carriers are concerned, and the conductivity type depends upon which species is in excess.

By a somewhat different process, *rate-grown* junctions are formed. It has been found that the extent to which *n*-forming impurities solidify out varies directly with the rate at which the crystal is gown, but the *p*-forming impurity solidification is independent of the growing rate. Thus, if the seed crystal is withdrawn very slowly from a melt containing both types of impurity, the impurities present are largely of the *p*-forming type. If the rate is then increased, *n*-forming impurities solidify more rapidly than those of the *p*-forming type, and the *n*-type region of the crystal is formed. This process lends itself readily to the production of a crystal consisting of a series of alternate *n*-type and *p*-type layers.

Grown junctions formed by either method have a rather gradual transition from the region of uniform acceptor concentration to that of uniform donor concentration, and are referred to as *graded junctions*. Since the junction area of a grown ingot is much larger than that usually required for a single device, the ingot is cut into rods along its length and leads are attached. If

* *P*-type single-crystal ingots may be formed by this method, beginning with a melt doped with 3-valence acceptor impurities.

the ingot is of the type that contains many junctions, each rod is cut into separate sections and leads are attached to each section.

For several years following the announcement of junction devices, processes for growing junctions were laboratory operations and grown-junction devices were produced in small quantities at high cost. Because of this, early commercial junction devices (diodes and transistors) were produced by the alloy process, the first process to be adapted to large-scale production. In this process, a wafer of either *p*-type or *n*-type material is coated or otherwise brought in contact with impurity material that forms semiconductor material of the opposite type when used as a doping agent. Typical examples are an *n*-type silicon wafer coated (by an evaporation process) with aluminum, a 3-valence *p*-forming element; a pellet of indium, a *p*-forming element, in contact with a wafer of *n*-type germanium; and a pellet of lead doped with arsenic, a 5-valence *n*-forming element, in contact with *p*-type germanium. The semiconductor-impurity assembly is then heated in a furnace to a temperature above the melting point of the impurity material but below that of the semiconductor. As a result, semiconductor material from the surface of the crystal dissolves into the molten impurity material to form a liquid solution or alloy. As the crystal is cooled, *heavily doped* semiconductor material crystallizes as an extension of the original crystal region and a *pn* junction exists at the surface that was the boundary between the solid and the melt. (The actual junction lies a fraction of a mil farther into the crystal, owing to the diffusion of impurity atoms.) Leads are attached to the impurity material, the outermost part of which is essentially pure, and to a contact made to the back surface of the wafer. It is important to recognize that the alloyed or regrown region is high-conductivity semiconductor material. For example, the *p*-region formed by alloying indium to *n*-type germanium has a resistivity at room temperature of the order of 0.01 ohm-cm, or less, compared to original-crystal resistivities of the order of 1 to 10 ohms-cm. Junctions formed by alloying have a very narrow transition from essentially uniformly doped *p*-type material to uniformly doped *n*-type material and are considered to be *abrupt junctions*.

A third method for forming *pn* junctions is the vapor-diffusion process. It is the most recent of the three and has largely replaced the other two methods. In this process, a heated slab of single-crystal *n*-type or *p*-type material is exposed to *p*-forming or *n*-forming impurities in vapor form, there being no liquid phase as in the other two methods. The impurity diffuses slowly into the surface of the material, the speed of the diffusion process being greatly increased by elevating the temperature, with the impurity concentration highest at the surface and decreasing exponentially inward. The inner region of the original crystal is unaffected by the diffusion action, but material near the surface becomes doped to the opposite conductivity type and a *graded junction* is formed just below the surface. Large-area uniform junctions can be formed in this way. If a large-area device is not required, the wafer can be scribed and broken into many small-area junctions. The vapor-diffusion process can be closely controlled and is extensively used to form diodes and transistors. It is also used in the fabrication of integrated circuits (Sec. 11-3).

We shall now direct our attention to the electrical nature of the *pn* junction. At absolute zero, the donated electrons of the *n*-region and the holes of the *p*-region are bound to their parent atoms, as shown in Fig. 2-1(a), in which immobile impurity atoms are represented by circled polarity signs and uncircled polarity signs represent holes and electrons. If the temperature of the crystal is now increased, say to room temperature, thermal lattice vibrations liberate almost all of the holes and electrons of the impurity atoms and create a few electron-hole pairs through valence-bond disruptions.

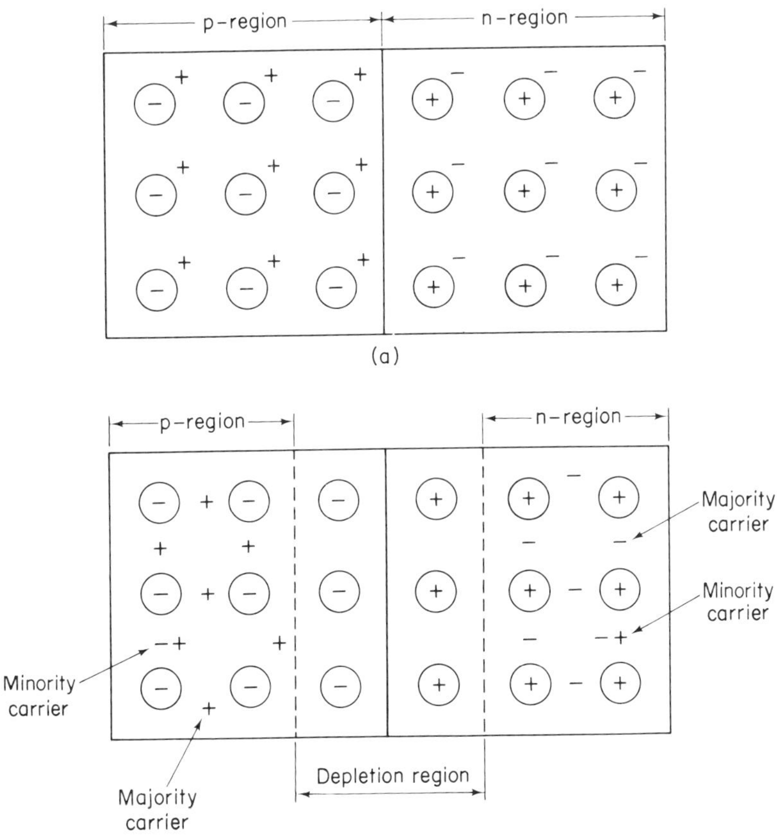

Fig. 2-1. Electronic composition of a *pn* junction. (a) Composition at absolute zero. (b) Composition at room temperature.

These liberated carriers are free to move throughout the crystal, and a diffusion flow of p-region holes into the n-region and of n-region electrons into the p-region begins. Prior to this flow the crystal was electrically neutral, but electron diffusion leaves unneutralized positive donor ions on the n-side of the junction and hole diffusion leaves unneutralized negative acceptor ions on the p-side. These ions create an electric field at the junction that opposes the diffusion action and the action ceases when the field is strong enough to oppose any further *net* diffusion. That is, under equilibrium conditions, positive charge on the n-side prevents the further entering of p-region holes and additional n-region electrons are repelled from the p-side by the negative charge therein. Thus we see that the net effect of the diffusion of carriers across the junction is the creation on each side of the junction of a thin layer that is depleted of carriers, as shown in Fig. 2-1(b).This is known as the *carrier-depletion* region, the *space-change* region, the *transition region*, or as simply the *junction region.*

The results of majority-carrier diffusion are illustrated in Fig. 2-2 for an abrupt junction having essentially the same doping level on each side of the junction. (As mentioned earlier, this *is not* a characteristic of an abrupt junction formed by alloying.) The x coordinate of each plot represents distance from the junction in a direction normal to the plane of the junction. Figure. 2-2(a) shows the diffusion tendencies of majority carriers, Fig. 2-2(b) is a plot of electrostatic potential in the junction region, and Fig. 2-2(c) shows the variation of space-charge density with distance from the junction. Since the difference in electrostatic potential between the n- and p-regions is indicative of the amount of energy required to move either a hole from the p-region to the n-region or an electron from the n-region to the p-region, this potential difference is known as the junction *barrier potential.* The direction of the electric field in the space-charge region, as shown in Fig. 2-2(b), is seen to be such that minority carriers drift across the junction. Under equilibrium conditions, this drift current is exactly equal to the opposite current of majority carriers having sufficient energy to surmount the potential barrier. In other words, the drift current of p-region electrons is equal to and opposite the diffusion current of electrons into the p-region, and the drift current of n-region holes is neutralized by the hole diffusion current.

Let us now consider the effects of applying an external voltage between terminals that serve as contacts to the two regions. If the polarity of the applied voltage makes the p-region positive with respect to the n-region, the height of the potential barrier is reduced, as shown in Fig. 2-3. This permits increased numbers of p-region holes and n-region electrons to cross the junction and a net current of majority carriers occurs. If the voltage drops in the contacts and in the p- and n-regions outside the carrier-depletion (space-charge) region are negligible, the potential-barrier reduction is equal to the applied voltage. Since the number of carriers having sufficient energy

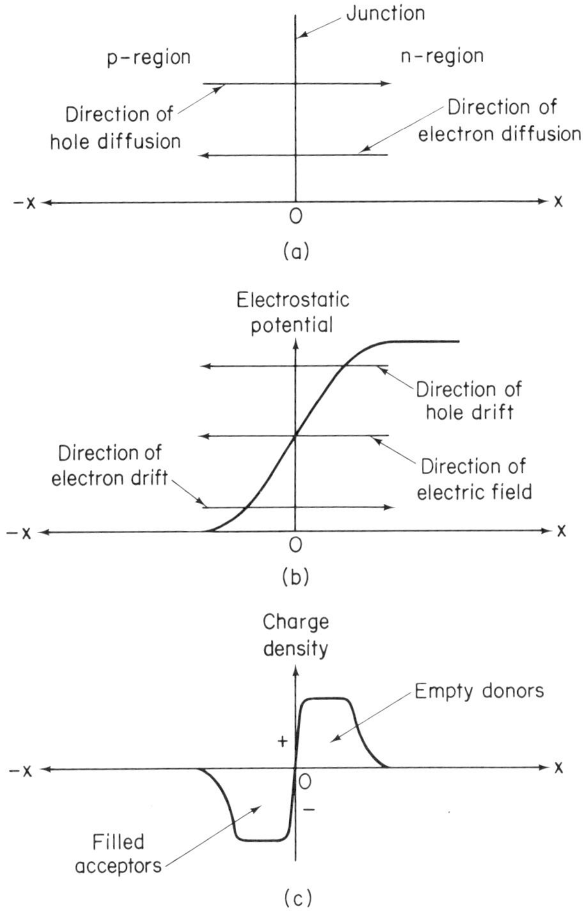

Fig. 2-2. The *pn* junction in equilibrium. (a) Carrier diffusion tendencies. (b) Potential barrier and carrier drift tendencies. (c) Space-charge layer.

to climb the barrier decreases exponentially with barrier height, a small change in height causes a large current change and junction current increases strongly with applied voltage. This is the low-resistance or *forward connection* of the *pn* junction, and the junction is said to be *forward biased.* The equilibrium barrier height is dependent upon the material and its doping, and is of the order of tenths of a volt, at most. Therefore, a comparatively small externally applied voltage produces appreciable conduction in the forward direction, In most *pn*-junction devices, an external voltage of the order of 1 V results in considerable forward current—so much so that most of the applied voltage appears as drops in the *p*- and *n*-regions and in the contacts, and *damage to the device may ensue*. The amount of current that can be safely

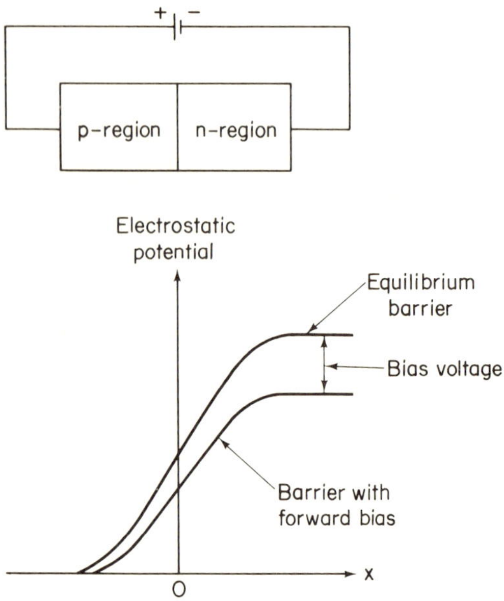

Fig. 2-3. The *pn* junction with applied forward bias.

passed depends upon the junction area, the cooling provisions, etc. Large-area devices have forward-current ratings of the order of hundreds of amperes.

Reversing the polarity of the applied voltage has a quite different effect on the behavior of the *pn* junction. When the *p*-region is negative with respect to the *n*-region, the potential-barrier height is increased, as shown in Fig. 2-4. This decreases the number of majority carriers having sufficient energy to climb the barrier, and a net current of minority carriers occurs. For voltages of the order of 0.1 V or greater, the majority-carrier current is so reduced that the junction current consists almost entirely of minority carriers and is essentially independent of voltage magnitude. This is known as the *reverse-saturation current* I_s, and it occurs because the minority-carrier current is independent of bias magnitude and polarity but is determined by the rate of thermal generation of carriers in *p*- and *n*-regions near the space-charge layer. These carriers "fall down" the potential barrier, constituting a current that is dependent upon their rate of arrival at the top of the barrier rather than upon barrier height. Because of the relatively small minority-carrier density, the junction current for this applied-voltage polarity is small compared to typical forward currents. Thus this is the high-resistance or *reverse connection* of the *pn* junction, and the junction is said to be *reverse biased.*

In summarizing the effects of bias on junction current, we have shown that the current is highly dependent on forward-bias voltage but relatively small and virtually independent of reverse-bias voltage for magnitudes greater than about 0.1 V (providing junction breakdown does not occur). The large difference in forward and reverse conduction characteristics almost ideally suits the *pn* junction for use as a rectifier, a device that converts alternating current into unidirectional current by permitting appreciable current flow in one direction only. Typical current-voltage characteristics of a small germanium junction rectifier (type 1N91) are shown in Fig. 2-5, current and voltage being represented as positive for forward bias. The large difference between the forward and reverse conduction characteristics is shown in Fig. 2-5(a). The much smaller current and voltage scales of Fig. 2-5(b) permit observation of the characteristics near zero bias. The existence of a reverse-saturation current may be seen and it is important to note the inadequacy of the device as a rectifier for small voltages, say of the order of 10 mV. (Nevertheless, the germanium diode is vastly superior to the high-vacuum tube as a low-voltage detector or rectifier.)

Junction breakdown. As the reverse-bias voltage of a *pn* junction is increased, the characteristic departs quite sharply from the saturation region, entering a region in which the current increases very strongly with voltage. This is the *reverse-breakdown region*, and the voltage at which it occurs is

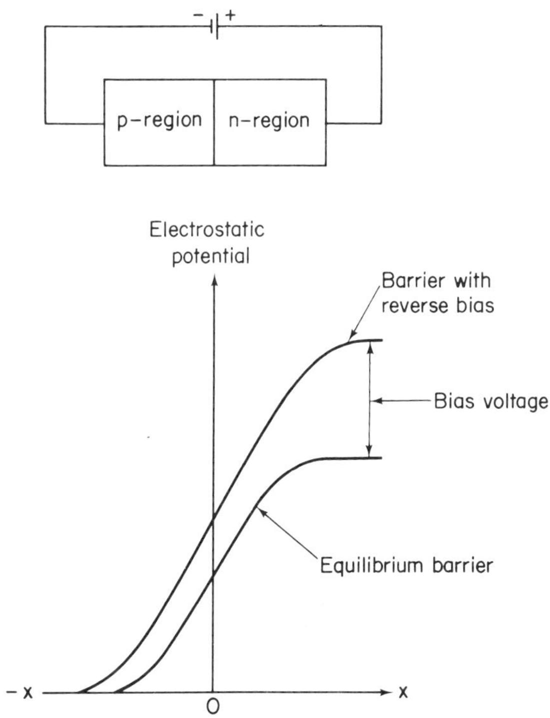

Fig. 2-4. The *pn* junction with applied reverse bias.

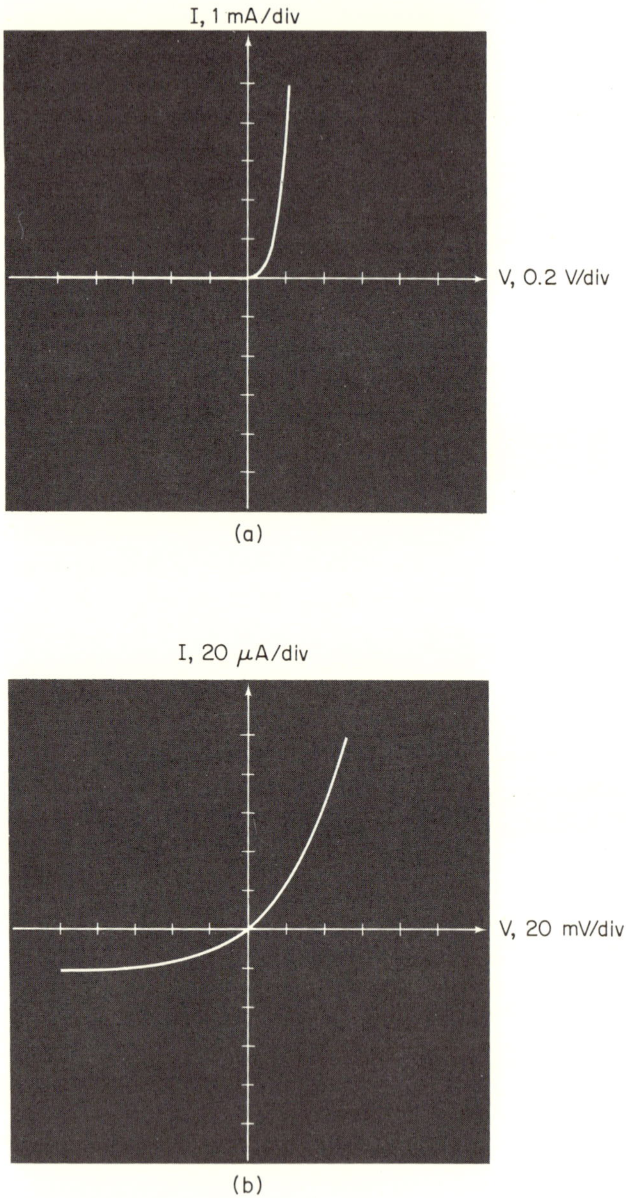

Fig. 2-5. Current–voltage characteristic of a germanium junction rectifier. (a) Normal characteristic. (b) Expanded scales showing behavior near the origin.

the *breakdown voltage*. The breakdown voltage ranges from a value of the order of 1 V to more than 1000 V, depending on the resistivities of the *p*- and *n*-regions and on the semiconductor material.

The effects of bias voltage on the carrier-depletion region are indicated in Figs. 2-3, and 2-4, in which it is seen that either decreasing forward bias or increasing reverse bias increases both the width of the region and the electric field (represented by the slope of the electrostatic-potential plot) therein. This will be shown by the expressions for depletion-region width and electric field derived in Sec. 2-6 in order to supplement the following qualitative consideration. If the electric field is sufficiently large, breakdown occurs by one of two possible mechanisms or by a combination of the two. These are *Zener breakdown* and *avalanche breakdown*. It is possible to distinguish between the two types of breakdown because, as we shall see, in Zener breakdown the breakdown voltage decreases as temperature increases, but in avalanche breakdown the breakdown voltage increases with temperature.

As mentioned earlier, donor and acceptor ions rigidly bound in the crystal lattice constitute a charge in the carrier-depletion region, where they are no longer neutralized by majority carriers. This is the charge that supports the reverse-bias voltage, the positive charge in the *n*-region being equal to the negative charge in the *p*-region, and almost the entire reverse voltage appears across the depletion region. Since a higher bias voltage requires a larger supporting charge, depletion-region thickness increases with bias voltage, varying for an abrupt junction as the *square root* of the voltage. With this manner of variation, the depletion-region electric field is seen to increase with bias voltage.

The depletion-region thickness required to provide the charge necessary to support a given reverse-bias voltage clearly depends on the density of ions in the region; the greater the density, the thinner the region. Therefore, the electric field in the region depends on the conductivities of the *p*- and *n*-regions, increasing with conductivity. From this we see that the higher the *p*- and *n*-region conductivities, the greater the electric field in the depletion region and the lower the voltage for which the electric field is high enough for breakdown.

Zener breakdown was first thought to be the mechanism of all junction breakdowns. If the *p*- and *n*-region conductivities are quite high, the carrier-depletion region is very thin, and a small reverse-bias voltage results in a large electric field. If the field is large enough, bound electrons vibrating about their mean positions in the lattice acquire from the field sufficient energy for the occurrence of valence-bond disruptions. This produces large quantities of electron-hole pairs in the depletion region, greatly enhancing the reverse current. An increase in temperature decreases the energy gap of the material (that is, the energy necessary to produce an electron-hole pair by valence-bond disruption), decreasing the breakdown voltage.

For wider junctions formed of lower conductivity material, it is found that reverse breakdown occurs at much lower voltages than expected on the basis of Zener-breakdown theory. This is believed to be due to *avalanche breakdown*, a phenomenon quite similar to the establishment of a self-maintained discharge between unheated electrodes in a gas. As p-region electrons and n-region holes enter the carrier-depletion region, they are accelerated by the electric field of the region. For normal reverse-bias operation, the kinetic energy gained by a carrier between collisions with semiconductor atoms, and lost during each collision, is less than that required for ionization of the atom. However, if the field is large enough, a carrier may gain sufficient energy in a single free path for an ionizing collision to occur and an electron-hole pair to be produced. This carrier-multiplication action takes place as the voltage approaches the breakdown value and is responsible for a significant increase in reverse current. This region of the reverse current-voltage characteristic is similar to the Townsend region of a gas discharge, and the processes responsible for the two regions are much the same. Since the primary carriers can go on to produce additional electron-hole pairs, and carriers generated in the depletion region may also experience ionizing collisions, further increase in voltage results in an avalanche of carriers and a current that is limited by the resistance of the p- and n-regions and that of the contacts. An increase in temperature decreases the minority-carrier mobility, requiring a higher electric field in order for electrons to acquire ionizing energy between collisions (even though this energy is somewhat less), and the breakdown voltage is increased. For *pn* junctions in silicon, an increase in temperature decreases the breakdown voltage of junctions that break down at less than about 5 V and increases the breakdown voltage of those breaking down at higher voltages. Presumably both Zener and avalanche phenomena occur in junctions breaking down at about 5 V.

The effect of p- and n-region conductivities on the avalanche-breakdown voltage V_a of abrupt *pn* junctions in silicon, and an indication of the order of magnitude of the resistivities of materials used in junction devices, are given by the relationship*

$$V_a = 39\rho_n + 8\rho_p \text{ volts} \tag{2-1}$$

where ρ_n and ρ_p are the resistivities in ohm-centimeters of the n- and p-regions, respectively. In the typical abrupt junction, the resistivity of one region is too low to have a significant effect on breakdown. When this is

* E. M. Baldwin and J. A. Mates, "Silicon Junction Diodes," *Proceedings of the 1955 Electronic Components Conference.*

true for the *p*-region, the approximate expression

$$V_a = 40\rho_n \text{ volts} \tag{2-2}$$

may be applied.

The reverse current-voltage characteristics of a typical germanium-junction diode at voltages up to the breakdown value is shown in Fig. 2-6. The current scale used permits examination of the characteristic at voltages below the breakdown value, showing the gradual increase in reverse current above the saturation value due to current leakage along the surface edges of the junction. Also shown is the steeper increase of current with voltage due to depletion-region carrier multiplication as the breakdown voltage is approached. Because of this effect and the increase of voltage with current due to *IR* drop in the *p*- and *n*-regions and in the contacts, the exact magnitude of the breakdown voltage is somewhat indefinite. In practice, the breakdown voltage is considered to be the voltage for a specified reverse current. For example, the breakdown voltage of a germanium *pn* junction having a reverse-saturation current of the order of microamperes might be specified as the voltage for which the reverse current is 100 μA, a current several times greater than the value at voltages significantly below the breakdown value. Nevertheless, voltage change with current in the breakdown region is small enough for the region to have great utility value, as may be seen from the reverse characteristic of a typical voltage-reference silicon diode (type 1N758) shown in Fig. 2-7. The much lower breakdown-region dynamic resistance of this device, compared to that of Fig. 2-6, is the result of the higher conductivity

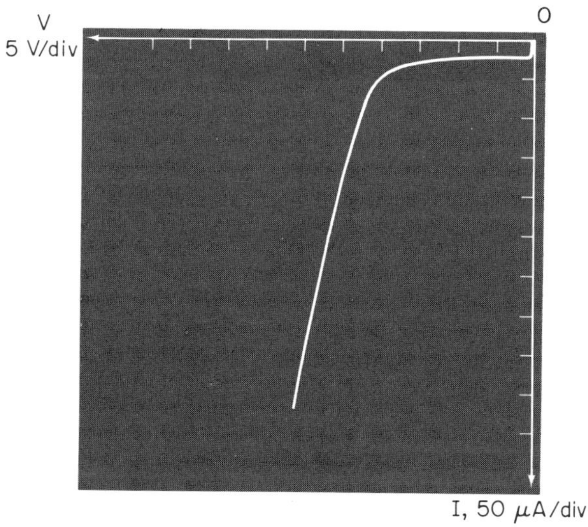

Fig. 2-6. Reverse characteristic of a typical germanium *pn* junction.

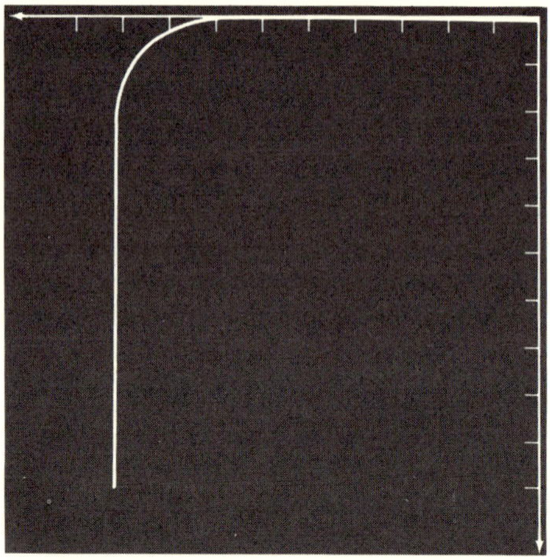

Fig. 2-7. Characteristic of a typical 1N758 voltage-reference silicon diode.

of its *p*- and *n*-regions. Commercially available voltage-reference silicon diodes, commonly referred to as Zener diodes, having breakdown-voltage ratings ranging from 2 V to several hundred volts are in widespread use as voltage references, limiters and regulators, and as wave clippers, surge arrestors, etc.

Junction breakdown may be either a *bulk property* or a *surface property* of a device. If it is a bulk property, breakdown usually occurs over much of the junction area, heat is generated uniformly throughout the junction, and the power-dissipation capability in reverse breakdown is essentially the same as for forward conduction. *Voltage-reference* (Zener) *diodes are bulk-breakdown devices.*

Avalanche breakdown often occurs at lower voltages than the value expected on the basis of the *p*- and *n*-region resistivities of the device. This is probably *surface breakdown*, in which breakdown occurs at a point or small-area spot along a surface edge of the junction. Because of the small area, the reverse-breakdown power-dissipation capability is very small compared to that for forward conduction and is quite uncertain. Therefore, the device is readily damaged by operation in the reverse-breakdown region, and no use can be made of this region. The junction breakdowns of most rectifiers and transistors in present use are surface properties of the devices. It is for this reason that when a high-voltage rectifier is formed by series-connecting lower-voltage units, it is usually necessary to shunt the units with resistors in order to insure a division of reverse voltage between them that will avoid reverse breakdown. It is possible to distinguish between the two forms of avalanche breakdown because, unlike bulk breakdown voltage, surface-breakdown voltage decreases as the temperature increases.

In a type of rectifier introduced comparatively recently, junction breakdown is a bulk property of the device. These are called *controlled-avalanche rectifiers* and have breakdown-voltage specifications similar to those of Zener diodes. That is, both upper and lower breakdown-voltage limits are specified, rather than the lower-limit specification of conventional rectifiers. Such rectifiers can be operated series-connected without the use of voltage-equalizing resistors, since they operate quite satisfactorily in reverse breakdown, and are less susceptible to damage by accidental breakdown. Perhaps eventually most rectifiers manufactured will be bulk-breakdown devices.

Reach-through. A phenomenon sometimes confused with junction breakdown is junction *reach-through* (also known as *punch-through*). This occurs when the reverse voltage is so high that the space-charge region extends all the way through one of the semiconductor regions to the metal contact and the electric field is said to have reached through the region. If this is the situation, no additional charge is available in the region to support an increase in applied voltage, and such an increase must appear as IR drop in the semiconductor regions and in the contacts. Therefore, when the reach through voltage is exceeded, the current increases strongly with reverse voltage as in reverse breakdown. Junction reach-through can usually be distinguished from breakdown because the current increase is more abrupt, there being no region in the current-voltage charcteristic comparable to the carrier-multiplication region near breakdown. Thus we see that the upper limit of junction reverse voltage is either the breakdown voltage or the reach-through voltage. Reach-through is an unlikely limitation of the reverse voltage of diodes and rectifiers, but often limits the collector voltage of transistors. In the most likely occurrence of reach-through in a transistor, the collector-base carrier-depletion region extends through the base to the high-conductivity emitter region, which acts as an electrical contact, resulting in a collector-emitter short for collector voltages greater than the reach-through value.

2-2 ENERGY-LEVEL REPRESENTATION OF *pn* JUNCTIONS

A quantitative view of the potential barrier at a junction and the effects of bias on the barrier may be obtained from consideration of electron energy-level diagrams. Such diagrams have been considered in Sec. 1-4.4 for intrinsic, *p*-type, and *n*-type materials, and we shall now extend this consideration to junctions of these materials.

The energy-level diagram of a *pn* junction in equilibrium is shown in Fig. 2-8. As pointed out in Sec. 1-4.4, the Fermi level lies below the middle of the energy gap E_g in the *p*-region and above the middle of the gap in the *n*-region. The Fermi level has these locations, *outside the junction region*, in the diagram of Fig. 2-8. Perhaps the most fundamental point related to energy-level diagrams for a system of solids in contact is that, *in a system in thermal equilibrium the Fermi-level energy E_f is constant throughout the entire system.* This is illustrated in Fig. 2-8 in which the Fermi levels of

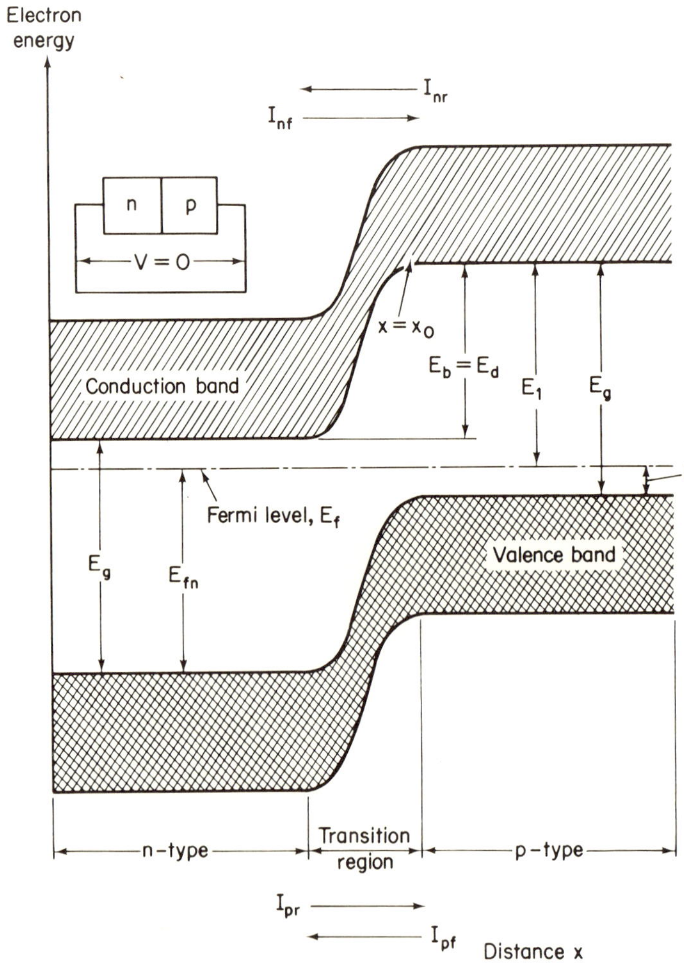

Fig. 2-8. Energy-level diagram of a *pn* junction in equilibrium (zero bias and zero net junction current). The Fermi levels of the *p*- and *n*-regions are aligned and the electron currents I_{nf} and I_{nr} are equal and opposite, as are the hole currents I_{pf} and I_{pr}.

the *p*- and *n*-regions coincide and run horizontally across the diagram. Fermi-level alignment is the result of the flow of carriers across the junction during the establishment of equilibrium conditions, as discussed in Sec. 2-1 for *pn* junctions, the argument for Fermi-level alignment in a *pn* junction being the same as that presented in Sec. 1-4.3 for solids in general.

We note that in Fig. 2-8, as a result of the alignment of the Fermi levels in the *p*- and *n*-regions and the locations of the conduction- and valence-band edges with respect to the Fermi levels outside the transition region, each band edge has a tapered step at the junction. The height of this step, E_b, in electron volts is equal to the contact potential in volts and is the magnitude of the junction equilibrium potential barrier referred to in Sec. 2-1. This magnitude is seen to be equal to $(E_{fn} - E_{fp})$, where E_{fn} is the Fermi level of the *n*-region and E_{fp} is the Fermi level of the *p*-region, each measured with respect to the top edge of the valence band in the appropriate region. The slope of the band edges in the junction region is equal to the magnitude of the electric field in the region. This field may have a magnitude of thousands or tens of thousands of volts per centimeter, depending on the doping of the *p*- and *n*-regions.

Since E_b is used as a general designation of barrier height, the height without bias is referred to as E_d, the barrier due to diffusion action, and E_b is indicated as equal to E_d in Fig. 2-8. The built-in junction potential difference is referred to as V_d, the diffusion voltage or contact potential, and V_b represents the total junction voltage. (The symbol V, rather than E, is used to represent potential and potential difference to avoid confusion with energy, which is represented by E.)

As pointed out in Sec. 2-1, all is not quiescent in the junction region following the establishment of equilibrium, and currents of electrons and holes continue to flow. Although it was stated in Sec. 2-1 that the net junction current is zero with zero bias, it might appear that the built-in potential difference V_d at the junction could be utilized as a source, that this source would supply a current upon short circuit (as with zero bias), and that a source voltage could be measured under open-circuit conditions. However, in applying electrodes to the *p*- and *n*-regions, and in making connections to these electrodes, additional contact potentials are introduced and the net circuit emf is zero *as long as all parts of the circuit are at the same temperature*. For this reason, the net junction current is zero with zero bias. Nevertheless, it is useful to consider the electron and hole currents under equilibrium conditions, because doing so will be helpful in determining the net junction current for conditions of forward or reverse bias.

Let us first consider the electron currents and examine them at the point $x = x_0$ in Fig. 2-8, the edge of the transition region on the *p*-side. The electron current that can flow to the left at $x = x_0$ is proportional to the number of electrons in the conduction band of the *p*-type material which, from Eq. (1-91) of Sec. 1-4.4, is proportional to $e^{-(E_g - E_f)/kT}$ evaluated on the *p*-side of the junction. From Fig. 2-8, this electron current is seen to be proportional to $e^{-E_1/kT}$ and is referred to as I_{nr}, the reverse electron current of the junction. Thus, I_{nr} is the current of minority electrons accelerated to the left "down the potential barrier" by the electric field as they diffuse into the transition region.

The electron current that flows to the right at $x = x_0$ consists of electrons that have climbed the potential barrier from the *n*-side of the junction. The number of electrons in the conduction band on the *n*-side is proportional to $e^{-(E_g - E_f)/kT}$, evaluated on the *n*-side, but the number having energies as high as the bottom level in the *p*-region conduction band is proportional to $e^{-E_b/kT}$, using the approximate Fermi-factor expression. Therefore, the number of electrons that can participate in current to the right at $x = x_0$ is proportional to

$$e^{-E_b/kT}e^{-(E_g - E_{fn})/kT} = e^{-E_1/kT} \tag{2-3}$$

since E_1 is the sum of E_b and the quantity $(E_g - E_{fn})$, as seen in Fig. 2-8, and this is the same as the number of electrons that contribute to I_{nr}. Thus the current of majority electrons climbing the barrier, referred to as the forward electron current I_{nf} of the junction, is equal to the reverse current I_{nr}, as indicated in Fig. 2-8, and the net electron current is zero.

Since the barrier to the flow of holes from the *p*-region to the *n*-region is the same as that for electron flow in the opposite direction, the arguments for hole currents are similar to those for electron currents. Therefore, the reverse current I_{pr} of *n*-region holes is equal to the foward-hole current I_{pf}, as indicated in Fig. 2-8, and the net hole current is zero.

As pointed out in Sec. 2-1 and indicated in Fig. 2-4, reverse-biasing the *pn* junction increases the potential barrier by the amount of the applied voltage. An energy diagram for this condition is shown by Fig. 2-9. Since the ordinate of the diagram is electron potential energy, making the *n*-region positive lowers the energy levels on that side of the junction. Outside the transition region the Fermi-level locations with respect to the band edges are unchanged by bias, since current flow does not significantly change the electron distribution in the *p*- and *n*-regions. Within the transition region the electron distribution is radically changed by the application of bias, and the Fermi level is no longer defined, but the *p*- and *n*-region Fermi levels are separated by an energy step of magnitude eV. The barrier height E_b with bias is given by

$$E_b = E_d - eV \tag{2-4}$$

where e is the electronic-charge *magnitude* and V is the voltage applied to the junction, being negative for reverse bias and positive for forward bias. The increased barrier height with reverse bias has no effect on the reverse electron and hole currents I_{nr} and I_{pr}, as explained in Sec. 2-1, but the

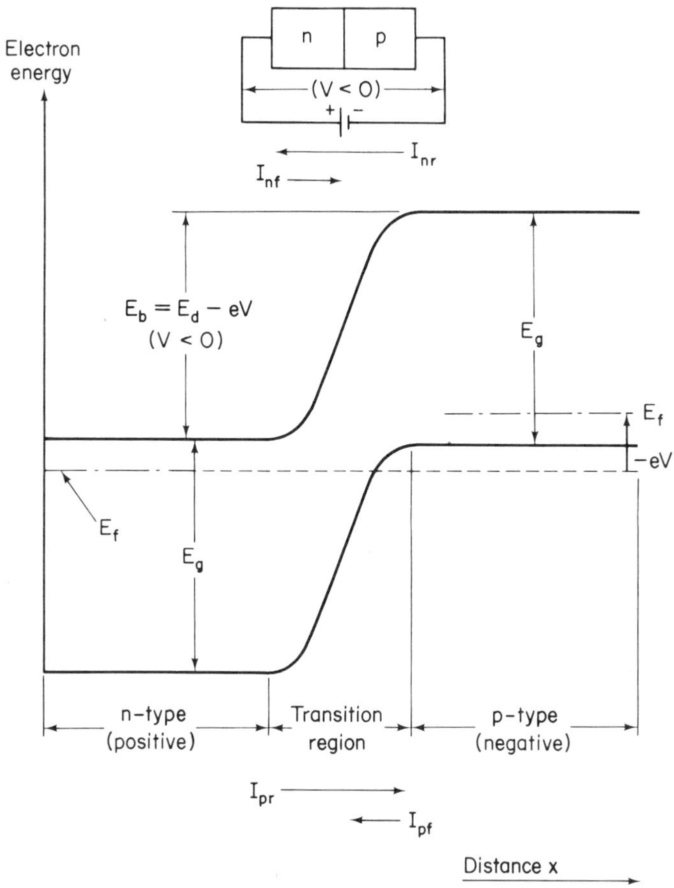

Fig. 2-9. Energy-level diagram of a *pn* junction with reverse bias ($V < 0$). The *n*-region Fermi level is lowered and the increased barrier height decreases the forward electron and hole currents.

forward electron and hole currents I_{nf} and I_{pf} are reduced, as indicated in Fig. 2-9.

From the same type of reasoning that led to Eq. (2-3), a change in barrier height of $-eV$ is seen to change the forward electron current I_{nf} by the factor $e^{-(-eV)/kT}$, which is less than unity for negative V. Since the equilibrium value of I_{nf} is I_{nr}, the value with reverse bias ($V < O$) is given by

$$I_{nf} = I_{nr}e^{eV/kT} \tag{2-5}$$

The net electron current to the right in Fig. 2-9 is then

$$I_n = I_{nr}e^{eV/kT} - I_{nr} = I_{nr}(e^{eV/kT} - 1) \tag{2-6}$$

From Eq. (2-6) we see that the electron current is negative for negative V, which is consistent with the established polarity definition that considers junction voltage and current to be positive for forward bias. The forward hole current I_{pf} with reverse bias is also decreased by the factor $e^{eV/kT}$, and the hole-current expressions are the same as Eqs. (2-5) and (2-6), except that the reverse-saturation hole current I_{pr} replaces the saturation electron current I_{nr}.

An electron-energy diagram for the condition of forward bias is shown in Fig. 2-10 and is similar to that for reverse bias, expect that making the p-region positive by the voltage V lowers the Fermi level in the p-region, causing it to be below the n-region level by the amount eV. This reduces the barrier height E_b (with respect to the equilibrium value E_d) by eV, which increases the forward electron and hole currents but has no effect on the reverse currents, as indicated in Fig. 2-10. As a result of the lowering of

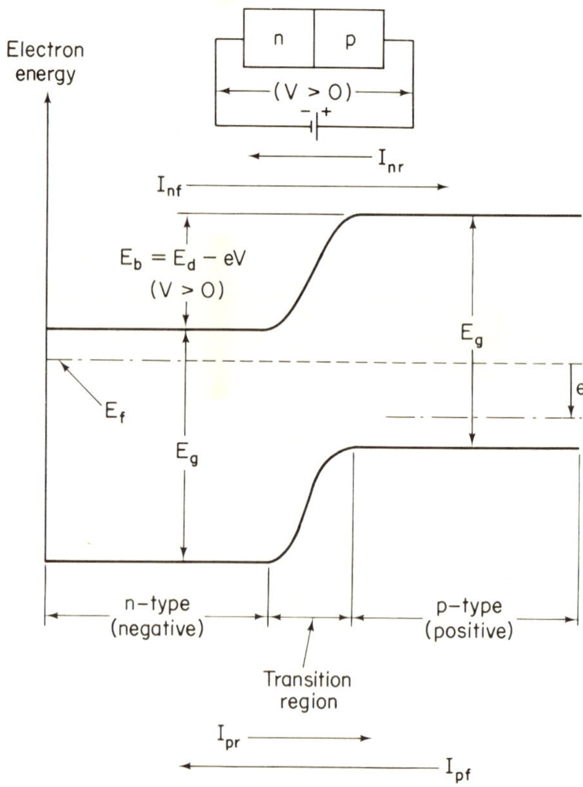

Fig. 2-10. Energy-level diagram of a *pn* junction with forward bias ($V > 0$). The p-region Fermi level is lowered and the decreased barrier height increases the forward electron and hole currents.

energy levels on the p-side of the junction by making the p-region positive, the concentration of n-region electrons having energies as great as the bottom level of the p-region conduction band is higher than the p-region electron concentration at this level. This concentration gradient acts as a driving force that produces a net electron current to the right in Fig. 2-10. The forward current is increased with respect to the equilibrium value by the factor $\mathrm{e}^{eV/kT}$, which is greater than unity for positive V, and Eqs. (2-5) and (2-6) apply for forward bias as well as for reverse bias. Since the total junction current is the sum of the net electron current and the net hole current and Eq. (2-5) applies to both, an expression for the current-voltage characteristic of a *pn* junction may be obtained from Eq. (2-5) by substituting I_s, the sum of the reverse-saturation currents for electrons and holes, for the electron saturation current I_{nr}. The result is

$$I = I_s(\mathrm{e}^{eV/kT} - 1) \tag{2-7}$$

which is known as the diode equation.

2-3 THE DIODE EQUATION

As shown in Sec. 2-2, the current-voltage characteristic of *pn* junction is given by

$$I = I_s(\mathrm{e}^{eV/kT} - 1) \tag{2-7}$$

where

I_s = saturation current for reverse bias.
e = electronic-change magnitude in coulombs.
V = voltage applied to the junction (positive values being forward).
k = Boltzmann's constant.
T = temperature, in degrees Kelvin.
e = Napierian base (2.718).

We shall examine this expression somewhat in detail and consider how the assumptions made in its derivation limit its applicability to practical junctions. These limitations will be further considered in the semiconductor-diode discussion of Chapter 4.

When V is in volts and T is 300°K, corresponding to room temperature, Eq. (2-7) becomes

$$I = I_s(\mathrm{e}^{39V} - 1) = I_s(10^{16.8V} - 1) \tag{2-8}$$

Thus, if V is -0.10 V,

$$I = I_s\left(\frac{1}{10^{1.68}} - 1\right) = -0.98 I_s$$

indicating that the current is essentially the reverse-saturation value for voltages more negative than about -0.1 V. In a practical *pn*-junction device, leakage current, carrier multiplication in the junction region, and junction breakdown cause the reverse characteristic to differ from that predicted by Eq. (2-7). For silicon junctions at room temperature, the leakage current at -0.10 V is usually several orders of magnitude greater than the reverse-saturation value, and reverse-current saturation is not observed.

For a forward bias of 0.10 V at room temperature,

$$I = I_s(10^{1.68} - 1) = I_s(48 - 1)$$

indicating that for $V \geq 0.1$ V the approximate forward characteristic is given by

$$I = I_s e^{39V} = I_s(10)^{16.8V} \qquad (2\text{-}9)$$

In forward operation of the practical device, IR drops in the *p*- and *n*-regions and in the contacts may significantly reduce the junction-applied voltage V below the terminal voltage, causing the expected forward current to be much less than that predicted by Eq. (2-7). No matter how high the applied forward voltage may be, *the junction potential barrier is never eliminated*, but only reduced, and the applied junction voltage V rarely exceeds a few tenths of a volt. Junction imperfections may also cause the forward current to be less than the expected value by reducing the coefficient of V in the exponential term of Eq. (2-7) below the value (e/kT) for an ideal junction. On the other hand, junction leakage increases the current with forward bias and may be the principal factor affecting the current in a silicon junction for forward voltages up to several tenths of a volt.

A very important assumption in the derivation of Eq. (2-7) is that negligible carrier recombination occurs in the carrier-depletion region. Thus it is assumed that *n*-region electrons in excess of the *p*-region thermal equilibrium density do not recombine in the carrier-depletion region but travel on into the *p*-region and that *p*-region holes travel through the transition region into the *n*-region. Unless the junction region is relatively free of recombination, no barrier exists and current flows equally well in either direction.

2-4 JUNCTION REVERSE-SATURATION CURRENT

With appreciable reverse bias, majority carriers are unable to cross the junction, and *p*-region electrons and *n*-region holes reaching the junction are accelerated through it by the electric field as rapidly as they arrive. Therefore, the minority-carrier density at the junction may be considered to be zero with appreciable reverse bias. This may also be seen from the expression for the electron density $n_p(0)$ in the p-region at the edge of the junction region, which is

$$n_p(0) = n_{po} e^{eV/kT} \tag{2-10}$$

Equation (2-10) is arrived at from the arguments that led to Eq. (2-5) and shows that $n_p(0)$ is negligible compared to the equilibrium value of the p-region electron density n_{po} for applied junction voltages V more negative than about -0.1 V. Similarly, the n-region hole density $p_n(0)$ at the edge of the junction region is given by

$$p_n(0) = p_{no} e^{eV/kT} \tag{2-11}$$

where p_{no} is the equilibrium value of the n-region hole density, which shows that $p_n(0)$ is also negligible for appreciable reverse bias.

Many diffusion lengths away from the junction, the minority-carrier density is the equilibrium value (n_{po} on the p-side and p_{no} on the n-side), being unaffected by the existence of the junction. An expression for minority-carrier density as a function of distance x from the edge of a reverse-biased abrupt junction may be obtained by solving the continuity equation. In reality, only the diffusion equation, given by Eq. (1-43), is required, since the electric field outside the junction region is negligible and there are no carrier sources or sinks in the p- and n-regions. Since the carrier density is not changing with time, dp_n/dt is zero, and Eq. (1-43) becomes

$$D_p \frac{d^2 p_n}{dx^2} - \frac{(p_n - p_{no})}{\tau_p} = 0 \tag{2-12}$$

For the boundary conditions

$$p_n = 0 \qquad \text{at } x = 0 \tag{2-13}$$

and

$$p_n = p_{no} \qquad \text{at } x = \infty \tag{2-14}$$

the solution to Eq. (2-12) is

$$p_n = p_{no}(1 - e^{-x/L_p}) \tag{2-15}$$

where L_p is the hole-diffusion length, which is defined as $L_p = (D_p \tau_p)^{1/2}$. By similar analysis, the p-region electron density n_p is given by

$$n_p = n_{po}(1 - e^{-x/L_n}) \tag{2-16}$$

The variation of minority-carrier density with distance x from the edges of an abrupt junction is shown in Fig. 2-11. From Eqs. (2-15) and (2-16), it is seen that the slopes of the $p_n(x)$ and $n_p(x)$ plots at $x = 0$ are such that a tangent drawn to each plot at this point reaches the equilibrium density at a distance equal to the diffusion length. This is indicated in Fig. 2-11.

The hole current I_{pn} to the right *toward* the junction in the n-region of Fig. 2-11 is a diffusion flow due to the density gradient in the region. This

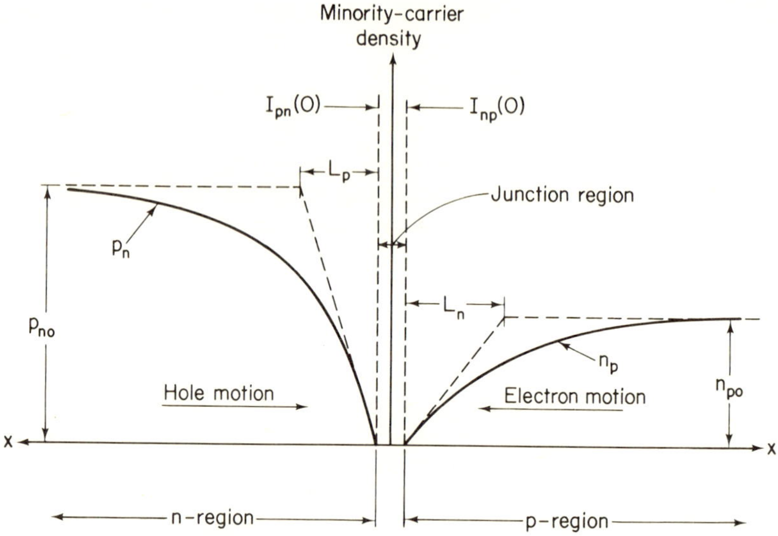

Fig. 2-11. Minority-carrier density in the p- and n-regions within a few diffusion lengths of a reverse-biased pn junction. (The junction-region width is greatly exaggerated compared to that of the p- and n-regions.) Hole-diffusion current $I_{pn}(0)$ at $x = 0$ and electron-diffusion current $I_{np}(0)$ at $x = 0$ are the reverse-saturation hole and electron currents.

current is a function of distance x in the n-region, depending on the hole-density gradient dp_n/dx, and has its maximum value $I_{pn}(0)$ at $x = 0$, where the gradient has its largest value. Current $I_{pn}(0)$ is the *excess* hole current at the junction edge, but for reverse-bias conditions such that the $p_n(x)$ plot of Fig. 2-11 is applicable, $I_{pn}(0)$ is the reverse-saturation hole current. Applying the diffusion-current expression of Eq. (1-33), we obtain

$$I_{pn}(0) = -AeD_p \frac{dp_n}{dx}\bigg|_{x=0} \tag{2-17}$$

where A is the junction area. From the slope of the $p_n(x)$ plot of Fig. 2-11 at $x = 0$, or by differentiating Eq. (2-15), we see that

$$\frac{dp_n}{dx}\bigg|_{x=0} = \frac{p_{no}}{L_p} \tag{2-18}$$

which, combined with Eq. (2-17), yields

$$I_{pn}(0) = -\frac{AeD_p p_{no}}{L_p} \tag{2-19}$$

the negative sign indicating that the hole flow is in the $-x$-direction or *toward* the junction.

If we treat electrons in the *p*-region in a similar manner, we find that the reverse-saturation electron current $I_{np}(0)$ flowing from the *p*-region into the junction is given by

$$I_{np}(0) = -\frac{AeD_n n_{po}}{L_n} \tag{2-20}$$

Since the magnitude of the reverse-saturation current I_s is the sum of the hole and electron saturation-current magnitudes, combining Eqs. (2-19) and (2-20) results in

$$I_s = Ae\left(\frac{D_p p_{no}}{L_p} + \frac{D_n n_{po}}{L_n}\right) \tag{2-21}$$

giving an expression for saturation current in terms of the physical properties of the bulk materials that make up the device. The saturation current may be expressed in terms of minority-carrier lifetime by using the relationships $L_n = (D_n \tau_n)^{1/2}$ and $L_p = (D_p \tau_p)^{1/2}$, as given by Eqs. (1-58) and (1-60), in conjunction with Eq. (2-21). The result of this combination is

$$I_s = Ae\left[p_{no}\left(\frac{D_p}{\tau_p}\right)^{1/2} + n_{po}\left(\frac{D_n}{\tau_n}\right)^{1/2}\right] \tag{2-22}$$

The reverse-saturation current I_s of a *pn*-junction device is a very important characteristic and should ordinarily be as low as possible. The important effect of minority-carrier lifetime on I_s is seen from Eq. (2-22), which shows the need for long-lifetime materials if I_s is to be low. The saturation current also varies directly with minority-carrier density, which increases exponentially with temperature. Thus I_s is very sensitive to temperature. At room temperature, the rate of change of I_s is about 10 percent per deg C for germanium junctions and 16 percent per deg C for silicon, I_s approximately doubling every 10°C in germanium and every 6°C in silicon.

At a given temperature, the minority-carrier density varies inversely with the majority-carrier density. Since the *p*- and *n*-region resistivities are inversely dependent upon their respective majority-carrier densities, we see that the saturation current varies directly with the resistivities of the bulk materials of the device. In the typical abrupt junction, the doping level of one region (the regrown region of an alloy junction) is so high that minority carriers in the region have negligible effect on the saturation current. In such a junction the saturation current varies directly with the resistivity of the low-conductivity region (that is, with the resistivity of the original crystal of an alloy junction), suggesting the use of low-resistivity material. However, the junction-breakdown voltage also varies directly with this resistivity, as indicated by Eq. (2-2). Therefore, the choice of the resistivity of the material used in a device may be a compromise based on these (and other) conflicting factors.

2-5 ELECTRON AND HOLE CURRENTS IN *p*- AND *n*-REGIONS

The current in regions far distant (many diffusion lengths) from a biased junction clearly is a majority-carrier current, since the minority-carrier density contributes little to the conductivity of the regions, and is a drift current, owing to the absence of density gradient in the regions. (The field responsible for the drift current is, of course, the result of IR drop in the regions.) However, in the p- and n-regions within a few diffusion lengths of the junction, the current consists of both diffusion components and drift components. This may be seen qualitatively for reverse bias by examination of Fig. 2-11 and quantitatively by obtaining expressions for electron and hole currents as a function of distance x from the junction region.

The current $I_{pn}(0)$ in Fig. 2-11 is a diffusion current of holes flowing into the reverse-biased junction at the edge of the n-region, as pointed out in Sec. 2-4. The junction-region field accelerates the holes through the junction to the edge of the p-region, where, in the absence of recombination in the junction region, they constitute a p-region hole current $I_{pp}(0)$ equal to $I_{pn}(0)$. Current $I_{pp}(0)$ is a majority-carrier drift current, and the total p-region current $I(0)$ at $x = 0$ (the junction edge) is the sum of $I_{pp}(0)$ and $I_{np}(0)$, the diffusion current of minority electrons flowing into the junction at the edge of the p-region. That is,

$$I(0) = I_{pp}(0) + I_{np}(0) = I_s$$

since the current at every plane cut through the device is the saturation value. At points within the p-region ($x > 0$) the electron-density gradient [that is, the slope of the $n_p(x)$ plot] is less than at $x = 0$, indicating that the minority-carrier diffusion current I_{np} decreases with distance x from the junction. An expression for I_{np} as a function of distance x may be obtained from the diffusion-current expression

$$J_n = -eD_n \frac{dn_p}{dx} \tag{1-32}$$

where

$$\frac{dn_p}{dx} = \frac{n_{po}}{L_n} e^{-x/L_n} \tag{2-23}$$

as obtained from the $n_p(x)$ expression of Eq. (2-16). From Eqs. (1-32) and (2-23),

$$I_{np} = -\frac{AeD_n n_{po}}{L_n} e^{-x/L_n} \tag{2-24}$$

indicating that electron flow toward the junction (in the $-x$-direction) decreases exponentially with distance from the junction. The decrease in minority-electron diffusion current I_{np} is accompanied by an increase in majority-hole current I_{pp}, since the sum of the two current magnitudes is constant. The variation of majority- and minority-carrier current through the n- and p-regions of a reverse-biased junction is shown in Fig. 2-12. The minority-carrier density plots are also shown for reference purposes. It is of interest to note the similarity of the $I_{np}(x)$ and $n_p(x)$ plots and of the plots of $I_{pn}(x)$ and $p_n(x)$, the plots having the same shape because the derivative of an exponential function is also an exponential.

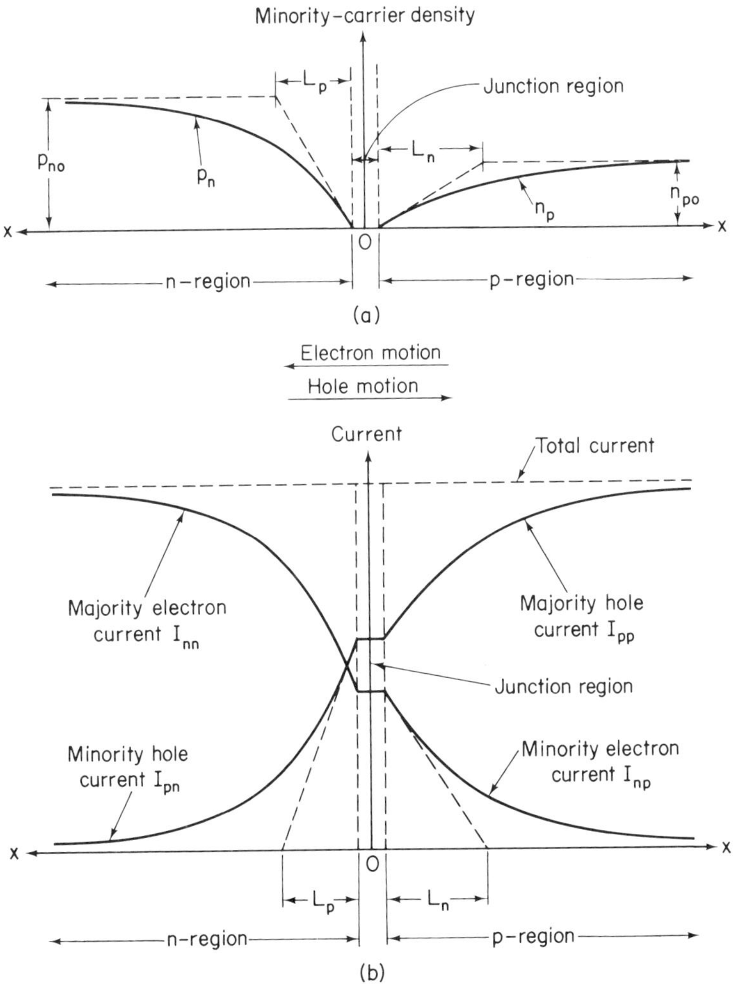

Fig. 2-12. (a) Minority-carrier densities and (b) hole and electron currents in the p- and n-regions near a reverse-biased junction, assuming negligible recombination in the junction region.

Majority-carrier diffusion current. The majority-carrier currents $I_{pp}(0)$ and $I_{nn}(0)$ at the junction edges of the p- and n-regions have been referred to as drift currents, but they are not the entire drift currents at $x = 0$, owing to the existence of majority-carrier diffusion currents in the regions. The reduction of minority-carrier density in the regions near the junction by reverse bias causes a corresponding reduction in majority-carrier density in order that electrical neutrality be maintained. Neutrality is established by majority carriers moving out of the regions (that is, repelled by the excess charge) until, at each point, the total positive-charge density (donor ions and holes) equals the total negative-charge density (acceptor ions and electrons). The majority-carrier distributions for reverse bias are shown in Fig. 2-13. Although the *deviations* from the equilibrium densities are the same

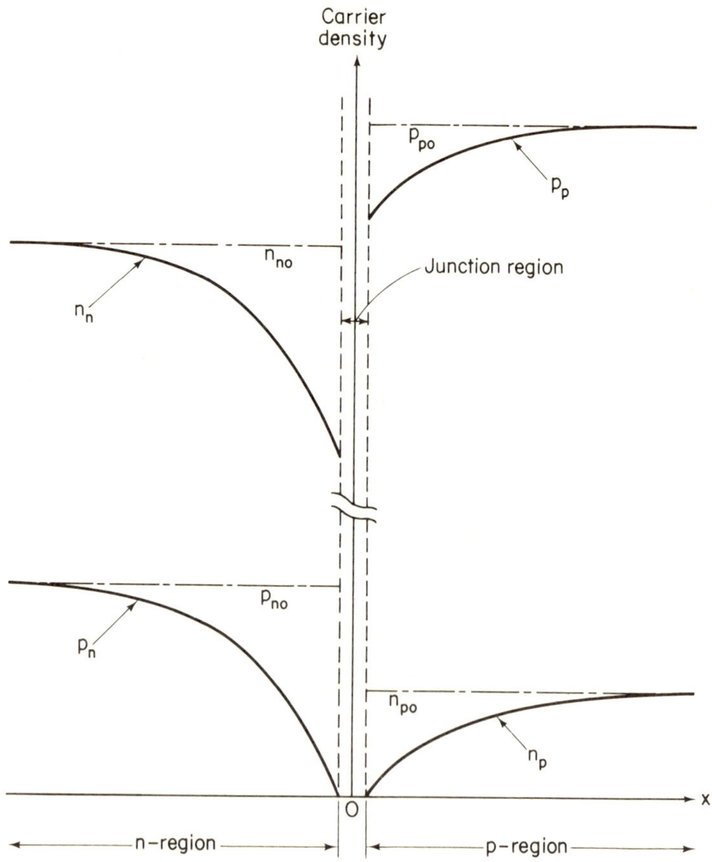

Fig. 2-13. Majority-carrier and minority-carrier densities in n- and p-regions near a reverse-biased junction. The vertical scale is broken to indicate the much larger majority-carrier densities.

(or very nearly so) for both minority and majority carriers, the *percentage* of change in majority-carrier density is very small, as suggested by the break in the linear vertical scale of Fig. 2-13. The majority-carrier-density gradient produces a diffusion flow of majority carriers *toward* the junction, tending to equalize the majority-carrier distribution. However, this equalization tendency increases the electric field in the *p*- and *n*-regions, resulting in an increase in drift flow of majority carriers *away* from the junction, which compensates for the diffusion flow. The resolution of *p*-region majority-carrier current into drift and diffusion components is shown by Fig. 2-14, in which flow away from the junction is considered positive. The minority-carrier diffusion current is shown for reference purposes and (for each x value) is represented as about double the majority value, since the diffusion constant for electrons is approximately double that for holes and the electron-density and hole-density gradients are approximately equal. The *p*-region

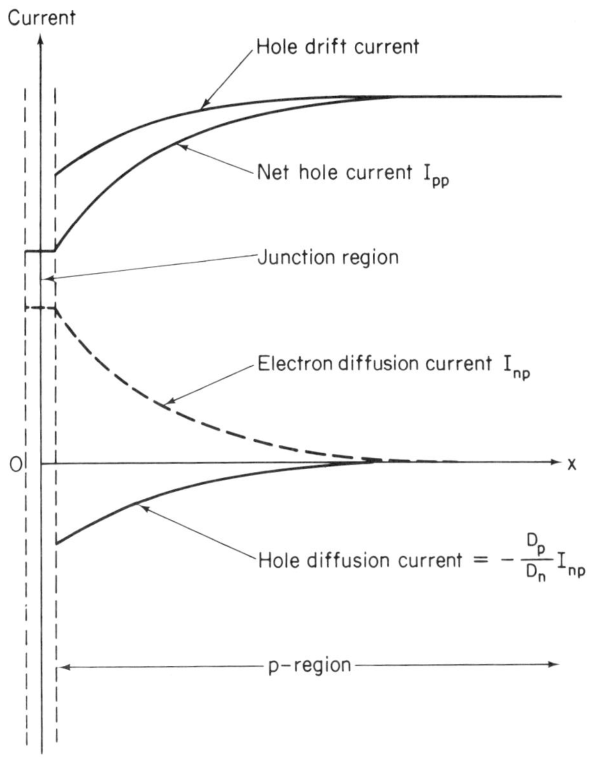

Fig. 2-14. Resolution of majority-carrier current into drift and diffusion components. The net hole current is the sum of the drift component and the opposite-flowing diffusion component. The minority-electron diffusion current is shown for purposes of reference.

current is represented as almost entirely drift current at distances several diffusion lengths from the junction.

Forward bias. From Eqs. (2-10) and (2-11), we see that forward bias ($V > 0$) increases the minority-carrier densities at the edges of the junction regions above the equilibrium values n_{po} and p_{no}. This is known as *minority-carrier injection* and results in a diffusion flow of minority carriers *away* from the junction *toward* the contacts. If we again consider the field in the p- and n-regions to have negligible effect on the minority-carrier distribution, an expression for minority-carrier density as a function of distance from the edge of the junction may be obtained by solving the diffusion equation. For steady-state conditions, this equation is

$$D_p \frac{d^2 p_n}{dx^2} - \left(\frac{p_n - p_{no}}{\tau_p}\right) = 0 \tag{2-12}$$

the same as for reverse bias. From Eq. (2-11), one boundary condition is

$$p_n = p_n(0) = p_{no} e^{eV/kT} \quad \text{at } x = 0 \tag{2-25}$$

and the other boundary condition is

$$p_n = p_{no} \quad \text{at } x = \infty \tag{2-26}$$

since recombination eliminates the excess-carrier density at great distances from the junction. The solution to Eq. (2-12) for these boundary conditions is

$$p_n = (p_{no} e^{eV/kT} - p_{no}) e^{-x/L_p} + p_{no} \tag{2-27}$$

and by similar analysis the p-region electron density n_p is given by

$$n_p = (n_{po} e^{eV/kT} - n_{po}) e^{-x/L_n} + n_{po} \tag{2-28}$$

In reality, Eqs. (2-27) and (2-28) apply whether V is positive or negative (that is, for both forward bias and reverse bias) and reduce to Eqs. (2-15) and (2-16), the expressions for minority-carrier density with the junction operating in reverse saturation, for V highly negative.

As with reverse bias, an expression for minority-electron diffusion current as a function of distance x from the junction may be obtained from the diffusion-current expression

$$J_n = -eD_n \frac{dn_p}{dx} \tag{1-32}$$

where

$$\frac{dn_p}{dx} = -\frac{n_{po}}{L_n}(e^{eV/kT} - 1) e^{-x/L_n} \tag{2-29}$$

as obtained from the $n_p(x)$ expression of Eq. (2-28). From Eqs. (1-32) and (2-29), the p-region electron current is given by

$$I_{np} = \frac{AeD_n n_{po}}{L_n}(e^{eV/kT} - 1) e^{-x/L_n} \tag{2-30}$$

where A is the area of the junction, indicating that electrons flow away from the junction (in the $+x$-direction) for forward bias ($V > 0$) and toward the junction for reverse bias ($V < 0$). Equation (2-30) reduces to Eq. (2-24), the $I_{np}(x)$ expression for reverse-saturation operation of the junction, for V highly negative. Similarly, treatment of holes in the n-region results in the hole-current expression

$$I_{pn} = \frac{AeD_p p_{no}}{L_p}(e^{eV/kT} - 1)e^{-x/L_p} \tag{2-31}$$

The variation of the minority-carrier densities in the p- and n-regions of a forward-biased junction are shown in Fig. 2-15(a). The minority-carrier diffusion current variation is as shown in Fig. 2-15(b), and, except for

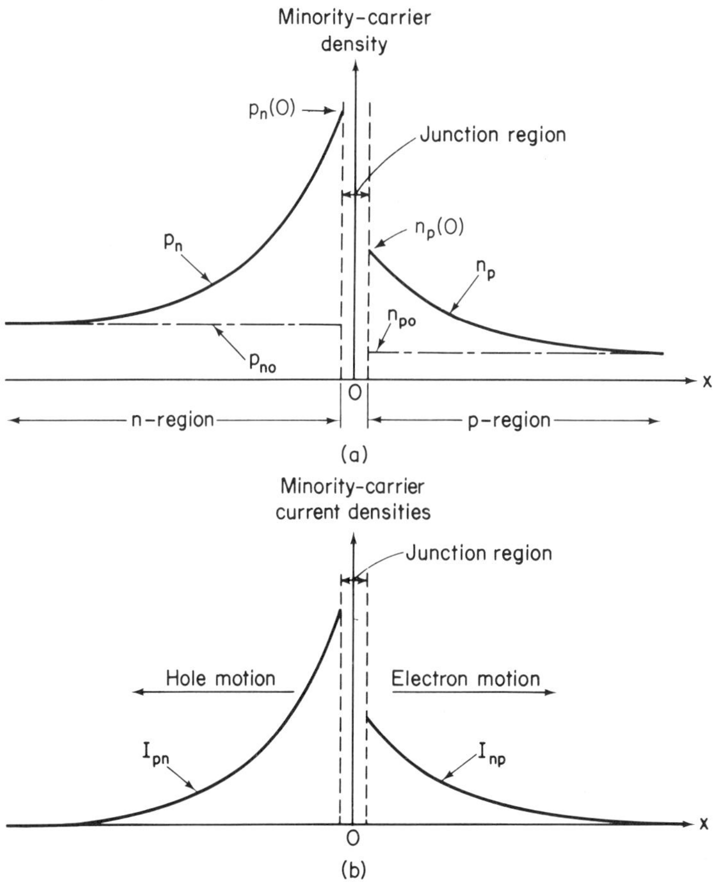

Fig. 2-15. (a) Minority-carrier densities and (b) minority-carrier current densities in the p- and n-regions near a forward-based junction.

magnitude and direction, the minority-carrier current plots are seen to be the same as those of Fig. 2-12(b) for reverse bias; therefore, the majority-current plots are also the same. The increase in minority-carrier density in the *p*- and *n*-regions near the junction results in a corresponding increase in majority-carrier density, excess minority carriers attracting majority carriers into the regions in order that electrical neutrality be maintained. This causes a majority-carrier diffusion current in each region, as explained for reverse-bias operation, and the majority-current resolution is as shown in Fig. (2-14) for reverse bias.

2-6 JUNCTION-REGION WIDTH AND CAPACITANCE

Our qualitative consideration of the carrier-depletion region and its dependence upon junction voltage and the conductivities of the *p*- and *n*-regions (Sec. 2-1) will now be extended. In so doing, we shall derive expressions for depletion-region width and for junction-region capacitance, an important junction property. This capacitance, also known as the transition, depletion-region or space-charge capacitance, is the capacitance associated with the space-charge change that results from the narrowing or widening of the depletion region when the junction voltage is changed.

Poisson's equation, which relates electrostatic potential and charge density, is the basis for quantitative consideration of the electrical properties of the carrier-depletion or space-charge region. This equation is

$$\frac{d^2V}{dx^2} = -\frac{Q}{\epsilon} \tag{2-32}$$

where V is the electrostatic potential, Q is the charge density, and ϵ is the dielectric permittivity of the medium. The application of Poisson's equation to the depletion region requires information regarding the variation of charge density within the region. This variation depends on the type of junction, and we shall consider both abrupt and graded junctions.

Abrupt junctions. Although the charge distribution in the carrier-depletion region of an abrupt junction is somewhat as shown in Fig. 2-2(c), results that agree well with experimental observations can be obtained by assuming the distribution to be as represented in Fig. 2-16(a). That is, it is assumed that the change from uniform donor concentration to uniform acceptor concentration takes place in negligible distance and that the depletion layers are completely void of mobile charge carriers and therefore end abruptly. The space-charge density in the region on the *n*-side of the metallurgical junction ($x \geq 0$) is then $Q_n = eN_d$ and on the *p*-side ($x \leq 0$) $Q_p = -eN_a$, where

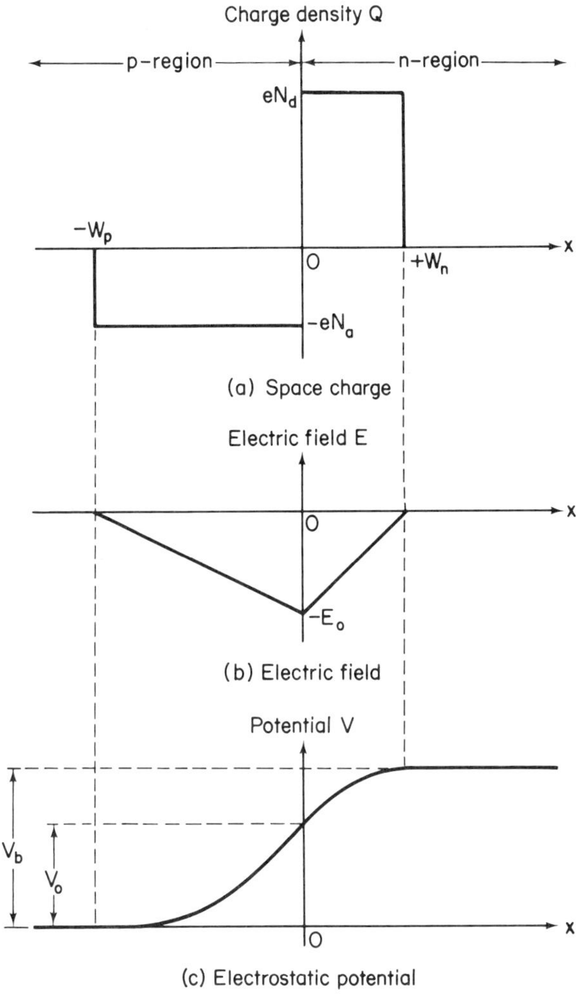

Fig. 2-16. Space-charge, electric-field, and electrostatic-potential distribution in an abrupt junction.

N_d and N_a are the *net* donor and acceptor ion densities. Applying Eq. (2-32) to the depletion region, we obtain

$$\frac{d^2V}{dx^2} = \frac{eN_a}{\epsilon} \quad \text{for } x \leq 0 \tag{2-33}$$

and

$$\frac{d^2V}{dx^2} = -\frac{eN_d}{\epsilon} \quad \text{for } x \geq 0 \tag{2-34}$$

Integrating Eqs. (2-33) and (2-34) yields

$$\frac{dV}{dx} = \frac{eN_a}{\epsilon}x + C \quad \text{for } x \leq 0 \tag{2-35}$$

and

$$\frac{dV}{dx} = \frac{eN_d}{\epsilon}x + C \quad \text{for } x \geq 0 \tag{2-36}$$

Since electric field E is defined as

$$E = -\frac{dV}{dx}$$

Eqs. (2-35) and (2-36) may be written as

$$E = -\frac{eN_a}{\epsilon}x + E_0 \quad \text{for } x \leq 0 \tag{2-37}$$

and

$$E = \frac{eN_d}{\epsilon}x + E_0 \quad \text{for } x \geq 0 \tag{2-38}$$

where E_0 is the electric field at the metallurgical junction ($x = 0$). Since the electric field is zero at the boundaries of the depletion region where the charge density is zero,*

$$E = 0 \quad \text{for } x = -W_p \tag{2-39}$$

and

$$E = 0 \quad \text{for } x = W_n \tag{2-40}$$

where W_n and W_p are the n-side and p-side depletion widths, as indicated in Fig. 2-16(a). From Eqs. (2-37) through (2-40), we obtain

$$E_0 = -\frac{eN_a}{\epsilon}W_p = -\frac{eN_d}{\epsilon}W_n \tag{2-41}$$

This relationship gives us two quite significant pieces of information. The p-region and n-region depletion widths are seen to be in the inverse ratio of the impurity densities; that is,

$$\frac{W_p}{W_n} = \frac{N_d}{N_a} \tag{2-42}$$

and the total positive charge eN_dW_n per unit area on the n-side of the junction is equal to the total negative charge eN_aW_p per unit area on the p-side. For this reason, the space charge in the depletion region is said to consist of a *dipole layer*.

* The small electric field responsible for majority-carrier drift current in the p- and n-regions (see Fig. 2-14) is considered negligible in this derivation.

Expressions for the electric field in the carrier-depletion region as a function of distance x from the metallurgical junction, as obtained from Eqs. (2-37) (2-38), and (2-41), are seen to be

$$E = -\frac{eN_a}{\epsilon}(x + W_p) \quad \text{for } x \leq 0 \tag{2-43}$$

and

$$E = \frac{eN_d}{\epsilon}(x - W_n) \qquad \text{for } x \geq 0 \tag{2-44}$$

and an $E(x)$ plot is shown in Fig. 2-16(b). The field is seen to have its maximum value E_0 at the junction ($x = 0$) and to decrease linearly with x from this negative maximum to zero at the boundaries of the depletion region, negative values indicating that the field is in the $-x$ direction.

The potential distribution in the carrier-depletion region may be obtained by integrating Eqs. (2-43) and (2-44), since

$$V = \int -E\,dx$$

Thus

$$V = \frac{eN_a}{\epsilon}\left(\frac{x^2}{2} + W_p x\right) + C \quad \text{for } x \leq 0 \tag{2-45}$$

and

$$V = \frac{eN_d}{\epsilon}\left(\frac{x^2}{2} - W_n x\right) + C \quad \text{for } x \geq 0 \tag{2-46}$$

where the integration constant C is the electrostatic potential V_0 for $x = 0$. If we arbitrarily let V be zero at the p-region boundary of the depletion region, that is, $V = 0$ when $x = -W_p$, it may be seen from Eq. (2-45) that

$$C = V_0 = \frac{eN_a}{2\epsilon}W_p^2 \tag{2-47}$$

The barrier potential V_b is the potential at the n-region boundary of the depletion region, and is obtained from Eqs. (2-46) and (2-47) for $x = W_n$. Thus

$$V_b = \frac{e}{2\epsilon}(N_d W_n^2 + N_a W_p^2) \tag{2-48}$$

Combining Eqs. (2-42) and (2-48) yields

$$V_b = \frac{eN_a W_p^2}{2\epsilon}\left(1 + \frac{N_a}{N_d}\right) \tag{2-49}$$

or

$$V_b = \frac{eN_d W_n^2}{2\epsilon}\left(1 + \frac{N_d}{N_a}\right) \tag{2-50}$$

From Eqs. (2-47) and (2-49), we see that

$$\frac{V_0}{V_b} = \frac{1}{(1 + N_a/N_d)} \tag{2-51}$$

showing that most of the potential barrier occurs in the region on the p-side of the junction when $N_d > N_a$, as in our example. This is indicated in the depletion-region potential-distribution plot of Fig. 2-16(c).

As pointed out in Sec. 2-2 [see Eq. (2-4)], the barrier potential V_b is given by

$$V_b = V_d - V \tag{2-52}$$

where V_d is the built-in junction potential and V is the voltage applied to the junction. Using this expression in Eqs. (2-49) and (2-50) and solving for depletion-region widths W_p and W_n, we obtain

$$W_p = \left[\frac{2\epsilon(V_d - V)}{eN_a(1 + N_a/N_d)}\right]^{1/2} \tag{2-53}$$

and

$$W_n = \left[\frac{2\epsilon(V_d - V)}{eN_d(1 + N_d/N_a)}\right]^{1/2} \tag{2-54}$$

From these equations, we see that the depletion-region width depends upon the square root of the junction voltage $(V_d - V)$.

In the typical abrupt junction (an alloy junction), the doping of one region is much greater than that of the other, and the depletion region lies almost entirely in the region of lower conductivity. For example, if $N_d \gg N_a$, we see from Eqs. (2-53) and (2-54) that $W_n \ll W_p$, and the depletion-region (or barrier) width W_b, given by

$$W_b = W_p + W_n \tag{2-55}$$

is

$$W_b \cong W_p \cong \left[\frac{2\epsilon(V_d - V)}{eN_a}\right]^{1/2} \tag{2-56}$$

Since the p-region conductivity σ_p is given by

$$\sigma_p = \mu_p e N_a$$

Eq. (2-56) may be written as

$$W_b \cong \left[\frac{2\epsilon\mu_p(V_d - V)}{\sigma_p}\right]^{1/2} \tag{2-57}$$

Therefore, for the alloy junction we can write the general relationship

$$W_b \cong \left[\frac{2\epsilon\mu(V_d - V)}{\sigma}\right]^{1/2} \tag{2-58}$$

where μ and σ are the carrier mobility and conductivity of the original crystal (the region of the device unaffected by alloying). As an indication of the order of magnitude of typical depletion-region widths, we can use this equation to determine W_b for a junction formed by alloying to n-type germanium having a resistivity of 1 ohm-cm (a conductivity of 1 mho/cm or 100 mho/meter). The dielectric constant of germanium is 15.8, μ_n is of the order of 0.39 m²/volt-sec (see Table 1-6), and the dielectric permittivity of free space is 8.85×10^{-12} farad/meter. Using these values in Eq. (2-58), we obtain

$$W_b \cong \left[\frac{2 \times 15.8 \times 8.85 \times 10^{-12} \times 0.39}{100}(V_d - V)\right]^{1/2}$$
$$\cong 10^{-6}(V_d - V)^{1/2} \text{ meters}$$

Thus the depletion-region thickness is of the order of 10^{-4} cm for a junction voltage of the order of 1 V.

We can determine the order of magnitude of the *average value* of the electric field in the depletion region by simply dividing the junction voltage by the depletion width. Thus, in our example, the electric field for a junction voltage of 1 V (obtained by using a reverse-bias voltage of a few tenths of a volt) has an average value of the order of 10,000 V/cm. The maximum value of the field may be obtained from Eq. (2-41), which can be written

$$E_0 = -\frac{eN_d}{\epsilon}W_n = -\frac{\sigma_n W_n}{\mu_n \epsilon} \tag{2-59}$$

Thus, in our example, the maximum value of the electric field for a junction voltage $(V_d - V)$ of 1 V is

$$E_0 = -\frac{100 \times 10^{-6}}{0.39 \times 15.8 \times 8.85 \times 10^{-12}} = -1.8 \times 10^6 \text{ V/m}$$

or 18,000 V/cm, suggesting an average value of 9,000 V/cm.

Junction-region capacitance. Since depletion-region width depends on junction voltage, a voltage change dV_b causes a change dQ_j in the charge Q_j stored in the depletion region on *one side* of the junction. This is caused by majority carriers, say n-region electrons, moving away from the junction through the n-region and the external circuit to the p-region contact where they neutralize p-region holes that have also moved away from the junction. Thus the charge change dQ in the external circuit is

$$dQ = -dQ_j \tag{2-60}$$

the negative sign being required because the sign of the mobile majority carriers that accomplish the charge change are opposite that of the immobile impurity ions that constitute the depletion-region charge. The current in the external circuit as a result of change in depletion-region charge Q_j is given by

$$i_j = \frac{dQ}{dt} \tag{2-61}$$

which can be written as

$$i_j = -\frac{dQ_j}{dt} = -\frac{dQ_j}{dV} \cdot \frac{dV}{dt} \tag{2-62}$$

where V is the voltage applied to the junction. This current is superimposed on the steady-state junction current predicted by the diode equation and may be the major component of total junction current for rapid changes in reverse-bias voltage; for forward bias it is likely to be masked by currents resulting from minority-carrier injection. Since charge flow does not occur within the depletion region, but takes place at the edges of the region, i_j is a displacement current and the associated capacitance C_j is given by

$$C_j = -\frac{dQ_j}{dV} \tag{2-63}$$

as seen from Eq. (2-62).

For the idealized abrupt junction shown in Fig. 2-16, we see that on the n-side of the junction

$$Q_j = AeN_dW_n \tag{2-64}$$

where W_n is a function of junction voltage. From Eq. (2-64) and the W_n expression of Eq. (2-54), we obtain

$$Q_j = A\left[\frac{2e\epsilon N_d(V_d - V)}{(1 + N_d/N_a)}\right]^{1/2} \tag{2-65}$$

From Eqs. (2-63) and (2-65), we see that

$$C_j = -\frac{dQ_j}{dV} = \frac{A}{2}\left[\frac{2e\epsilon N_d}{(V_d - V)(1 + N_d/N_a)}\right]^{1/2} \tag{2-66}$$

For an abrupt junction formed by alloying to n-type material, $N_a \gg N_d$ and Eq. (2-66) becomes

$$C_j \cong \frac{A}{2}\left[\frac{2e\epsilon N_d}{(V_d - V)}\right]^{1/2} = A\left[\frac{\sigma_n\epsilon}{2\mu_n(V_d - V)}\right]^{1/2} \tag{2-67}$$

and the general expression for the capacitance of an alloy junction is

$$C_j = A\left[\frac{\sigma\epsilon}{2\mu(V_d - V)}\right]^{1/2} \tag{2-68}$$

where σ and μ are the conductivity and carrier mobility of the original crystal. For reverse-bias voltages large compared to contact potential (diffusion voltage) V_d, the capacitance of an abrupt junction is seen to vary as $(-V)^{-1/2}$. For forward bias, the capacitance does not become excessively large compared to the reverse-bias value, since V is always appreciably less than V_d. The property of junction capacitance dependence upon voltage is utilized in the voltage-variable silicon capacitor, a device having such trade names as Varicap, Semicap, etc. Silicon is used in such devices because the reverse current is several orders of magnitude lower than that of germanium junctions, making relatively high-Q capacitors possible.

Junction capacitance differs from the capacitance of a parallel-plate capacitor in that the stored charge is distributed through the dielectric rather than concentrated at the edges. However, the charge change for a small change in junction voltage occurs entirely at the edges of the depletion layer. Therefore, the junction capacitance is the same as that of a parallel-plate capacitor of area A, spacing W_b, and dielectric permittivity ϵ. That is,

$$C_j = \frac{\epsilon A}{W_b} \tag{2-69}$$

Using the W_b expression given by Eq. (2-58) for an alloy junction, We find that Eq. (2-69) becomes

$$C_j = \epsilon A\left[\frac{\sigma}{2\epsilon\mu(V_d - V)}\right]^{1/2} = A\left[\frac{\sigma\epsilon}{2\mu(V_d - V)}\right]^{1/2} \tag{2-68}$$

which is identical to Eq. (2-68).

Junction capacitance is ordinarily an undesired junction property, voltage-variable capacitors being a notable exception, and should usually be as low as possible. Equation (2-68) indicates that junction capacitance can be kept low by using low-conductivity material. However, the effects of conductivity on reverse-saturation current and junction-breakdown voltage must also be considered in the choice of the conductivity of device material, and compromises are inevitable.

Graded junctions. Expressions for the depletion-region width and capacitance of a graded junction can be obtained through analysis similar to that for the abrupt junction. This analysis requires information regarding the impurity distribution in the *p*- and *n*-regions; as an example, let us consider a device in which the impurity density changes linearly with position, as shown by the plot of Fig. 2-17(a). If we again assume the depletion layers to be completely void of mobile charge carriers, the space-charge density plot is also linear, as shown in Fig. 2-17(b), and can be expressed as

$$Q = eax \tag{2-70}$$

where

$$a = \frac{N_a - N_d}{W_i} \tag{2-71}$$

and W_i is the width of the graded region. If we continue the analysis as for the abrupt junction, we obtain the depletion-region width and junction capacitance expressions

$$W_b \equiv W_n + W_p = 2W_n = \left[\frac{12\epsilon(V_d - V)}{ae}\right]^{1/3} \tag{2-72}$$

and

$$C_j = A\epsilon^{2/3}\left[\frac{ae}{12(V_d - V)}\right]^{1/3} \tag{2-73}$$

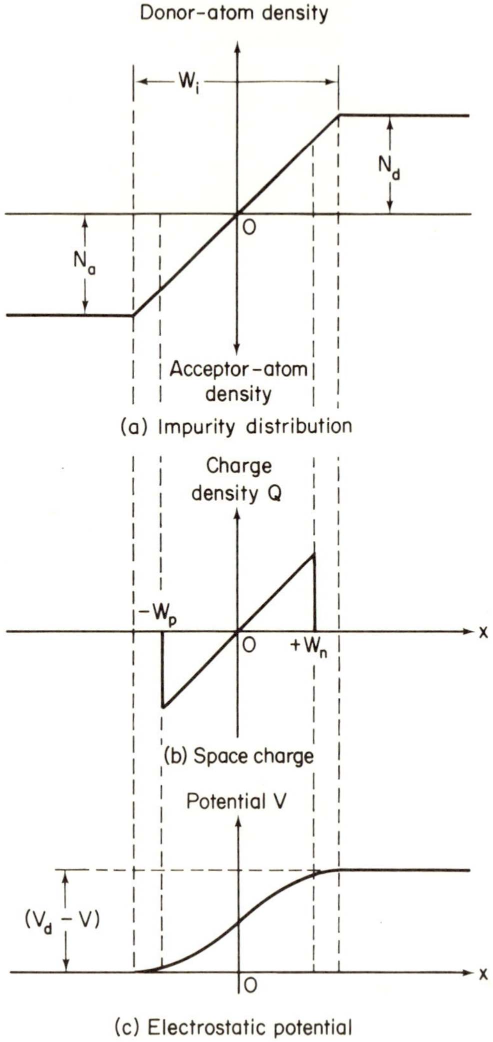

Fig. 2-17. Impurity distribution, space-charge, and electrostatic potential for a graded junction.

from which we see that the depletion region is symmetrical and the junction capacitance varies inversely as the cube root of the applied voltage for reverse-bias voltages large compared to the contact potential V_d.

As in the analysis of the abrupt junction, the derivation of Eqs. (2-72) and (2-73) is based on the assumption that the electric field is zero at the edges of the carrier-depletion region. It is also assumed that the depletion region does not extend into the uniformly doped regions. In reality, an electric field exists in the undepleted graded region because of the *impurity-concentration gradient*. With such a gradient, a gradient of majority-carrier density also exists, and majority carriers flow by diffusion toward the more lightly doped regions. As a result, unneutralized impurity ions exist in the more heavily doped regions, and there is an excess of majority-carrier charge in the regions to which the carriers have diffused. This separation of charge sets up an electric field in the graded region that causes drift flows of majority carriers opposite the directions of the diffusion flows. Under equilibrium conditions, the magnitude of the electric field is such that the drift flows equal the diffusion flows.* Therefore, a relatively small field (compared to the average depletion-region field) exists in the graded p- and n-regions of Fig. 2-17, and the potential barrier extends between the regions of uniform doping, as shown by the potential-distribution plot of Fig. 2-17(c).

2-7 DIFFUSION CAPACITANCE

Although forward-biasing a *pn* junction increases the depletion-region capacitance, it is also responsible for a much more important capacitive effect that swamps out the depletion-region capacitance when the forward current is appreciably larger than the reverse-saturation current. This is the storage or diffusion capacitance, and it is the capacitance associated with the charge stored in the p- and n-regions as a result of minority-carrier injection.

The existence of an additional capacitive effect with forward bias may be seen from the following qualitative consideration. Suppose that a step voltage is applied to a junction in the forward direction, resulting in a p-region electron density

$$n_{po}e^{eV/kT}$$

at the edge of the carrier-depletion region, as indicated by Eq. (2-10). Because of the large electric field in the very narrow depletion region, the time required for an electron to cross the region and appear on the p-side is very short. This time has been determined to be of the order of 10^{-11} sec, at most, and is much less than the time required for electrons to diffuse an appreciable distance, say a diffusion length, into the p-region. Therefore, an enormous electron-density gradient exists in the p-region almost immediately following the application of the step voltage, as shown by the "early" $n_p(x)$ plot of Fig. 2-18, causing an equally enormous electron-diffusion current. As electrons diffuse farther into the p-region, the electron-density gradient at the

* The concept of establishing an electric field in a semiconductor through donor or acceptor gradients, or both, is of considerable importance in device considerations and will be referred to later.

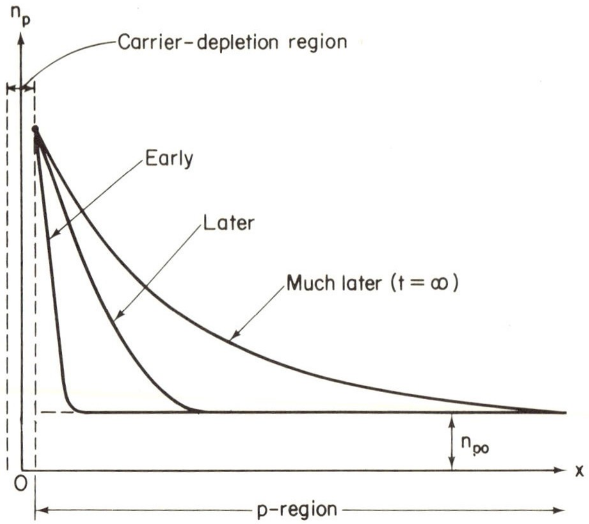

Fig. 2-18. Electron density in the p-region of a pn junction following the application of a forward-voltage step. The early values of the electron-density gradient dn_p/dx, and the resulting electron-diffusion currents, are much greater than the steady-state values.

depletion-region edge ($x = 0$) decreases, as shown by the "later" $n_p(x)$ plot of Fig. 2-18, decreasing the diffusion current. Eventually the steady-state electron-density distribution, given by Eq. (2-28), becomes established, and the electron-diffusion current at the depletion-region edge ($x = 0$) becomes the steady-state forward value given by

$$I_{np}(0) = \frac{AeD_n n_{po}}{L_n}(e^{eV/kT} - 1) \tag{2-74}$$

as obtained from Eq. (2-30). A similar variation in n-region hole-density gradient and hole-diffusion current occurs following the application of a forward voltage step. The decaying component of junction current following a change in forward voltage suggests the existence of a capacitance in shunt with the junction. In reality, it is difficult to change the junction voltage rapidly, and the application of a forward-voltage step is virtually impossible, owing to the enormous current flow that would occur.*

To determine the diffusion capacitance, we must find the charge stored in the p- and n-regions when the junction is forward biased. The excess electron density n'_p in the p-region of the forward-biased junction is given by

$$n'_p = n_{po}(e^{eV/kT} - 1)e^{-x/L_n} \tag{2-75}$$

* Note that this statement refers to *junction voltage*. A forward-voltage step may be readily applied to the *terminals* of a pn-junction device.

as obtained from Eq. (2-28). The total excess electron change Q_n in the p-region is obtained by integrating the $n'_p(x)$ expression. Thus

$$\begin{aligned} Q_n &= -eA \int_0^\infty n'_p(x)\,dx \\ &= -eA \int_0^\infty n_{po}(e^{eV/kT} - 1)e^{-x/L_n}\,dx \\ &= \Big[_0^\infty -eAn_{po}L_n(e^{eV/kT} - 1)e^{-x/L_n} \\ &= Aen_{po}L_n(e^{eV/kT} - 1) \end{aligned} \tag{2-76}$$

If the voltage is varied, the differential capacitance C_n associated with the change in p-region electron charge is defined as

$$C_n = \frac{dQ_n}{dV} \tag{2-77}$$

From Fqs. (2-76) and (2-77).

$$C_n = \frac{e^2 An_{po}L_n}{kT} e^{eV/kT} \tag{2-78}$$

There is a similar term for the capacitance C_p associated with holes stored in the n-region. Therefore,

$$C_p = \frac{e^2 Ap_{no}L_p}{kT} e^{eV/kT} \tag{2-79}$$

and the storage or diffusion capacitance C_D is given by

$$C_D = C_n + C_p = \frac{e^2 A}{kT}(n_{po}L_n + p_{no}L_p)e^{eV/kT} \tag{2-80}$$

and is seen to increase strongly with forward bias.

Since the exponential term of Eq. (2-78) also appears in the diode equation, we can express C_n in terms of the electron current at the junction. From Eq. (2-6),

$$I_n + I_{nr} = I_{nr}e^{eV/kT} \tag{2-81}$$

where I_n is the net forward electron current and I_{nr} is the reverse-saturation electron current. From Eq. (2-20), the magnitude of I_{nr} is given by

$$I_{nr} = \frac{eAD_n n_{po}}{L_n} \tag{2-82}$$

Substituting this I_{nr} expression in Eq. (2-81) yields

$$I_n + I_{nr} = \frac{eAD_n n_{po}}{L_n} e^{eV/kT} \tag{2-83}$$

which may be written

$$I_n + I_{nr} = \frac{kTD_n}{eL_n^2} \times \frac{e^2 An_{po}L_n}{kT} e^{eV/kT} \tag{2-84}$$

From Eqs. (2-78) and (2-84),

$$I_n + I_{nr} = \frac{kTD_n}{eL_n^2} C_n \tag{2-85}$$

or

$$C_n = \frac{eL_n^2}{kTD_n}(I_n + I_{nr}) \tag{2-86}$$

Similarly,

$$C_p = \frac{eL_p^2}{kTD_p}(I_p + I_{pr}) \tag{2-87}$$

where I_p is the net forward hole current and I_{pr} is the reverse-saturation hole current. Since the forward current is the sum of I_n and I_p, we see from Eqs. (2-86) and (2-87) that the diffusion capacitance is directly proportional to forward current for values large compared to the reverse-saturation current. Quite often either the electron current or the hole current predominates at a junction, and the diffusion capacitance is given by either Eq. (2-86) or Eq. (2-87). Thus for an alloy junction

$$C_D = \frac{eL^2}{kTD}(I + I_s) = \frac{e\tau}{kT}(I + I_s) \tag{2-88}$$

where I is the junction current, I_s is the reverse-saturation current, and τ, L, and D are the minority-carrier lifetime, diffusion length, and diffusion constant of the original crystal. From Eq. (2-88) we see that the diffusion capacitance is of the order of microfarads for τ of the order of 10 μs, a typical value, and $(I + I_s)$ of the order of milliamperes. This is a capacitance several orders of magnitude greater than the depletion-region capacitance, which might be of the order of 10^{-4} μF, or less.

2-8 JUNCTION RESISTANCE

An important property of a *pn*-junction device, such as a diode, is its resistance. Since the diode is a nonlinear device, its resistance depends upon operating current, and we are interested in two resistances—the d-c resistance and the dynamic or small-signal a-c resistance. The diode may be considered to consist of an ideal semiconductor diode, whose characteristics are defined by the diode equation, in series with the bulk resistance R_b of the *p*- and *n*-regions. The d-c resistance R_{dc} of the series combination is given by

$$R_{dc} = R_b + \frac{V}{I} \tag{2-89}$$

where V is the voltage applied to the junction and I is the junction current.

Applying the diode equation, we find that Eq. (2-89) becomes

$$R_{dc} = R_b + \frac{V}{I_s(e^{eV/kT} - 1)} \tag{2-90}$$

It is convenient to calculate the a-c conductance g_{ac}, where

$$g_{ac} = \frac{dI}{dV} \tag{2-91}$$

From this expression and the diode equation, given by

$$I = I_s(e^{eV/kT} - 1) \tag{2-92}$$

we obtain

$$g_{ac} = \frac{e}{kT} I_s e^{eV/kT} \tag{2-93}$$

which can be written

$$g_{ac} = \frac{e}{kT}(I + I_s) \tag{2-94}$$

as seen from Eq. (2-92). Since the a-c resistance r_{ac} of the ideal diode is the reciprocal of the a-c conductance, we see that

$$r_{ac} = \frac{kT}{e(I + I_s)} \tag{2-95}$$

and the diode a-c resistance R_{ac} is given by

$$R_{ac} = R_b + \frac{kT}{e(I + I_s)} \tag{2-96}$$

At room temperature, the value of r_{ac} is

$$r_{ac} = \frac{25}{I + I_s} \text{ ohms} \tag{2-97}$$

if I and I_s are in milliamperes. From this relationship, we see that the forward ideal-diode resistance at high currents becomes negligible, compared to the bulk resistance of the *p*- and *n*-regions, and the diode characteristic is ohmic. For low forward currents and for reverse-bias operation, the ideal-diode resistance becomes so high that the bulk resistance is negligible. For reverse-bias (negative) currents approaching I_s, the ideal diode resistance becomes meaningless, owing to the leakage resistance in shunt with the junction.

From this consideration and that of Secs. 2-6 and 2-7, we see that for small signals superimposed upon the d-c bias voltage and current, the semiconductor diode may be represented by the circuit model shown in Fig. 2-19. In this circuit, C_s is the capacitance between the connecting leads, R_b is the bulk-region resistance, r_j is the dynamic resistance of the junction, and C_j is the sum of the depletion-region and diffusion capacitances, r_j and

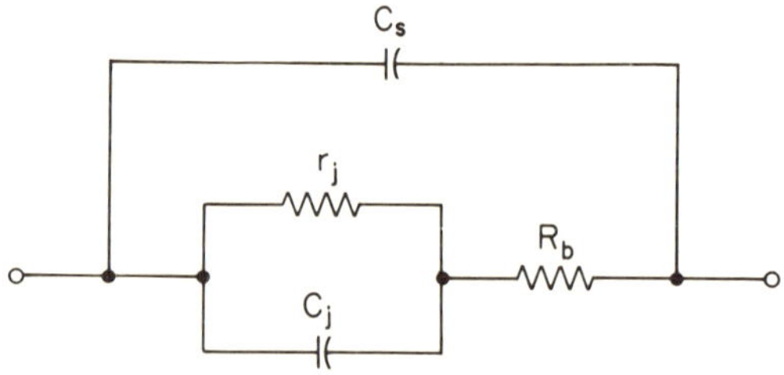

Fig. 2-19. Small-signal circuit model of a *pn*-junction diode. Dynamic resistance r_j and junction capacitance C_j are highly dependent upon d-c current and voltage, and bulk-region resistance R_b may be dependent on forward current.

C_j being highly dependent upon d-c bias conditions. Bulk-region resistance R_b may also be bias dependent, depending on the forward current. This is the effect of *conductivity modulation*, which is discussed in Sec. 4-5.

PROBLEMS

2-1. An alloy-junction silicon diode is formed by alloying to 10-ohm-cm *n*-type silicon. What reverse-breakdown voltage might be expected of the diode? Account for a lower value that might be observed.

2-2. An abrupt junction is formed by alloying to a wafer of *n*-type germanium having a resistivity of 5 ohm-cm and a minority-carrier lifetime of 10 μs. If the junction area is 0.01 cm^2, what reverse-saturation current might be expected at a temperature of 300°K?

2-3. For what reverse-bias voltage might the current in a *pn* junction reach 90 percent of its saturation value?

2-4. Determine the ratio of forward-to-reverse current that might be expected for a *pn* junction at bias voltages of 0.02 V, 0.05 V, and 0.10 V in magnitude at a temperature of 300°K.

2-5. Two identical diodes are connected series-opposing (back-to-back). If a d-c voltage of 1 V is applied to the series combination, what voltage might be expected across each diode at a temperature of 300°K?

2-6. The reverse-saturation current of a germanium *pn* junction is 1 μA at a temperature of 25°C. What value might be expected at 100°C?

2-7. Repeat Prob. 2-6 for a silicon *pn* junction.

2-8. A germanium *pn* junction having an area of 0.010 cm^2 has an *n*-region majority-carrier density of 10^{15} cm^{-3} and a *p*-region majority-carrier density of 10^{16} cm^{-3}. The minority-carrier lifetime of each region is 10 μs.

(a) Determine the *n*-region hole density $p_n(0)$ at the edge of the carrier-depletion (space-charge) region for a forward-bias voltage of 0.10 V at a temperature of 300°K.

(b) Determine the value of the hole current and that of the electron current *at the junction* for the conditions of (a).
(c) The *n*-region majority-carrier current for the conditions of (a) may be resolved into a drift component and a diffusion component. Determine each component at the edge of the carrier-depletion region.
(d) For the conditions of (a), determine the hole and electron currents in the *n*-region at a distance from the junction of one hole diffusion length and resolve the electron current into diffusion and drift components.

2-9. The transition-region current of a forward-biased *pn* junction consists of an electron current of 1.0 mA and a hole current of 2.0 mA. Determine the approximate magnitude of each of the following *n*-region currents at the transition-region edge and indicate whether the carriers move toward or away from the junction: (a) Hole diffusion current, (b) electron diffusion current, and (c) electron drift current.

2-10. A forward-biased p^+n junction (a junction having a *p*-region very heavily doped relative to the *n*-region doping) has a current of 100 mA. The *n*-region hole current through a plane parallel to the junction is 75 mA. Determine the drift and diffusion components of electron current through the plane and indicate whether the current carriers constituting the components move toward or away from the junction.

2-11. An abrupt *pn* junction is formed by alloying to a wafer of *n*-type germanium having a donor-atom concentration of 8.7×10^{14} atoms/cm^3. Assuming the *p*-region Fermi level to be at the bottom of the energy gap (that is, $E_{fp} = 0$) and using the *n*-region Fermi level determined in Prob. 1-15 ($E_{fn} = 0.43$ eV), determine the depletion-region thickness for zero bias and the maximum value of the electric field in the transition region.

2-12. Repeat Prob. 2-11 for a reverse voltage of 100 V.

2-13. An abrupt-junction diode formed by alloying to 5-ohm-cm *p*-type germanium has a *p*-region thickness of 1 mil (0.001″). What *p*-region reach-through (punch-through) voltage might be expected?

2-14. An abrupt *pn* junction has *p*- and *n*-regions of equal resistivities. If the reverse bias is such that 0.001 in. of the *p*-region is depleted of majority carriers, what is the approximate value of the total depletion width?

2-15. An abrupt *pn* junction is formed by alloying to a wafer of *n*-type germanium having a resistivity of 5-ohm-cm. If the junction has an area of 0.01 cm^2, what transition capacitance might be expected for a reverse voltage of 10 V?

2-16. Repeat Prob. 2-15 for *p*-type germanium.

2-17. If the *n*-region of the junction of Prob. 2-16 has a minority-carrier lifetime of 10 μs, and the junction reverse-saturation current is 5 μA, determine the approximate parallel combination of dynamic (a-c) resistance and capacitance that might be expected to represent the forward-biased junction when the forward current is 100 μA.

2-18. Show that the width of the carrier-depletion region and the transition capacitance of a graded junction are given by Eqs. (2-72) and (2-73).

3

Metal-Semiconductor Contacts

The theory of metal-semiconductor contacts, commonly referred to simply as contacts, is of importance because of their rectifying properties and because contacts are required in the connections to *pn* junctions. There are, therefore, two general types of contacts; these are rectifying contacts and nonrectifying or ohmic contacts. Contacts to the *p*- and *n*-regions of *pn* junctions are (or should be) ohmic contacts, but rectifying contacts also have important applications. Finally, there are two kinds of rectifying metal-semiconductor contacts: those to *n*-type semiconductors and those to *p*-type semiconductors.

3-1 RECTIFYING n-TYPE SEMICONDUCTOR CONTACTS*

Energy-level diagrams for a metal and an *n*-type semiconductor are shown in Fig. 3-1(a). The two materials are considered to have parallel-plane faces in close proximity to each other and to be electrically neutral so that the space between their faces is field-free. The Fermi level in the metal is at

* J. N. Shive, *The Properties, Physics and Design of Semiconductor Devices*. Princeton, N.J.: D. Van Nostrand Co., Inc., 1959.

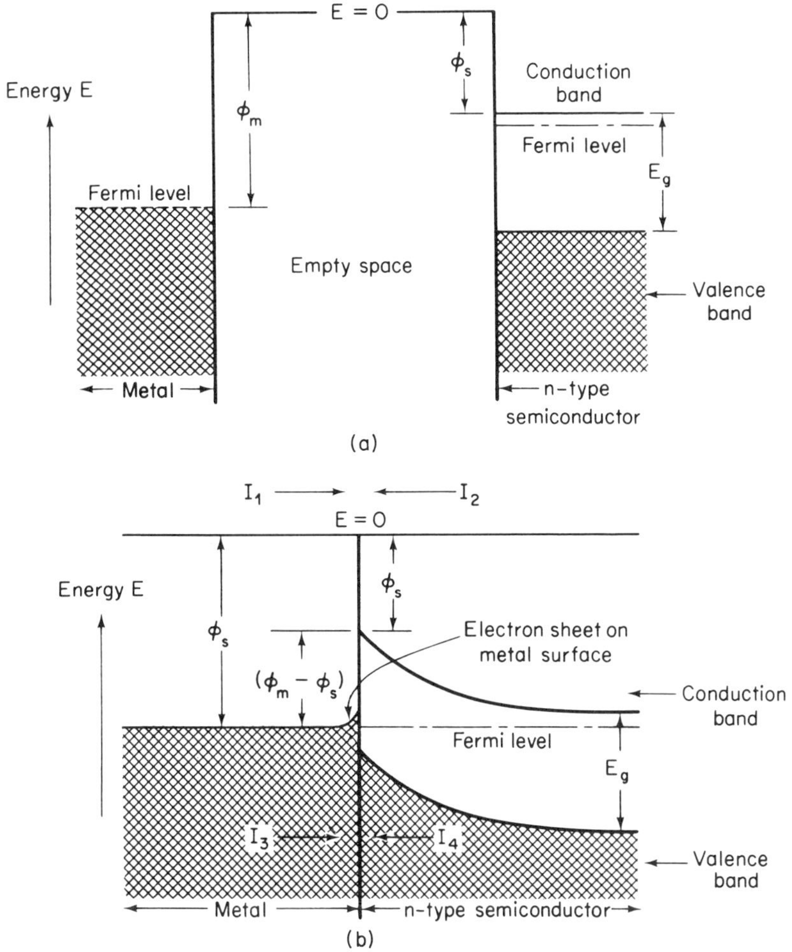

Fig. 3-1. Energy-level diagrams for a metal-to-*n*-type semiconductor contact. (a) Specimens not connected. (b) Specimen faces in contact and equilibrium conditions established.

the top of the filled portion of the conduction band and is ϕ_m below the reference level $E = 0$, the emission level outside the metal surface, where ϕ_m is the work function of the metal.* The work function ϕ_s of the semiconductor is approximately the vertical distance between the reference level ($E = 0$) and the bottom of the conduction band, where the majority of the conduction-band electrons are located.

* The work function of a material is the minimum amount of energy that, when supplied to an electron, will effect the emission of an electron from the material.

If an external connection is made between the metal and an ohmic contact on the semiconductor, electrons will flow from the semiconductor to the metal until the excess negative charge on the metal and the excess positive charge on the semiconductor bring the two Fermi levels into alignment. The excess negative charge on the metal is in the form of a thin electron sheet on its surface and the excess positive charge on the semiconductor is a depleted layer of ionized donor atoms. As a result of this dipole layer of charge, an electric field exists in the space between the opposing metal-semiconductor faces, and the potential difference between the faces is the contact potential. If the two opposing faces are brought closer together, the electric field in the space between them becomes greater and the dipole-charge layer becomes larger. This charge behavior is the result of increasing capacitance betwen the two surfaces as they are brought closer together. Since the capacitance is charged to a fixed voltage, the contact potential, the charge increases directly with capacitance. When the surfaces are brought into physical contact, with zero externally applied voltage, the situation is as illustrated by the energy-level diagram of Fig. 3-1(b), and the field due to contact potential now lies entirely within the semiconductor. Since an electron at the semiconductor surface continues to require an energy increase equal to the work-function energy ϕ_s to reach the reference level ($E = 0$), the n-region conduction band is bent upward near the contact, as shown in Fig. 3-1(b). This bending results in a barrier of approximately $(\phi_m - \phi_s)$ to the flow of n-region conduction-band electrons toward the contact and to the flow of electrons from the metal to the semiconductor. In order to cross the barrier, an electron in either the metal or the semiconductor must have an energy of $(\phi_m - \phi_s)$ above the Fermi energy. Therefore, the current I_1 of electrons in the metal that are able to surmount the barrier is equal to the current I_2 of n-region electrons that are also able to do so, and the net electron current is zero. Since the same barriers exist for hole flow, the current I_3 of holes in the metal that are able to cross the contact is equal to the current I_4 of n-region holes that are able to reach the metal, and the net hole current is zero.

The effect of making the n-type semiconductor negative with respect to the metal is shown by the energy-level diagram of Fig. 3-2. With the semiconductor negative by the voltage V, the potential energy of conduction-band electons deep in the n-region is increased, the Fermi level being shifted upward by the energy eV. However, there is no change in the potential energy of n-region conduction-band electrons at the contact, an energy increase ϕ_s continuing to be required by each surface electron in order to reach the reference level ($E = 0$), and the barrier to the flow of n-region electrons toward the contact has been reduced, as indicated in Fig. 3-2. This reduction increases the current I_2 of n-region electrons that are able to surmount

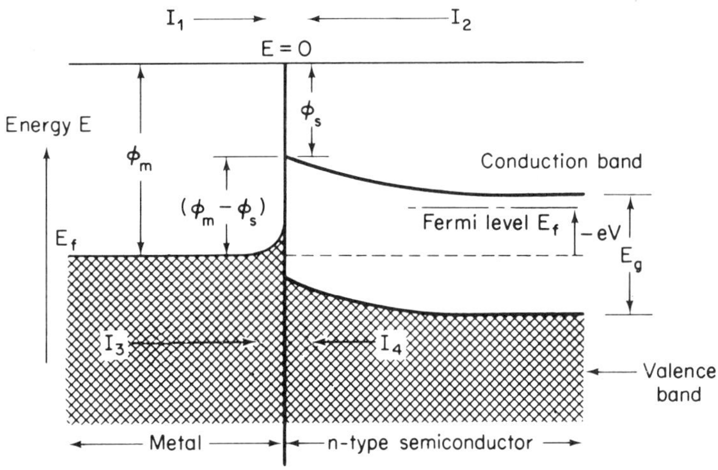

Fig. 3-2. Energy-level diagram for a forward-biased metal-to-*n*-type semiconductor contact. The *n*-region is negative by the voltage V, reducing the barrier to *n*-region electrons and to metal-region holes and increasing electron current I_2 and hole current I_3 above the equilibrium values.

the barrier, but the current I_1 of metal-region electrons crossing the contact is unchanged, since the barrier to their flow continues to be approximately $(\phi_m - \phi_s)$, the value for zero bias. Similarly, the barrier to the flow of holes from the metal to the semiconductor is decreased, increasing hole current I_3, but the barrier to the flow of *n*-region holes into the metal is unchanged and hole current I_4 continues to be the value for zero bias. Since the zero-bias barrier may be as small as a fraction of an electron-volt, making the *n*-region negative by a relatively small voltage, say a few tenths of a volt, results in a large current through the contact, and the contact is forward biased.

Figure 3-3 shows the rectifying barrier with the *n*-type semiconductor positive by the voltage V with respect to the metal. As with forward bias, the barrier to the flow of metal-region electrons and *n*-region holes across the contact is unchanged by bias, but the barrier to *n*-region electrons and metal-region holes is increased, since the potential energy of conduction-band electrons deep in the *n*-region is decreased, the Fermi level being shifted downward by the energy eV. Therefore, the metal-region electron current I_1 and the *n*-region hole current I_4 across the contact are unaffected by the bias voltage, but the *n*-region electron-current I_2 and the metal-region hole current I_3 are reduced. If the bias voltage is sufficiently large, say greater

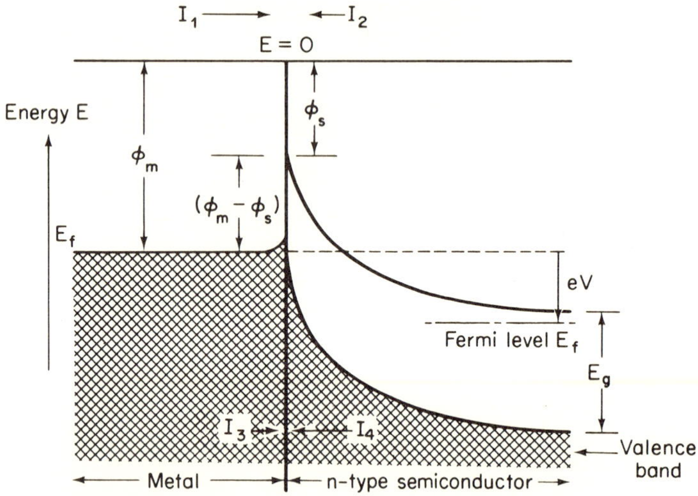

Fig. 3-3. Energy-level diagram for a reverse-biased metal-to-*n*-type semiconductor contact. The *n*-region is positive by the voltage V, increasing the barrier to *n*-region electrons and to metal-region holes and decreasing electron current I_2 and hole current I_3.

than about 0.1 V, currents I_2 and I_3 become negligible and the net contact current is $I_1 + I_4$, a current unaffected by bias voltage. Thus, the contact current saturates as the voltage is increased, and the contact is reverse biased by making the *n*-region positive.

We see that a rectifying barrier *may* exist at an *n*-type semiconductor contact if the thermionic work function ϕ_m of the metal is greater than the semiconductor work function ϕ_s. Like the *pn* junction, a rectifying contact cannot usually be formed by merely pressing together ordinary specimens of material, an ohmic contact being the usual result of such an endeavor. At the very least, a good rectifying contact requires that the surface of the semiconductor specimen be polished or etched in order to remove imperfections produced in cutting the specimen from the original ingot. Otherwise, the very short minority-carrier lifetime in the surface region of the semiconductor precludes rectifying action. When a rectifying *n*-type contact exists, the polarities of the applied voltages for forward and reverse bias, and the actions occuring at the contact are similar to those for a *pn* junction, the metal of the contact acting somewhat as a junction *p*-region. That is, the contact is forward biased when the metal is positive and reverse biased when it is negative, and the current-voltage characteristic is defined by the diode equation.

3-2 RECTIFYING *p*-TYPE SEMICONDUCTOR CONTACTS

Figure 3-4 shows an energy-level diagram for a *p*-type semiconductor contact under equilibrium conditions. As in the energy diagram for the *n*-type contact, ϕ_m is the work function of the metal and ϕ_s is that of the semiconductor, ϕ_s being approximately the vertical distance between the reference level ($E = 0$) and the top of the valence band. With the Fermi levels aligned for zero applied voltage, deep in the *p*-region the top of the valence band is just below the Fermi level, but at the contact, it is ϕ_s below the reference level ($E = 0$). As a result, a barrier of $(\phi_s - \phi_m)$ exists to electron flow from the metal to the *p*-region and to hole flow from the *p*-region to the metal. Unlike the *n*-type contact, no barrier exists to the flow across the contact of metal-region holes and *p*-region electrons. As in the junction, the current I_1 of metal-region electrons able to surmount the barrier is equal to the current I_2 of *p*-region electrons "falling down" the barrier, and the net electron current is zero. Similarly, current I_3 of metal-region holes falling down the barrier is equal and opposite current I_4 of holes climbing the barrier, and the net contact current is zero.

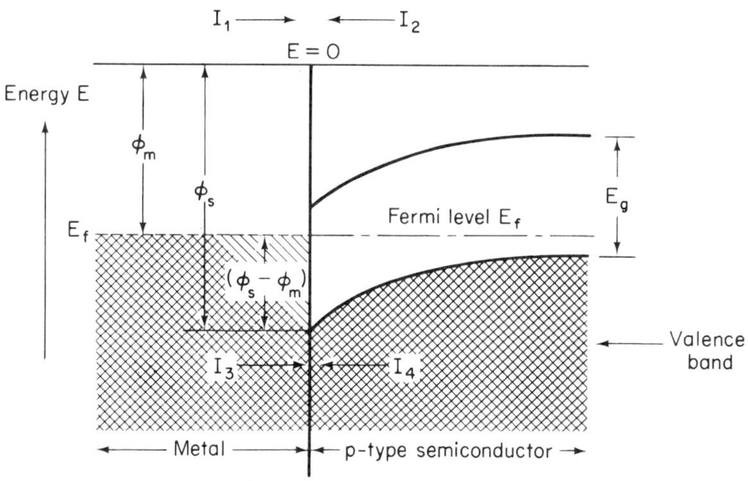

Fig. 3-4. Energy-level diagram for a metal-to-*p*-type semiconductor contact under equilibrium conditions (zero bias). A barrier of $(\phi_s - \phi_m)$ exists at the contact for metal-region electrons and *p*-region holes. Majority and minority electron currents I_1 and I_2 are equal and opposite as are minority and majority hole currents I_3 and I_4.

The effect of making the *p*-type semiconductor negative with respect to the metal is shown by the energy-level diagram of Fig. 3-5. Unlike the effect on the *n*-type contact, this polarity of applied voltage increases the barrier to the flow of majority carriers across the contact, and the contact is reverse biased. Figure 3-6 shows the reduction of the barrier to majority-carrier flow when the *p*-region is positive with respect to the metal, forward-biasing the contact.

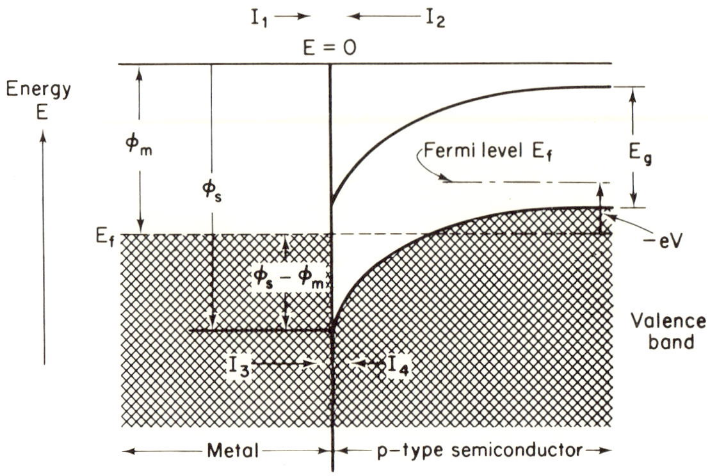

Fig. 3-5. Energy-level diagram for a reverse-biased metal-to-*p*-type semiconductor contact. The *p*-region is negative by the voltage V, increasing the barrier to metal-region electrons and to *p*-region holes and reducing majority-electron current I_1 and majority-hole current I_4 below the equilibrium values. Minority-electron current I_2 and minority-hole current I_3 are unaffected by bias.

We see that a rectifying barrier *may* exist at a *p*-type semiconductor contact if the thermionic work function ϕ_s of the semiconductor is greater than the metal work function ϕ_m. It is of interest to note that the polarity of the applied voltage for forward operation of the *p*-type contact is just the opposite to that of the *n*-type contact. This difference was the first indication of the existence of two semiconductor types, and the distinguishing designations, *n* and *p*, as originally used referred to the polarity of the semiconductor for easy flow of current. This serves as a method for determining the type of a semiconductor, but it is not a particularly good one, owing to the difficulty in obtaining a rectifying contact. A better arrangement is the hot-probe method described in Sec. 1-1.3.

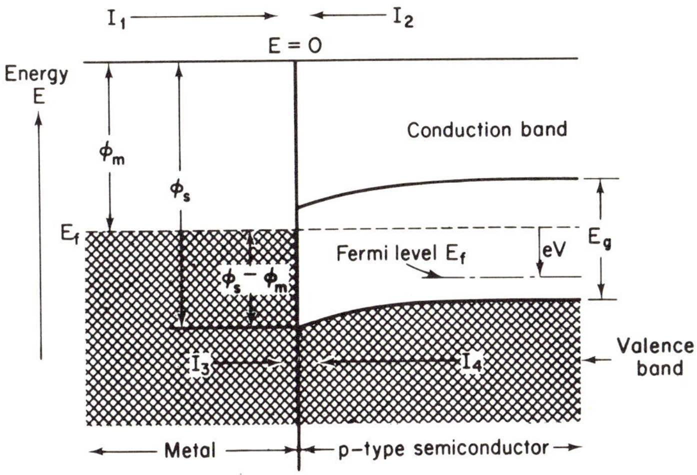

Fig. 3-6. Energy-level diagram for a forward-biased metal-to-*p*-type semiconductor. The *p*-region is positive by the voltage V, decreasing the barrier to metal-region electrons and to *p*-region holes and increasing majority-electron current I_1 and majority-hole current I_4 above the equilibrium values. Minority-electron current I_2 and minority-hole current $\mathbf{I}_3$ are unaffected by bias.

3-3 OHMIC CONTACTS

Since the presentation of the last two sections was based on the conditions that $\phi_m > \phi_s$ for the *n*-type contact and $\phi_s > \phi_m$ for the *p*-type contact, let us now consider contacts in which these conditions are not met. Figure 3-7 shows an energy-level digram for the contact between a metal and an *n*-type semiconductor for which $\phi_s > \phi_m$. The diagram is for zero applied voltage and depicts the situation the instant after contact is established. We observe that, near the metal, the lower edge of the semiconductor conduction band lies *below* the Fermi level, producing the depression A. Electrons immediately flow into this depression, just as a liquid would flow into a depression in its surface, and the depression is obliterated. Thereafter, electrons may flow unimpeded in either direction in the semiconductor conduction band, and we have a nonrectifying or ohmic contact. Similar reasoning shows the absence of a rectifying barrier at a contact between a metal and a *p*-type semiconductor for which $\phi_s < \phi_m$.

Such contacts are ideally resistanceless, the only resistance present

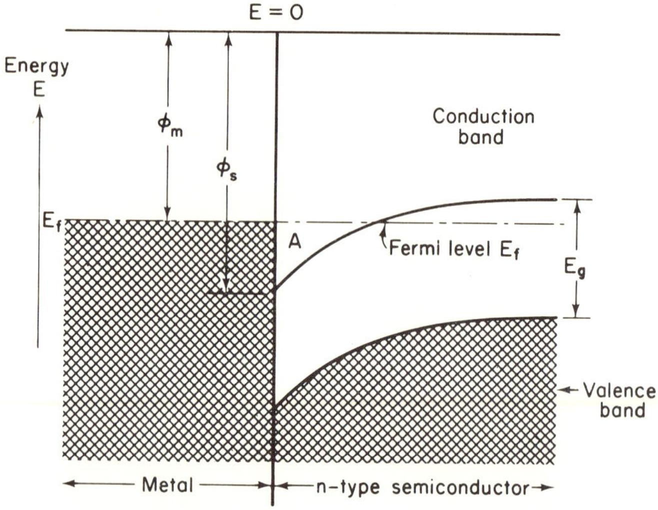

Fig. 3-7. Energy-level diagram for a contact between a metal and an n-type semiconductor for which $\phi_s < \phi_m$. Conditions are depicted immediately after contact. Depression A subsequently fills up with electrons and the contact becomes resistanceless.

being that of the metal and semiconductor specimens and their connections. Thus, we see that an ohmic contact to a semiconductor can be obtained by choosing metals such that $\phi_m < \phi_s$ for n-type metarial and $\phi_m > \phi_s$ for p-type material. As mentioned in Sec. 3-1, ohmic contacts may also exist as a result of semiconductor surface imperfections that reduce the minority-carrier lifetime and, for this reason, the crystalline structure at the surface of a semiconductor may be deliberately damaged, say by sandblasting, when an ohmic contact is to be made. Soldering or welding leads to a semiconductor usually results in a ohmic contact, owing to surface damage. When the minority-carrier diffusion length is short compared to the width of the transition region, almost all carrier recombination takes place in the transition region, and no change occurs in the carrier density on either the p-side or the n-side of the contact. (That is, no minority-carrier injection occurs upon the application of a voltage of the polarity that would forward-bias a rectifying contact.) The currents in the metal and semiconductor regions near a forward-biased contact are shown in Fig. 3-8(a) for a rectifying contact, in which some recombination occurs in the transition region, and in Fig. 3-8(b) for an ohmic contact. Because of the high conductivity of the metal, the hole current in the metal region is represented as being negligible. As pointed out earlier, an ohmic contact is the usual result of pressing a metal against a semiconductor specimen, or of pressing together specimens of

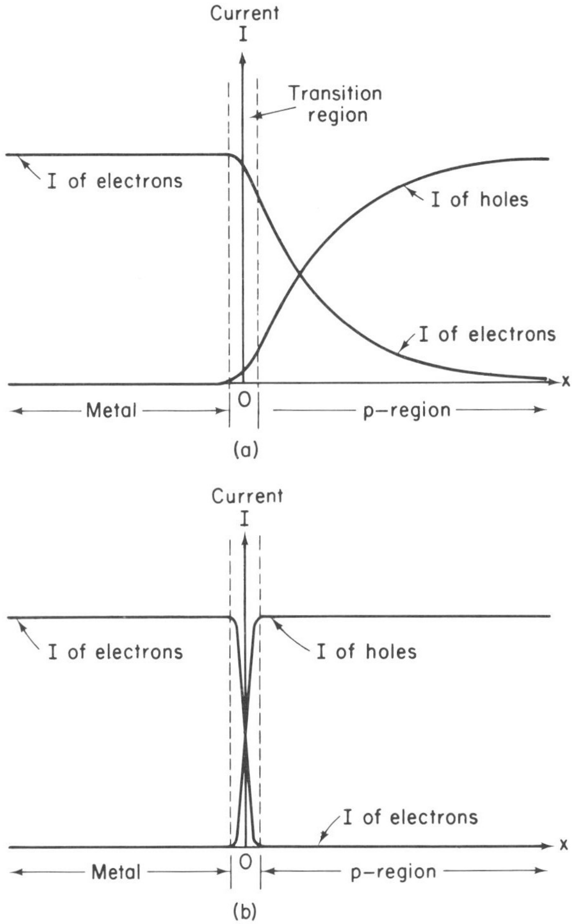

Fig. 3-8. Hole and electron currents in the metal- and *p*-regions near a forward-biased metal-*p*-type contact. (a) A rectifying contact for which diffusion length $\gg$ width of transition region. (b) Ohmic contact for which diffusion length $\ll$ width of transition region.

p-type and *n*-type material, owing to imperfections in the semiconductor surface.

An ohmic contact can also be thought of as a contact having a very large reverse-saturation current, owing to the short minority-carrier diffusion length in the semiconductor region. According to the diode equation, there is negligible difference in the magnitudes of the forward and reverse voltages for forward and reverse currents of equal magnitude, providing the magnitude is small compared to the reverse-saturation current. Therefore, rectifying contacts and *pn* junctions appear to be ohmic for very small currents and, conversely, contacts that appear to be ohmic for small currents may show nonohmic characteristics for large currents.

PROBLEMS

3-1. Draw an energy-level diagram for a contact between a metal and a *p*-type semiconductor for which $\phi_s < \phi_m$, and show that such a contact is resistanceless.

3-2. A rectifying metal-semiconductor contact has a reverse-saturation current of 10 mA. Determine the forward and reverse voltages that might be expected for forward and reverse currents of 1.0 mA.

4

Semiconductor Diodes

A semiconductor diode is defined as "a two-electrode semiconductor device having an asymmetrical voltage-current characteristic."* Thus, we see that a diode is formed from a *pn* junction by making ohmic contacts to the *p*- and *n*-regions, the *n*-region being the cathode and the *p*-region acting as the anode. Diodes of this type are *junction diodes*, and many of their characteristics were discussed in Chapter 2. Other types of semiconductor diodes utilize the rectification properties of metal-semiconductor contacts. These include *point-contact* diodes, *hot-carrier* diodes, and *metallic rectifiers*. Semiconductor diodes are often referred to as *crystal diodes*.

4-1 JUNCTION DIODES

Junction diodes may be produced by any of the methods described in Sec. 2-1, but most are either alloy-junction or diffused-junction devices. These techniques may also be used to produce multilayer diodes such as *pin*, *npn*, or *pnpn* diodes which will be discussed later (Chapter 12). We restrict our present consideration to simple *pn* diodes.

* Institute of Radio Engineers, "Standards on Electron Devices, Definitions of Semiconductor Terms." New York, 1954.

Since silicon and germanium are the most widely used semiconductor materials, it is of interest to compare the conduction characteristics of silicon and germanium diodes. The characteristics of the two device types are quite different, owing to the different energy-gap widths of silicon and germanium (that is, the difference in the energy required for the disruption of a valence bond). Because of the large energy gap of silicon, the reverse-saturation current of the silicon diode is so small at room temprature that it is masked by the reverse-leakage current, and the reverse current shows little evidence of saturation. This can be seen from the reverse characteristic of a typical type 1N461 silicon diode shown in Fig. 4-1. On the other hand, the reverse-saturation current of a typical germanium diode is likely to be the major component of its reverse current, owing to the small energy gap of germanium. Therefore, the reverse current of a silicon diode is several orders of magnitude lower than that of a germanium diode of comparable current and voltage ratings, as can be seen by comparison of the reverse characteristics of Fig. 4-1. The reverse-conduction characteristics of silicon junctions are also superior to those of germanium in that higher junction-breakdown voltages are obtainable in silicon. Single-junction silicon rectifiers having maximum reverse-voltage ratings as high as 1500 V are available, whereas the upper limit for germanium rectifiers is of the order of a few hundred volts. As a matter of fact, germanium power rectifiers are not in widespread use, owing to their inferior reverse characteristics.

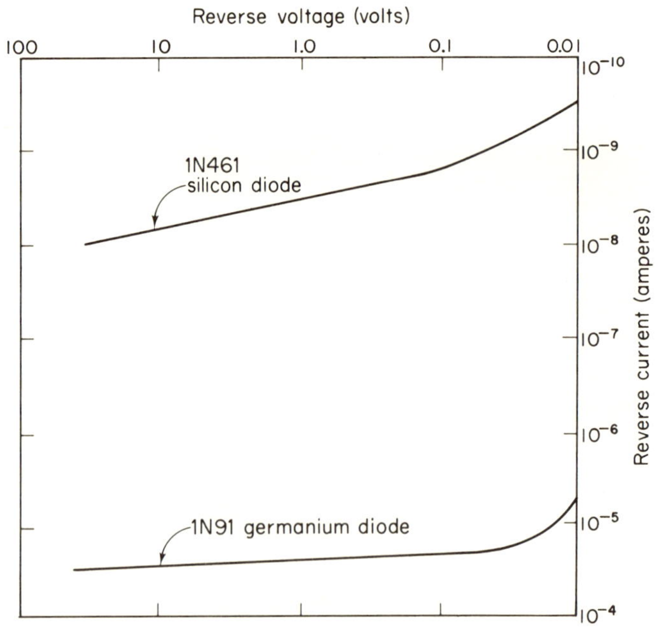

Fig. 4-1. Reverse conduction characteristics of comparable silicon and germanium diodes at room temperature. Note the differences in the reverse-current magnitudes and in the effect of voltage on reverse current.

Although the higher energy gap of silicon is responsible for the superiority of the reverse characteristics of silicon diodes, the effect on the forward-conduction characteristic is just the opposite. Since the barrier to the flow of majority carriers across the junction is greater in the higher energy-gap material, the forward voltage required to lower the barrier sufficiently for the flow of appreciable forward current is larger for a silicon junction than for a germanium junction. Therefore, the forward conduction characteristic of a silicon diode is inferior to that of a germanium diode, as may be seen from Fig. 4-2. In fact, at room temperature, the forward current of a silicon diode is negligible for most applications unless the forward voltage is greater than about 0.4 V. Therefore, a forward-biased silicon diode, or a series combination of diodes, may be used in voltage-limiting or regulating applications at voltages below those of Zener diodes. Diodes designed for this use are known as *stabistors.* The barrier height and the corresponding voltage required for appreciable forward current are also dependent upon the doping levels of *p*- and *n*-regions. This can be seen from the forward characteristics of Fig. 4-3, in which the relative doping levels are suggested by the rated junction-breakdown voltages, the more heavily doped devices having the lower breakdown voltages, as indicated by Eq. (2-1). Thus, the voltages of available single-junction stabistors range over some 0.2 V.

Although the forward conduction characteristic of a silicon diode is inferior to that of a germanium device, it is vastly superior to that of any existing electron tube. Silicon rectifiers may pass forward currents of the order of hundreds of amperes with conduction drops of the order of 1 V. This is an order of magnitude below the conduction drop of the best electron-

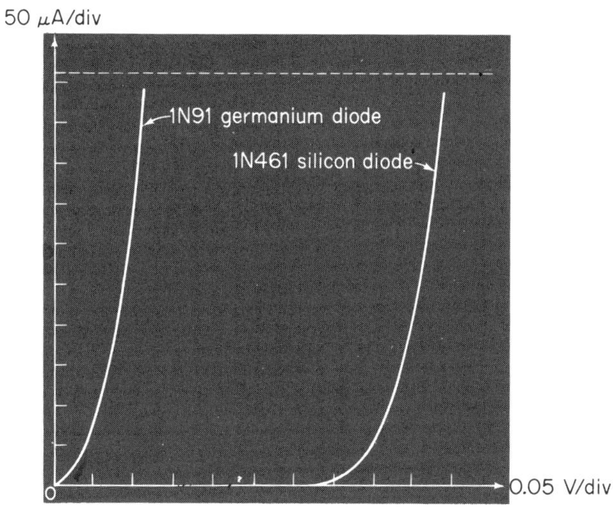

Fig. 4-2. Forward conduction characteristics of typical silicon and germanium junction diodes. Note the superiority of the type 1N91 germanium diode over the type 1N461 silicon diode.

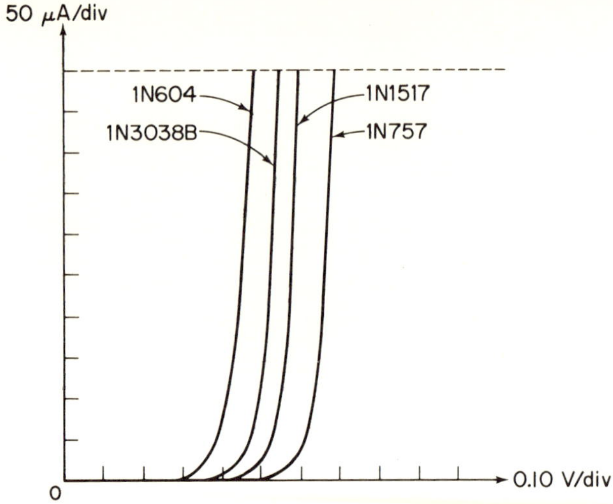

Fig. 4-3. Effect of doping level on the forward conduction characteristics of silicon diodes. Voltage-reference diodes having lower breakdown voltages V_a have more heavily doped *p*- and *n*-regions and poorer forward characteristics. The rated breakdown voltages for the types 1N757, 1N1517, and 1N3038B diodes are 9.1 V, 25 V, and 56 V, respectively and the type 1N604 is a rectifier having a maximum reverse-voltage rating of 400 V.

tube rectifiers. Where higher-current rectifier units are required, either the junction area is increased or units are paralleled, there being some advantage to paralleling units owing to problems associated with large-area devices. For a conduction drop of 1 V, silicon and germanium devices pass currents of the order of 200 and 500 $A/(cm)^2$, respectively. Thus, large current capacities can be obtained with small physical size.

From Fig. 4-1, the reverse current of the typical 1N461 silicon diode, a general-purpose diode having an average-current rating of 60 mA, is seen to be of the order of nanoamperes (10^{-9} amperes) at room temperature. These currents are of the order of the magnitude of the reverse currents of comparable *high-vacuum rectifiers* at the same voltages. In fact, silicon-diode and high-vacuum-rectifier reverse currents are of the same general nature, both being leakage currents.

When higher reverse voltages are required, silicon rectifiers may be connected in series. However, it may be necessary to shunt the units with resistors in order to equalize the reverse voltages, as pointed out in Sec. 2-1. Shunt capacitors may also be required to provide voltage equalization in the event of reverse-voltage surges. Because of their small size and high efficiency, junction rectifiers have replaced other rectifier types in most applications.

Diode equation for silicon junctions. Since leakage current is likely to be the major factor contributing to the reverse current of a silicon *pn* junction, the leakage path should be considered in arriving at an expression for the $I(V)$ characteristic of the junction. On the basis of a junction model that considers the leakage film to make an ohmic contact with the semiconductor material on one side of the junction and a recifying contact with the bulk region on the other side, Cutler and Bath derived an expression for the leakage current of an alloy junction.* By considering the total current to be the sum of this leakage current and the current predicted by the diode equation, they show the $I(V)$ characteristic to be given by

$$I = I_s(e^{eV/kT} - 1) \pm I_b\left(e^{eV/kT} - \frac{eV}{kT} - 1\right)^{1/2} \tag{4-1}$$

where I_s is the reverse-saturation current, I_b is the leakage-current factor, the plus sign is used for forward bias (positive V), and the minus sign is used for reverse bias (negative V). At room temperature, Eq. (4-1) becomes approximately

$$I = I_s(e^{40V} - 1) \pm I_b(e^{40V} - 40V - 1)^{1/2} \tag{4-2}$$

It turns out that the leakage-current factor I_b is usually several orders of magnitude larger than the reverse-saturation current I_s, typical room-temperature values being of the order of 10^{-10} and 10^{-14} amperes, respectively, for a relatively low-power diode. Therefore, the reverse characteristic is approximately

$$I = -I_b\left(e^{eV/kT} - \frac{eV}{kT} - 1\right)^{1/2} \tag{4-3}$$

which becomes

$$I \cong -I_b\left(-\frac{eV}{kT} - 1\right)^{1/2} \quad \text{for } V < -0.1 \text{ V} \tag{4-4}$$

At room temperature for reverse voltages more negative than about -0.1 V, Eq. (4-3) becomes

$$I \cong -I_b(-40V)^{1/2} \tag{4-5}$$

the dependence of reverse current on V being the same as that of the width of the carrier-depletion region.

For forward voltages greater than about 0.1 V, Eq. (4-1) becomes

$$I \cong I_s e^{eV/kT} + I_b e^{eV/2kT} \tag{4-6}$$

Because $I_b \gg I_s$, the forward current is dominated by the leakage component for several tenths of a volt and, for typical alloy-junction diodes at room temperature,

$$I \cong I_b e^{20V} \quad \text{for } 0.1 \text{ V} < V < 0.4 \text{ V to } 0.5 \text{ V} \tag{4-7}$$

* M. Cutler and H. M. Bath, "Surface Leakage Current in Silicon Fused-Junction Diodes," *Proc. IRE*, Vol. 45 (January 1957), pp. 39–43.

At higher forward voltages, the saturation component overtakes the leakage component and, at room temperature,

$$I \cong I_s e^{40V} \quad \text{for } V > 0.4 \text{ V to } 0.5 \text{ V} \tag{4-8}$$

These two distinct regions of the forward characteristic are evident in the plot for a type 1N758 alloy-junction diode shown in Fig. 4-4.

Unlike the germanium *pn* junction, the saturation current I_s of a silicon junction cannot be measured directly, owing to leakage current. However, the values of I_s and I_b for an abrupt junction can be obtained by utilizing Eq. (4-6). If both sides of Eq. (4-6) are multiplied by $e^{-eV/2kT}$, we obtain

$$e^{-eV/2kT} I = I_s e^{eV/2kT} + I_b \quad \text{for } V > 0.1 \text{ V} \tag{4-9}$$

Therefore, a plot of $I_f e^{-eV/2kT}$ versus $e^{eV/2kT}$ has an intercept of I_b and a slope of I_s.

The $I(V)$ characteristic of a diode is often expressed as

$$I = I_s(e^{eV/nkT} - 1) \tag{4-10}$$

where n is unity for an ideal junction or rectifying contact. Factor n depends on the device and its operating point. Thus, for a typical alloy-junction diode, n is about 2 for forward voltages from 0.1 V to about 0.5 V, and decreases to about 1 for higher forward voltages.

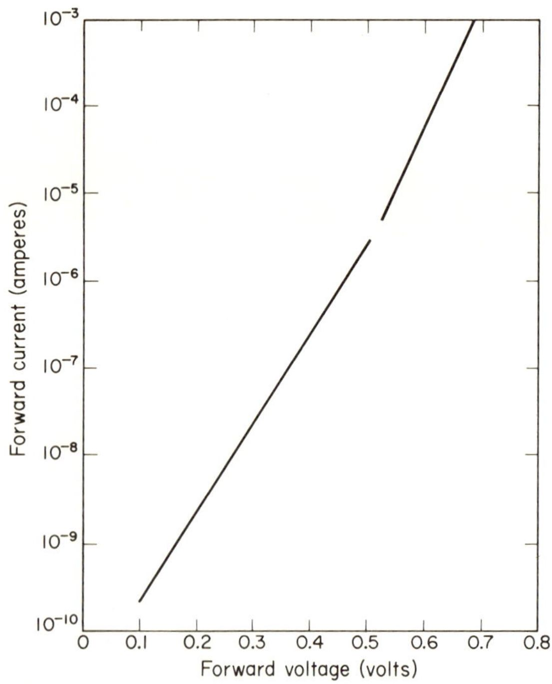

Fig. 4-4. Forward conduction characteristic of a typical type 1N758 alloy-junction silicon diode at room temperature. Note the difference in the characteristic that occurs at about 0.5 V.

Temperature effects. As pointed out in Secs. 2-1 and 2-4, both the reverse-breakdown voltage and the reverse current of a *pn* junction are sensitive to temperature. Since the forward conduction characteristic depends on the reverse-saturation current, according to the diode equation, it too is temperature dependent.

An increase in temperature may either increase or decrease the breakdown voltage of a diode, as may be seen from Table 4-1, which is compiled from manufacturer's specifications for Zener diodes. Breakdown voltage V_z is the voltage for a specified current in the breakdown region. This current depends on the power-dissipation rating of the diode and the breakdown voltage. The temperature coefficient of V_z is negative for low-voltage diodes, for which the breakdown mechanism is Zener in nature, and positive for higher-voltage diodes, which are avalanche-breakdown devices, as explained in Sec. 2-1. For diodes having breakdown voltages well above 5 V, the breakdown voltage increases roughly 0.1 percent per deg C. Because of the low temperature coefficient of diodes having breakdown voltages of the order of 5 V, higher reference voltages are sometimes obtained by using series-connected 5-V diodes instead of single diodes.

Table 4-1 Temperature coefficient of Zener diodes

Type	*Reference Voltage* V_z *(volts)*	*Temperature* (%/°C)	*Coefficient of* V_z *(Millivolts/°C)*
1N748	3.9	−0.05	−1.9
1N750	4.7	−0.02	−0.9
1N751	5.1	−0.01	−0.5
1N752	5.6	+0.005	+0.3
1N756	8.2	0.05	4.1
1N759	12	0.06	7.2
1N968	20	0.075	15.
1N980	62	0.09	56.
1N992	200	0.10	200.

The forward voltage of a silicon *pn* junction typically decreases about 2 mV per deg C.* This temperature effect has certain utility value, an example being the *temperature-compensated voltage-reference diode*. Such a "diode" is a series combination of one or more forward-connected diodes, depending on the number required for compensation, and a positive temperature-coefficient reverse-breakdown diode. The reference voltage is the voltage across the series combination and can have an extremely low temperature coefficient. For example, the type 1N945 voltage-reference diode has a V_z specification

* For high currents at high temperatures, the forward voltage of a silicon diode may increase with temperature, owing to the effect of temperature on bulk resistance. The decrease in forward voltage with increase in temperature is due to barrier-height reduction caused by *p*- and *n*-region Fermi-level shifts.

of from 11.12 V to 12.28 V and a maximum temperature coefficient of 0.0005 percent per deg C. Voltages obtained by using such "diodes" are much less temperature dependent than voltages supplied by batteries.

In Sec. 2-4, the effect of temperature on reverse-saturation current I_s of a diode at room temperature was given as about 10 percent per deg C for germanium and 16 percent per deg C for silicon. That is, I_s approximately doubles every 10°C in germanium and every 6°C in silicon. However, the reverse current of a typical silicon diode at room temperature is largely leakage current, which has a temperature sensitivity much less than that of the saturation current, as may be seen from the reverse current-temperature plots of Fig. 4.5. At higher temperatures, the reverse-saturation current of the silicon diode becomes significant, and reverse-current increase with temperature becomes stronger than that of the germanium diode. Even though its reverse-saturation current is more sensitive to temperature than that of the germanium diode, the silicon diode is usable at higher tempera-

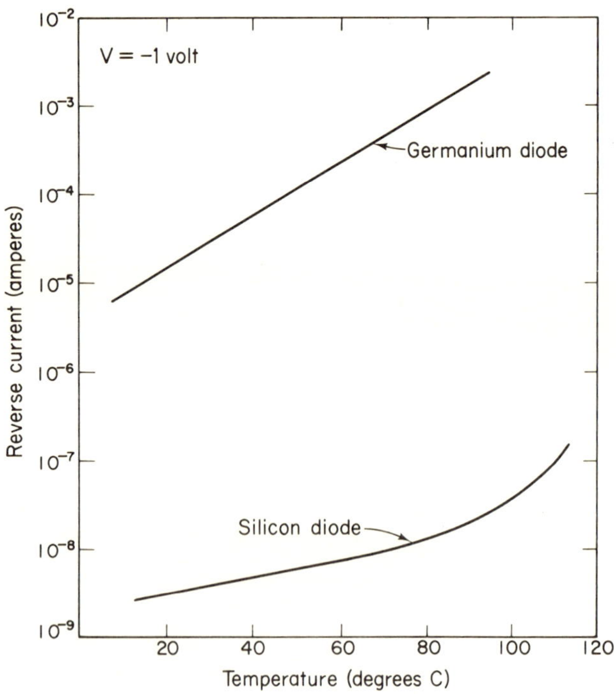

Fig. 4-5. Effect of temperature on the reverse current of germanium and silicon diodes having comparable average-current ratings. Note the differences in the reverse-current magnitudes and in the effects of temperature on reverse current.

tures, owing to its much lower reverse current at room temperature.* This is an important advantage of silicon devices.

4-2 THE POINT-CONTACT DIODE

The point-contact diode is the oldest of the semiconductor diodes. The first patent for such a diode in the form we know today—a fine wire called a *cat whisker* in contact with the crystal—was filed in 1906, three years after the discovery of the vacuum diode. Crystals of galena (lead sulphide) were commonly used with cat whiskers that could be moved about in order to locate a "sensitive spot" on the crystal.

The rectification action at the point contact is either that of a metal-semiconductor contact (Chapter 3) or that of a *pn* junction (Chapter 2) or a combination of the two. Junctions occur naturally in certain crystals, owing to the segregation of impurities upon solidification, and may exist at the sensitive spots on the surface of the galena crystal. At present, point-contact diodes usually use crystals of *n*-type germanium or silicon with whiskers of phosphor-bronze wire. The pointed wire is brought into *permanent contact* with the crystal to form a rectifying contact, as shown in the practical structure of Fig. 4-6.

The rectifying property of a point-contact diode can usually be enhanced by briefly heating the contact region in a process called *forming*. Forming may be accomplished by charging a capacitor and then discharging it through the contact in the low-resistance direction, using a small series resistance to

Fig. 4-6. Examples of practical point-contact diode structures.

* The upper limit of the junction temperature is about 100°C for germanium devices and 200°C for those of silicon.

limit the discharge current. Although many different combinations of voltage, capacitance, and resistance have been used, the classic arrangement uses a 1 μF capacitor charged to 45 V and discharged through the contact and a 100-ohm series resistor. In manufacturing practice, the forming-current pulse is usually supplied by a pulse generator, the typical forming pulse having a magnitude of the order of 100 mA and a duration of the order of 1 ms. What happens at the contact during the forming process depends on the amount of heat developed and the material of the metal whisker. The surface oxide may be merely broken through to result in a metal-semiconductor interface. On the other hand, the heating may be sufficient to cause the formation of *p*-type material under the whisker by alloying or by the diffusion of acceptor impurities into the semiconductor. That is, a *pn* junction

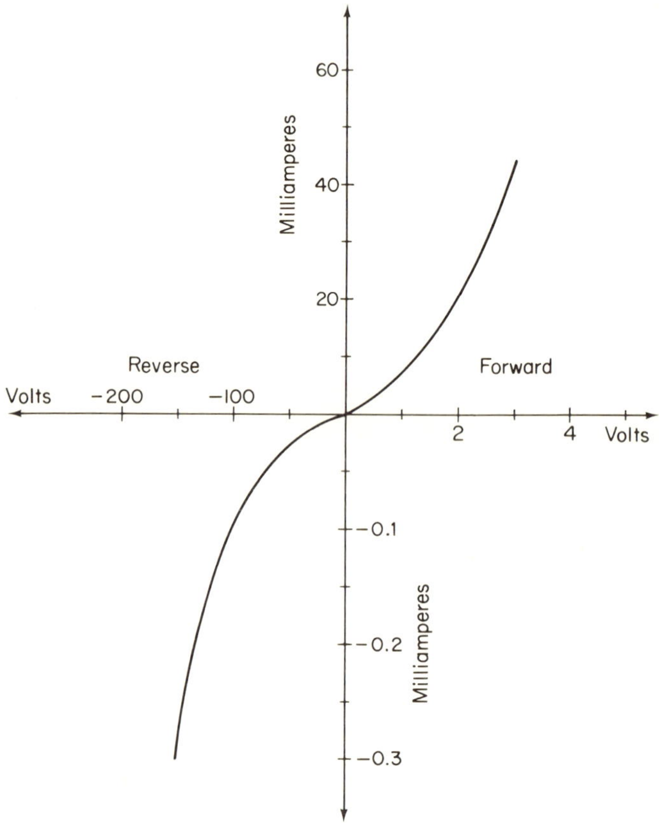

Fig. 4-7. Current–voltage characteristic of a typical type 1N55A point-contact germanium diode. Note the different current and voltage scales for the reverse characteristic.

may be established under the whisker. If the whisker contains both donor and acceptor impurities, an *npn* device may exist after forming.

The forward and reverse conduction characteristics of a typical point-contact germanium diode (type 1N55A) are shown in Fig. 4-7. Because of the small contact area, the forward characteristic is inferior to that of the typical junction diode, and the average-current rating usually does not exceed 50 mA. Because of leakage and thermal effects, the reverse current continues to increase with voltage, rather than saturating as in the germanium junction diode, and the maximum reverse-voltage rating is lower than that obtainable with junction diodes, usually not exceeding 200 V. When used as a rectifier, the 1N55A has forward-current ratings of 150 mA peak and 50 mA average and a maximum reverse-voltage rating of 150 V.

Even with its inferior conduction characteristics, the point-contact diode continued to be widely used after the advent of the junction diode, owing to its superior high-frequency characteristics. This superiority is due to its small junction capacitance and extremely rapid recovery of high reverse resistance following forward conduction (see Sec. 4-5), owing to the nature of the contact. For this reason, most germanium diodes are of the point-contact type. In recent years, silicon junction diodes having vastly improved high-frequency characteristics have been devleoped and have replaced point-contact diodes in many high-frequency applications. Nevertheless, junction diodes continue to be inadequate for the highest frequency applications of point-contact diodes. However, the point-contact diode is being replaced in these applications by the *hot-carrier* diode, a device having frequency characteristics comparable to those of the point-contact diode. In addition, the hot-carrier diode has superior electrical performance and greater mechanical ruggedness.

4-3 THE HOT-CARRIER DIODE

The hot-carrier or Schottky-barrier diode is an improved metal-semiconductor diode made possible by recent advances in semiconductor technology. The term "hot carrier" stems from the fact that in a forward-biased metal-*n*-type semiconductor contact, electrons are injected into the metal at much higher energy levels than those of the existing free electrons of the metal. This is shown by the energy-level diagram of Fig. 3-2. Thus, we see that a point-contact diode *may be* a hot-electron device. However, current use of the term "hot carrier" implies devices produced by a different type of technology.

The structure of a typical hot-carrier diode is shown by the cross-section

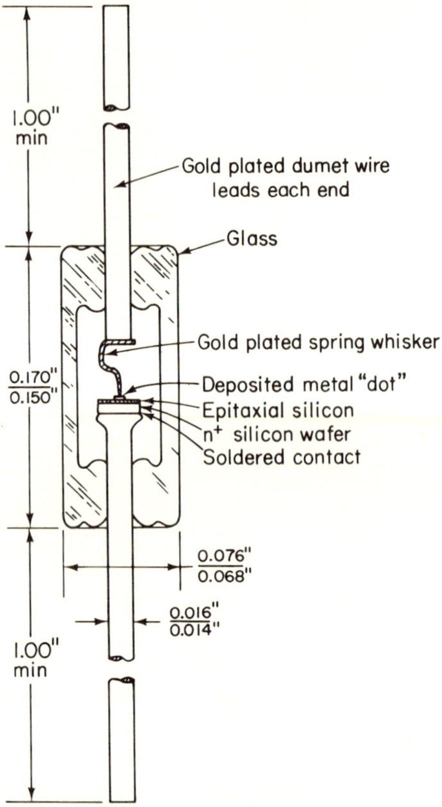

Fig. 4-8. Cross-section of a hot-carrier silicon diode. (From Hewlett-Packard Associates Application Note 907, *The Hot-Carrier Diode; Theory, Design and Application.*)

view of Fig. 4-8. The device is fabricated by using a thin, lightly doped, *epitaxial*, *n*-type silicon layer on a heavily doped (n^+) silicon substrate. The term "epitaxial" means that atoms of the layer are aligned with those of the crystalline structure upon which it is formed. The use of such a layer is a means of working with an extremely thin wafer of material, with the high-conductivity substrate providing a low-resistance connection to the wafer. The rectifying contact with the epitaxial layer is formed by depositing a metal "dot" thereon.

Hot-carrier diodes have a larger contact area with larger capacitance than point-contact diodes, but they also have the ability to handle greater power and have greater mechanical stability. Their electrical characteristics are more reproducible and are more nearly those of the ideal rectifying junction or contact, n of Eq. (4-10) being about 1.08 or less compared to 1.3 or more for the point-contact diode. Since they are *majority-carrier devices*, their recovery of reverse resistance following forward conduction (see Sec. 4-5) is superior to that of the highest-speed junction diodes and comparable to the performance of the fastest point-contact diodes. Current-voltage characteristics of comparable hot-carrier and point-contact diodes are shown by Fig. 4-9. The superiority of the hot-carrier device is clearly evident.

4-4 METALLIC RECTIFIERS

As explained in Sec. 3-2, a rectifying barrier may exist at the contact between a *p*-type semiconductor and a metal, the metal acting as though it were high-conductivity *n*-type material. By making an ohmic contact to the semiconductor, a diode is formed with the metal acting as the cathode and the semiconductor as the anode. Large-area diodes of this type are known as *metallic rectifiers;* commonly used types are copper-oxide rectifiers and selenium rectifiers. In these devices, rectifying contacts are formed by chemical or electrolytic processes in order to obtain high minority-carrier lifetime in the contact region.

A copper-oxide rectifier cell consists of a layer of cuprous oxide, a *p*-type semiconductor, on copper. The current-voltage characteristic of a typical

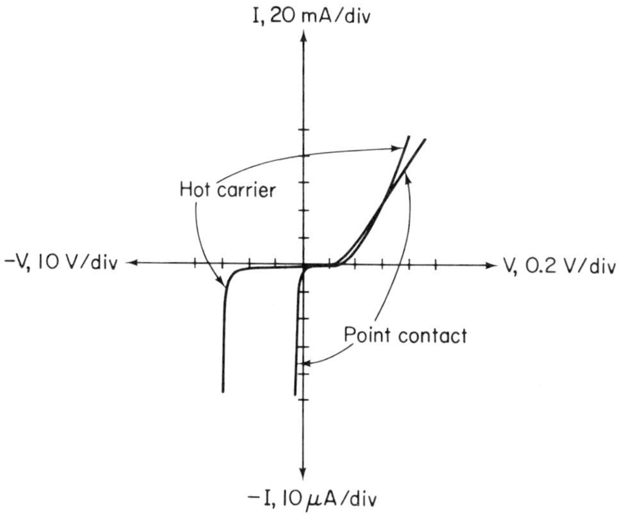

Fig. 4-9. I(V) characteristics of comparable hot-carrier and point-contact silicon diodes. (From Hewlett Packard Journal, Vol. 17, No. 4, December 1965.)

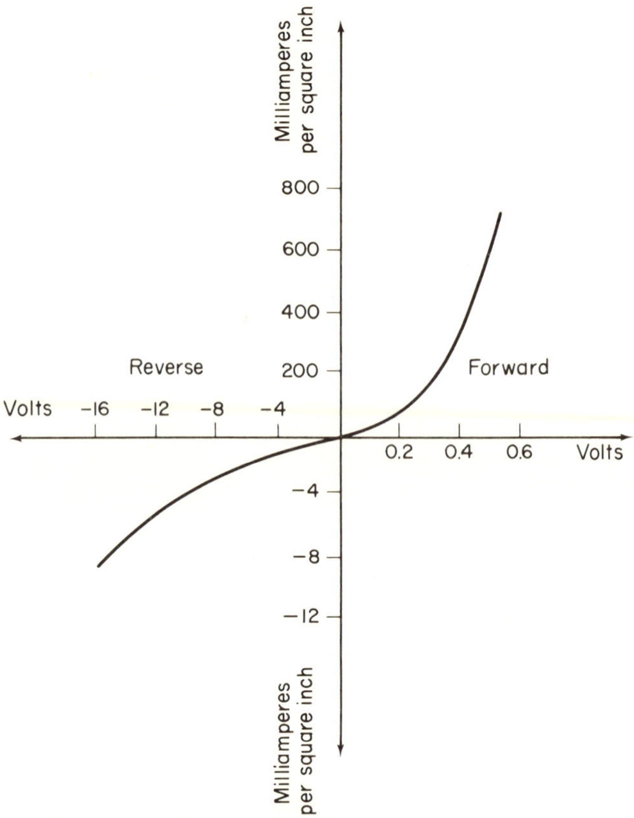

Fig. 4-10. Current-voltage characteristic of a copper-oxide rectifier cell operating at a temperature of 40°C. Note the different current and voltage scales for the reverse characteristic.

cell is shown in Fig. 4-10. The rather high reverse conduction restricts the reverse voltage to an upper limit of about 10 V. Although cells are connected in series to obtain higher back-voltage ratings, the low back voltage per cell causes the copper-oxide rectifier to be essentially a low-voltage device. By connecting cells in parallel or by using large contact areas, high forward-current capacities are obtained. Copper-oxide rectifiers have been used commercially since the 1920's; typical applications are in battery chargers and as instrument rectifiers.

In the late 1930's, metallic rectifiers using selenium as the *p*-type semiconductor were introduced commercially. The current-voltage characteristic of a typical selenium rectifier cell is shown in Fig. 4-11. The forward characteristic is seen to be similar to that of the copper-oxide cell, but the reverse

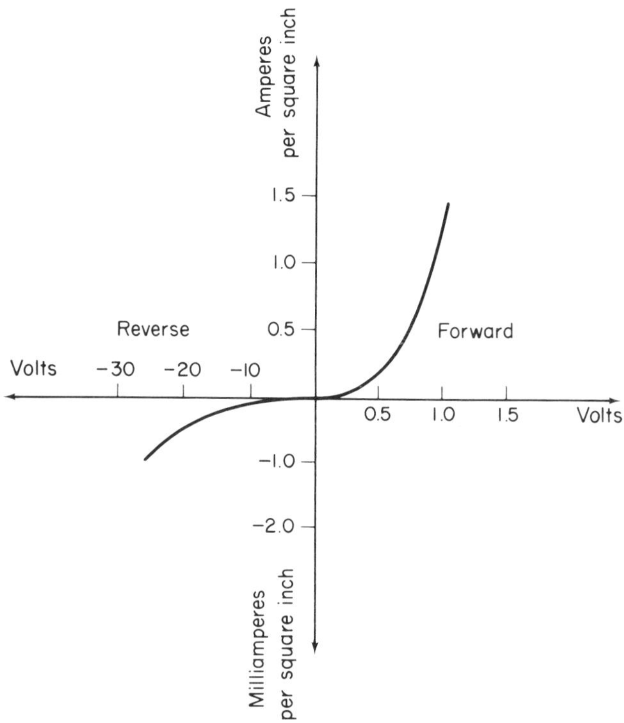

Fig. 4-11. Current–voltage characteristic of a selenium rectifier cell operating at a temperature of 50°C. Note the different current and voltage scales for the reverse characteristic.

characteristic is significantly superior. The cell can operate at back voltages up to 25 V (as compared to 10 V for the copper-oxide cell), passing appreciably less reverse current at higher operating temperatures.

Selenium rectifiers are quite versatile, having been used at "voltages as high as 500,000 volts supplying a current of a few microamperes and at currents of over 100,000 amperes at 9 volts."* They have replaced copper-oxide rectifiers and electron-tube rectifiers in many applications, a typical example being their wide use in power supplies of radio and television receivers. In recent years, they have been replaced to a considerable extent by silicon-junction rectifiers, which have higher back voltages per cell and superior forward-conduction characteristics, but their manufacture and use has continued because they are less costly than silicon rectifiers in low-voltage applications.

* Gerneral Electric Company Bulletin GET-2350, "Selenium Rectifier Stacks."

4-5 CONDUCTIVITY MODULATION

As higher reverse voltages are required of *pn* junction rectifiers, higher-resistivity semiconductor material must be used, as indicated by the avalanche-voltage expression [Eq. (2-1)]. In addition, the high-resistivity region of higher-voltage rectifiers must be wider in order to avoid reach-through (see Sec. 2-1). As a result, the forward-conduction characteristic is seriously degraded because of *IR* drop in the high-resistivity region. For example, the type 1N1130 silicon-junction rectifier, with a maximum reverse-voltage rating of 1500 V may have a forward voltage as high as 15 V at a current of 300 mA. For some time, it seemed inevitable that the forward conduction characteristic of a junction be compromised in order to obtain a high reverse voltage, but this difficulty has been resolved in the *conductivity-modulation* rectifier.

Consider a *pn* junction device consisting of a low-conductivity *n*-region sandwiched between a high-conductivity *p*-region and a high-conductivity *n*-region, the high-conductivity regions being referred to as the p^+ and n^+ regions. In such a single-crystal device, rectification takes place at the $n - p^+$ interface, but the $n - n^+$ *interface* has no electrical effect on the performance of the device. During forward conduction, almost all of the current results from the injection of holes into the low-conductivity *n*-region. For large forward currents, holes flood the *n*-region, raising the minority-carrier density far above the *equilibrium majority-carrier* density. The building up in the region of a large space charge of minority holes is prevented by a corresponding flow of *extra* majority electrons from the n^+ region. This flow occurs in order that electrical neutrality of the *n*-region be maintained, and the influx of extra carriers into the region so increases the conductivity that it may become almost metallic for large forward currents. The bulk-region resistance of the device (which is largely that of the *n*-region at low current) may thus drop to a value several orders of magnitude below the low-current value. Because the conductivity of the *n*-region is thus modulated by the passage of forward current, the device is called a *conductivity-modulation rectifier.* Even in relatively low-voltage rectifiers, forward conduction can be vastly improved by conductivity modulation. Since the low-conductivity region is so lightly doped, it is sometimes referred to as an intrinsic region and the device called a *pin* rectifier. If the low-conductivity region happens to be *p*-type, conductivity modulation of the region occurs during forward conduction as a result of holes from the p^+ region that neutralize electrons injected by the n^+ region, rectification occuring at the $p - n^+$ interface. If the rectifying junction is formed in a low-conductivity epitaxial layer on a high-conductivity substrate of the same material type such as that of Fig. 4-8, the device is known as an *epitaxial rectifier.*

It should not be concluded from the preceding discussion that an adjacent high-conductivity region of the same type is necessary for conductivity modulation of the low-conductivity region of a *pn* junction. If the low-conductivity region is as narrow as a diffusion length, or less (as it should be in epitaxial and *pin* devices), minority carriers injected into the region with large forward current and the influx of neutralizing majority carriers from the ohmic contact greatly reduce the resistance of the region. In any event, large forward currents result in conductivity modulation of the portion of the low-conducitivity region that is in close proximity to the transition region. Thus, we see that minority-carrier lifetime has an important effect on the forward-conduction characteristic of a *pn* junction device. Because of conductivity modulation, the bulk-region resistance of a diode may be an inverse function of forward current with the bulk-region *IR* drop essentially independent of current for large forward values.

4-6 DIODE SWITCHING TRANSIENTS

Most practical applications of a diode require that its conduction state change rather quickly. For example, a back-biased diode may suddenly become forward biased, or a forward-biased diode may experience a sudden increase in bias level, giving rise to a *turn-on transient* and a *forward-recovery time*. In a more important example, a forward-biased diode becomes reverse biased, resulting in a *turn-off transient* and a *reverse-recovery time*. The forward- and reverse-recovery times of a diode are defined as follows:* The forward-recovery time is "the time measured from the time ($t = 0$) the voltage or current has increased by a specified part of the change between the initial and final steady-state values to the time at which a specified forward voltage or current is reached." The reverse-recovery time is "the time measured from the time ($t = 0$) the forward voltage across or the forward current through the diode has changed by a specified part of the total change occurring in the rise-time portion of the monitored waveform to the time at which a specified reverse voltage across or current thorugh the diode is reached." These definitions rather meticulously point out that recovery time is not the total time required for equilibrium to become established after a change begins, owing to the uncertainty of the duration of exponential changes. For example, the time might be that required for a change of 80 percent of the total change, beginning the time measurement when 10 percent of the total change has occurred and terminating it at the 90 percent point.

* "IEEE Standard Test Procedure for Semiconductor Diodes." *IEEE Transactions on Electron Devices*, Vol. ED-11, No. 8 (August 1964).

Diode recovery times, particularly reverse-recovery time, are important diode specifications.

Turn-on transient. Let us consider the circuit shown in Fig. 4-12. The symbol representing the diode is standard notation, in which the arrow points in the direction of easy current flow, that is, from the p-region to the n-region of a junction diode. This symbol is often printed on the surface of a commercial diode in order to designate its terminals; in diodes too small for this designation, the cathode terminal is usually identified by the bar of the symbol or by some other significant marking such as the color-code indication of the diode type. Voltage V_f of Fig. 4-12 represents the generated voltage of the transient-driving source and resistance R is the Thevenin-equivalent output resistance of the source. In the practical test circuit, switch S is usually a mercury relay and may be a single-pole double-throw switch for changing from one active bias condition to another. Diode voltage v is monitored during the turn-on interval.

For $t < 0$, switch S of Fig. 4-12 is open and the diode is in equilibrium at zero bias. When S closes at $t = 0$, the diode current immediately increases to a value I_f determined by the resistance of the circuit, owing to the diffusion capacitance of the diode. Since source resistance R is usually quite large compared to the bulk-region resistance of the diode, we assume the diode forward voltage to be small compared to V_f and the current I_f for $t \geq 0$ to be approximately equal to V_f/R. If the forward current is assumed to result entirely from holes injected into the low-conductivity n-region, the hole-diffusion current at the n-region edge of the junction region is equal to I_f and the excess-hole-density gradient at this edge (that is, at $x = 0$) is given by*

$$\left.\frac{dp'_n}{dx}\right|_{x=0} = -\frac{I_f}{eAD_p} \tag{4-12}$$

as dictated by Eq. (1-33). The hole-density gradient becomes this value at $t = 0$ and stays constant thereafter, owing to the constant I_f. The distribution of the n-region hole density p_n at various intervals of time after $t = 0$ is shown in Fig. 4-13. Very early in the turn-on interval, the constant

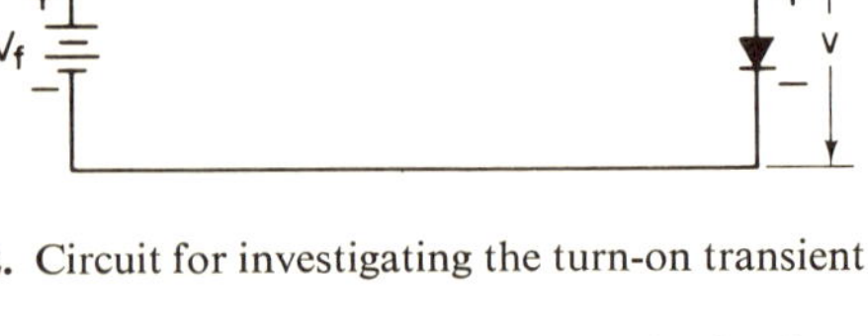

Fig. 4-12. Circuit for investigating the turn-on transient of a diode.

* If a significant fraction of I_f results from electron injection into the p-region, I_f in Eq. (4-11) becomes kI_f, k being less than unity.

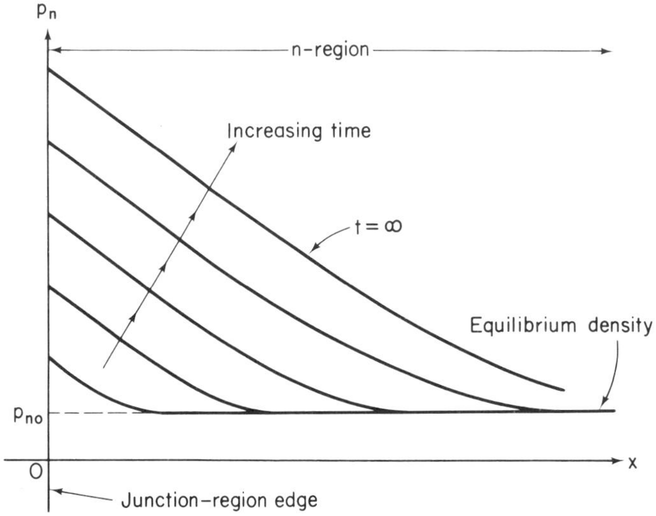

Fig. 4-13. The hole distribution in the n-region of a diode following the application of constant forward current. Note the constant $p_n(x)$ slope at $x = 0$.

current of holes flowing into the n-region increases $p_n(0)$ (the p_n value for $x = 0$) above the equilibrium density p_{no}, but p_n for $x > 0$ is virtually unaffected, owing to the time required for the diffusion process. Because the p_n values are small during the early stages of the turn-on process, the recombination term $(p_n - p_{no})/\tau_p$ is small and most of the holes flowing into the n-region are stored therein. As turn-on continues, $p_n(0)$ continues to increase, increasing p_n for $x > 0$. This increase in the store of n-region holes results in more of the input hole current supplying recombination, and the fraction going into storage decreases, becoming zero when the hole store reaches its static value and the entire input current supplies recombination. The voltage v across the diode depends on $p_n(0)$ through the relationship

$$p_n(0) = p_{no}e^{eV/kT} \tag{4-13}$$

as shown in Sec. 2-4. Thus, v is zero at $t = 0$, and increases rapidly for initial values of $t > 0$, owing to the rapid initial increase in $p_n(0)$ for an input-current step, gradually approaching an equilibrium value as $p_n(0)$ approaches its static value. A plot of $v(t)$ is shown in Fig. 4-14(b) and may be obtained analytically by applying Eq. (4-13) to the solution of the diffusion equation. From the diode equation, the equilibrium value of the diode voltage is seen to be

$$v = \frac{kT}{e} \ln\left(\frac{I_f + I_s}{I_s}\right) \tag{4-14}$$

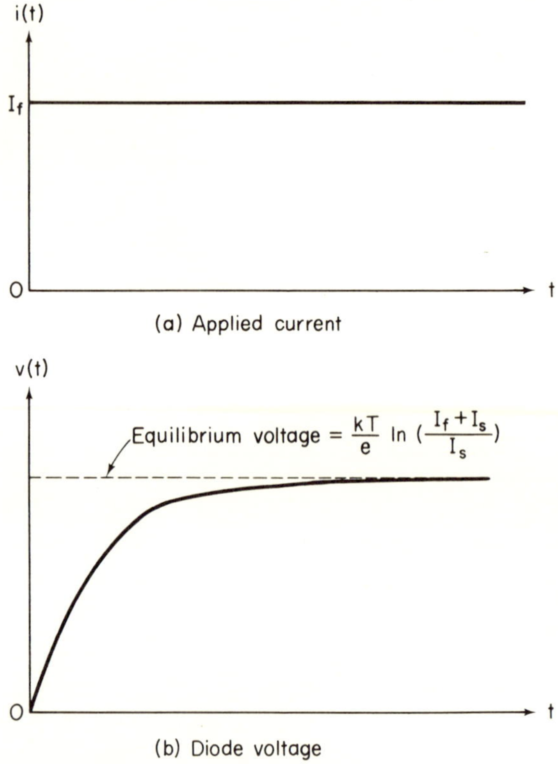

Fig. 4-14. The turn-on transient for an idealized junction diode operated in the circuit of Fig. 4-12.

Although the turn-on transient has been described in terms of minority-hole flow, we know from the discussion of Sec. 2.5 that an electron current flows in the *n*-region from the contact toward the junction in order to supply recombination and to neutralize the charge of the increasing store of holes in the region near the junction.

The $v(t)$ plot of Fig. 4-14 is for an idealized diode having negligible bulk-region resistance. The presence of bulk resistance R_b adds ar instantaneous step of voltage of magnitude $I_f R_b$ to the voltages of the plot. Perhaps even more important is the effect on diode voltage of a forward current that results in a minority-carrier-injection level high enough for the occurrence of conductivity modulation. Conductivity-modulation diodes show turn-on voltage transients such as those of Fig. 4-15. Initially, the $v(t)$ plot for a large forward current is due almost entirely to bulk-resistance *IR* drop, which decreases as injected-minority-carrier penetration of the *n*-region decreases the bulk-resistance. For small forward currents, the bulk-resistance *IR* drop has little effect on the $v(t)$ plot.

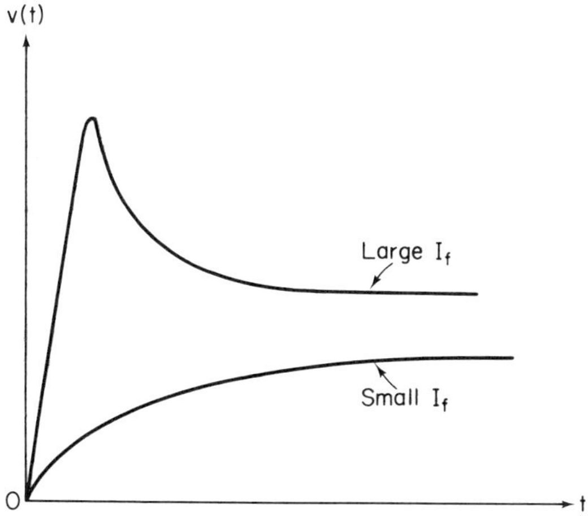

Fig. 4-15. Effect of conductivity modulation on the turn-on transient of a junction diode operated in the circuit of Fig. 4-12. For small I_f values, diode bulk resistance has little effect on the $v(t)$ plot, but large I_f values result in an IR drop that decreases with time, owing to conductivity modulation. The initial rate of rise of the $v(t)$ display for large I_f values is a characteristic of the test equipment rather than of the diode.

The experimental determination of turn-on $v(t)$ plots such as those of Fig. 4-15 requires the use of fast-rise-time pulse generators and oscilloscopes since, typically, equilibrium values of diode voltages are reached in times of the order of a few nanoseconds (10^{-9}s).

Turn-off transient. Suppose that our idealized junction diode (a resistanceless diode consisting of a high-conductivity p-region and a low-conductivity n-region) is operated in the circuit of Fig. 4-16. In this circuit, the diode may be either forward biased or reverse biased, depending on the position of switch S. With S in position 1, the diode is forward biased and V_f and R are of such magnitudes that $i = I_f \cong V_f/R$. If S is quickly changed from position 1 to position 2, holes stored in the n-region during forward-bias

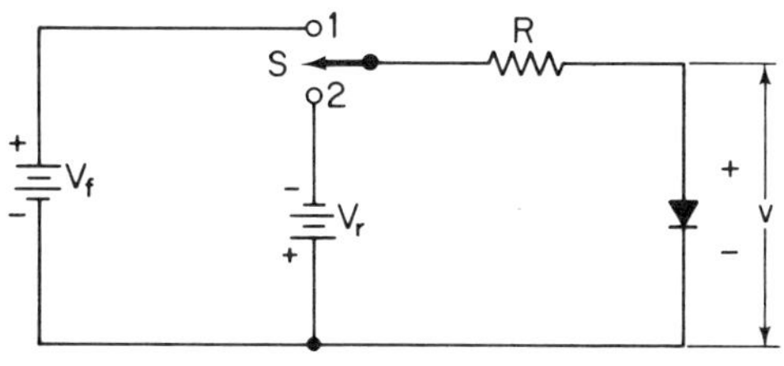

Fig. 4-16. Circuit for investigating the turn-off transient of a diode.

operation prevent the junction from immediately becoming reverse biased. The voltage v across the diode continues to depend on $p_n(0)$ through the relationship

$$p_n(0) = p_{no}e^{eV/kT} \tag{4-13}$$

which indicates that v is positive and the diode is forward biased as long as $p_n(0) > p_{no}$. Initially, $p_n(0)$ changes little, and the diode voltage continues to be the relatively small forward-bias value for some time after the switch is changed to position 2. During this time, the reverse current is constrained by the circuit, current i being given by

$$i = -\frac{V_r + v}{R} \cong -\frac{V_r}{R} \tag{4-15}$$

for $V_r \gg v$. This reverse current is the result of holes being withdrawn from the n-region into the p-region, and this, along with the carrier recombination, reduces the n-region hole store and decreases $p_n(0)$. Eventually $p_n(0)$ falls to the equilibrium density p_{no}, and the junction voltage becomes zero. As $p_n(0)$ is further reduced, the junction becomes reverse biased, diode voltage v reverses, and the reverse current begins to decrease, as indicated by Eq. (4-15). At this point, the n-region hole store has been largely removed, and the reverse current rapidly falls to its normal saturation value I_s as diode voltage $-v$ rises to the applied voltage V_r.

If we consider switch S of Fig. 4-16 to be in position 1 (and the diode to be in its d-c steady state) for $t < 0$ and to be changed to position 2 at $t = 0$, the circuit current i is given by

$$i = I_f \cong \frac{V_f}{R} \quad \text{for } t < 0 \tag{4-16}$$

and

$$i = I_f \cong -\frac{V_f}{R} \quad \text{for } t_s > t \geq 0 \tag{4-17}$$

where t_s is the period of time during which the diode voltage remains positive for reverse conduction. Time t_s is known as the *storage-delay time*. The variation of diode current i and voltage v with time for this manner of diode operation is shown in Fig. 4-17. For the practical diode, a step in the $v(t)$ plot occurs at $t = 0$, owing to the reversal of the IR drop in the bulk resistance of the diode.

The hole-density distribution in the n-region of the diode during the turn-off transient is shown in Fig. 4-18. The only instantaneous change in p_n at $t = 0$ is a change of dp'_n/dx at $x = 0$ from $-I_f/eAD_p$ [as given by Eq. (4-12)] to I_r/eAD_p, the hole-density slope at $x = 0$ reversing to provide the reverse diffusion current I_r. This slope remains fixed as turn-off proceeds, owing to the constant reverse current, but the hole density decreases because of hole removal across the junction and because of recombination at

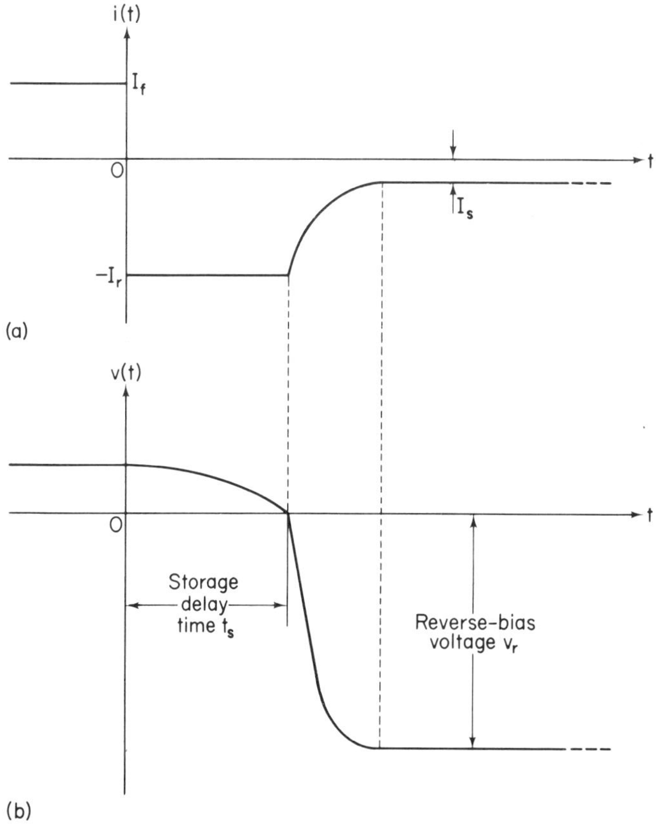

Fig. 4-17. The turn-off transient for an idealized junction diode operated in the circuit of Fig. 4-16. (a) Diode current $i(t)$; (b) diode terminal voltage $v(t)$.

the rate $(p_n - p_{no})/\tau_p$. When the n-region hole store becomes too small to support the hole-density slope I_r/eAD_p at $x = 0$, I_r can no longer be maintained, and the reverse current begins to decrease.

At the end of the turn-off interval, the n-region hole density assumes the normal distribution for reverse bias, as shown by the $p_n(x)$ plot for $t = \infty$, and the reverse current becomes the saturation value I_s.

The reverse-recovery time of a junction diode is usually much greater than the forward-recovery time, owing to the storage-delay time, and is an exceedingly important characteristic of diodes used in fast-switching applications. Hot-carrier diodes and *junctionless* point-contact diodes have storage-delay times much shorter than those of junction diodes, since they are majority-carrier devices. Hot electrons injected into the metal during forward bias give up their excess energy in a fraction of a picosecond (10^{-12} s), a time

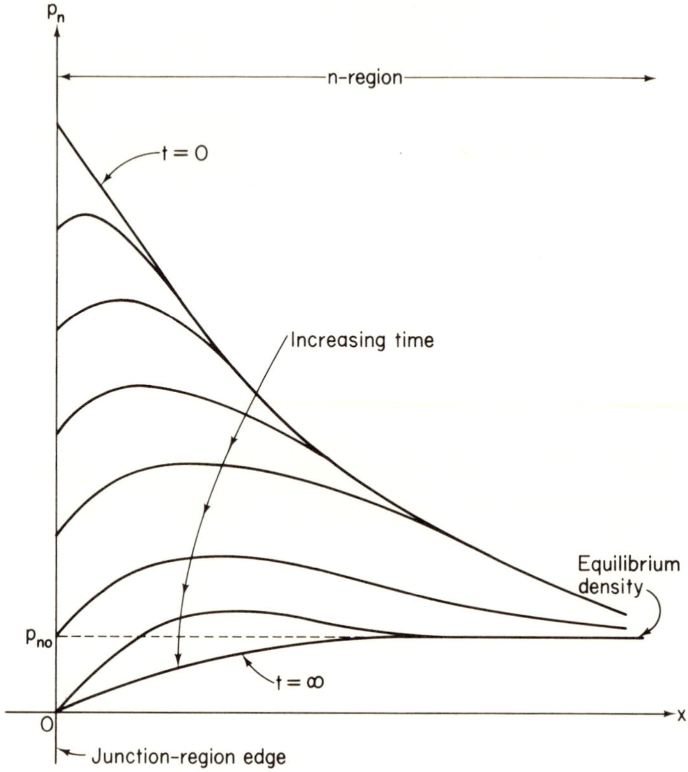

Fig. 4-18. The hole distribution in the *n*-region of a diode during the turn-off transient following the application of constant reverse current. Note the constant $p_n(x)$ slope at $x = 0$ for $t < 0$.

much shorter than the recombination lifetime of injected minority carriers. After losing their excess energy, injected electrons become part of the sea of free electrons in the metal. At present, hot-carrier diodes are said to be too fast for reverse-recovery time measurement by conventional methods. Storage-delay times for junction devices range from nanoseconds, or less, in high-speed diodes to tens of microseconds, or more, in low-speed rectifiers, depending on the forward- and reverse-current magnitudes and the minority-carrier lifetimes. The storage-delay time is reduced by removing the stored carriers faster, and this can be effected by increasing the reverse current I_r and by decreasing the minority-carrier lifetimes. Minority-carrier lifetimes are beyond the control of the circuit designer in most instances, but they may be deliberately reduced in the fabrication of high-speed junction devices. When this occurs, decreased turn-off time is obtained by degrading the forward and reverse conduction characteristics, as pointed out in Secs. 2-4 and 4-5.

4-7 LIGHT-SENSITIVE AND LIGHT-EMITTING DIODES

It was pointed out in Sec. 1-3 that electron-hole pairs can be generated in a semiconductor crystal by shining light of an appropriate wavelength on it; it was also pointed out that the process is reversible and light emission may occur when the carriers recombine. These phenomenona are utilized in *photodiodes* and *solar cells* and in *light-emitting diodes* and *junction lasers*.

4-7.1 Photodiodes and Solar Cells

Figure 4-19 shows the construction of an early-type photodiode. Light falling on the device increases, above their equilibrium values, the densities of holes and electrons in both the *p*- and *n*-regions of the diode near the upper surface. As a result, a diffusion flow of both holes and electrons occurs in both regions, the flow being in all directions away from the region of carrier generation. Carriers generated within a diffusion length of the junction, and diffusing toward it, are likely to reach the junction before recombining. Majority carriers doing so are turned away from the junction by the repelling action of the electric field of the transition region, but minority carriers are swept across the junction by the field.

The current of minority carriers crossing the junction *as a result of illumination* is the photo current I_{ph}. This current is independent of junction bias, since an electric field exists in the transition region regardless of bias. If the photodiode is short-circuited, the circuit current is I_{ph}, providing that *diode bulk-resistance IR drop* is negligible; if significant, the *IR* drop forward-biases the junction, reducing the *circuit current* below I_{ph}. A more

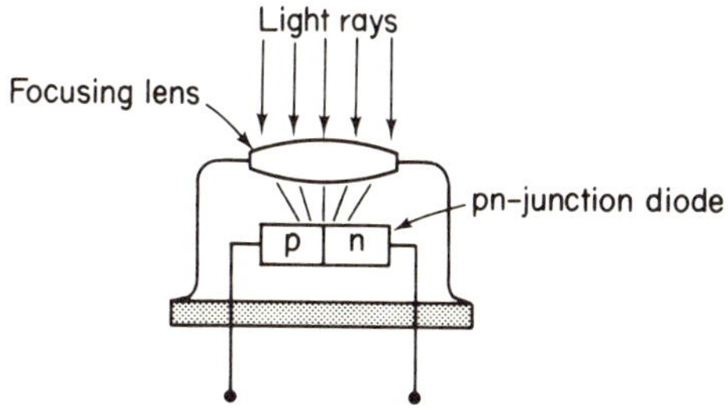

Fig. 4-19. Structure of a grown-junction photodiode.

accurate indication of I_{ph} is the increase in reverse-bias current with illumination. If the photodiode is open-circuited, minority carriers crossing the junction build up a negative charge in the *n*-region and a positive charge in the *p*-region. As a result, the junction becomes forward biased, and a forward current I_f of majority carriers occurs. Under equilibrium conditions, the forward current just equals the photo current; that is,

$$I_{ph} = I_f \tag{4-18}$$

and the voltage V_{ph} across the *open-circuited* diode terminals is such that this condition is satisfied. Thus,

$$V_{ph} = V_f \tag{4-19}$$

From the diode equation

$$I_f = I_s(e^{eV_f/kT} - 1)$$

and Eqs. (4-18) and (4-19), we obtain

$$V_{ph} = \frac{kT}{e} \ln \frac{(I_{ph} + I_s)}{I_s} \tag{4-20}$$

where V_{ph} is the *generated voltage* of the photodiode. V_{ph} is sometimes called the *photo-voltaic voltage* of the diode.

The current-voltage characteristics of a photodiode with and without illumination are shown by Fig. 4-20(a). The *i-v* characteristic with illumination is seen to be that of a normal diode, except that it is displaced on the current scale by an amount I_{ph}; this suggests the circuit model of Fig. 4-20(b).

Solar cells are large-area photodiodes fabricated for the purpose of converting light from the sun into electrical energy. Solar cells may be of either the *pn*-junction or the metal-semiconductor type. The construction of a typical *pn*-junction cell is shown in Fig. 4-21. It consists of a thin diffused *p*-layer on a wafer of *n*-type material. The thin *p*-region permits light penetration to the *pn* junction and beyond. A ring contact to the *p*-region maximizes the area available for illumination.

Metal-semiconductor solar cells have been used for many years, an

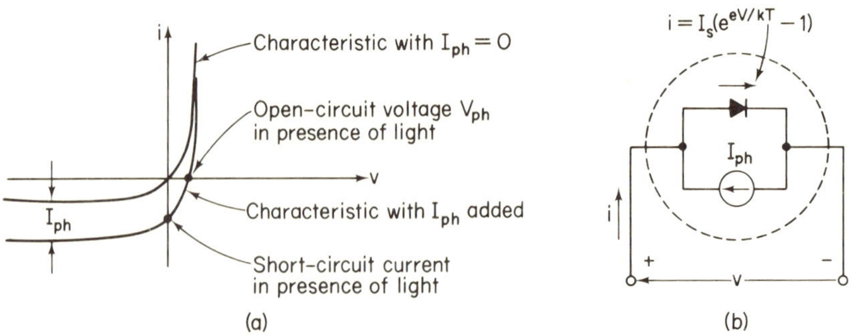

Fig. 4-20. (a) The *i-v* characteristics of a photodiode; (b) circuit model for a photodiode.

example being the familiar light-intensity meter. Such a cell consists of a rectifying contact between a thin semiconductor region and an appropriate metal, as shown in Fig. 4-22. Minority carriers generated in the semiconductor region by illumination and reaching the contact by diffusion are swept across the contact region by the electric field therein. As a result, the contact becomes forward biased and a voltage V_{ph} is generated precisely as in the *pn*-junction cell.

We are now in position to make a few observations regarding the characteristics of photodiodes and solar cells. In most of these devices, photo current I_{ph} is almost exactly proportional to illumination over a wide range of light fluxes. Since reverse-saturation current I_s can be very small, the ratio of photo current to dark current can be quite large. From the photo-voltage expression of Eq. (4-20), we see that this ratio may be more important than the magnitude of I_{ph}. Since kT/e is about 0.025 V at room temperature, the logarithmic expression for V_{ph} results in V_{ph} values of less than a volt for enormous I_{ph}/I_s ratios. Therefore, solar cells are inherently low-voltage sources.

4-7.2 Light-Emitting Diodes and Junction Lasers

When an electron and a hole recombine in a semiconductor, an amount of energy of approximately E_g is released, where E_g is the energy gap of the material (that is, E_g is the energy required for disruption of a valence bond). If the recombination event involves a recombination center, the released

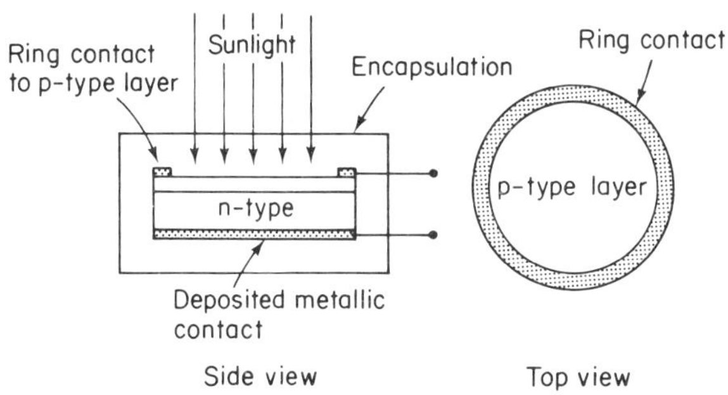

Fig. 4-21. Construction of a *pn*-junction solar cell.

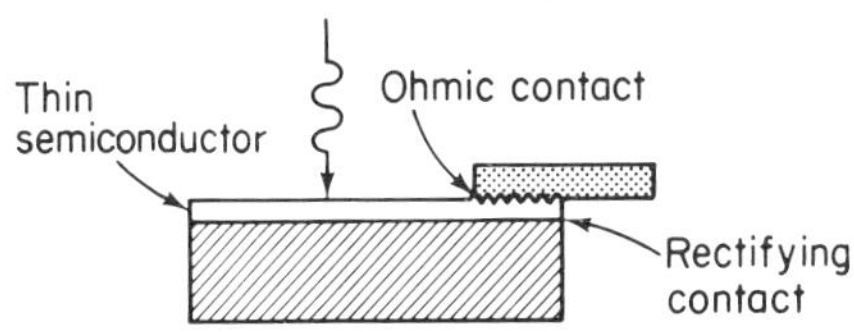

Fig. 4-22. Construction of a metal-semiconductor solar cell.

energy is delivered to the crystal as heat; if the recombination is direct, a photon of energy is emitted. In germanium and silicon, recombination via recombination centers is the dominant mechanism, and very little recombination light is emitted by germanium and silicon diodes. On the other hand, significant direct recombination does occur in certain semiconductors that are compunds of 3-valence and 5-valence materials. Gallium arsenide and gallium phosphide are such semiconductors, and junction diodes can be fabricated from these materials in such a way that they have the capability of emitting an appreciable amount of light. Such devices are called *light-emitting diodes.*

Our consideration of Sec. 2.5 has shown that all current that flows in a a *pn*-junction diode can be accounted for by recombination in the device. Thus, the larger the current, the greater the recombination rate, suggesting that light-emitting diodes be forward biased. Light is also emitted from diodes operated in the reverse-breakdown region, but this method of obtaining minority-carrier injection is clearly less efficient than by forward bias. The output of light-emitting diodes is approximately proportional to forward current over a very large current range.

Light-emitting diode emission occurs over a narrow band of frequencies centered approximately about the value E_g/h, where h is Planck's constant. This range is the result of the involvement of donor and acceptor energy levels in the recombination processes, resulting in the emission of photons of energy less than E_g, and because recombining carriers can have thermal energies of the order of kT resulting in photons having energies as much as a few kT above E_g. The output of gallium-arsenide diodes is in the infrared region of the specturm, but diodes emitting visible light also exist. Gallium-phosphide diodes emit green light, and red light can be obtained from diodes fabricated from alloys of gallium-arsenide and gallium-phosphide. An advantage of light-emitting diodes over other light sources is the fast rate at which they can be turned on and off, typical diodes having on-off response times of the order of 10 ns.

Junction lasers. Electromagnetic radiation may be expressed mathematically by using an electric-field vector of the form

$$E = A \sin (2\pi f t + \phi) \tag{4-21}$$

If A and ϕ are constants, the intensity of the radiation at some future time can be predicted from its value at a particular time of measurement. Light having this characteristic is known as *coherent light.* If A and ϕ are random functions of time, the light is *incoherent.* Althoughly principally monochromatic, the emission of light-emitting diodes is incoherent, since recombination events occur at random times. If the emission from a *pn* junction can be made coherent, the device is called a junction laser. The word *laser* is an abbreviation of "Light Amplification by Stimulated Emission."

For reasons beyond the scope of our consideration, it turns out that the probability of a photon-producing recombination event depends upon the presence nearby of another photon of the same energy. These photon-assisted recombinations result in emission known as *stimulated emission*. If this type of recombination predominates, the emission is largely coherent, since new photons are produced regularly by photons travelling past recombination sites. This is achieved when large numbers of carriers are available for recombinations and the diode surfaces are treated to keep most of the generated photons within the device.

The construction of a junction laser is somewhat as shown by Fig. 4-23. Both flat surfaces are reflecting in order to retain generated light, with one surface made less so in order for emission to occur. If the distances between the reflecting surfaces is an integral number of wavelengths, the wave is reinforced with each pass. As a result, a standing wave is developed as well as a travelling wave, which leaves the diode through the surface having less reflection. The emitted light comes out in a flat sheet along the plane of the junction. Thus, we see that a forward-biased *pn*-junction diode may produce either incoherent light or both coherent and incoherent light. For illumination purposes, incoherent light is quite adequate, but other applications (communication systems, for example) require coherent light.

4-8 TUNNEL DIODES

As explained in Sec. 2-1, the breakdown voltage of a *pn* junction decreases with increased doping of the *p*- and *n*-regions. In 1957, Esaki* discovered that the doping level could be made high enough to reduce the breakdown voltage to zero. That is, the junction is a short circuit for zero bias. This short circuit exists for all reverse-bias voltages and up to a small forward-bias voltage, where the characteristic reverts to a normal forward characteristic. Current flow in the short-circuit region of the characteristic is by the phenomenon used in Sec. 2-1 to explain Zener breakdown. This is

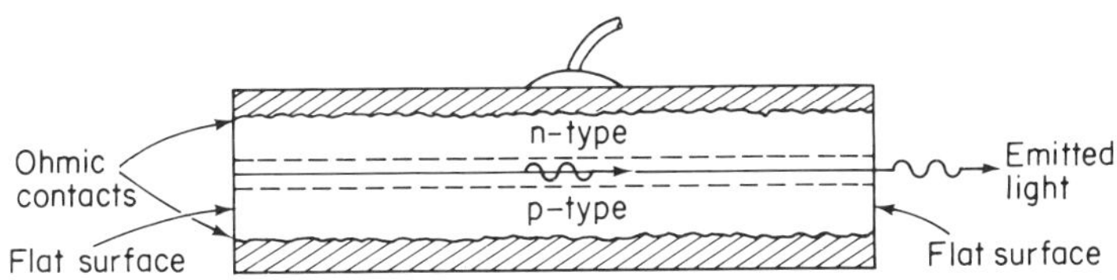

Fig. 4-23. Construction of a semiconductor *pn*-junction laser.

* Leo Esaki, "New Phenomenon in Narrow Germanium P-N Junctions," *Phys. Rev.*, Vol. 109 (April 1, 1958) pp. 603–604.

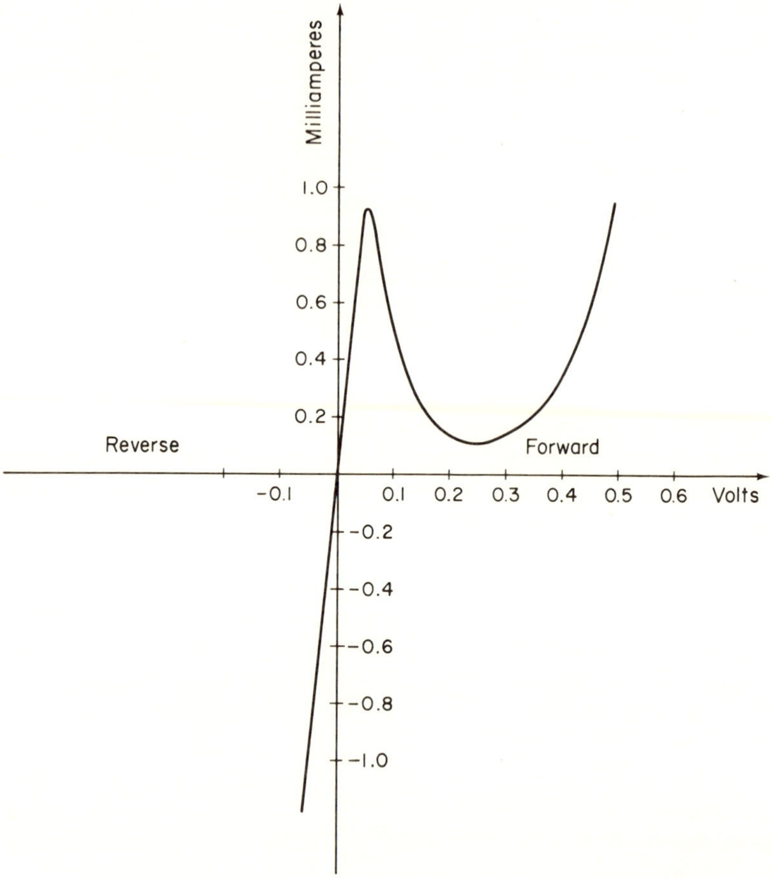

Fig. 4-24. Current-voltage characteristic of a typical germanium tunnel diode.

known as quantum-mechanical tunneling; hence the term *tunnel diode*. The device is also known as the Esaki diode, in deference to its discoverer.

The current-voltage characteristic of a typical germanium tunnel diode is shown in Fig. 4-24. Its principal feature is the negative-resistance region between the region of the short-circuit characteristic and that of the normal forward characteristic. This negative-resistance region is utilized in high-frequency amplification and in high-speed switching applications. The voltage magnitudes of the normal forward characteristic are seen to be significantly larger than those of the conventional germanium junction diode, owing to the heavily doped p- and n-regions.

In order for tunneling to occur, the bottom of the n-region conduction band must be lower in energy than the top of the p-region valence band. This condition is known as "overlapping bands," and when it exists, occupied energy levels (electrons) in one band are directly across the barrier from unoccupied energy levels in the other band. But for the barrier, electrons could move freely, without change in energy, from one type of material to the other; if the barrier is such that they do so, they are said to "tunnel through" the barrier. The *n*-region conduction band and the *p*-region valence band overlap in any *pn* junction that is reverse biased by a few tenths of a volt, or more, but electron tunneling does not occur unless the barrier region is quite thin. If the barrier is thin enough, usually no thicker than 100 Å, the random motion of an electron may enable it to move through the barrier to occupy the same energy level in the adjacent energy band. Thus, the two requirements for electron tunneling are that the barrier between the *p*- and *n*-regions be thin and that there be unoccupied energy levels in one region at the same energies as those of electrons in the other region.

For the tunnel diode, the *n*-region conduction band and the *p*-region valence band overlap with zero bias, owing to the heavily doped *p*- and *n*-regions, and the transition from one type of impurity to the other is very sharp in order that the barrier region be thin. In germanium, impurity concentrations of the order of 10^{19} atoms per cubic centimeter are used, and the material is said to be doped to *degeneracy*, meaning that its electrical behavior over a wide temperature range is essentially metallic in nature. Overlapping bands with zero bias means that the Fermi level lies in the conduction band of the *n*-region and in the valence band of the *p*-region, as indicated in the energy-level diagram of Fig. 4-25(a). Thus, the lower part of the *n*-region conduction band is essentially filled and the upper part of the *p*-region valence band is essentially empty. This is indicated in Fig. 4-25(a), and in subsequent energy-level diagrams, by cross-hatching energy-level regions having a probability of occupancy greater than 50 percent. With zero bias, electrons are able to tunnel equally well in either direction, and the net junction current is zero.

An energy-level diagram for the tunnel diode for the condition of reverse bias is shown in Fig. 4-25(b). With the *p*-region negative, the Fermi level of the region is shifted upward with respect to the Fermi level of the *n*-region. As a result, *p*-region valence-band electrons tunnel through the barrier to fill unoccupied energy levels in the *n*-region conduction band, and the current is limited by the resistance of the *p*- and *n*-regions and of the contacts, as shown by the reverse characteristic of Fig. 4-24.

The effect on the tunnel diode of a small forward-bias voltage is shown by the energy-level diagram of Fig. 4-26(a). With the *n*-region negative, the Fermi level of the region is shifted upward with respect to the Fermi level of the *p*-region. As a result, *n*-region conduction-band electrons tunnel through the barrier to fill unoccupied energy levels in the *p*-region valence band and, as for reverse bias, the junction is a short circuit.

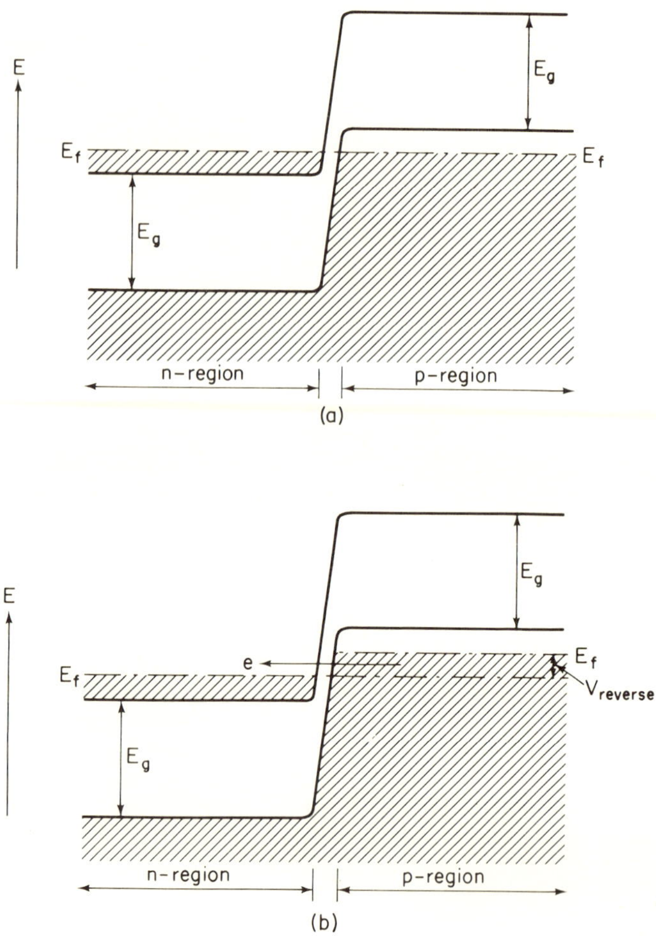

Fig. 4-25. Energy-level diagrams of a tunnel diode: (a) for zero bias; (b) for reverse bias. With reverse bias *p*-region valence-band electrons tunnel through the barrier to the *n*-region conduction band.

As the forward voltage is increased above the value represented by the energy-level diagram of Fig. 4-26(a), the higher-energy *n*-region conduction-band electrons are unable to tunnel through the barrier to the *p*-region, owing to the absence of energy levels in the *p*-region energy gap. Further increase in forward voltage so reduces the number of tunneling electrons that the forward current no longer increases with voltage, but reaches a maximum value and begins to decrease, causing the negative-resistance region of the forward $I(V)$ characteristic. This condition is represented by the energy-level diagram of

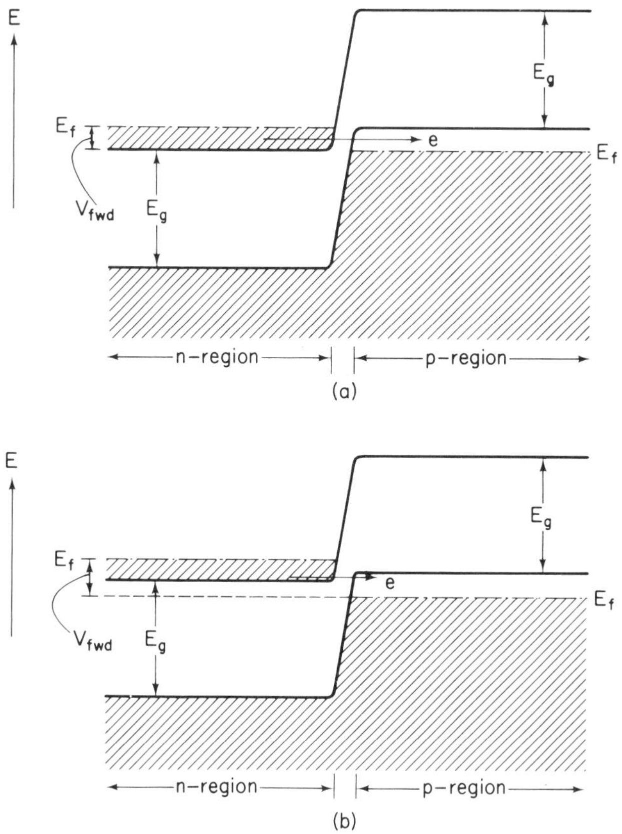

Fig. 4-26. Energy-level diagrams of a forward-biased tunnel diode: (a) *n*-region conduction-band electrons tunnel through the barrier to the *p*-region. (b) With increased bias, tunneling is suppressed since few of the *n*-region conduction-band electrons "see" available energy levels across the barrier.

Fig. 4-26(b) in which overlapping of the *n*-region conduction band and the *p*-region valence band has almost ceased to exist.

A further increase in forward voltage above the value of Fig. 4-26(b) entirely eliminates band overlapping and tunneling is completely suppressed, *n*-region conduction-band electrons being unable to "see" available energy levels across the barrier. This condition is represented by the energy-level diagram of Fig. 4-27(a). Because of the heavily doped *p*- and *n*-regions, the barrier continues to be so high that few *n*-region electrons and *p*-region holes are able to cross it, and the forward current is quite low. This condition exists at the current minimum or valley of the forward $I(V)$

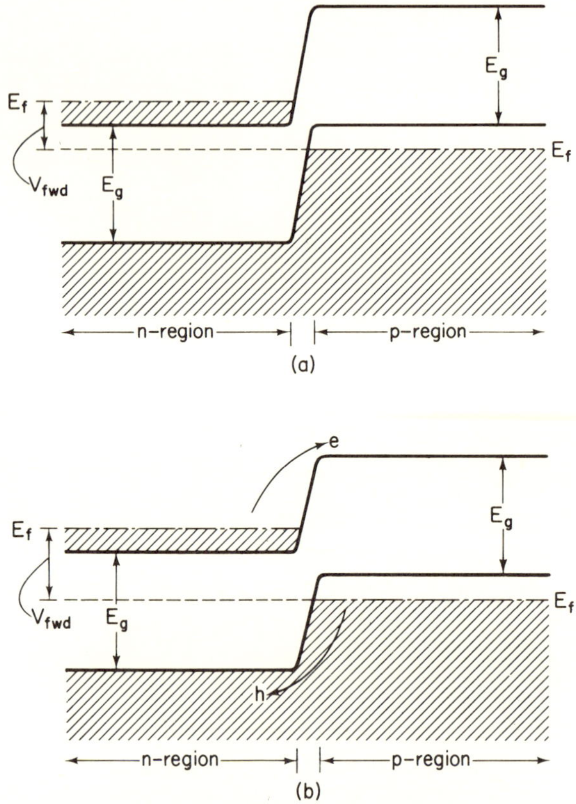

Fig. 4-27. Energy-level diagrams of a tunnel diode so forward biased that electron tunneling is entirely suppressed. (a) Tunneling suppressed but barrier too high for significant numbers of majority carriers to cross the junction. (b) Barrier so reduced that large quantities of electrons (*e*) and holes (*h*) cross the junction.

characteristic of Fig. 4-24. As the forward voltage is increased above the valley-point value, the barrier becomes low enough for significant numbers of majority carriers to cross the junction, and the current increases strongly with voltage. This is the normal forward-characteristic region of the $I(V)$ characteristic of Fig. 4-24, and this condition is represented by the energy-level diagram of Fig. 4-27(b).

Important tunnel-diode specifications are the peak-point current I_p, the valley-point current I_v (or the peak-to-valley current ratio I_p/I_v), the forward peak-point-current voltage V_{fp}, V_{fp} being the voltage in the normal-forward-characteristic region of the $I(V)$ characteristic for which the current is the forward-peak value, and the diode capacitance C and series resistance R_s when biased in the negative-resistance region. Additional specifications are the maximum forward and reverse current ratings and typical values (rather than specified upper and lower limits) of the peak-point voltage V_p, the valley-point voltage V_v, and the negative conductance. The peak-point current depends upon the junction area and ranges from a fraction of a milliampere to as high as several amperes. The peak-to-valley current ratio is a property of the semiconductor material, typical values being of the order of $\frac{10}{1}$ for germanium diodes and $\frac{4}{1}$ for silicon diodes, as may be seen from the forward characteristics of Fig. 4-28. Figure 4-28 also shows the valley-point voltage and the forward peak-point-current voltage to be higher for the silicon diode, owing to the larger energy gap of silcion. The V_{fp} specification is important in switching applications, since it is an indication of the maximum possible diode-voltage change that is obtainable. Thus, we see that the choice of germanium or silicon tunnel diodes may depend upon whether current change or voltage change is more important in the application. Tunnel diodes of gallium-arsenide have I_p/I_v ratios comparable to those of germanium diodes and V_{fp} values somewhat higher than those of silicon diodes.

In general, tunnel diodes are used primarily in applications for which there are no suitable transistors. The tunnel diode is superior to the transistor in the following ways: It is basically a higher-frequency device, because its electrical characteristics depend upon the propagation of electric waves

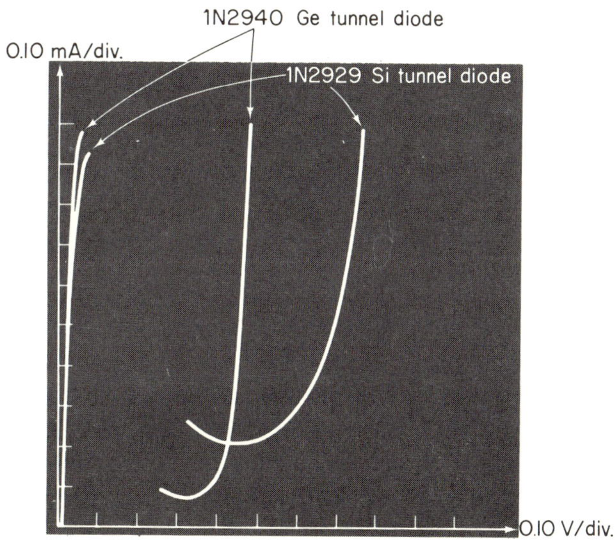

Fig. 4-28. Forward conduction characteristic of a typical type 1N2929 silicon tunnel diode and of a typical type 1N2940 germanium tunnel diode. (The negative resistance portion of each characteristic is missing, owing to the type of curve tracer used to obtain the display.) Note the higher valley-point and forward-peak-current voltages of the silicon diode.

across a space-change layer rather than upon the drift and diffusion of carriers. It is not dependent upon minority-carrier lifetime and is, therefore, less sensitive to surface conditions and to nuclear radiation, often being operated unencapsualted, and it can be fabricated from low-lifetime materials unsuitable for other devices. It can be operated over a broader temperature range owing to the heavier doping, and it is basically less costly than the transistor.

The disadvantages of the tunnel diode that restrict its use are as follows: It is a two-terminal device requiring special circuitry, owing to the difficulty in cascading amplifier stages, etc. Its output voltages are quite low, being of the order of tenths of a volt at most. Although these voltages are large enough to switch transistors having base-emitter voltages of the same order of magnitude, they cause the tunnel diode to be basically a low-power device. Another disadvantage is the problem of obtaining a suitable low-voltage power supply. Such a supply must have an extremely low internal resistance in order that the diode operate in the negative-resistance region and be supplied appreciable power at such a low voltage.

4-9 BACKWARD DIODES

As mentioned earlier, conventional semiconductor diodes, particularly silicon diodes, do not work very well near the origin of the conduction characteristic, silicon diodes at room temperature acting as virtual open circuits to applied voltages of either polarity unless the magnitude exceeds several hundred millivolts. This difficulty is alleviated to a considerable degree in a version of the tunnel diode that consists of a *pn* junction between two regions doped just short of degeneracy. If the doping is such that the *n*-region Fermi level is located at the bottom edge of the conduction band and the *p*-region Fermi level is at the top edge of the valence band, the bands overlap with the smallest reverse bias, but no overlapping occurs with either zero bias or forward bias. When such a device has a barrier thin enough for the occurrence of tunneling, its reverse conduction characteristic is that of the tunnel diode, but it has a normal forward characteristic, owing to the absence of tunneling. Thus, the device conducts extremely well for one polarity of applied voltage, but quite poorly for voltages of the opposite polarity, unless the voltage magnitude exceeds several tenths of a volt, and it can be used as a rectifier or detector of relatively small voltages. Since it operates with its normal anode and cathode regions interchanged, the device is known as a *backward diode*.

The forward and reverse conduction characteristics of a germanium backward diode (type BD-3) are shown in Fig. 4-29(a); also shown for comparison

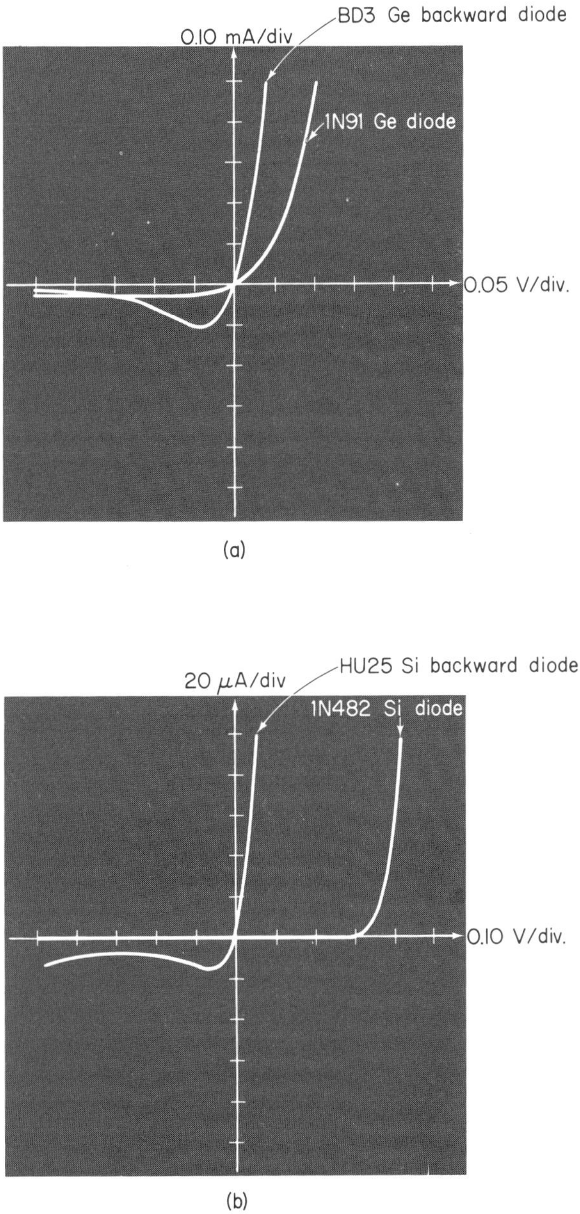

Fig. 4-29. Comparison of the conduction characteristics of backward diodes with those of conventional junction diodes. (a) Germanium backward diode (type BD-3) and germanium conventional diode (type 1N91). (b) Silicon backward diode (type HU25) and silicon conventional diode (type 1N482).

are the characteristics of a conventional germanium junction rectifier (type 1N91). Figure 4-29(b) compares the conduction characteristics of a silicon backward diode (type HU25) with those of a conventional silicon junction diode (high-conductance type 1N482). The current scales of Fig. 4-29 are chosen so that the reverse characteristics of the devices may be compared. Since some tunneling does occur when their *pn* junctions are forward biased, the "reverse characteristics" of backward diodes are inferior to those of conventional diodes, but the backward-diode "forward characteristics" are decidedly superior. This is particularly evident in the silicon-diode comparison of Fig. 4-29(b). The voltage scales chosen in Fig. 4-29 do not permit comparison of the maximum reverse voltages of the devices, but, as has been implied, the "reverse voltage" of the backward diode is limited to a value of the order of 1 V. Another important feature of the backward diode is the absence of appreciable minority-carrier storage effects.

4-10 DIODE RECTIFIER CIRCUITS

A rectifier is a device that converts alternating current into unidirectional current by virtue of a current-voltage (i–v) characteristic that permits appreciable current flow in one direction only. Although d-c power represents a significant fraction of the total electric power used, very little is generated; instead, almost all is obtained through rectification processes. In this section, we consider circuits using silicon *pn*-junction rectifiers operated at frequencies low enough that the effects of capacitance, turn-on and turn-off times, etc. may be neglected. That is, our consideration is based on diode i–v characteristics, and approximate ones at that.

Approximate i–v characteristics. The i–v characteristic of an ideal diode (ideal from the point of view of its use as a rectifier) is shown in Fig. 4-30(a).

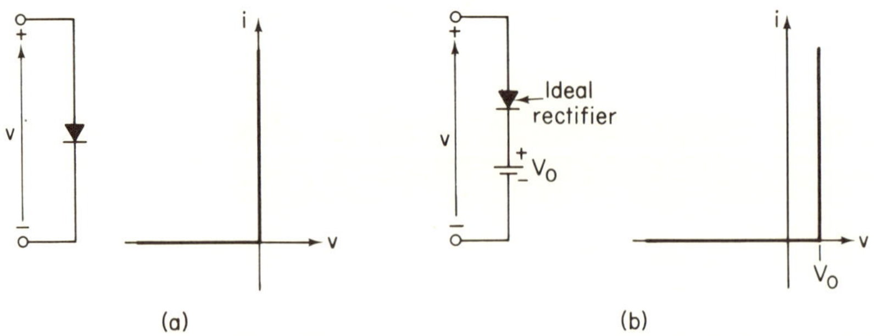

Fig. 4-30. Approximate diode *i-v* characteristics. (a) Ideal rectifier. (b) Ideal rectifier and battery.

As its characteristic indicates, the ideal rectifier acts like a short circuit in the forward direction and an open circuit in the reverse direction. This characteristic is a good approximation for that of a silicon *pn* diode operating at moderate temperatures in circuits having applied voltages large compared to 1 V but less than the reverse-breakdown value. This is because the forward voltage is usually less than 1 V, owing to conductivity modulation, and the reverse current is quite small.

For circuits in which the diode forward voltage is significant relative to the applied voltage, a silicon diode may be represented by the series combination of an ideal diode and a battery V_0, as shown in Fig. 4-30(b). Voltage V_0 is the threshold voltage for appreciable forward conduction. The i–v characteristics of Fig. 4-30 are said to be *piecewise* linear, since the characteristics consist of separate linear pieces.

Figure 4-31(a) shows the i–v characteristic of an ideal voltage-reference diode (commonly known as a breakdown or Zener diode). If the idealized characteristic is an inadequate representation, the *reverse region* of the characteristic may be represented by the series combination of an ideal diode (not an ideal voltage-reference diode), a resistor r_z, and a battery V_z, as shown in Fig. 4-31(b). The ideal diode and battery prevent current flow for reverse voltages having magnitudes less than the breakdown value V_z, and r_z is the dynamic resistance dv/di of the diode in the breakdown region. For the ideal voltage-reference diode, $r_z = 0$.

4-10.1 The Half-Wave Rectifier

The circuit shown in Fig. 4-32(a) is known as a half-wave rectifier because just half of the supply-voltage wave is transmitted to load resistor R_L, provided diode D is an ideal rectifier. The voltage source supplies a sinusoidal voltage

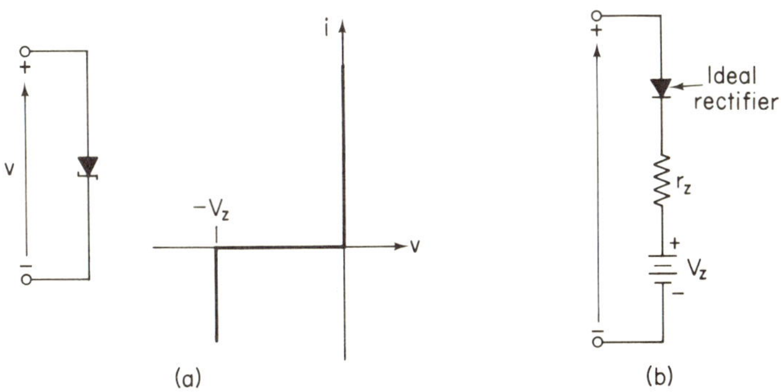

Fig. 4-31. Approximate voltage-reference diode representations. (a) Ideal *i-v* characteristic. (b) Circuit-model representation of the reverse region of a practical diode.

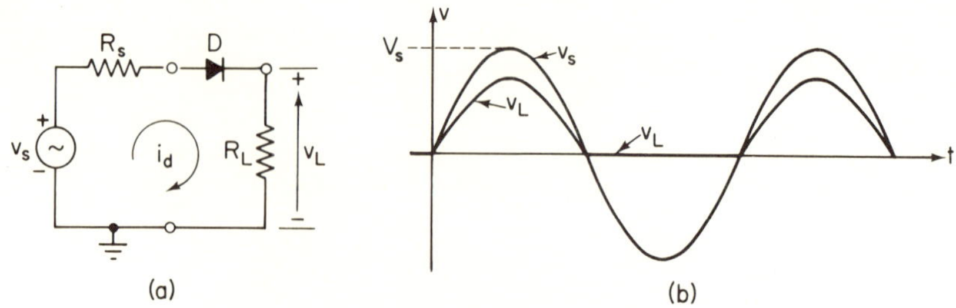

Fig. 4-32. A half-wave diode rectifier. (a) Circuit. (b) Voltage waveforms.

$$v_s = V_s \sin \omega_s t \tag{4-22}$$

and has a source resistance R_s. If D is assumed ideal, it is considered to be a short circuit when v_s is positive and an open circuit for negative values of v_s. Therefore, load voltage v_L can be expressed as

$$v_L = \frac{R_L}{R_s + R_L} v_s \quad \text{for } v_s > 0 \tag{4-23}$$

$$v_L = 0 \quad \text{for } v_s < 0 \tag{4-24}$$

and the wave forms of v_s and v_L are as shown in Fig. 4-32(b).

The load voltage developed by the half-wave rectifier consists of half-sine-wave pulses and is periodic in time. Therefore, v_L can be expressed by a Fourier series as

$$v_L = V_L\left(\frac{1}{\pi} + \frac{1}{2}\sin \omega_s t - \frac{2}{3\pi}\cos 2\omega_s t - \frac{2}{15\pi}\cos 4\omega_s t - \cdots\right) \tag{4-25}$$

where V_L is the maximum instantaneous value of v_L. Thus v_L consists of a d-c component

$$V_{L,\text{dc}} = \frac{V_L}{\pi} \tag{4-26}$$

and sinusoidal components at the fundamental radian frequency ω_s and at integer multiples of ω_s. The existence in the load voltage of frequencies *not present* in the applied voltage is a consequence of the nonlinearity of the diode. This is a general characteristic of nonlinear circuits.

The rms value $V_{L,\text{rms}}$ of the load voltage is of interest, since it determines the heating power in R_L, and may be obtained from the expression

$$V_{L,\text{rms}} = \sqrt{\frac{1}{2\pi}\int_0^{2\pi} v_L^2 \, d\omega t} = \sqrt{\frac{V_L}{2\pi}\int_0^{\pi} \sin^2 \omega t \, d\omega t} \tag{4-27}$$

from which

$$V_{L,\text{rms}} = \frac{V_L}{2} \tag{4-28}$$

The ratio of the rms and average values of load voltage is known as the *form factor F* of the wave form; that is,

$$F = \frac{V_{L,\mathrm{rms}}}{V_{L,\mathrm{dc}}} \tag{4-29}$$

From Eqs. (4-26) through (4-29), the form factor of a half-wave-rectified sine wave is seen to be $\pi/2$. A form factor greater than unity indicates the presence of fluctuation (a-c) components in a d-c wave form.

Ripple factor. The output voltage of a rectifier may be considered to consist of two components—the d-c component $V_{L,\mathrm{dc}}$ and the a-c component $V_{L,\mathrm{ac}}$. Then the effective (rms) value of the voltage is given by

$$V_{L,\mathrm{rms}} = \sqrt{V_{L,\mathrm{dc}}^2 + V_{L,\mathrm{ac}}^2} \tag{4-30}$$

and

$$V_{L,\mathrm{ac}} = \sqrt{V_{L,\mathrm{rms}}^2 - V_{L,\mathrm{dc}}^2} \tag{4-31}$$

where $V_{L,\mathrm{ac}}$ is the effective value of the a-c components. The ratio of the d-c and a-c components is known as the ripple factor of the wave form; that is,

$$\text{Ripple factor} = \frac{V_{L,\mathrm{ac}}}{V_{L,\mathrm{dc}}} \tag{4-32}$$

From Eqs. (4-31) and (4-32), we obtain

$$\text{Ripple factor} = \frac{\sqrt{V_{L,\mathrm{rms}}^2 - V_{L,\mathrm{dc}}^2}}{V_{L,\mathrm{dc}}} = \sqrt{\frac{V_{L,\mathrm{rms}}^2}{V_{L,\mathrm{dc}}^2} - 1} = \sqrt{F^2 - 1} \tag{4-33}$$

where F is the form factor of the wave form, as given by Eq. (4-29). For the half-wave rectifier supplying a resistance load,

$$\text{Ripple factor} = \sqrt{\left(\frac{\pi}{2}\right)^2 - 1} = 1.21 \tag{4-34}$$

indicating that the a-c component in the output exceeds the d-c component.

Rectification efficiency. The efficiency of conversion of a-c rectifier-input power to d-c power is known as the rectification efficiency η_r. For ideal diodes, the rectifier input is the heating power in the load resistor; therefore,

$$\eta_r = \left(\frac{V_{L,\mathrm{dc}}}{V_{L,\mathrm{rms}}}\right)^2 \tag{4-35}$$

Applying this expression to the half-wave rectifier yields

$$\eta_r = \left(\frac{V_L/\pi}{V_L/2}\right)^2 = \frac{4}{\pi^2}$$

which has a value of 40.6 percent.

Diode threshold voltage. If the forward threshold voltage V_0 of the diode is significant relative to the peak input voltage V_s, the approximate i–v characteristic of Fig. 4-30(b) may be used in the load-voltage determination. The effect of V_0 is to delay the flow of diode current i_d until v_s rises to the value V_0 at $\omega t = \alpha_1$ and to cause i_d to cease when v_s falls to V_0 at $\omega t = \alpha_2$, as shown in Fig. 4-33(b). Thus i_d consists of a series of pulses of duration less than 180 deg, each pulse being the upper portion of a sinusoid.

From Fig. 4-33, we see that i_d can be expressed as

$$i_d = \frac{V_s \sin \omega t - V_0}{R_s + R_L} \quad \text{for } \alpha_1 < \omega t < \alpha_2 \tag{4-36}$$

$$i_d = 0 \quad \text{for } \alpha_2 < \omega t < 2\pi + \alpha_1 \tag{4-37}$$

The average and rms value of i_d may be obtained from

$$I_{d,\text{dc}} = \frac{1}{2\pi}\int_0^{2\pi} i_d \, d\omega t \tag{4-38}$$

and

$$I_{d,\text{rms}} = \sqrt{\frac{1}{2\pi}\int_0^{2\pi} i_d^2 \, d\omega t} \tag{4-39}$$

using the i_d expressions of Eqs. (4-36) and (4-37).

4-10.2 The Full-Wave Rectifier

The circuit of a full-wave rectifier supplying a resistive load is shown in Fig. 4-34(a). It is usually called a *biphase rectifier*, because it consists of two half-wave rectifiers connected to a single load resistor and supplied

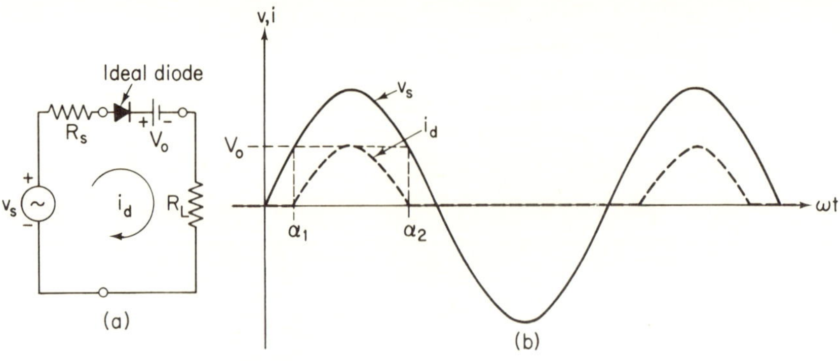

Fig. 4-33. A half-wave diode rectifier with significant diode forward-threshold voltage V_0. (a) Circuit showing the diode model of Fig. 4-30(b). (b) Waveforms.

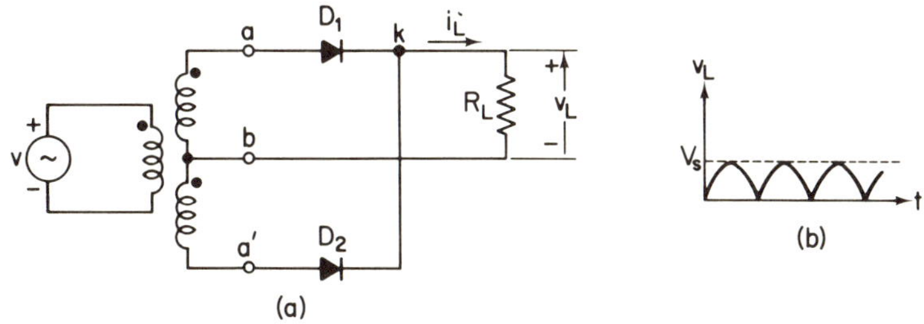

Fig. 4-34. Full-wave (biphase) rectifier with resistance load. (a) Circuit. (b) Output-voltage waveform.

with sinusoidal input voltages that are equal in magnitude but opposite in phase; that is,

$$v_{ab} = V_s \sin \omega t \tag{4-40}$$

and

$$v_{a'b} = -v_{ab} = V_s \sin (\omega t + \pi) \tag{4-41}$$

If the diodes are considered to be ideal rectifiers, the circuit operates as follows. During positive half-cycles of v_{ab}, D_1 acts as a short circuit, D_2 acts as an open circuit, and $v_L = v_{ab}$. During negative half-cycles of v_{ab}, the diode conditions are interchanged and $v_L = v_{a'b}$. Thus the diodes act as switches that connect R_L to either a or a', whichever is positive, and the wave form of v_L is as shown in Fig. 4-33(b).

The load-voltage developed by the full-wave rectifier can be expressed by a Fourier series as

$$v_L = V_s\left(\frac{2}{\pi} - \frac{4}{3\pi}\cos 2\omega t - \frac{4}{15\pi}\cos 4\omega t - \cdots\right) \tag{4-42}$$

Thus v_L consists of a d-c component

$$V_{L,\text{dc}} = \frac{2V_s}{\pi} \tag{4-43}$$

which is the average of each half-cycle of v_{ab}, and a set of sinusoidal components having radian frequencies that are even multiples of ω_s. The rms value of v_L is seen to be the same as that of v_{ab}. Thus

$$V_{L,\text{rms}} = \frac{V_s}{\sqrt{2}} \tag{4-44}$$

and the form factor of the v_L wave form is

$$F = \frac{\pi}{2\sqrt{2}} = 1.11 \tag{4-45}$$

From Eqs. (4-33) and (4-35),

$$\text{Ripple factor} = \sqrt{\left(\frac{\pi}{2\sqrt{2}}\right)^2 - 1} \cong 0.48 \tag{4-46}$$

and the rectification efficiency η_r is

$$\eta_r = \left(\frac{2V_s/\pi}{V_s/\sqrt{2}}\right)^2 = \frac{8}{\pi^2} \tag{4-47}$$

which has a value of 81.2 percent. The higher rectification efficiency and lower ripple factor, relative to values for the half-wave rectifier, are indicative of the superiority of the full-wave circuit.

Peak inverse voltage. The reverse voltage applied to a rectifier diode during its nonconducting period is called an inverse voltage, and the peak inverse voltage (PIV) is an important consideration, since the PIV rating should not be exceeded. From Fig. 4-32(b), the PIV of a half-wave rectifier is seen to be the peak source voltage V_s. In the full-wave rectifier, the sum of the voltages across the two diodes is equal to the full transformer voltage 2 v_{ab}. However, at any instant of time one of the diodes acts as a short circuit and the other is an open circuit. Therefore, the voltage across the "open circuit" is 2 v_{ab} and each diode has a PIV of $2V_s$; that is, the PIV is the peak value of the full transformer voltage.

4-10.3 The Bridge Rectifier

Another type of circuit that may be used for rectifying single-phase alternating current is the *bridge circuit* shown in Fig. 4-35. During positive half-cycles of v_s, current i_1 flows through D_1, R_L and D_3, but no current flows through D_2 and D_4, owing to their reverse voltages. When v_s is negative, current i_2 flows through D_4, R_L and D_2, while D_1 and D_3 are nonconducting. Since currents i_1 and i_2 flow through R_L in the same direction, full-wave

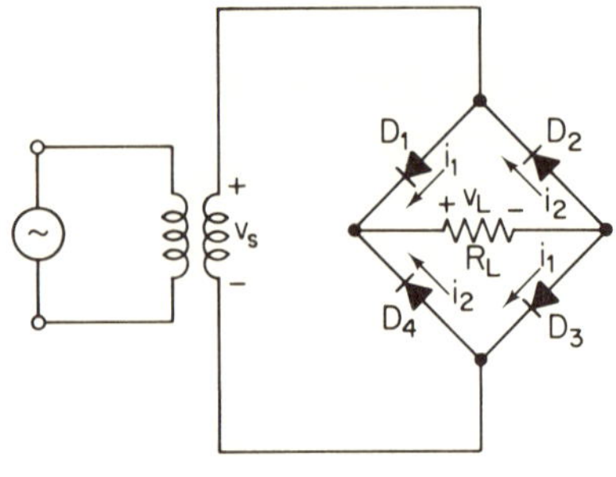

Fig. 4-35. A bridge-rectifier circuit.

rectification occurs, the wave form of the load voltage is the same as that shown in Fig. 4-34(b), and v_L is given by Eq. (4-34).

Like the diodes of the biphase circuit of Fig. 4-34(a), each bridge-rectifier diode has a PIV equal to the peak value of the full transformer voltage. This is because series-connected diode pairs (D_1-D_4 and D_2-D_3) are subjected to the transformer voltage, and one diode of each pair acts as a short circuit at any instant of time. The principal advantage of the bridge rectifier, relative to the biphase circuit, is that rectification of the full transformer voltages halves the voltage required for a paricular load voltage. The cost saving through use of a lower-voltage transformer and lower PIV-rated diodes more than compensates for the additional diodes required by the bridge circuit. Therefore, there seems little reason for using semiconductor-diode biphase rectifiers.

4-11 POLYPHASE RECTIFIERS

Almost all electric power is generated three-phase and distributed by three-phase circuits. Since large single-phase loads are undesirable, single-phase rectifiers become impractical when large quantities of d-c power are required; instead, three-phase rectifier circuits or circuits derived from a three-phase system are preferred. Other important advantages are the higher rectification efficiency and lower output-ripple factor inherent with the polyphase rectifier, the improvement in both rectification efficiency and ripple reduction increasing with the number of phases. For this reason, the choice of the number of phases often depends upon the maximum permissible ripple factor.

Three-phase half-wave rectifier. The simplest of the polyphase rectifiers is the three-phase half-wave circuit shown in Fig. 4-36. This rectifier is seen to consist of three single-phase half-wave rectifiers that send current in the same direction through the common load R_L.

The source voltages for the individual diodes are balanced three-phase voltages, as shown in Fig. 4-37(a). At time t_1, v_{ao} is positive and phase voltages v_{bo} and v_{co} are negative. Therefore, only D_1 is conducting at t_1 and, if the diodes are assumed to be ideal rectifiers, load voltage v_{ko} equals v_{ao}. The anode-cathode voltage v_{bk} of D_2 is

$$v_{bk} = v_{bo} - v_{ko} \tag{4-48}$$

and during conduction of D_1

$$v_{bk} = v_{bo} - v_{ao} \tag{4-49}$$

Thus the anode of D_2 is made negative by load voltage v_{ko}. At later time

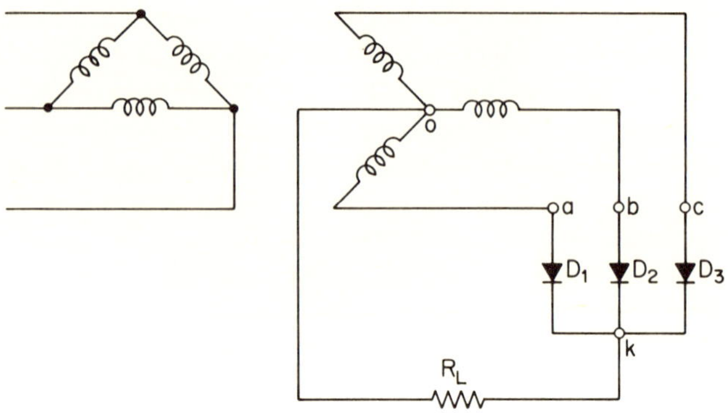

Fig. 4-36. Three-phase half-wave rectifier.

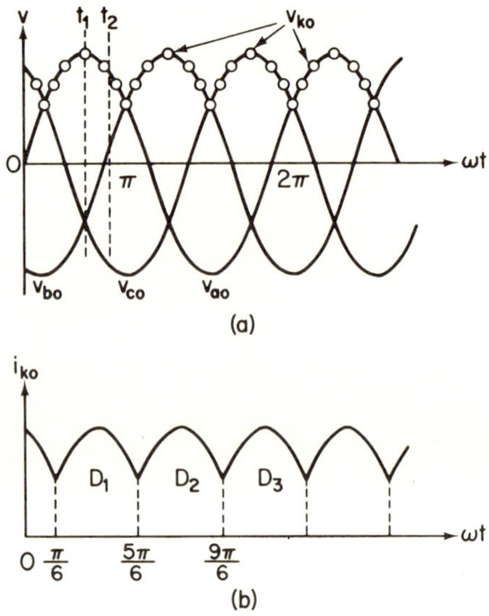

Fig. 4-37. Phase and load voltages and currents of the three-phase rectifier of Fig. 4-36.

t_2, both v_{ao} and v_{bo} are positive, but v_{ao} is greater than v_{bo}. Therefore, D_1 continues to conduct, and, as indicated by Eq. (4-49), the anode-cathode voltage v_{bk} of D_2 continues to be negative. This process continues until $\omega t = 5\pi/6$, when v_{ao} is equal to v_{bo} and v_{bk} becomes equal to zero. When v_{bo} exceeds v_{ao}, D_2 begins to conduct and v_{ko} becomes equal to v_{bo}. The anode-cathode voltage v_{ak} of D_1 then becomes negative, being given by

$$v_{ak} = v_{ao} - v_{ko} = v_{ao} - v_{bo} \tag{4-50}$$

and D_1 ceases conduction, the load current having transferred, or commutated, to D_2. A similar process takes place at $\omega t = 9\pi/6$ when the current commutates between D_2 and D_3, and at $\pi/6$ when the current is transferred to D_1. The presence of lead inductance prevents this commutating action from taking place in zero time, as implied here, but at power frequencies the commutation time may be negligible compared to the period of the waves.

For ideal rectifiers, the load voltage v_{ko} is equal to the supply voltage of the conducting phase. If the supply voltage for D_1 is

$$v_{ao} = V_s \sin \omega t \tag{4-51}$$

the load voltage during the conduction of D_1 is

$$v_{ko} = V_s \sin \omega t \quad \text{for } \frac{\pi}{6} < \omega t < \frac{5\pi}{6} \tag{4-52}$$

Since the wave form of the load voltage repeats itself every third of a cycle, the average load voltage $V_{L,\text{dc}}$ is

$$V_{L,\text{dc}} = \frac{3}{2\pi} \int_{\pi/6}^{5\pi/6} V_s \sin \omega t \, d\omega t = 0.827 V_s \tag{4-53}$$

The effective value $V_{L,\text{rms}}$ of the load voltage is

$$V_{L,\text{rms}} = \sqrt{\frac{3}{2\pi} \int_{\pi/6}^{5\pi/6} V_s^2 \sin^2 \omega t \, d\omega t} = 0.838 V_s \tag{4-54}$$

The form factor of the load voltage is

$$F = \frac{V_{L,\text{rms}}}{V_{L,\text{dc}}} = \frac{0.838}{0.827} = 1.013 \tag{4-55}$$

the ripple factor is

$$\sqrt{F^2 - 1} = \sqrt{(1.013)^2 - 1} = 0.17 = 17 \text{ percent} \tag{4-56}$$

and the rectification efficiency is $(0.827/0.838)^2$, which is equal to 96.5 percent. Thus we see considerable improvement over the single-phase full-wave rectifier with its ripple factor of 48 percent and rectification efficiency of 81.2 percent.

Zigzag-connected three-phase half-wave rectifier. The three-phase half-wave circuit of Fig. 4-36 finds little use, because of d-c magnetization of the transformer core by the d-c component of each anode current. This may result in saturation of the core and a high exciting current. The zigzag connection of Fig. 4-38 is a practical arrangement for eliminating d-c magnetization. Each transformer is seen to have two secondary windings; windings of the same transformer are indicated by being drawn parallel to each other. The secondary windings of each transformer carry d-c components in opposite directions, resulting in zero net d-c compoent of magnetizing force.

The supply voltage for each diode is the sum of two equal winding voltages having a phase difference of 60 deg. Thus the phase voltage is 1.732 times the winding voltage. To provide the same supply voltage for each diode as that with the straight three-phase connection (windings of the same transformer connected in series), each winding voltage must be about 15 percent higher. Such a transformer may be considerably less expensive than one having the lower winding voltage but able to operate satisfactorily with an appreciable d-c component of magnetizing current.

Six-phase half-wave rectifier. The secondary windings of the transformers of the zigzag connection of Fig. 4-38 may be arranged to provide phase voltages 60 deg apart to supply a six-phase half-wave rectifier, as shown by Fig. 4-39. Except for the number of phases, the operation of this circuit differs little from that of the three-phase half-wave rectifier. The principal difference is the absence of d-c magnetization of the transformer cores, since each transformer supplies two phases and the d-c magnetizing components add to zero.

The voltage wave forms for the six-phase rectifier of Fig. 4-39 are shown by Fig. 4-40. Reference to these wave forms shows that the anode voltage of a given diode is positive for only 60 deg of the cycle and that six pulses of load current occur each cycle. An analysis similar to that of the three-phase circuit shows that

$$V_{L,\mathrm{dc}} = 0.955V_s \tag{4-57}$$

$$V_{L,\mathrm{rms}} = 0.956V_s \tag{4-58}$$

$$F = 1.001 \tag{4-59}$$

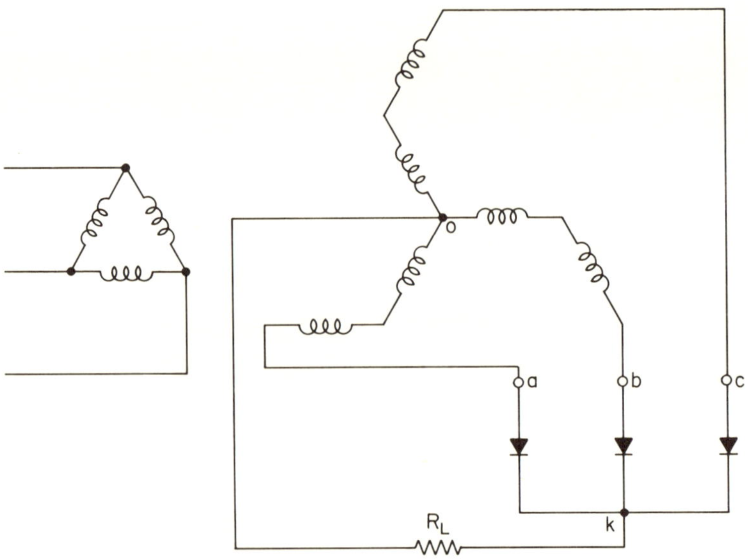

Fig. 4-38. Zigzag connection for a three-phase half-wave rectifier.

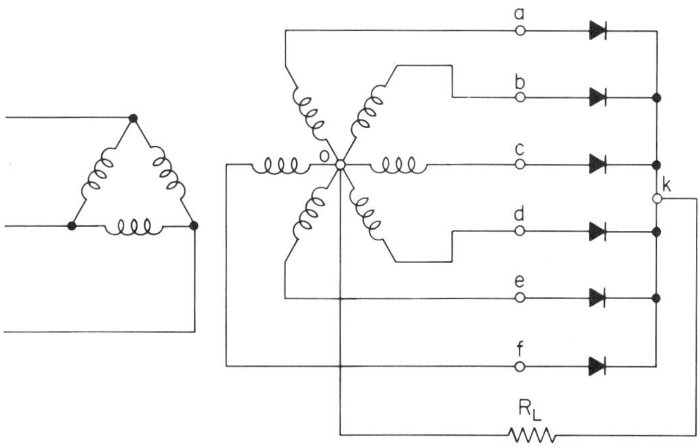

Fig. 4-39. Six-phase half-wave rectifier.

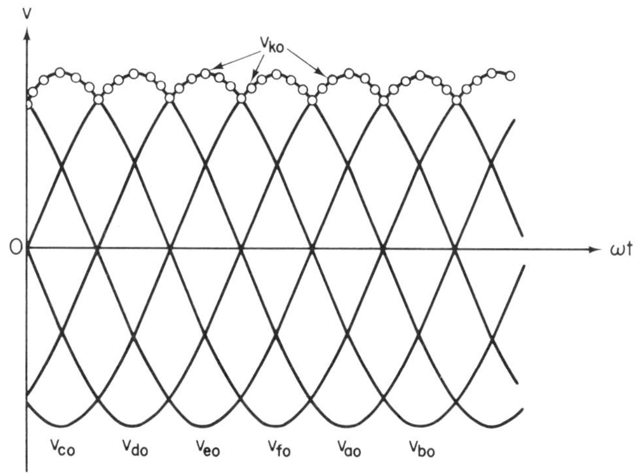

Fig. 4-40. Voltage waveforms for the six-phase half-wave rectifier of Fig. 4-39.

$$\text{Ripple factor} = 4 \text{ percent}$$

and

$$\eta_r = 99.8 \text{ percent} \tag{4-60}$$

These results show a significant improvement in ripple factor and rectification efficiency, relative to the three-phase circuit.

Three-phase full-wave rectifier. The performance of the three-phase full-wave rectifier shown in Fig. 4-41 is similar to that of the six-phase half-wave

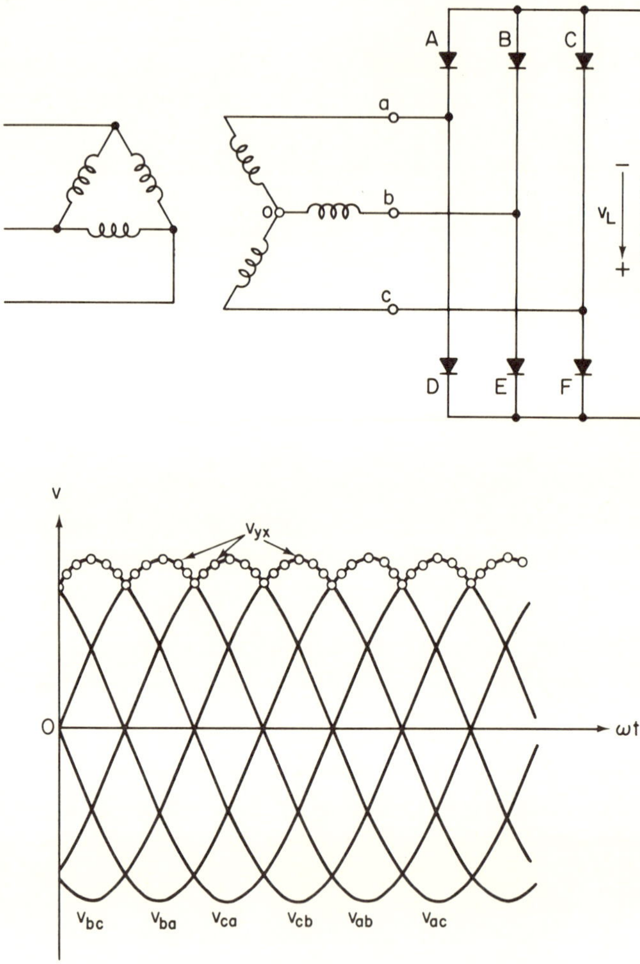

Fig. 4-41. Three-phase full-wave rectifier.

circuit, the principal difference being that each transformer secondary carries current during both half-cycles. Thus circuit operation is similar to that of the bridge rectifier. There is further similarity in that two diodes must conduct in order for current to flow through load R. Thus we see why the circuit is also known as the *three-phase bridge rectifier*.

Wave forms of transformer-secondary voltage and load voltage are also shown by Fig. 4-41. The load-voltage wave form is seen to be identical to the wave form of the six-phase half-wave rectifier; therefore, the rectification efficiency, ripple factor, etc., are the same. It should be noted, however, that the d-c load voltage is proportional to line-to-line, rather than phase, voltage. Thus, if V_s is the peak value of the phase voltage,

$$V_{L,\mathrm{dc}} = 0.955\sqrt{3}\ V_s = 1.65V_s \tag{4-61}$$

which is $2\sqrt{3}$ times larger than $V_{L,\mathrm{dc}}$ of the six-phase half-wave circuit using the same transformers, provided that secondary windings of individual transformers are connected series-aiding. Therefore, the six-phase circuit has little to recommend it when semiconductor rectifiers are used.

4-12 THE SHUNT-CAPACITOR FILTER

Many devices such as amplifiers and radio receivers require d-c power sources having a ripple factor much smaller than that obtainable directly from a rectifier; consequently, filter circuits of some sort are employed. Filter circuits not only reduce ripple, but they also increase rectification efficiency by preventing a-c components of current from flowing in the load resistance.

One of the simplest and most widely used filter types consists of a capacitor shunted across the load resistance, as shown in Fig. 4-42 for a half-wave rectifier. The capacitor stores energy obtained from the source during the conduction period of the rectifier and delivers energy to the load during the nonconduction period. In this way, the time during which current flows through the load is longer than if the capacitor were not present; in fact, if the capacitance is sufficiently large, the load current is continuous. This prolonging of the current represents a reduced ripple factor and results in increased d-c output current and voltage.

In the analysis of the rectifier-filter combination, we continue to assume the diode to be an ideal rectifier. Thus, when current flows through diode D, points a and k in Fig. 4-42(a) are considered to be at the same potential, and instantaneous output voltage v_{kb} is equal to instantaneous supply voltage

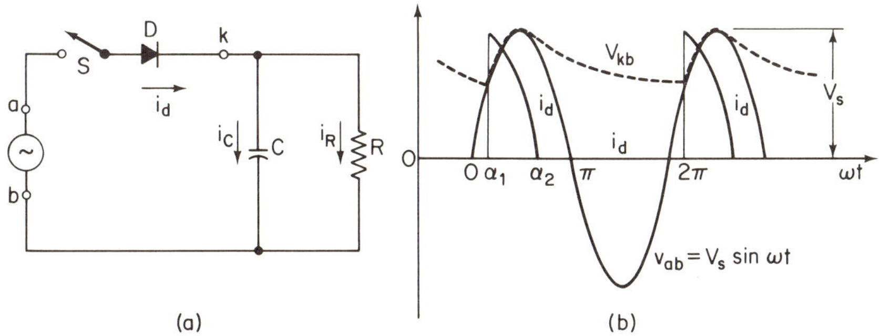

Fig. 4-42. (a) Half-wave rectifier with capacitor filter. (b) Voltage- and diode-current waveforms in the circuit of (a).

v_{ab}. Suppose that switch S is closed at time t equal to zero with the capacitor uncharged. As the supply voltage, given by

$$v_{ab} = V_s \sin \omega t \tag{4-62}$$

increases, current i_d flows through the diode, and the output voltage v_{kb} is equal to the supply voltage v_{ab}. The capacitor is charged, the current i_C through the capacitor being given by

$$i_C = C\frac{dV_{ab}}{dt} = \omega C V_s \cos \omega t \tag{4-63}$$

At $\omega t = \pi/2$, the capacitor current is zero and the capacitor is charged to V_s, the maximum value of the supply voltage. As v_{ab} decreases from the value V_m, the capacitor discharges through load resistor R and i_C reverses. When a capacitor discharges through a resistance, the maximum discharge rate occurs at the beginning of the discharge. Since, at $\omega t = \pi/2$, $dv_{ab}/dt = 0$, the capacitor is able to discharge rapidly enough for v_{kb} to follow v_{ab} downward for some time after $\omega t = \pi/2$, as shown in Fig. 4-42(b). At $\omega t = \alpha_2$, C is no longer able to discharge rapidly enough for its voltage to follow the supply voltage, and v_{ab} begins to fall below v_{kb}. When v_{kb} is greater than v_{ab} (point k more positive than point a with respect to point b), the cathode of the diode is positive with respect to the anode, and conduction ceases. With i_d equal to zero, the load R is disconnected from the supply voltage, and load current i_R is supplied entirely by the capacitor discharge.

Thus, from $\omega t = \alpha_2$ to the value of ωt at which the diode again begins to conduct, the load voltage v_{kb} is the familiar capacitor-discharge wave. The discharge rate of the capacitor varies inversely as the RC product, and if R and C are sufficiently large, the capacitor will not be completely discharged at $\omega t = 2\pi$. In this case, i_d does not begin flowing when v_{ab} first becomes positive, since no current can flow as long as v_{ab} is less than v_{kb}. At $\omega t = \alpha_1 + 2\pi$, v_{ab} equals v_{kb}, the diode begins to conduct, the capacitor begins to recharge, and the process is repeated.

Perhaps the most important fundamental point to be recognized in the operation of the circuit of Fig. 4-42(a) is that the potential of point k is determined either by the potential of point a or by the voltage to which the capacitor is charged, depending upon which is the larger. During conduction, the potential of k is determined by that of a, since v_{ab} is larger than the voltage across the capacitor. But when v_{ab} falls below the capacitor voltage, the diode current ceases and the potential of k depends upon the capacitor charge.

The general nature of the wave forms of the currents defined in Fig. 4-42(a) is shown by Fig. 4-43. From the circuit of Fig. 4-42(a), two current relations are apparent. First, during the conduction operiod of the rectifier,

$$i_d = i_C + i_R \tag{4-64}$$

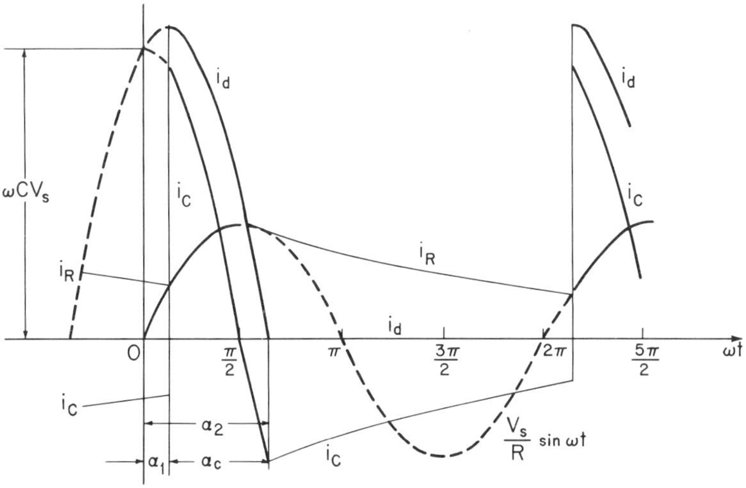

Fig. 4-43. Typical waveforms in the circuit of Fig. 4-42(a).

and second, during the nonconduction period

$$i_d = 0 \tag{4-65}$$

and therefore

$$i_C = -i_R \tag{4-66}$$

During the conduction period, the diode is considered to be a short circuit, and the currents are given by

$$i_R = \frac{V_s}{R} \sin \omega t \tag{4-67}$$

$$i_C = \omega C V_s \cos \omega t \tag{4-68}$$

and

$$i_d = \frac{V_s}{R} (\sin \omega t + \omega RC \cos \omega t) \tag{4-69}$$

$$i_d = \frac{V_s}{R} \sqrt{1 + \omega^2 R^2 C^2} \sin (\omega t + \tan^{-1} \omega RC) \tag{4-70}$$

It should be recognized that Eqs. (4-67) through (4-70) apply only between $\omega t = \alpha_1$, the ignition agnle, and $\omega t = \alpha_2$, the extinction angle. The extinction angle α_2 may be determined by applying Eq. (4-70) at the end of the conduction period when $i_d = 0$. For $i_d = 0$,

$$\omega t + \tan^{-1} \omega CR = \pi \tag{4-71}$$

but $\omega t = \alpha_2$ at $i_d = 0$; therefore,

$$\alpha_2 = \pi - \tan^{-1} \omega CR \tag{4-72}$$

Since ω, C, and R are all positive quantities, α_2 must lie in either the second or fourth quadrant. However, for $C = 0$, $\alpha_2 = 180$ deg, the value for a resistance load. Therefore,

$$90° < \alpha_2 < 180° \tag{4-73}$$

Combining Eqs. (4-70) and (4-72), we obtain

$$i_d = \frac{V_s}{R}\sqrt{1 + \omega^2R^2C^2}\sin(\omega t + \pi - \alpha_2) \tag{4-74}$$

from which

$$i_d = \frac{V_s}{R}\sqrt{1 + \omega^2R^2C^2}\sin(\alpha_2 - \omega t) \qquad \alpha_1 < \omega t < \alpha_2 \tag{4-75}$$

From this equation, and by inspection of Fig. 4-43, we see that the maximum diode current $i_{d,\max}$ occurs at $\omega t = \alpha_2 - \pi/2$, if this value of ωt lies within the angle of conduction (that is, if $\alpha_2 - \alpha_1 \geq \pi/2$); otherwise, $i_{d,\max}$ occurs at the beginning of conduction. Therefore,

$$i_{d,\max} \leq \frac{V_s}{R}\sqrt{1 + \omega^2R^2C^2} \tag{4-76}$$

This equation may be used to determine the required peak-current rating of the diode, although it results in a conservative rating if the conduction angle α_c is less than $\pi/2$. In a reasonably good filter, α_c is less than 90 deg, $i_{d,\max}$ occurs at $\omega t = \alpha_1$, and from Eq. (4-74) and by inspection of Fig. 4-43 we see that

$$i_{d,\max} = \frac{V_s}{R}\sqrt{1 + \omega^2R^2C^2}\sin(\alpha_2 - \alpha_1) \tag{4-77}$$

Complete quantitative anlysis of the operation of the rectifier-filter combination requires determining the equation for the capacitor-discharge current during the nonconduction period and using it in conjunction with Eqs. (4-64) through (4-77) to determine expressions for α_1, d-c output voltage and current, ripple factor, etc. However, such detailed consideration is hardly justified, since these factors may be readily determined by the use of approximations which yield results that are quite satisfactory for most design purposes.

Approximate analysis. If the output ripple voltage is to be small, say a few percent of V_s at most, several simplifying approximations may be made. In addition to considering the diode to be ideal and the a-c supply to have negligible impedance, we make the following assumptions.

1. The extinction angle α_2 is $\pi/2$. That is, the capacitor begins supplying the load current when the supply voltage falls below V_s. Equation (4-72) shows that this is a good approximation when ωCR is large.

2. The capacitor voltage v_{kb} is assumed to decrease linearly, at its initial rate of change, during the nonconduction period. Applying approximation 1, during the nonconduction period we have

$$v_{kb} = V_s e^{-t/RC} \tag{4-78}$$

and

$$\left| \frac{dv_{kb}}{dt} \right|_{t=0} = -\frac{V_s}{RC} \tag{4-79}$$

Thus the output-voltage change Δv_{kb} is assumed to be given by

$$\Delta v_{kb} = \frac{V_s}{RC} t \tag{4-80}$$

where t is the period of the capacitor discharge.

3. For the purpose of determining output ripple voltage and d-c output voltage, the ignition angle α_1 is assumed to be $\pi/2$. That is, the discharge period of the capacitor is assumed to be $1/f$, where f is the supply frequency, and the charging period is assumed to be zero.

Based upon these assumptions, the output-voltage wave form is approximately represented by the saw-toothed wave shown in Fig. 4-44. From Eq. (4-80),

$$\Delta v_{kb} = \frac{V_s}{fRC} \tag{4-81}$$

Inspection of Fig. 4-44 shows that

$$V_{L,\mathrm{dc}} = V_s - \frac{\Delta v_{kb}}{2} \tag{4-82}$$

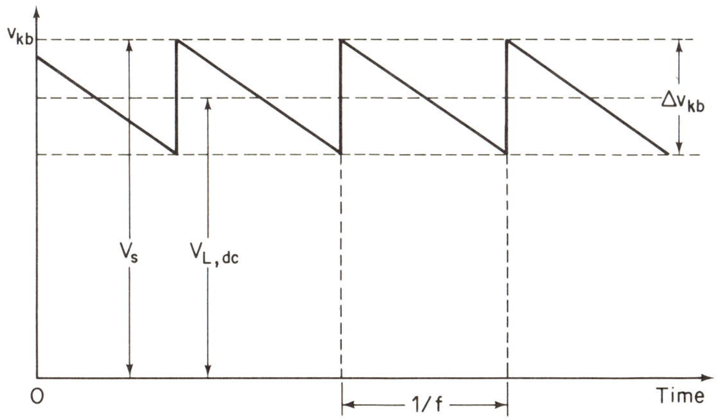

Fig. 4-44. Approximate representation of the output-voltage waveform of the shunt-capacitor filter of Fig. 4-42(a).

By Fourier analysis the idealized output-voltage wave form of Fig. 4-44 may be represented by the series

$$v_{kb} = f(t) = V_{L,\mathrm{dc}} + \frac{\Delta v_{kb}}{\pi}\left(-\sin \omega t + \frac{1}{2}\sin 2\omega t - \frac{1}{3}\sin 3\omega t + \cdots\right) \tag{4-83}$$

from which the output ripple voltage may be determined. The rms value of the a-c components of the idealized output-voltage wave may be shown to be $\Delta v_{kb}/2\sqrt{3}$, or about 128 percent of the fundamental component. It is seen that the approximations result in a ripple voltage somewhat larger than the actual value.

The approximate value of the ignition angle α_1 may be determined by assuming the supply voltage to be $V_s - \Delta v_{kb}$ at $\omega t = \alpha_1$. This is seen to be a fairly accurate assumption if ωRC is large. If α_1 is known, the peak value of the diode current can be determined from Eq. (4-77).

Diode surge current. The peak diode current given by Eq. (4-76) recurs each cycle and should not exceed the *recurrent-peak-current* rating of the diode. However, this is not the largest current that may pass through the diode. When switch S of Fig. 4-44 is closed, v_{kb} is zero owing to the absence of charge in capacitor C. Since v_{kb} cannot change instantly, C acts as a short circuit at the instant of closing S and the current through an ideal diode would be the value of v_{ab} at this instant divided by the source resistance R_s. Therefore, the diode current may be as large as V_s/R_s, depending on the point in the supply-voltage cycle at which S happens to be closed, and a resistance should be connected in series with the diode so that its *surge-current* rating is not exceeded.

Use of a full-wave rectifier. If a full-wave rectifier is used, the capacitor is charged twice per supply-voltage cycle, and the discharge time between periods is approximatly half that for a half-wave rectifier. Thus for a given RC combination, the ripple factor is lower; or for a given load and ripple factor, the filter capacitor is smaller and less costly. However, for a given d-c output voltage, the supply voltage of a biphase rectifier must be approximately twice as large as for the half-wave rectifier. For this reason, it is usually more economical to use a half-wave rectifier with a larger capacitor, and the biphase circuit is little used with this type of filter. As diodes have become less costly, the half-wave rectifier has yielded to the full-wave-bridge circuit, since the smaller capacitor requirement often justifies the additional diodes. A bridge rectifier-shunt capacitor filter combination is shown in Fig. 4-45(a), and the rectifier current and voltage wave forms are illustrated in Fig. 4-45(b).

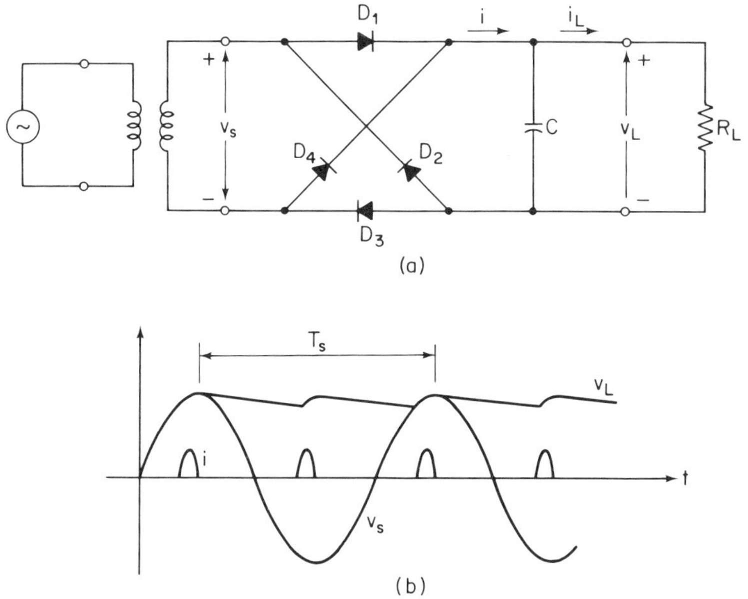

Fig. 4-45. Bridge rectifier with a shunt-capacitor filter. (a) Circuit. (b) Waveforms.

4-13 REGULATED POWER SUPPLIES

The rectifier-capacitor type of d-c power supply just considered is best suited for a fixed load, owing to variation of d-c output voltage and output ripple voltage with load. It is also desirable that load current be relatively low, say of the order of milliamperes at most, in order that the ripple voltage be small. This is seen from Eq. (4-81), in which V_s/R is the approximate load current. Since few loads meet these specifications, an electronic voltage regulator is often used to stabilize d-c output voltage and reduce output ripple voltage.

The simplest type of electonic regulator uses a Zener diode to fix the output voltage at a value equal to the breakdown voltage of the diode. A regulator circuit of this type is shown in Fig. 4-46. As in the circuit of Fig. 4-45(a), the bridge rectifier supplies pulses of charge to the capacitor each half-cycle. Between pulses, current is drawn from the capacitor by the load which consists of R_1, R_L, and regulator diode D_5. Thus, the capacitor voltage v_C has the wave form shown for v_L in Fig. 4-45.

If we consider the Zener diode to be ideal, having the i–v characteristic

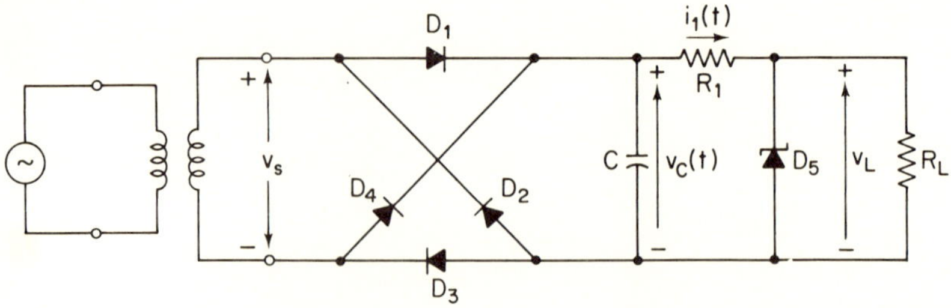

Fig. 4-46. Zener-diode regulated d-c power supply.

shown in Fig. 4-31(a), its regulating action is as follows. As long as $v_C(t)$ is large enough, D_5 will be broken down, and the voltage across it is the breakdown value V_B. Fluctuations in $v_C(t)$ then produce variations in the current $i_1(t)$ which are absorbed by D_5. As a result, a constant current equal to V_B/R_L flows in the load; that is, the output ripple voltage is zero. If the load current is changed (that is, if R_L changes), an equal but opposite change occurs in the average current thorugh D_5, provided that

$$\frac{R_L v_{C,\min}}{R_1 + R_L} \geqq V_B \tag{4-84}$$

$i_1(t)$ is unaffected, and the output voltage continues to have an average value of V_B and zero ripple.

As a practical matter, the Zener diode has a body (bulk-region) resistance ranging from a few ohms to several tens of ohms, depending on the power-dissipation rating and breakdown voltage, and may be represented by the circuit model of Fig. 4-31(b). As a result, the output ripple voltage Δv_L is not zero, but is given by

$$\Delta v_L = \frac{r_z}{R_1 + r_z} \Delta v_C \tag{4-85}$$

provided $r_z \ll R_L$, and the average value of v_L changes somewhat with load current owing to the effect of r_z. For this reason, two-stage Zener-diode regulators are often found in power supplies, as shown in Fig. 4-47.*

In this circuit, the voltage provided at C_1 by the bridge rectifier and surge-current-limiting resistor R_1 is 150 V d-c with 11.5 V peak-to-peak ripple. The resistor R_2 and double-anode regulator diode† D_5 provide additional filtering, reducing the ripple voltage by a factor of 180 while reducing the d-c voltage to 47 V. The d-c current through D_5 is approximately 50 mA. The second regulator stage provides further filtering, reducing the

* S. Schwartz, *Selected Semiconductor Circuits Handbook*. New York: John Wiley & Sons, Inc., 1960, pp. 8–42.

† A double-anode regulator diode has essentially the same breakdown voltage for either polarity of applied voltage.

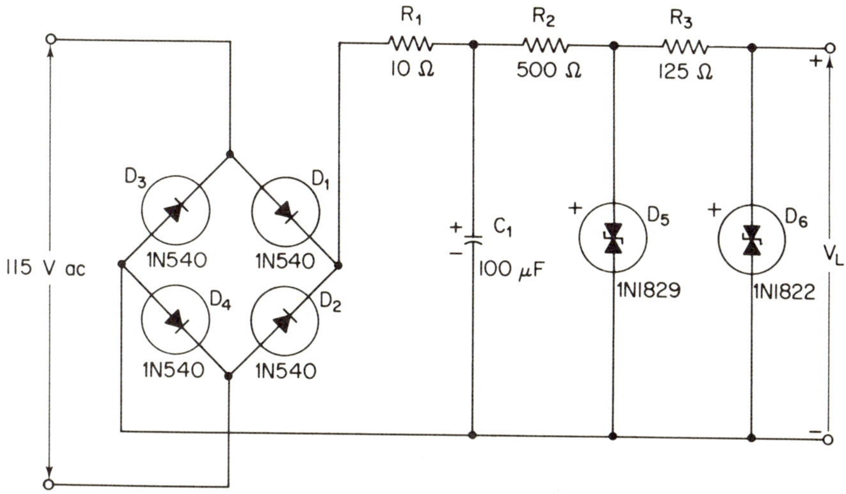

Fig. 4-47. Two-stage Zener-diode regulated power supply.

ripple by a factor of 16 at full load while regulating the output at 24 V for a load-current change of from zero to 150 mA. The output changes by about 2 percent for the 150-mA change in load, resulting in an output resistance $\Delta V_L/\Delta I_L$ of approximately 3 ohms. The output is also stabilized for changes in a-c input voltage, changing little for input voltages ranging from 100 V to 140 V.

Voltage regulators much more elaborate than those of Figs. 4-46 and 4-47 are used when the ultimate in regulation is required. These regulators combine Zener diodes and amplifying devices such as bipolar and field-effect transistors that are yet to be considered.

4-14 THE DIODE LIMITER

The circuit of Fig. 4-48(a) is a diode limiter or clipper. If the diodes are ideal, both are reverse-biased and nonconducting for v_s values ranging between V_1 and $-V_2$ and output voltage v_o is equal to v_s. When v_s is greater than V_1, diode D_1 conducts and $v_o = V_1$. Similarly, $v_o = -V_2$ for v_s more negative than $-V_2$, owing to conduction of D_2. Thus, no matter how wide the range of v_s variation, v_o is restricted to the range between the values of V_1 and $-V_2$. The variation of v_o as a function of v_s, for no load connected to the output terminals, is shown by the voltage transfer characteristic of Fig. 4-48(b).

The diode limiter can be used to protect a circuit against the application

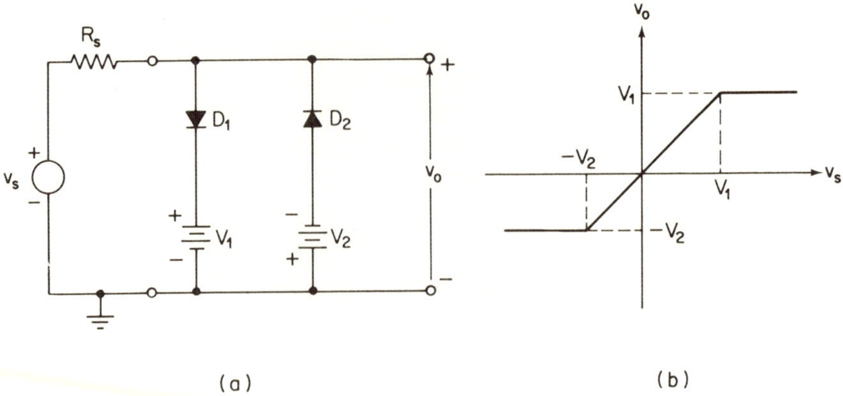

Fig. 4-48. A diode limiter. (a) Circuit. (b) Transfer characteristic for no load connected to the output terminals.

of excessive voltage. For example, R_s in Fig. 4-48 might be the multiplier resistance of a conventional voltmeter and v_o the voltage across the meter element. The limiter can also be used to alter the wave form of a signal voltage. For example, if $V_1 = V_2$ and v_s is a sinusoid having an amplitude much larger than V_1, v_o has very nearly a square wave form.

The battery symbols in Fig. 4-48(a) indicate that V_1 and V_2 are d-c voltages, but batteries are not necessarily used. Figure 4-49 shows diode limiter circuits without batteries. In Fig. 4-49(a), D_1 and D_2 are silicon diodes, and v_o is limited to the range between $+V_t$ and $-V_t$, where V_t is the forward threshold voltage of the diodes. In Fig. 4-49(b), D_1 and D_2 are Zener diodes in series opposing, and v_o may range between $+(V_B + V_t)$ and $-(V_B + V_t)$, where V_B and V_t are the breakdown and forward threshold voltages of the diodes. Operation of the circuit of Fig. 4-49(c) is identical to that of Fig. 4-49(b), since the double-anode voltage-regulator diode is the equivalent of two series-opposing Zener diodes.

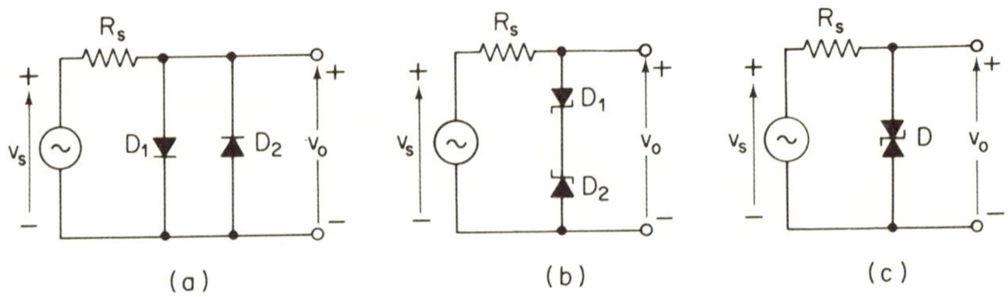

Fig. 4-49. Diode limiter circuits using: (a) conventional silicon diodes, (b) Zener diodes, (c) a double-anode regulator diode.

4-15 DIODE CLAMPING CIRCUITS AND a-c ELECTRONIC VOLTMETERS

It is often desirable that the positive or negative extremity of a periodic wave form be established at some constant reference level V_R. When the voltage is at this level, it is said to be *clamped* to V_R, and the circuit performing this function is known as a *clamping circuit*.

A diode clamping circuit is shown in Fig. 4-50. The circuit is identical to the half-wave rectifier-capacitor filter arrangement, except that the positions of the capacitor and the diode are interchanged and no load is shown. For an input voltage $v_s = V_s \sin \omega t$, a capacitor C that is uncharged at $t = 0$, and an ideal rectifier D, C charges through D to V_s during the first quarter-cycle. From then on, C remains charged to $v_C = V_s$, since D prevents C from discharging, and the output voltage is $v_o = v_s - V_s$. Thus the wave form of the output voltage, after the first quarter-cycle, is the same as that of the input voltage, except that the output wave is shifted downward by an amount equal to the peak value V_s of the input voltage, as shown in Fig. 4-50(b). Since the output voltage v_o cannot rise above zero, the circuit is said to clamp the positive peak of v_o at zero volts.

Diode clamping circuits have a number of useful applications. For example, it is necessary that signal voltages at certain points in TV and radar receivers have their peak values clamped at specified levels. Another important application is in the peak-responding a-c electronic voltmeter. Since

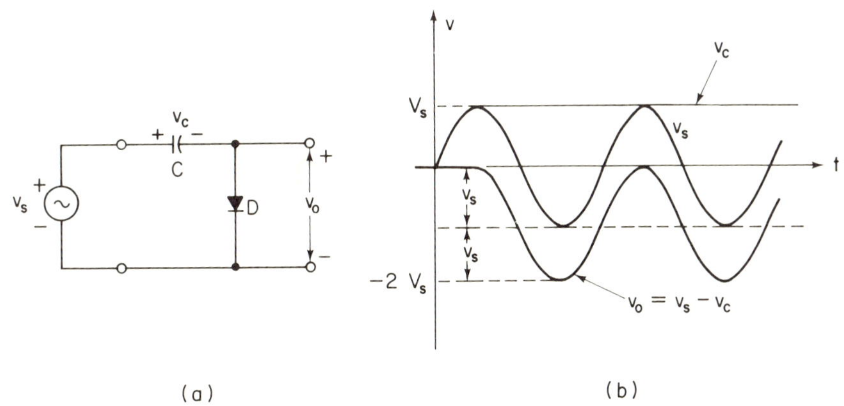

Fig. 4-50. (a) Diode clamping circuit. (b) Waveforms of the circuit of (a).

the average value of v_o in Fig. 4-50 is equal to the peak value V_s of input voltage v_s, measuring v_o by means of a high-impedance d-c voltmeter provides a measurement of V_s. The meter scale of such a voltmeter is calibrated to read $V_s/\sqrt{2}$ so that it indicates the rms value of a sinusoid. Thus the peak-responding electronic voltmeter consists of the circuit of Fig. 4-50 with a high-impedance d-c voltmeter as its load.* Advantages of the electronic voltmeter are its very high input impedance, since the clamping circuit draws no current once C has been charged, and its broad frequency range (up to hundreds of MHz).

Since the output-voltage wave form in Fig. 4-50 is the same as that of the input voltage but for the presence of the d-c component, the circuit is also called a *d-c inserter*. The more common terms for the circuit are *d-c restorer* and *reinserter*, which are used because the signal probably had a d-c component that was removed by a capacative coupling network. The latter terms are somewhat misleading, because they imply that the d-c component introduced is identical to that lost in transmission; such is unlikely. All three terms are used interchangeably in practice.

By reversing the diode in Fig. 4-50, the negative peak of a signal wave form may be clamped. This is shown in Fig. 4-51 for a rectangular wave form of input voltage having a d-c component. When input voltage v_s is negative, the capacitor is charged to the maximum negative value of v_s with the polarity shown. Thus the output voltage is $v_o = V_2 + v_s$, as shown in Fig.

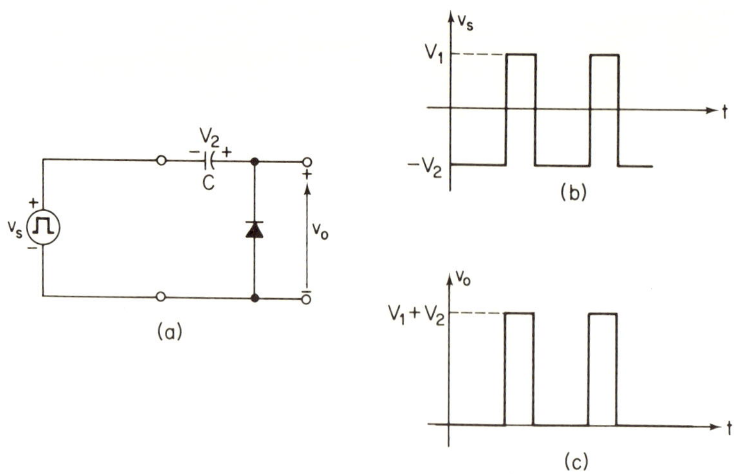

Fig. 4-51. Negative-peak clamping circuit. (a) Circuit. (b) Input voltage. (c) Output voltage.

* This is somewhat of an oversimplification, since a filter is also included to reduce the a-c voltage applied to the d-c voltmeter.

4-51(b), and the wave form of the output is the same as that of input, except for an upward shift equal to the negative peak of the input voltage. Therefore, the negative peak of the output signal is clamped at zero volts.

It is of interest to note how the d-c component of v_o in Fig. 4-51 relates to the input voltage. If $V_1 = V_2 = V$ and the duration of the negative region of the input wave is double that of the positive region, we see that the average value of v_o is $2V/3$ and that of v_s is $-V/3$. Thus the d-c component of v_o equals the difference between the peak (negative) and average values of v_s. This shows that the peak-responding a-c electronic voltmeter indicates the peak value of a voltage with respect to its average value.

If the connections to v_s in Fig. 4-51, are interchanged, the negative peak v_o continues to be clamped at 0 V, but the average value is $4V/3$. This voltage is seen to be the difference between the positive peak value of v_s and its average value. Thus we see that a peak-responding a-c electronic voltmeter may indicate differently when its connections are interchanged. When this occurs, the voltmeter is said to have *turn-over error.* "Turn-over error" is somewhat of a misnomer, because the difference in readings is not the fault of the voltmeter. However, it is a meaningful term, since an error does indeed occur when it is present, if either voltmeter reading is considered to be the rms value of the a-c voltage. The existence of turn-over error indicates the presence of a nonsinusoidal a-c voltage, but its absence does not ensure that the voltage is sinusoidal; it merely indicates the equality of the positive and negative peaks.

The response to peak-about-average voltage is a useful feature of the a-c electronic voltmeter, since it can be used to measure a-c voltages superimposed on relatively large d-c voltages—the output ripple voltage of a d-c power supply, for example. Because of the difficulty in rectifying small a-c voltages, the peak-responding voltmeter is usually restricted to the measurement of voltages greater than about 0.1 V. Smaller a-c voltages may be measured by means of an *average-responding* electronic voltmeter that blocks the d-c voltage upon which the a-c signal may be superimposed, and amplifies the signal prior to rectification. The full-wave rectified voltage is measured by means of a d-c voltmeter that is calibrated to indicate the rms value of a sinusoid. This type of voltmeter may be used to measure voltages as low as a fraction of a millivolt, but the frequency range is much less than that of the peak-responding instrument. Most a-c electronic voltmeters are either peak- or average-responding instruments; however, rms-responding electronic voltmeters are in existence.

Effect of source resistance. Although source resistance has not been considered in the circuits of Figs. 4-50(a), and 4-51(a), this is not the practical situation. Source resistance and diode forward resistance prevent the capacitor from charging to V_s during the first positive half-cycle of v_s, but the diode conducts and C receives additional charge during the next positive half-cycle. This process continues, with C receiving additional charge during each cycle of v_s, until C becomes charged to V_s and the diode current becomes zero. Therefore, after several cycles of operation, the clamping circuits perform as though source resistance were nonexistent.

Effect of signal-level change. Suppose that, after steady-state conditions are established in the circuit of Fig. 4-50(a), the amplitude of v_s is increased. When this occurs, there will again be a time interval during which the diode conducts and the d-c voltage across C increases as required to clamp the positive peak of v_o to 0 V. After this interval, the circuit performs as before. But suppose the input signal decreases. In this case, clamping action requires that the d-c voltage across C also decrease, but there is no mechanism in the circuit of Fig. 4-50(a) for such a discharge. Therefore, it is necessary to shunt a resistor across either C or the diode to permit a decrease in capacitor voltage. Such a resistor R is shown in Fig. 4-52(a). In this arrangement C will discharge through the series combination of R and the resistance of the source.

Figure 4-52(b) shows the wave form of v_o of Fig. 4-52(a) for a sinusoidal v_s that has an abrupt decrease in amplitude at time t_1. The first two cycles

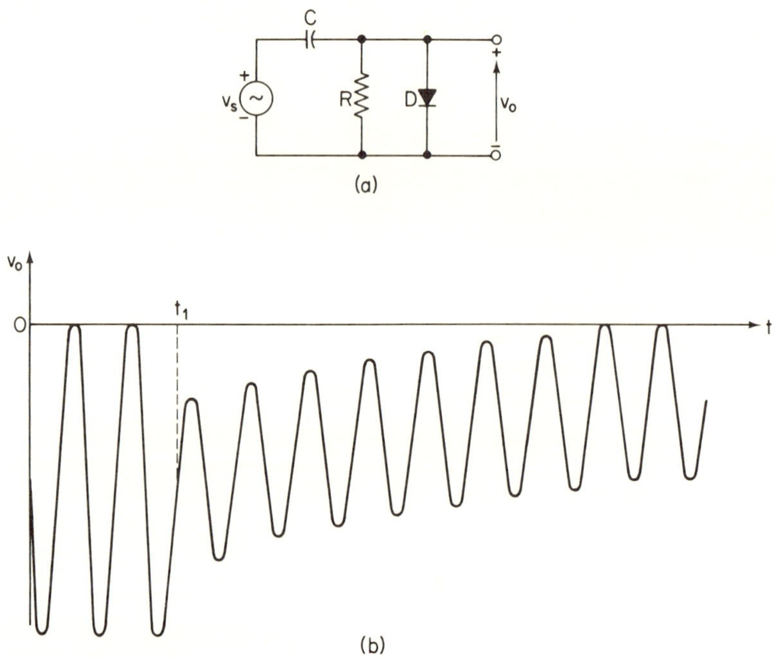

Fig. 4-52. (a) Practical clamping circuit having a resistor that permits the capacitor to discharge. (b) Effect of an abrupt decrease in input signal.

of v_o indicate the steady-state condition established after the large-amplitude signal has been applied. Since the capacitor cannot discharge abruptly at t_1, the positive peaks of the sinusoid fall short of reaching 0 V immediately after t_1. As the capacitor discharges, its voltage decreases exponentially, and after several cycles the positive peaks again reach zero. Unlike the situation with R absent, C continues to discharge through R even through the positive peaks reach zero. Therefore, the diode conducts during each positive half-cycle of v_s to replace the charge lost by C during the nonconducting interval. In most applications, the diode-conduction interval is but a small part of the cycle, since it is not usually necessary that clamping be reestablished immediately after a reduction in signal amplitude, permitting R to be relatively large.

Other clamping levels. Signal wave forms can be clamped at levels other than zero by adding a battery in series with the diode, as shown in Fig. 4-53. In this arrangement the positive extremity of v_o is clamped to V_R.

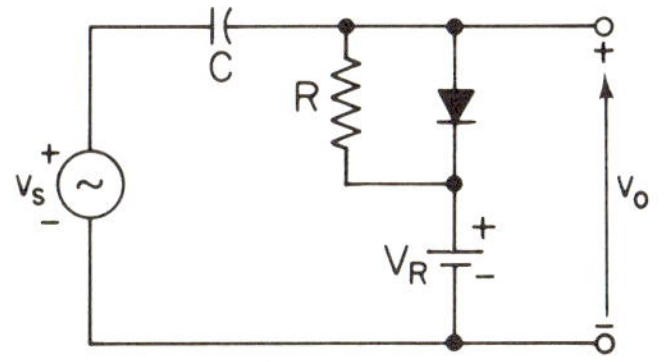

Fig. 4-53. A circuit that clamps the positive extremity of v_O to V_R.

PROBLEMS

4-1. For a particular current, the forward-bias voltage of a silicon *pn*-junction diode is 0.450 V at a temperature of 25°C. What voltage might be expected for the same forward current at 125°C?

4-2. An illuminated *pn*-junction germanium photodiode having a reverse-saturation current of 10 μA has a reverse current of 100 μA for a reverse voltage of 1.0 V. If the temperature is 300°K, what open-circuit voltage might the diode be expected to have with this illumination?

4-3. In the half-wave rectifier circuit shown in Fig. P4-3, input voltage v_s is a sinusoidal voltage and the average (d-c) load current is 1.0 A. Determine the approximate rms value of v_s.

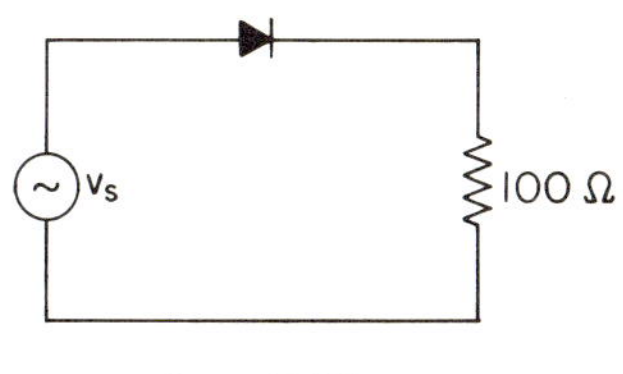

Figure P4-3

4-4. Repeat Prob. 4-3 for the bridge rectifier of Fig. P4-4.

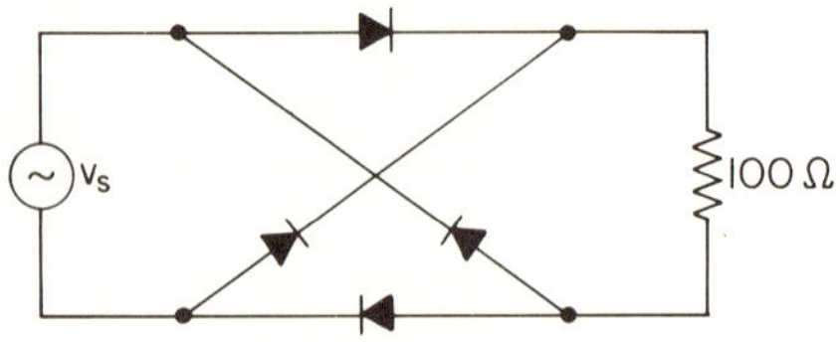

Figure P4-4

4-5. A half-wave rectifier is used to charge a 6-V battery as shown in Fig. P4-5. The supply voltage is $v_s = 12 \sin 377t$. V_o in the diode model of Fig. 4-30(b) may be considered to be 0.5 V.

(a) What value of resistance R is required to limit the peak current to 1.0 A?
(b) What percentage of the time does the diode conduct?
(c) What is the diode peak-inverse voltage?

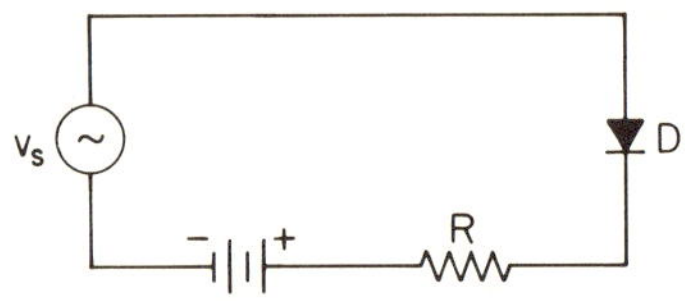

Figure P4-5

4-6. Determine the average (d-c) current in the circuit of Prob. 4-5 and the average power dissipated in the diode.

4-7. A half-wave rectifier, operating from a 60-cycle supply line, and a shunt-capacitor filter supply a d-c voltage of 100 V at a d-c load current of 10 mA. The filter consists of a 20 μF capacitor. Determine the approximate value of each of the following: (a) rectifier a-c supply voltage, (b) peak-to-peak output ripple voltage, (c) extinction angle α_2, (d) ignition angle α_1, (e) peak diode current, and (f) rectifier peak-inverse voltage.

4-8. Repeat Prob. 4-7 for a full-wave bridge rectifier.

4-9. The Zener-diode regulator of Fig. P4-9 is to supply a load current that ranges frm 0 to 20 mA. For good regulation, the diode current should not be less than 1.0 mA. Input voltage v_i has an average value of 24 V and a peak-to-peak ripple voltage of 2 V. Determine: (a) the maximum value of R, (b) the

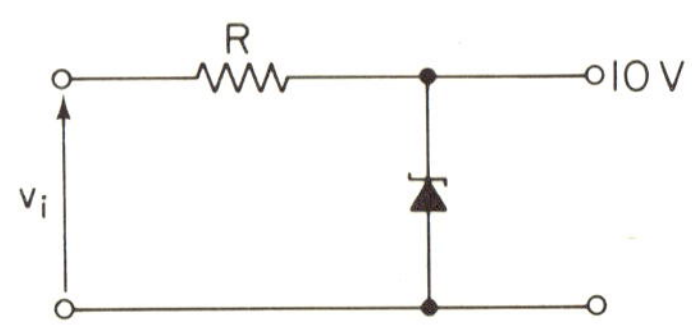

Figure P4-9

ripple current through the diode for the R value of (a), and (c) the peak-to-peak output ripple voltage for the R value of (a) if the Zener diode has a dynamic resistance of 10 Ω.

4-10. If the Zener diode of Prob. 4-9 has a power-dissipation rating of 250 mW, determine the range that the average value of v_i can have for regulation of a load current of 10 mA.

4-11. Using the experimentally determined information given in the text for the two-stage Zener-diode regulated power supply of Fig. 4-47, determine each of the following: (a) the d-c current of each diode when the load current is 150 mA, (b) the largest average power that each regulator diode may be required to dissipate, (c) the approximate dynamic resistance of each regulator diode, and (d) the approximate average voltage and peak-to-peak ripple voltage that might have been expected at the regulator input terminals (using approximations suggested in the text). Account for the difference between the values of (d) and those given in the text.

4-12. The output voltage v_L of the full-wave rectifier of Fig. 4-34 is applied to a peak-responding a-c electronic voltmeter. What voltmeter readings are expected if $V_s = 100$ V?

4-13. The output voltage of a pulse generator is shown in Fig. P4-13. If we measure voltage using a peak-responding a-c electronic voltmeter, what readings are expected?

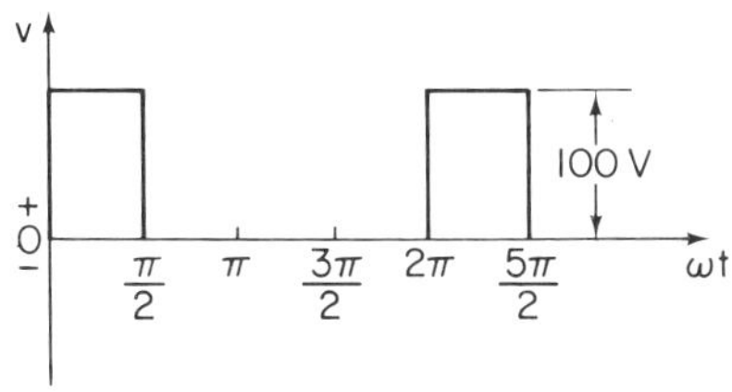

Figure P4-13

4-14. Repeat Prob. 4-13 if an average-responding a-c electronic voltmeter is used.

5

The Bipolar Transistor

The term *transistor*, a contraction of the words transfer resistor, was first applied in 1948 to a semiconductor control device discovered by Shockley, Bardeen and Brattain.* In this device, the current of one electrode is controlled by that of a second electrode and the device is the dual of a vacuum control device, being current controlled rather than voltage controlled. This type of transistor is referred to as a *bipolar transistor*, since its operation involves the movement of both majority and minority carriers, and it is the conventional transistor that has become so widely used. Other transistor types include the unipolar or *field-effect transistor* (FET) and the *unijunction transistor* (UJT), considered in Chapters 10 and 12. In current usage, the term *transistor* implies the bipolar device unless designated otherwise.

The bipolar transistor works on the principle that a forward-biased *pn* junction or metal-semiconductor contact, acting as a control element, injects excess minority carriers into the region adjacent to it, and a reverse-biased junction in (or contact to) the same material collects the injected carriers. Transistor classifications are similar to those of semiconductor diodes in that there are both *point-contact* and *junction* types, with junction devices classified according to the process used in forming their junctions. Early bipolar transistors were of the point-contact type, in which carrier

* J. Bardeen and W. H. Brattain, "The Transistor—A Semiconductor Triode," *Physical Review*, Vol. 74 (1948), p. 230.

injection and collection occurred through the use of closely spaced metallic whiskers in contact with a pellet of *n*-type material. However, this type was soon made obsolete by the development of the bipolar junction transistor (BJT), and we shall give point-contact transistors no further consideration.

The BJT consists of a layer of either *p*- or *n*-type semiconductor sandwiched between layers of opposite conductivity material, in either the *npn* or *pnp* configuration, with a lead making an ohmic contact to each layer. The outer layer in which the minority carriers originate is called the *emitter* region, the inner layer is the *base* region, and the other outer layer is the *collector* region. The base region through which injected minority carriers pass must be very thin, usually less than 10^{-3} cm, in order for effective transistor action to occur.

5-1 TYPES OF BIPOLAR JUNCTION TRANSISTORS (BJTs)

Bipolar junction transistors are produced by extension of the technologies for forming *pn* junctions. Thus we have *grown-junction* BJTs, *alloy-junction* BJTs, and *diffused-junction* BJTs.

Grown-junction BJTs are of the *double-doped* type, the *grown-diffused* type, or the *rate-grown* type and have a structure somewhat like that shown by Fig. 5-1(a). In producing double-doped grown BJTs, a crystal is pulled from a semiconductor melt that is doped first to form *n*-type material, then to form *p*-type material, and finally to re-form material of the *n* type. This method is obsolete, because of the difficulty in obtaining thin base regions, but grown-diffused BJTs are produced by a modification of the process.

In the grown-diffused process, the collector region of the crystal is grown from a lightly doped *n*-type melt. After the collector region is formed, both *p*-forming and *n*-forming impurities are added to the melt, and the rest of the crystal is grown. The region of the crystal grown last is the emitter and it is heavily doped *n*-type material because of the relatively high concentration of *n*-forming impurity atoms. Although the concentration of *p*-forming impurity atoms in the emitter region is too small to be significant, it is large compared to the impurity-atom concentration of the lightly doped *n*-region. Therefore, *p*-forming impurities diffuse across the n^+-n interface of the hot crystal and produce a thin *p*-region. This region is the BJT base.

Rate-grown BJT crystals are produced by first using a pulling rate that causes the impurity concentration in the crystal to be, say, of the *n*-forming type. Then the rate is changed to produce *p*-type material, and finally is changed to again form *n*-type material. After grown-junction BJT crystals are produced, they are sliced and diced, each die being a BJT.

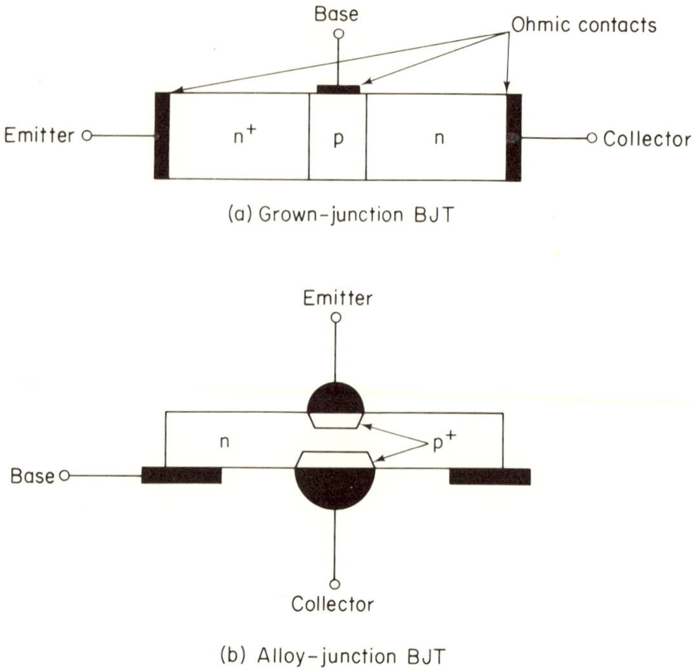

Fig. 5-1. Cross-sectional views of grown and alloy BJT structures.

Unlike grown-junction BJTs, alloy-junction units are fabricated individually. Each BJT is formed in a wafer of, say, lightly doped *n*-type material by placing each side of the wafer in contact with *p*-forming impurity material. When the assembly is placed in an alloying furnace, high-conductivity *p*-type material is formed on each side of the wafer with a thin *n*-region lying between the two p^+ regions. Alloy-junction *npn* BJTs are formed by using wafers of *p*-type semiconductor in contact with *n*-forming material. An important difference between alloy BJTs and those produced by other processes is that the alloy unit inherently has a high-conductivity collector region. The structure of the alloy BJT is somewhat like that shown by Fig. 5-1(b).

Diffused-junction BJTs are formed in a wafer of, say, lightly doped *n*-type material by exposing the heated wafer first to *p*-forming impurities in vapor form and then to *n*-forming impurities in the same form. The first diffusion process forms a thin *p*-type skin, the outer part of which is converted to high-conductivity *n*-type material by the second diffusion process. The wafer-surface areas affected by each diffusion process are controlled by masking techniques such as those described in Sec. 11-3 for integrated-circuit fabrication, and hundreds of BJTs can be produced simultaneously in a single wafer approximately 1 in. square. Individual devices are obtained by scribing and breaking the wafer into chips.* A cross-sectional view of the structure of a typical diffused-junction BJT is somewhat as shown by Fig. 5-2(a). This is a *passivated planar* structure, the term *passivated* meaning that junction surfaces are protected by a silicon-dioxide coating in order that the device not be contaminated by its environment. The term *planar* is

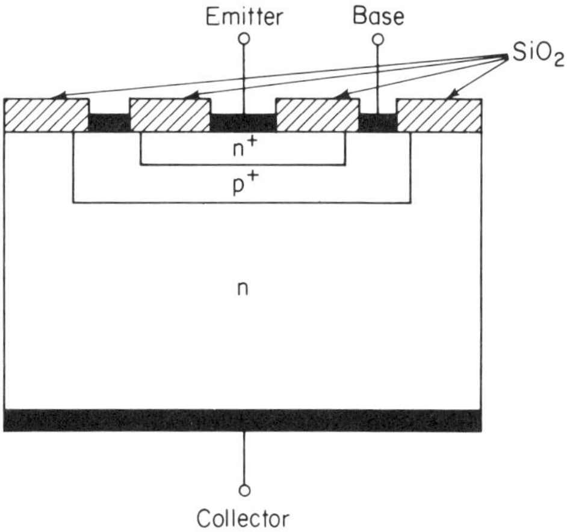

(a) Diffused-junction BJT

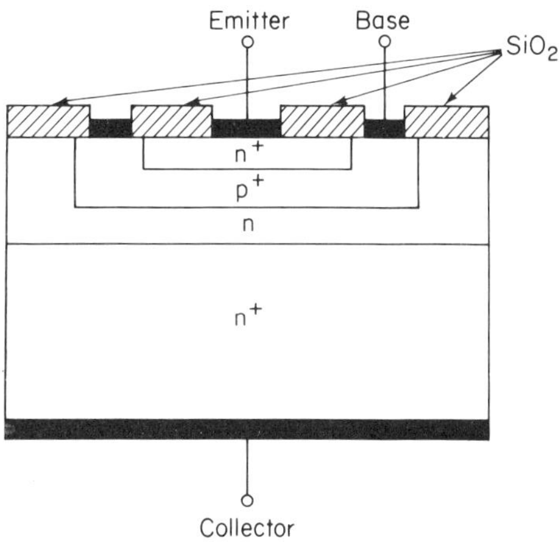

(b) Epitaxial diffused-junction BJT

Fig. 5-2. Cross-sectional views of diffused-junction BJT structures.

* The area of a BJT chip depends on the desired power dissipation rating, but it may be as small as 20 mils by 20 mils (a mil is 0.001 in.), this being the smallest chip that can be handled conveniently.

derived from the fact that the edges of both the emitter-base and the collector-base junctions lie in the same geometrical plane.

An inherent disadvantage of both the grown-junction BJT and the conventional diffused-junction BJT is the relatively high bulk resistance of the low-conductivity collector region. This is due to the impracticality of working with extremely thin semiconductor slices. This difficulty is alleviated in the diffused-epitaxial BJT shown in Fig. 5-2(b). In the epitaxial structure, the *npn* unit (or *pnp* as the case may be) is formed by vapor diffusion in a thin, lightly doped semiconductor layer, known as an *epitaxial* layer because the atoms are aligned in a continuation of the original crystalline structure, on a high-conductivity substrate. Another way of achieving low collector resistance is by forming the n^+ layer of Fig. 5-2(b) by diffusion from the bottom. Such devices are called *triple-diffused BJTs.*

5-2 BJT CHARACTERISTICS

Although the technologies for producing BJTs do not work equally well for *npn* and *pnp* types or for germanium and silicon devices, commerical *npn* and *pnp* BJTs in both germanium and silicon are available. In general, grown BJTs of both germanium and silicon are of the *npn* type; alloy germanium BJTs are either *npn* or *pnp*; alloy silicon BJTs are *pnp*; diffused germanium BJTs are *pnp*, and diffused silicon BJTs are either *npn* or *pnp* devices. The latter situation is particularly fortunate, since the diffusion technology is the most promising of the three and silicon devices are preferred in most applications.*

5-2.1 Basic Operating Principles

The normal bias polarities for amplifier operation of *npn* and *pnp* BJTs are shown in Fig. 5-3. In order for the emitter-base junction to be forward biased, the emitter is made negative with respect to the base for an *npn* BJT and positive for a BJT of the *pnp* type. Conversely, the back-biased collector-base junction requires that the collector of an *npn* BJT be positive with respect to the base, but the collector of a *pnp* BJT must be negative. In accordance with convention, BJT-electrode current is considered to be positive when *conventional current* flows from the external circuit into the electrode. Thus the directions of positive transistor currents are as indicated by the arrows

* Of the new transistor types introduced in recent years, almost all are diffused BJTs, and most are silicon devices.

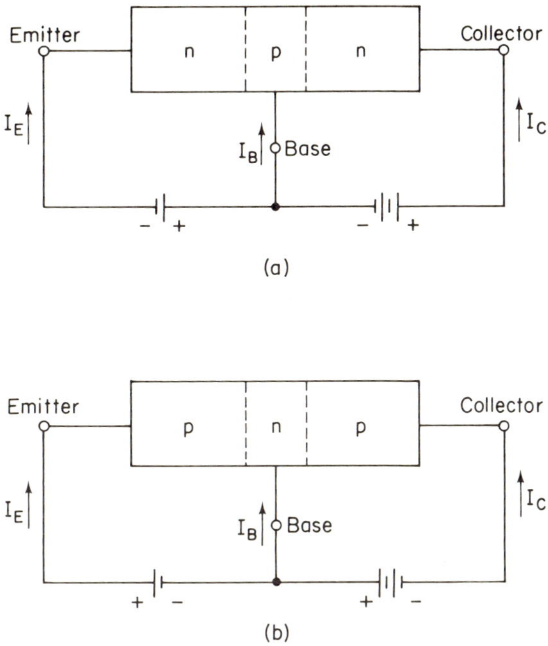

Fig. 5-3. Normal bias polarities for amplifier operation of (a) an *npn* BJT and (b) a *pnp* BJT. The directions of positive electrode currents are shown by arrows. From the polarities of the bias voltages, it is seen that the emitter-bias current I_E is negative for the *npn* BJT and positive for the *pnp* BJT. Conversely, the collector-bias current I_C is positive for *npn* BJTs and negative for *pnp* BJTs.

in Fig. 5-3. The d-c emitter current I_E is seen to be negative for an *npn* type. Conversely, the d-c collector current I_C is positive for an *npn* BJT and negative for a *pnp* device. From the circuits of Fig. 5-3, we see that direction of base-bias current I_B depends upon the relative magnitudes of I_E and I_C. For amplifier operation, I_E *usually* has the greater magnitude, making I_B positive for an *npn* BJT and negative for a *pnp* BJT.

BJT model. In our initial consideration of BJT operation, we shall use the one-dimensional model of an *npn* BJT shown by Fig. 5-4. Each of the semiconductor regions is assumed to be homogeneous and extrinsic, and the junctions are considered to be abrupt.* This model preserves the principal features of the BJT structures of Figs. 5-1 and 5-2, in which minority-carrier flow is *essentially normal* to the planes of the junctions. Although it neglects effects related to current flow in the transverse direction and those due to

* From the discussion of Sec. 2-1, we recognize that abrupt junctions are characteristic of alloy BJTs, but grown and diffused BJTs have graded junctions.

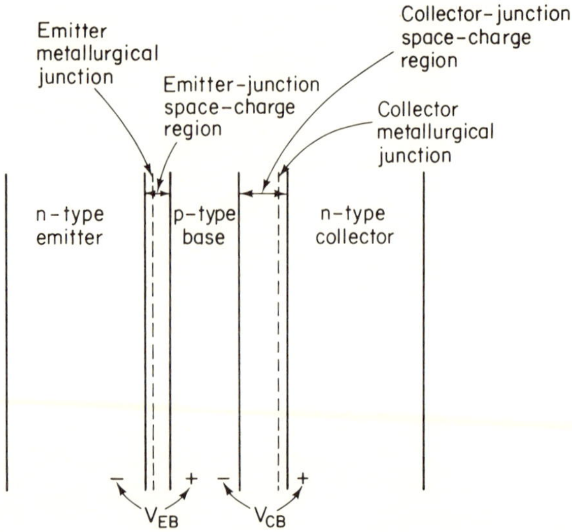

Fig. 5-4. An idealized physical model of an *npn* BJT showing polarities of terminal voltages for amplifier operation.

graded junctions, the one-dimensional assumption facilitates consideration of the mechanisms that are basic to all types of bipolar transistors. Effects neglected because of our choice of transistor model will be considered later.

Active-mode operation. As mentioned earlier, when a BJT is used as an amplifier, the emitter junction is forward biased and the collector junction is reverse biased. This operating condition is called *active-mode* operation. Other modes of operation relating to open-switch and closed-switch states will be discussed later. In active-mode operation, the minority-carrier concentrations at the edges of the emitter-junction region are increased above the equilibrium values, owing to the forward bias. Conversely, reverse bias causes the minority-carrier concentrations at the edges of the collector-junction region to be decreased to a negligible fraction of the corresponding equilibrium values. Thus the base minority-carrier concentration is higher at the emitter edge than at the collector edge, and the resulting concentration gradient causes electrons (for our *npn* BJT) to diffuse from emitter to collector. Upon reaching the collector edge of the base, diffusing electrons enter the collector-junction field and are swept through the junction region into the collector, causing the collector current to be larger than the reverse-saturation value. Since the electron concentration at the emitter edge of the base is strongly dependent upon the emitter-base voltage, as shown by Eq. (2-10), the base electron-concentration gradient and resulting electron-diffusion current are controlled by this voltage. Thus the collector current is controlled by the emitter-base voltage and the BJT can be used as an amplifier, since a large change in collector current can be produced by a small change in emitter-base voltage.

Minority-carrier distributions. The minority-carrier distributions in the three neutral regions of our idealized *npn* BJT of Fig. 5-4 have the general form shown by Fig. 5-5. The excess hole densities in the emitter and collector regions vary exponentially with distance from the junction regions, exactly as shown by Figs. 2-11 and 2-15 for a junction diode, if the emitter and collector contacts are several diffusion lengths away from their respective junctions. However, the electron distribution in the base is essentially linear, since the base width W is a very small fraction of a diffusion length and few electrons disappear by recombination while diffusing through the base. With negligible recombination, the electron-current density is the same throughout the base, requiring a uniform electron-density gradient and a linear electron-density distribution.

The notation used in Fig. 5-5 is a modification of that used for diodes, in that the subscript of the carrier symbol refers to the region of the BJT rather than to the material type. It should also be recognized that Fig. 5-5 is not drawn to scale, in that the equilibrium minority-carrier density p_{eo} in the emitter is usually a very small fraction of the base equilibrium-density n_{bo}, owing to the heavily doped emitter region. Thus the emitter minority-carrier density $p_e(0)$ at the junction edge is correspondingly smaller than the base minority-carrier density $n_b(0)$ at the emitter edge. Also, for typical operating emitter-base voltages, $p_e(0)$ and $n_b(0)$ may be several orders of magnitude greater than the corresponding equilibrium densities p_{eo} and n_{bo}.

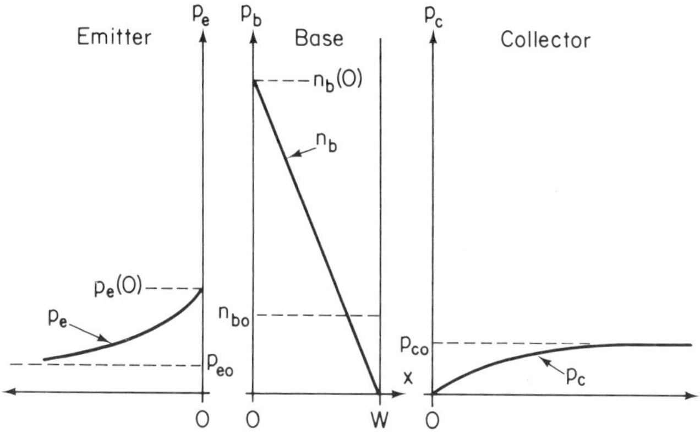

Fig. 5-5. Minority-carrier concentrations in an *npn* BJT biased in the active mode. The concentrations are above the equilibrium values at the edges of the forward-biased emitter-base junction and essentially zero at the edges of the reverse-biased collector-base junction.

The collector equilibrium minority-carrier density p_{co} may be larger or smaller than the base equilibrium density n_{bo}, depending on the technology used in producing the BJT.

Composition of terminal currents. In our initial consideration of BJT operation, we assume that the neutral *p*- and *n*-regions are field-free, and that no recombination or generation of carriers occurs in the space-charge layers. Therefore, electron and hole current flow in neutral regions at the edges of a space-charge layer occurs only by diffusion, and the electron and hole components of junction current are uniform within the layer. Thus, the emitter-junction current i_E is the sum of the diffusion current i_{pe} of holes injected into the emitter region and the diffusion current i_{ne} of electrons injected into the base. That is,

$$i_E = i_{pe} + i_{ne} \tag{5-1}$$

Because the base electron-density gradient is so much larger than the emitter hole-density gradient, current i_{ne} is much larger than i_{pe}, and the emitter-junction current consists primarily of electrons injected into the base. The ratio i_{ne}/i_E is an indication of the effectiveness of the emitter in injecting electrons into the base and is known as the *emitter efficiency* γ. That is,

$$\gamma \equiv \frac{i_{ne}}{i_E} = \frac{i_{ne}}{i_{ne} + i_{pe}} = \frac{1}{1 + \dfrac{i_{pe}}{i_{ne}}} \tag{5-2}$$

Typical BJTs have emitter efficiencies of the order of 0.99. A quantitative expression for emitter efficiency in terms of bulk properties of the emitter and base regions is derived in Sec. 5-3.

The collector-junction current i_C is also the sum of minority-carrier currents, consisting of holes extracted from the *n*-type collector region and electrons from the *p*-type base. With zero emitter current, the collector current is the collector-base junction reverse-saturation value I_{CBO}. For an emitter current i_E, i_C is increased by the current i_{nc} of electrons that are injected into the base and reach the collector junction by diffusion. Since the base width is a very small fraction of a diffusion length, as mentioned earlier, very few electrons disappear by recombination while diffusing through the base, and current i_{nc} is only slightly less than i_{ne}. The ratio i_{nc}/i_{ne} is an indication of the effectiveness of the base in transporting injected carriers and is known as the *base-transport factor* β. That is,

$$\beta = \frac{i_{nc}}{i_{ne}} \tag{5-3}$$

Typical BJTs have base-transport factors greater than 0.99. A quantitative expression for base-transport factor in terms of bulk properties of the base region is considered in Sec. 5-3.

Since both the emitter efficiency and the base-transport factor are very close to unity, the increase in collector current i_C above the I_{CBO} value is only slightly less than the emitter current i_E, and it is convenient to express i_C in terms of i_E. In this expression, the change in i_C is considered to have the magnitude $\alpha_{fb} i_E$, and i_C is given by

$$i_C = I_{CBO} + |\alpha_{fb} i_E| \tag{5-4}$$

The parameter α_{fb} is the *common-base forward current-transfer ratio* and is often referred to simply as the alpha of the BJT. Strictly speaking, α_{fb} is defined as the negative of the ratio of incremental changes in collector current and emitter current, with constant collector-base voltage; that is,

$$\alpha_{fb} = -\left.\frac{di_C}{di_E}\right|_{v_{CB}\text{ fixed}} \tag{5-5}$$

Incremental current changes are used in the α_{fb} definition because the effect on i_C of a change in i_E is somewhat dependent on the level of i_E, and the collector-base voltage v_{CB} is specified as constant bacause i_C and i_E are not completely independent of v_{CB}. Also, specifying constant v_{CB} implies that the current changes are initiated in the emitter circuit. The negative sign is necessary in Eq. (5-5), since α_{fb} is a positive quantity and i_C and i_E are of opposite algebraic sign. With α_{fb} defined as in Eq. (5-5), we see that Eq. (5-4) must be considered an approximate relationship. Although we have introduced α_{fb} at this point to facilitate discussion of the relationship between electrode currents, "alpha" is an important small-signal linear equivalent-circuit parameter that is considered in greater detail in Sec. 5-3.

The base current i_B of a BJT biased in the active mode is usually very small compared with either the emitter current or the collector current. This is so because the magnitude of i_B is the difference between the magnitudes of i_E and i_C, which are nearly equal. Although relatively small, i_B is an extremely important electrode current, and an understanding of the mechanisms that produce it is essential to understanding junction transistors.

Unlike emitter and collector currents, which are dominated by minority carriers that are injected at the emitter and diffuse through the base, base current results from the flow of majority carriers in the base region. In the active mode, the base current of an *npn* BJT is usually *positive*, corresponding to the flow of electrons *out* of the base region at the base terminal as holes are created at the base-region contact. The flow of holes into the base region from the base contact has three origins:

1. Some holes are injected into the emitter, resulting in the i_{pe} component of emitter current of Eq. (5-2).

2. Holes are required to combine with the electrons that disappear by recombination while diffusing through the base.

3. Holes are extracted from the *n*-type collector because the collector junction is reverse biased.

The components of base current resulting from holes injected into the emitter and those that are extracted from the collector can be determined by the static *pn*-junction diode theory of Secs. 2-4 and 2-5. The hole current resulting from recombination in the base can also be evaluated, since the approximate form of the electron distribution in the base is known. Clearly, this component of base current varies directly with the quantity of electrons stored in the base, as controlled by the emitter-base voltage, and is inversely proportional to the base-region lifetime.

BJT notation. Although our discussion so far of BJT operation is almost entirely in terms of the *npn* configuration, the operation of *pnp* devices is similar in every respect expect that the roles of electrons and holes are interchanged. The conventional symbol for each type of configuration is shown by Fig. 5-6, in which *e* designates the emitter electrode, *c* the collector, and *b* the base. The arrow used to represent the emitter electrode is drawn in the direction of forward-conventional emitter current. Thus the arrow is directed toward the base for the *pnp* BJT and away from the base for the *npn* type. Symbols for instantaneous electrode currents and voltages are shown in Fig. 5-6, and the conventional positive reference directions are indicated. In accordance with standard notation: (1) Lower-case letters denote either total instantaneous quantities or instantaneous values of variational (a-c) components, depending on the case of the subscript. (2) Upper-case letters denote either average (d-c) quantities or rms values of variational components, depending on the case of the subscript. (3) Upper-case subscripts denote either total instantaneous quantities or average quantities, depending on the case of the symbol. (4) Lower-case subscripts denote variational components. Since *e* is used as a subscript, confusion is avoided by using the letter *v* to denote voltage. Electrode voltages are best designated by double-subscript notation, since the reference electrode depends on how the BJT is used, although the second subscript may be dropped when the reference is understood. The additional subscript *M* or *m* is

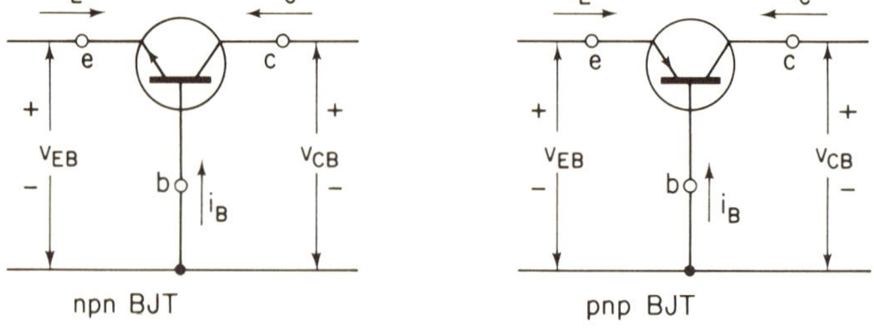

Fig. 5-6. Conventional symbols for BJTs including positive reference directions of electrode instantaneous voltages and currents.

used to represent maximum values, and the addition of subscript (AV) indicates an average value. Supply voltages are indicated by repeating the terminal subscript, and a third subscript may be used to distinguish the reference terminal. The following typical symbols indicate the system of notation.*

i_C	instantaneous total collector current
i_c	instantaneous value of variational (a-c) component of collector current
I_C	average (d-c) value of collector current
I_c	effective (rms) value of variational (a-c) component of collector current
$I_{C(\mathrm{AV})}$	average (d-c) value of collector current with signal applied
v_{CB}	instantaneous total collector-base voltage
v_{cb}	instantaneous value of variational (a-c) component of collector-base voltage
V_{CB}	average (d-c) value of collector-base voltage
V_{cb}	effective (rms) value of variational (a-c) component of collector-base voltage
V_{CC}	collector-circuit supply voltage
V_{CCB}	collector-base circuit supply voltage

As an example of this notation, for a BJT operating as a linear amplifier under sinusoidal conditions we may write

$$i_C = I_C + i_c = I_{C(\mathrm{AV})} + i_{cm} \sin \omega t = I_C + \sqrt{2}\, I_c \sin \omega t \tag{5-6}$$

and

$$\begin{aligned} v_{CB} &= V_{CB} + v_{cb} = V_{CB(\mathrm{AV})} + v_{cbm} \sin (\omega t + \phi) \\ &= V_{CB} + \sqrt{2}\, V_{cb} \sin (\omega t + \phi) \end{aligned} \tag{5-7}$$

5-2.2 Common-Base Characteristics

The relations between electrode voltages and currents of junction transistors may be shown by static characteristics. A family of static characteristics may be obtained by plotting any electrode current or voltage against any other, holding one of the remaining variables constant. Curves of collector current i_C versus collector-base voltage v_{CB} for either constant emitter current i_E or constant emitter-base voltage v_{EB} are known as common-base *collector characteristics.* Constant-emitter-current collector characteristics of typical *npn* BJTs are shown by Figs. 5-7 and 5-8.

* See *Standard Letter Symbols for Semiconductor Devices, IEEE No. 255.* New York: The Institute of Electrical and Electronics Engineers, December 1963.

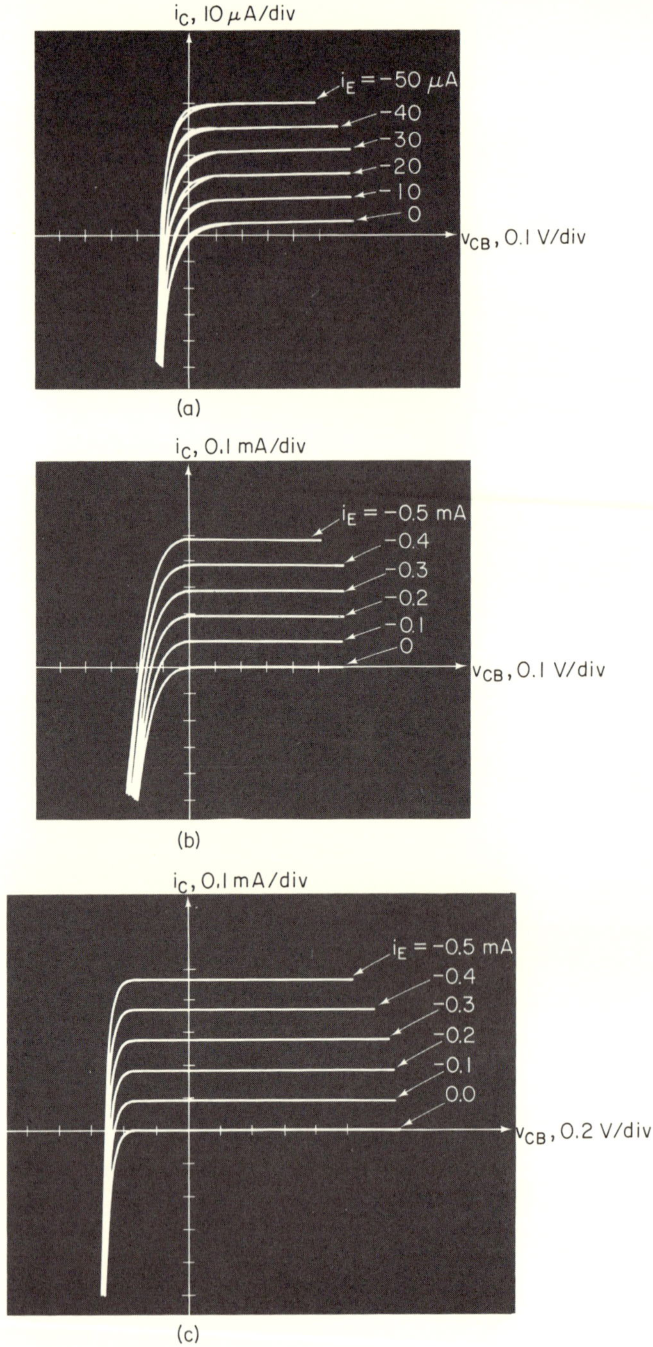

Fig. 5-7. Constant-emitter-current collector characteristics of typical *npn* transistors. (a) Typical 2N634A germanium BJT for collector currents comparable to the reverse-saturation value I_{CBO}. (b) BJT of (a) for collector currents large compared to I_{CBO}. (c) Typical 2N706 silicon BJT.

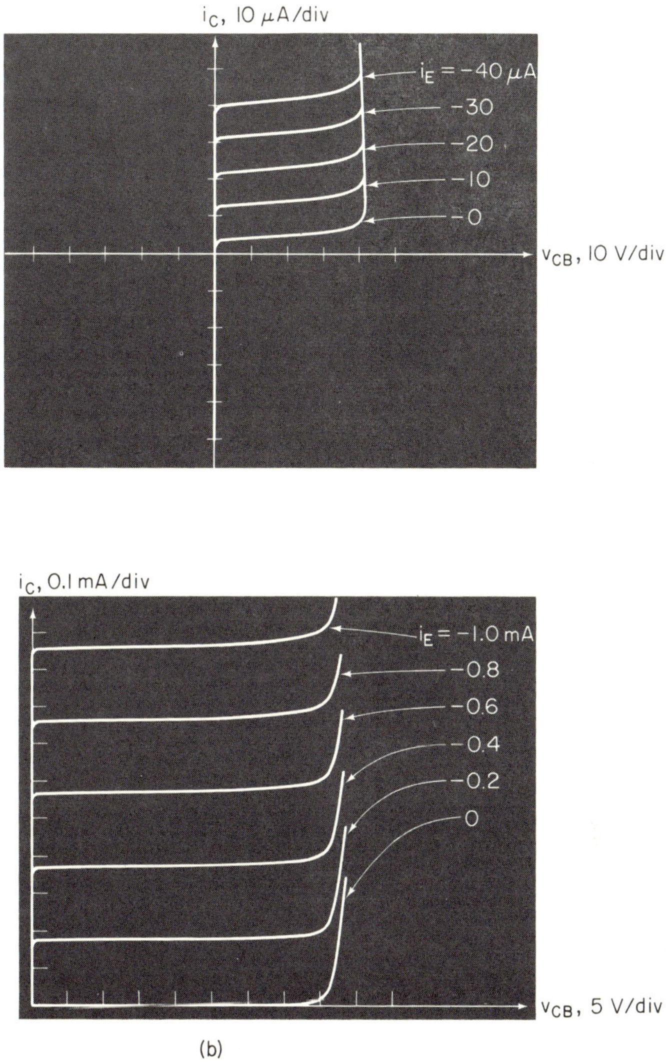

Fig. 5-8. Constant-emitter-current collector characteristics of a typical (type 2N634A) *npn* germanium BJT.

The curve family of Fig. 5-7(a) shows operation near the origin of the constant-i_E collector characteristic. The curve for $i_E = 0$ (the emitter terminal open-circuited) is the normal characteristic of a germanium *pn* junction. For negative collector voltages, the collector-base junction is forward biased and the current is strongly dependent upon v_{CB}. For positive voltages, the normal bias for active-mode operation, the junction is reverse biased and the collector current is essentially independent of v_{CB} for values greater than about 0.1 V, being the reverse-saturation value referred to as I_{CBO}. For the BJT of Fig. 5-7(a), I_{CBO} is seen to be about 5 μA. For negative i_E values, electrons injected into the base by the forward-biased emitter-base junction increase the collector current, and i_C is seen to be approximately $I_{CBO} + |i_E|$ for positive values of v_{CB}.* That is, in Fig. 5-7(a), 10-μA increments in $-i_E$ produce i_C increments of approximately 10 μA.

Figures 5-7(b) and 5-7(c) show constant-i_E collector characteristics near the v_{CB} origin for $-i_E$ values large compared to I_{CBO}. Again we see approximately equal increments of $-i_E$ and i_C, but of particular interest is the fact that, for fixed i_E, i_C is essentially independent of v_{CB} for values ranging down to zero and slightly below. Since a collector-base-junction field exists regardless of bias, being merely increased or decreased by bias, electrons diffusing through the base produce collector current even though v_{CB} is zero, and reducing i_C to zero necessitates forward-biasing the collector junction and injecting collector-region electrons into the base in order to produce an opposing electron flow. Thus we see that a forward-bias voltage of the order of tenths of a volt is required for zero i_C in Figs. 5-7(b) and 5-7(c). The linearity of collector characteristics at low voltages is an important attribute of the BJT. For example, BJT circuits commonly operate from a single $1\frac{1}{2}$-V supply battery.

The lack of complete independence of collector current i_C from collector voltage v_{CB} in active-mode operation is shown by Fig. 5-8. The increase of I_{CBO} with v_{CB}, as shown by the zero-i_E curve of Fig. 5-8(a), is due to collector-junction leakage current, and this is reflected in each of the fixed i_E curves of the figure. However, the increase in i_C with v_{CB} of the curves of Fig. 5-8(b) cannot be attributed to change in I_{CBO}, since I_{CBO} is negligible compared to i_C. Both sets of curves of Fig. 5-8 show abrupt increases in i_C when the collector-junction avalanche-breakdown voltage is reached, but the i_C increase with v_{CB} of Fig. 5-8(b) below avalanche breakdown is the result of decreased carrier recombination in the base.

The effects of collector voltage on base-charge storage and carrier recombination of an *npn* BJT operating with fixed emitter current may be seen from the base minority-carrier-concentration plots of Fig. 5-9. As we saw in Fig. 5-5, the base electron density n_b decreases approximately linearly from the value $n_b(0)$ at the edge of the emitter space-charge layer ($x = 0$) to virtually zero at the edge of the collector-charge layer ($x = W$). For a relatively low value V_1 of collector voltage v_{CB}, the edge of the collector-charge layer occurs at $x = W_1$, the effective base width for $v_{CB} = V_1$. As

* The discrepancy in Fig. 5-7(a) between i_C values and the values $I_{CBO} + |i_E|$ is less than the uncertainty of the measurements.

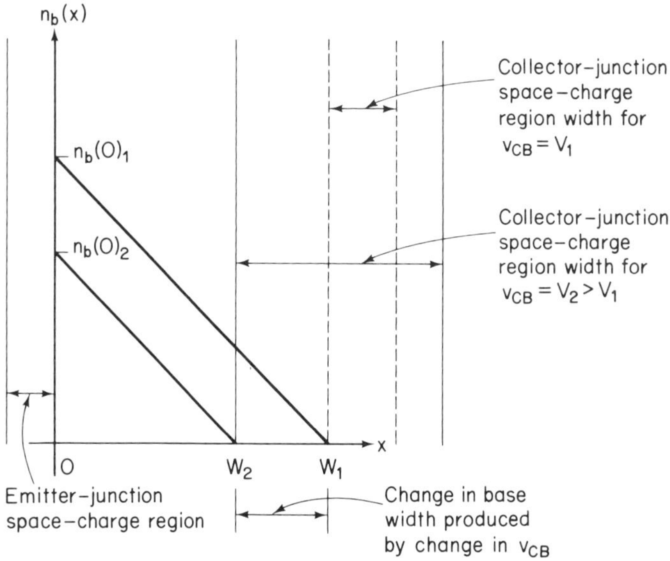

Fig. 5-9. The effect of collector-base voltage v_{CB} on the base width and the base minority-carrier distribution of an *npn* BJT operated with fixed emitter current i_E. Since the major component of i_E is proportional to the base electron-concentration gradient dn_b/dx, the slope of the $n_b(x)$ is essentially independent of v_{CB}.

v_{CB} is increased to V_2, the increased collector-charge-layer thickness reduces the base width to W_2, but the slope of the $n_b(x)$ plot is virtually unchanged, since the major component of i_E is proportional to the base electron-concentration gradient dn_b/dx and i_E is fixed. Therefore, $n_b(0)$ decreases to $n_b(0)_2$ and decreased electron storage in the base results in less carrier recombination. The change in W with collector voltage is known as *base-width modulation* and is often referred to as the Early effect.* The increase in i_C with v_{CB} as a result of base-width modulation is rather small, when i_E is fixed, but the decrease in i_B may be quite pronounced.

Another important effect of base-width modulation occurs if the *p*-region of an *npn* BJT (or the *n*-region of a *pnp* device) is so thin that the effective base width is reduced to zero by a collector voltage lower than the avalanche-breakdown value. The voltage that results in this condition is known as the reach-through voltage $V_{(RT)}$ (sometimes referred to as the punch-through voltage), since the collector space-charge region extends to the space-charge

* J. M. Early, "Effects of Space-Charge Layer Widening in Junction Transistors," *Proc. IRE*, Vol. 40 (1952), pp 1401–1406.

region of the emitter-base junction. The collector-emitter voltage of a BJT may be limited by the reach-through value, since direct collector-emitter conduction occurs for voltages larger than this value.

The decrease in the $n_b(0)$ value required for a particular emitter current, with increased collector voltage, suggests that emitter-junction forward conduction increases with v_{CB}, since

$$n_b(0) = n_{bo}e^{ev_{BE}/kT} \tag{5-8}$$

as may be deduced from Eq. (2-10). This may also be seen from the base minority-carrier concentration plots of Fig. 5-10 for an *npn* BJT operated with fixed emitter-base-junction voltage v_{EB}. For v_{EB} fixed, the $n_b(0)$ intercept of the $n_b(x)$ plots is unaffected by collector voltage and the decreased base width with higher v_{CB} results in a steeper $n_b(x)$ plot and a correspondingly larger electron-diffusion current. Thus the electron component i_{ne} of i_E increases with v_{CB}, but the hole-current component i_{pe} is unchanged.

Common-base emitter characteristic. The effect of the collector junction on the emitter-base-junction conduction characteristic of a typical *npn* germanium BJT is shown by Fig. 5-11. Figure 5-11(a) shows operation near the origin of the i_E versus v_{EB} curve family. The curve for $i_C = 0$ (the collector terminal open-circuited) is the normal characteristic of a germanium *pn* junction, the junction being reverse biased for positive v_{EB} and forward biased when v_{EB} is negative (normal emitter bias for active-mode operation).

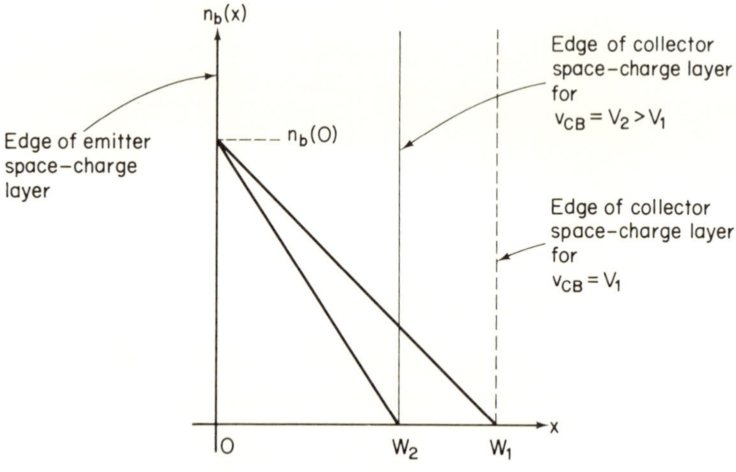

Fig. 5-10. The effect of collector-base voltage v_{CB} on the emitter current i_E of an *npn* BJT operated with fixed emitter-base voltage. The higher base-region electron-concentration gradient dn_b/dx, owing to lower base-width W_2, with higher v_{CB} results in a larger electron-diffusion-current component of i_E.

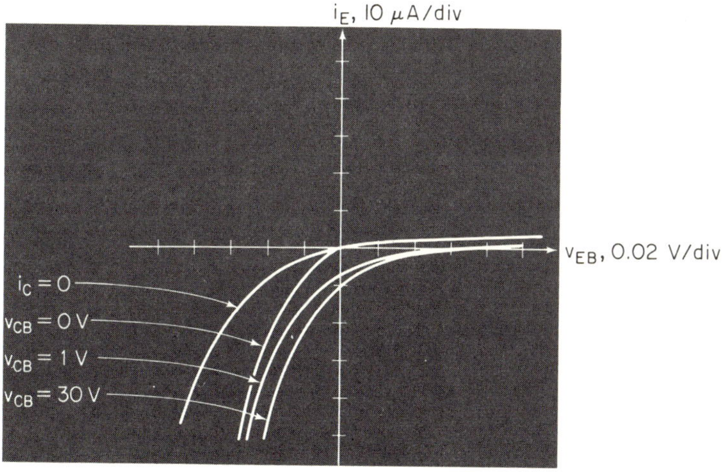

(a)

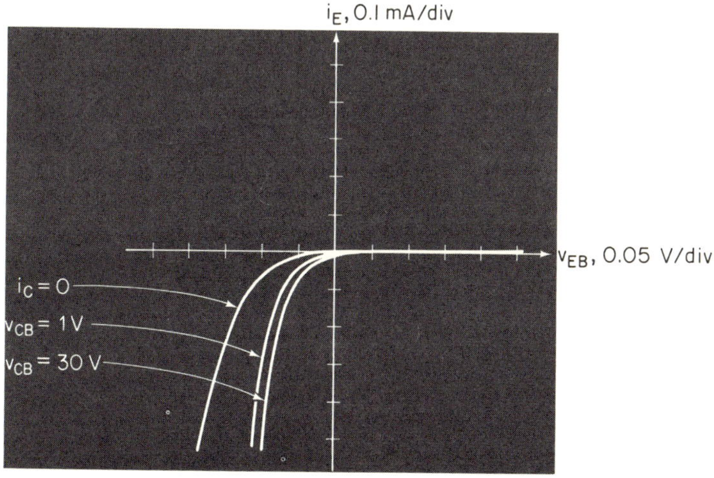

(b)

Fig. 5-11. The effect of the collector junction on the emitter-base-junction conduction characteristic (common-base emitter characteristic) of a typical (type 2N634A) *npn* germanium BJT.

With $v_{CB} = 0$ (the collector-base terminals short-circuited) the reverse-bias characteristic is seen to be the same as that for $i_C = 0$, but the forward-bias conduction characteristic is significantly improved, since most of electrons injected into the base leave by way of the nearby collector junction rather than through the relatively high transverse base resistance to the distant base contact. With v_{CB} positive (normal bias for active-mode operation) the emitter current is seen to have a negative intercept, owing to the effect of part of the collector reverse-saturation current flowing through the emitter junction with $v_{EB} = 0$, and the larger v_{CB}, the better the forward conduction characteristic becomes. Figure 5-11(b) shows i_E versus v_{EB} characteristic curves for i_E values large compared to the emitter-base junction reverse-saturation current.

A family of constant-collector-voltage i_E versus v_{EB} curves for a BJT is known as its common-base *emitter characteristic*. From the emitter characteristic of Fig. 5-11(b) the emitter-base voltage of a germanium BJT is seen to be of the order of a tenth of a volt. Silicon BJTs have somewhat higher emitter voltages, but the emitter voltages of both silicon and germanium devices are small enough that the power dissipated in the emitter junction is not usually an important consideration and the power rating of a BJT is the maximum collector-junction power that can be dissipated without overheating.

It is of interest to compare the constant-v_{EB} collector-characteristic curves of Fig. 5-12 with the constant-i_E curves of the same BJT. Differences of particular significance in the active-mode region (positive v_{CB}) are

1. Larger i_C for $v_{EB} = 0$ (emitter-base terminals short-circuited) than for $i_E = 0$ (emitter terminal open-circuited) as seen by comparing Figs. 5-7(a) and 5-12(a).
2. The dependence of i_C upon i_E is more nearly linear than upon v_{EB}, as seen by comparing the vertical spacings of the characteristic curves.
3. The effect of v_{CB} on i_C for fixed v_{EB} is appreciably greater than for fixed i_E, as seen by comparing the slopes of the curves of Figs. 5-12(b) and 5-8(b).

The increase in i_C above the I_{CBO} value (the value with the emitter open-circuited) when the emitter is connected to the base ($v_{EB} = 0$) is the result of part of the collector reverse-saturation current flowing through the emitter-base junction. This causes the emitter junction to be slightly forward-biased and minority carriers injected into the base enhance the collector current. This effect is very pronounced if the base is open-circuited so that the entire collector current flows through the emitter junction, as we shall see when *common-emitter collector characteristics* are considered.

Comparison of the effects on i_C of i_E and v_{EB} may be best shown by *common-base transfer characteristics* such as those of Fig. 5-13. Since i_C approximately equals i_E, which varies exponentially with the emitter-base junction voltage according to the diode equation, the i_C verse i_E characteristic is more nearly linear than the i_C versus v_{EB} characteristic. It is for this reason that the BJT is often considered to be current controlled rather than voltage

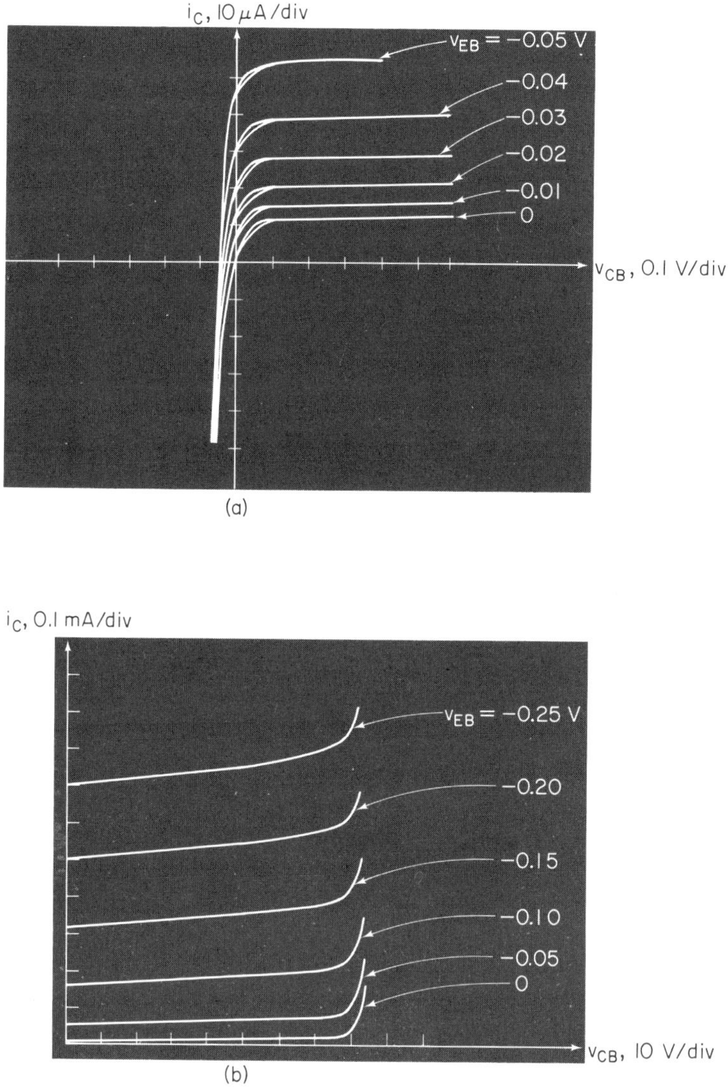

Fig. 5-12. Constant-emitter-voltage collector characteristics of the (type 2N634A) *npn* germanium BJT of Figs. 5-7 and 5-8.

controlled. The difference in the nonlinearities of the two transfer characteristics is seen to be less pronounced for higher collector currents, which require emitter currents large enough that the majority of v_{EB} is IR drop rather than junction voltage.

The slopes of the constant-v_{EB} collector characteristic curves of Fig. 5-12(b) (that is, the increases of i_C with v_{CB}) are greater than those of the

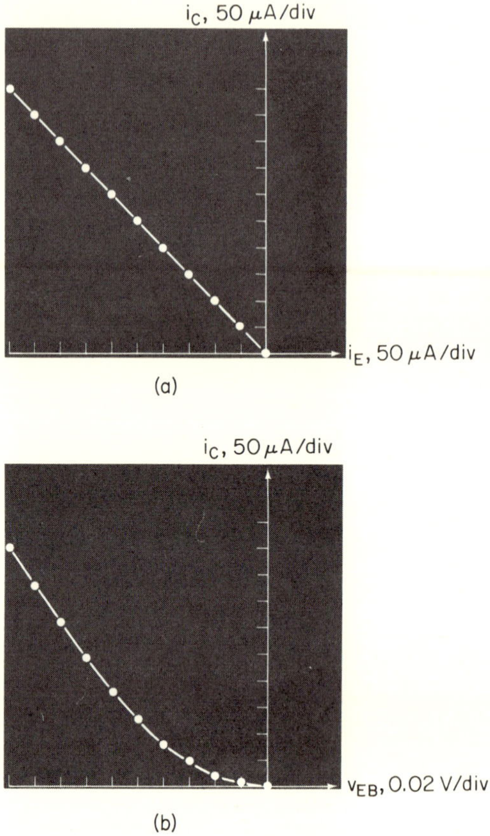

Fig. 5-13. Common-base transfer characteristics of the (type 2N634A) *npn* germanium BJT of Figs. 5-7 and 5-12. The i_C vs. i_E characteristic of (a) is virtually linear but the i_C vs. v_{EB} characteristic of (b) is quite nonlinear for low values of v_{EB}.

constant-i_E curves of Fig. 5-8(b), because i_E increases with v_{CB} when v_{EB} is fixed, as shown by the emitter-characteristic curves of Fig. 5-11.

Elementary BJT amplifier. The circuit shown in Fig. 5-14 is known as a *common-base amplifier*, since the base is common to the input and output

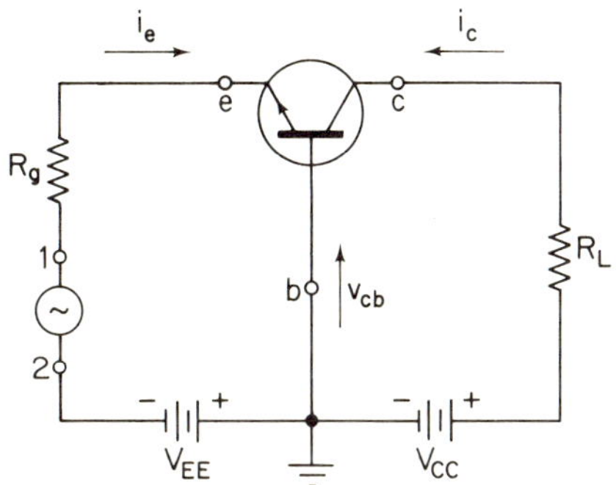

Fig. 5-14. Elementary common-base (grounded-base) amplifier circuit.

circuits. An *npn* BJT is shown; for the *pnp* type, the polarities of the d-c supplies are reversed. In this circuit, signal voltage v_{12} superimposed on the emitter-supply voltage V_{EE} produces an a-c component i_e of emitter current that causes an a-c component i_c of collector current. Alternating current i_c flowing through load resistance R_L results in an a-c component v_{cb} of collector voltage given by

$$v_{cb} = -i_c R_L \tag{5-9}$$

For a junction transistor, the amplitude of i_c is slightly less than that of i_e and a small current loss occurs. However, if R_L is appreciably larger than the load resistance of v_{12} (that is, larger than the sum of resistance R_g and the a-c resistance between the emitter-base terminals), v_{cb} is greater than v_{12} and voltage amplification occurs. In addition, the a-c power developed in R_L is greater than that delivered by the signal source, and a power gain occurs.

Determination of the quiescent operating point, given V_{EE}, V_{CC}, R_g, R_L, and the BJT characteristics, is accomplished by drawing appropriate load lines on the emitter and collector characteristics. Frequently only the collector characteristic is supplied by the manufacturer, and the approximate d-c emitter current I_E can be determined by assuming the emitter-base voltage negligible compared to V_{EE}; that is,

$$I_E \cong \frac{V_{EE}}{R_g} \tag{5-10}$$

When the a-c emitter-current component i_e is given, the a-c components of collector current and voltage can be obtained from the collector operating line. If a-c signal voltage v_{12} is given, the approximate value of i_e can be obtained by neglecting the BJT input impedance; that is,

$$i_e \simeq \frac{v_{12}}{R_g} \tag{5-11}$$

If required, the BJT input impedance can be obtained from the emitter characteristic.

5-2.3 Common-Emitter Characteristics

Since a current loss occurs in the common-base amplifier circuit of Fig. 5-14, power gain is dependent on the existence of voltage gain, and this occurs only if the load resistance is larger than the input resistance. In the interest of highest power gain, it is clear that both current gain and voltage gain are desired properties of amplifier stages. Both can be achieved in the common-emitter arrangement of Fig. 5-15, in which an input signal is simulated by an adjustable base-supply voltage V_{BB}.

The usual directions* of the *conventional currents* of the electrodes of an *npn* BJT operating in the active mode are shown in Fig. 5-15, indicating

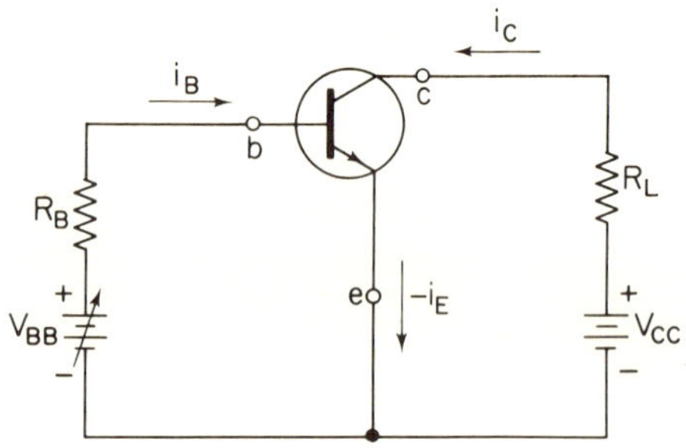

Fig. 5-15. Common-emitter BJT connection. The usual directions of the *conventional currents* of the electrodes of the *npn* BJT are indicated.

that the emitter current i_E is negative, the collector current i_C and base current i_B are positive, and

$$-i_E = i_B + i_C \tag{5-12}$$

* For small values of i_C, say an order of magnitude greater than I_{CBO}, the i_B of a high-alpha *npn* BJT may be negative.

Since i_E is negative, we obtain from Eq. (5-4)

$$i_C \cong I_{CBO} - \alpha_{fb} i_E \tag{5-13}$$

Substituting for i_E, we find that Eq. (5-13) becomes

$$i_C \cong I_{CBO} + \alpha_{fb}(i_B + i_C) \tag{5-14}$$

from which

$$i_C \cong \frac{I_{CBO}}{1 - \alpha_{fb}} + \frac{\alpha_{fb}}{1 - \alpha_{fb}} i_B \tag{5-15}$$

Since α_{fb} is slightly less than unity for a reasonably good BJT, the quantity $\alpha_{fb}/(1 - \alpha_{fb})$ is large compared to unity. Therefore, a small change in base current results in a relatively large change in collector current, suggesting that amplifier-input current be supplied to the base.

Equation (5-15) may be written as

$$i_C \cong I_{CEO} + \alpha_{fe} i_B \tag{5-16}$$

where I_{CEO} is the collector-to-emitter current with the base open-circuited, being given by

$$I_{CEO} \cong \frac{I_{CBO}}{1 - \alpha_{fb}} \tag{5-17}$$

and α_{fe} is the *common-emitter forward-current-transfer ratio* given by

$$\alpha_{fe} = \frac{\alpha_{fb}}{1 - \alpha_{fb}} \tag{5-18}$$

Since α_{fb} of commercial BJTs is very close to unity, deviating by no more than a few percent,

$$\alpha_{fe} \cong \frac{1}{1 - \alpha_{fb}} \tag{5-19}$$

and

$$I_{CEO} \cong \alpha_{fe} I_{CBO} \tag{5-20}$$

Equation (5-20) is the more useful I_{CEO} expression, since α_{fe} is more readily determined than $(1 - \alpha_{fb})$.

Curves of collector current i_C versus collector-emitter voltage v_{CE} for fixed base current i_B are shown for a typical *npn* germanium BJT by Fig. 5-16. This curve family is a *common-emitter collector characteristic* and is for the BJT whose common-base collector characteristic is given by Fig. 5-7(a). For v_{CE} values greater than a few tenths of a volt, the common-emitter characteristic is essentially as predicted by Eq. (5-16) in that a base current i_B causes an i_C increase that is approximately proportional to i_B.

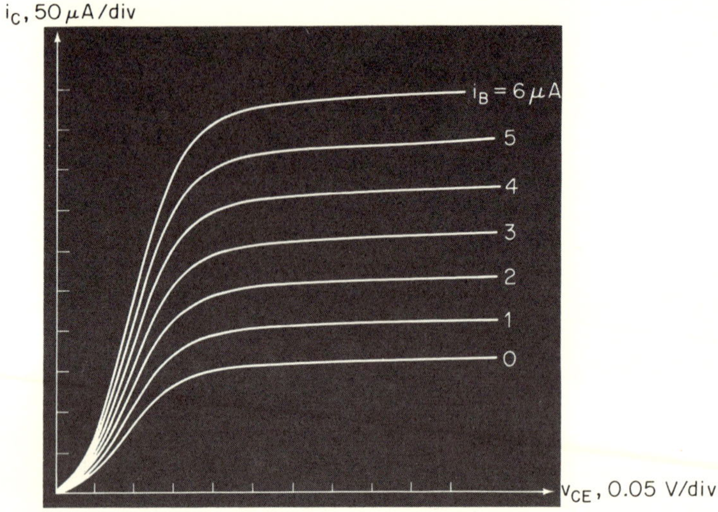

Fig. 5-16. Constant-base-current collector characteristic of the (type 2N634A) *npn* germanium BJT of Fig. 5-7(a).

Because of the effect of v_{CE} on i_C, the forward current-transfer ratio α_{fe}* is defined by

$$\alpha_{fe} = \left.\frac{di_C}{di_B}\right|_{v_{CE}\text{ fixed}} \tag{5-21}$$

and is seen to be about 40 for the BJT of Fig. 5-16 at $v_{CE} = 0.5$ V and an i_C of the order of I_{CEO}. Using this value of α_{fe} and the I_{CBO} value obtained from Fig. 5-8 (say 4 μA), we verify the I_{CEO} (I_{CBO}) relationship of Eq. (5-20).

At this point, let us look beyond the derivation of Eq. (5-16) in an effort to see why i_C is so much larger with the base open-circuited than with the emitter open and why i_C is so much more sensitive to i_B than to i_E. With the base open, the collector and emitter junctions are in series and a voltage v_{CE} divides between them, reverse-biasing the collector junction and forward-biasing the emitter junction. The emitter junction is forward biased because it must pass the collector reverse-saturation current I_{CBO}. The forward-biased emitter junction injects electrons into the base (for an *npn* transistor), enhancing the collector current, which increases the forward bias. This regenerative action continues until i_C reaches $I_{CBO}/(1 - \alpha_{fb})$, at which point stabilization occurs. Figure 5-17 shows the effect of permitting electrons to enter the base region by way of the base contact; that is, the effect of operating the BJT with negative base current. This reduces the collector current, i_C becoming

* Although not an approved IEEE symbol, β is widely used to designate the common-emitter forward current-transfer ratio. Other symbols frequently used are h_{21e} and h_{fe}.

I_{CBO} for $i_B = -I_{CBO}$. It is of interest that active-mode operation can occur with negative base current and that i_C is approximately proportional to i_B for negative magnitudes less than I_{CBO}. Operation in this region of the collector characteristic is perfectly acceptable for small base-input signals.

The strong dependence of i_C upon i_B is understood if we recognize that one of the major components of i_B is the hole component i_{pe} of emitter current. Since i_{pe} is a small fraction of i_E, and this fraction is fixed if the collector voltage is unchanged, a small change in i_{pe} is accompanied by a much larger change in i_E and in the resulting i_C. From another point of view, a change in i_B produces a change in electron-density gradient in the base and in the resulting diffusion flow of electrons through the base, since a significant part of any change in i_B is the change in the base-recombination component.

Further comparison of the common-base collector characteristics of Fig. 5-7 and the common-emitter characteristic curves discloses that the knees of the common-emitter curves occur at positive collector voltages, since v_{CE} includes the base-emitter voltage v_{BE}. For i_C values below the knee, the polarity of v_{CE} is controlled by that of v_{BE}, since the collector-base junction is forward biased (as shown by the negative-v_{CB} region of Fig. 5-7). That is,

$$v_{CE} = v_{BE} - |v_{CB}| \tag{5-22}$$

with v_{BE} being the larger of the two forward-bias voltages owing to $i_E > i_C$. This is known as the *saturation region* of the characteristic, and the voltage at the knee is known as the saturation voltage $V_{CE,\text{sat}}$. The upper limit of

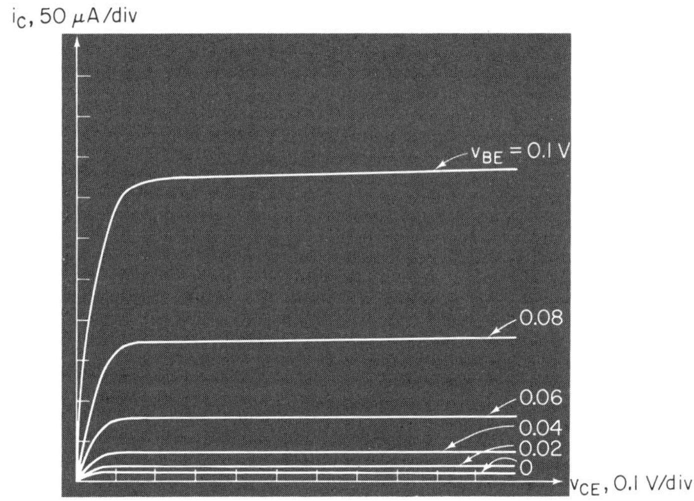

Fig. 5-17. Constant-base-current collector characteristic of the type 2N634A *npn* germanium BJT of Fig. 5-16 for both positive and negative base currents.

$V_{CE,\text{sat}}$, for specified values of i_B and i_C, is an important BJT specification, since it is the edge of the *active region.*

Another important difference between the common-base and common-emitter collector characteristics is the greater effect of collector voltage on active-region collector current in the common-emitter configuration, as may be seen by comparing the slopes of the curves of Fig. 5-18(a) with those of Fig. 5-8. With i_B fixed, any increase in i_C with v_{CE} produces an equal increase in i_E with further enhancement of i_C. Thus i_C is more dependent upon collector voltage when i_B is fixed than for fixed i_E.

Collector avalanche breakdown. Comparison of the collector characteristics of Figs. 5-8(b) and 5-18(b) shows that the beginning of the *collector-breakdown region* (upper-voltage edge of the active region) occurs at lower collector voltages for BJT operation in the common-emitter configuration. This is also shown in Fig. 5-19 for a silicon BJT. Collector breakdown in the common-base configuration is simply avalanche breakdown of the collector-base junction (Sec. 2-1), but common-emitter breakdown is a phenomenon that involves transistor action. Like diode reverse-breakdown, collector-base breakdown may be a small-area effect, and BJT characteristics may be permanently altered by breakdown currents much smaller than those that result in rated power dissipation. The *emitter-open* ($i_E = 0$) collector-base breakdown voltage BV_{CBO} is often considered to be the v_{CB} value for an i_C of the order of ten times I_{CBO}, since such an i_C value is not likely to damage the BJT.

A sharp increase in common-emitter i_C when v_{CE} is raised to a critical value could be due to the condition of *reach-through* (Secs. 2-1 and 5-2.2), but this is not the situation in the BJTs of Figs. 5-18(b) and 5-19(b), since breakdown voltage decreases with an increase in i_B. Instead, breakdown occurs because carrier generation in the collector space-charge region causes α_{fb} to increase to unity. The effect of unity α_{fb} may be seen from Eq. (5-17), which indicates that BJT operation in the common-emitter configuration is unstable if $\alpha_{fb} = 1$. The carrier generation responsible for unity α_{fb} is the phenomenon that causes collector-base avalanche breakdown, but it occurs to a lesser degree for $\alpha_{fb} = 1$. Collector-base avalanche breakdown may be though of as occurring when α_{fb} increases to infinity. The decrease in common-emitter breakdown voltage with increase in i_B is the result of α_{fb} increasing with i_C (Sec. 5-3.5) and requiring a smaller collector-junction field to become unity. Unlike common-base breakdown, common-emitter breakdown is not likely to damage the BJT unless the power-dissipation rating of the device is exceeded.

Because of the effect of i_B on collector breakdown, the breakdown voltage BV_{CEO} is specified for BJT operation with the base open ($i_B = 0$). If appreciable breakdown current flows v_{CE} may fall below BV_{CEO}, as in Fig. 5-19(b),

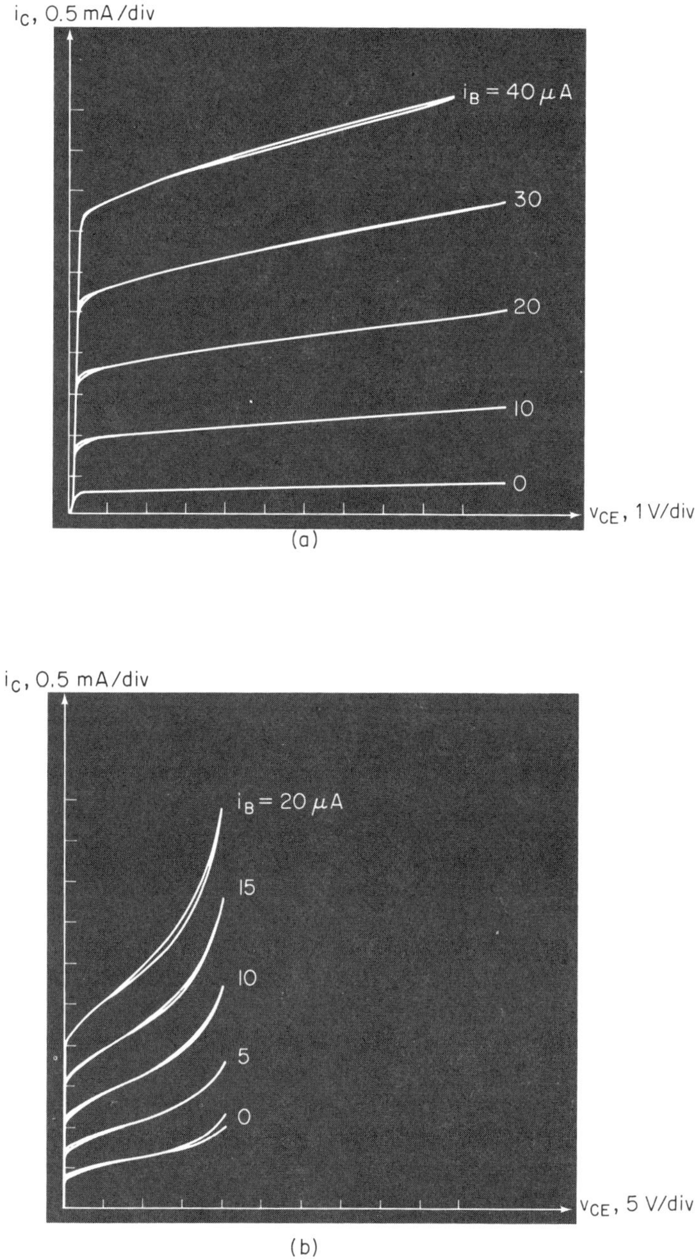

Fig. 5-18. Common-emitter collector characteristics of the (type 2N634A) *npn* BJT of Fig. 5-8. The v_{CE} scale is chosen so that the effect of v_{CE} on i_C may be seen.

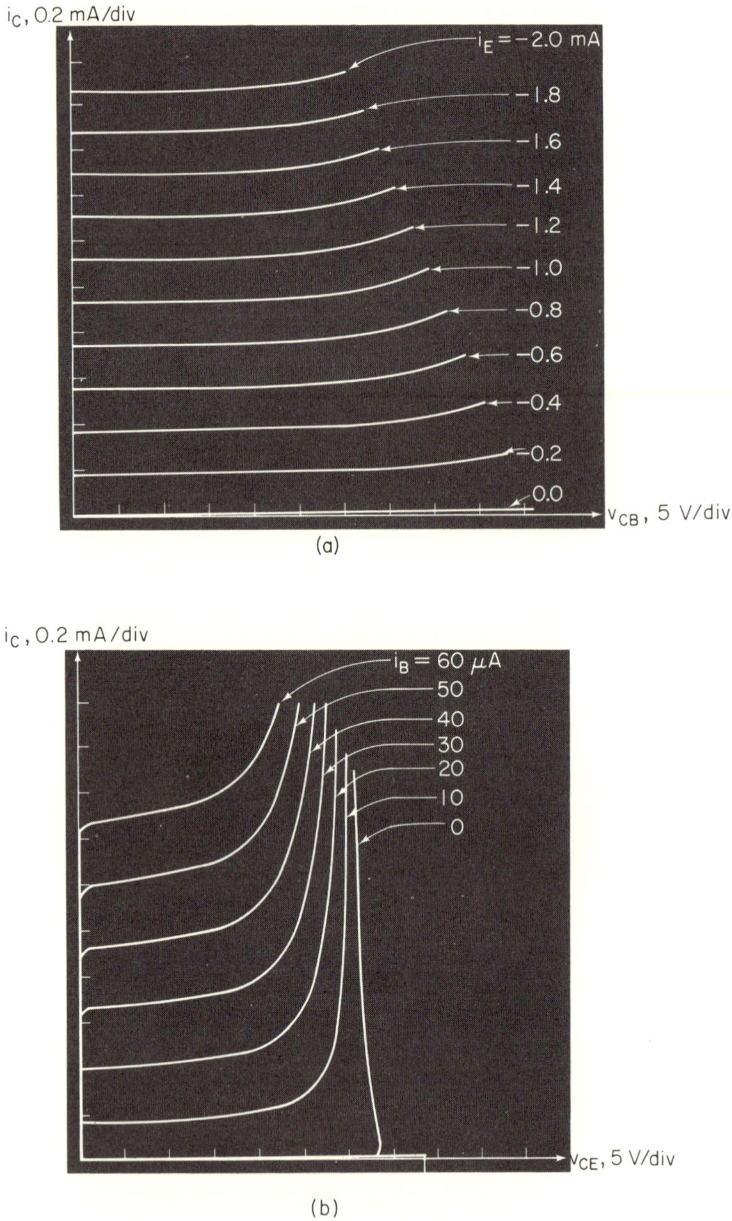

Fig. 5-19. Collector characteristics of the (type 2N706) *npn* silicon BJT of Fig. 5-7(c). (a) Common-base characteristic. (b) Common-emitter characteristic. Because I_{CEO} is indistinguishable from the v_{CE} axis in (b), the $i_B = 0$ curve indicates operation after breakdown.

owing to α_{fb} increasing with i_C and requiring a smaller v_{CE} value for unity. The collector voltage in the breakdown region is known as the sustaining voltage $V_{CE,\text{sus}}$. The difference between BV_{CEO} and $V_{CE,\text{sus}}$ depends on the sensitivity of α_{fb} to i_C at low currents, say of the order of I_{CEO}.* For common-emitter operation with $i_B > 0$ (that is, for active-mode operation), i_C is observed to increase sharply with v_{CE} as $V_{CE,\text{sus}}$ is approached [see Figs. 5-18(b) and 5-19(b)].

Base reach-through. Unlike the devices of Figs. 5-18 and 5-19, BJTs may have collector-emitter voltages limited by base-region reach-through. In order to test for reach-through, one measures the open-circuit emitter-base voltage when a voltage is applied between the collector and the base. If v_{CB} is greater than the reach-through voltage V_{RT}, an emitter-base voltage v_{BE} is observed and

$$v_{BE} = v_{CB} - V_{RT} \tag{5-23}$$

Thus the collector-emitter voltage is limited by either avalanche breakdown or reach-through, whichever voltage is lower.

Constant-v_{BE} characteristic. Figure 5-20 shows common-emitter collector-characteristic curves of a typical *npn* germanium BJT operated with fixed values of base-emitter voltage. Comparison of this characteristic with the

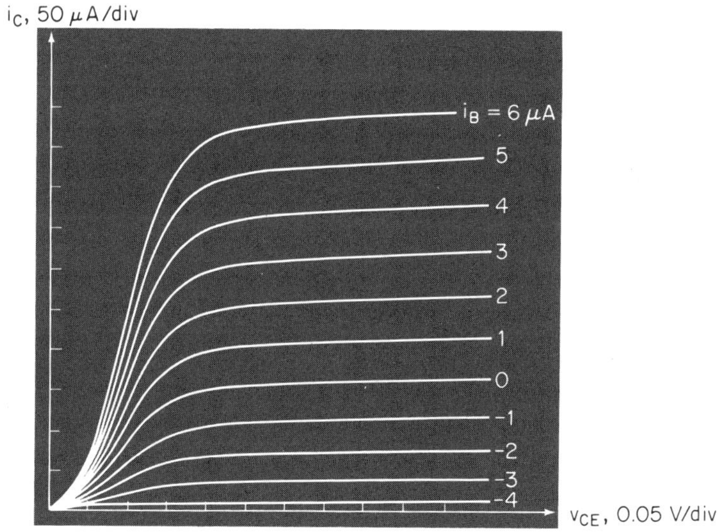

Fig. 5-20. Constant-base-voltage common-emitter collector characteristic of the type 2N634A *npn* germanium transistor of Fig. 5-17.

* Because I_{CEO} is indistinguishable from the v_{CE} axis in Fig. 5-19(b), the zero-i_B curve indicates operation after breakdown.

constant-i_B curves of Fig. 5-17 shows i_C to be a more nearly linear function of i_B than of v_{BE}. This is also seen by comparing the i_C versus i_B and i_C versus v_{BE} common-emitter transfer characteristics of Fig. 5-21.

The common-emitter base characteristic. The common-emitter base characteristic consists of a family of i_B versus v_{BE} curves for fixed v_{CE} values, as shown in Fig. 5-22 for a typical *npn* germanium BJT. Like the common-base emitter characteristic, the base characteristic is dependent upon collector voltage, but the effect is in the opposite direction. For v_{CE} equal to zero (collector-emitter terminals short-circuited), the base-conduction characteristic is that of the base-emitter and base-collector diodes in parallel with i_B dividing between the two junctions. When v_{CE} is positive (the normal bias for active-mode operation), v_{BE} has a positive intercept, since the base-emitter junction is forward biased with zero i_B as a result of collector current I_{CEO}. Because of this intercept, and because the emitter-junction current is much larger than

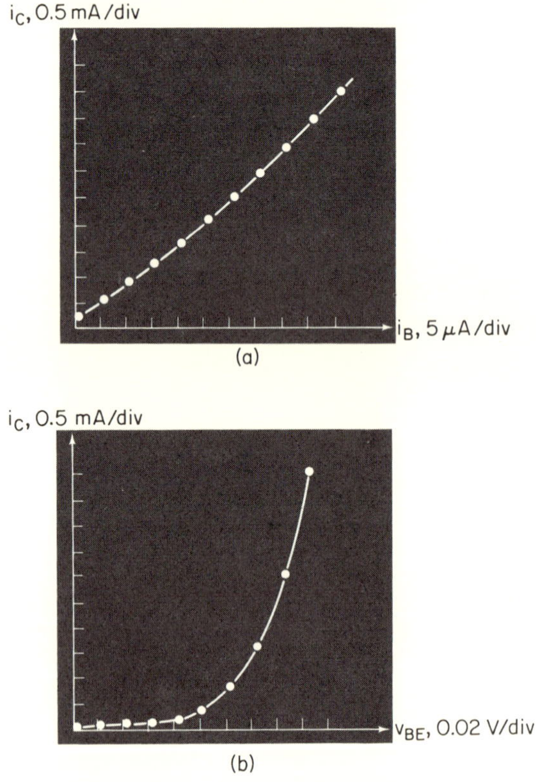

Fig. 5-21. Common-emitter transfer characteristics of the (type 2N634A) *npn* germanium transistor of Figs. 5-17 and 5-20. (a) $i_C - i_B$ characteristic. (b) $i_C - v_{BE}$ characteristic.

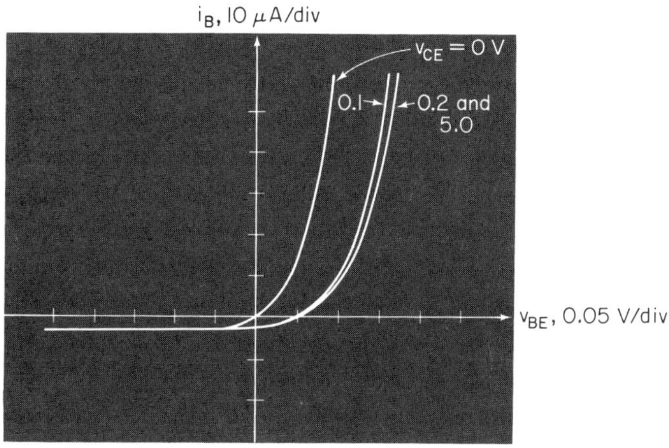

Fig. 5-22. Common-emitter base characteristic of a typical (type 2N634A) *npn* germanium transistor.

i_B when v_{CE} becomes positive, a significantly larger v_{BE} value is required for a particular i_B than when v_{CE} is zero. As v_{CE} becomes more positive, decreased recombination in the base with larger collector voltages further degrades the base-conduction characteristic, since a particular i_B value results in larger values of i_E and v_{BE}, but the effect is too small to be seen in Fig. 5-22 for v_{CE} values greater than a few tenths of a volt.*

The common-emitter amplifier. The elementary common-emitter amplifier stage shown in Fig. 5-23 can be analyzed graphically by constructing load

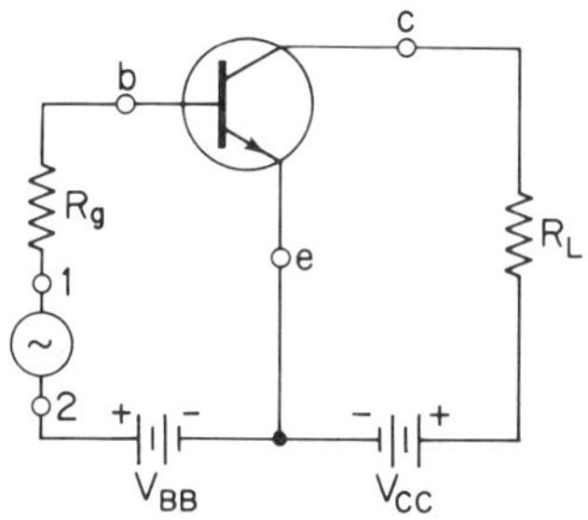

Fig. 5-23. An *npn* BJT operated as a common-emitter amplifier.

lines on the common-emitter base and collector characteristics. If the base characteristic is not given, the d-c base current I_B can be determined approximately by

* For a typical BJT operating at a v_{CE} of the order of volts, a change in v_{CE} of 1 V requires a v_{BE} change of the order of 1 mV for a particular value of i_B.

$$I_B \cong \frac{V_{BB}}{R_g} \tag{5-24}$$

provided $V_{BB} \gg V_{BE}$, and the a-c input current i_b by

$$i_b \cong \frac{e_{12}}{R_g} \tag{5-25}$$

if R_g is large compared to the transistor input impedance. Comparison of the common-base and common-emitter collector characteristics shows that the common-emitter arrangement requires a much smaller a-c input current for given a-c conditions in the collector circuit. It is also seen from the characteristics that, for operation in the linear region, the current gain is much greater than unity with the common-emitter arrangement and always less than unity for common-base operation. Since the two arrangements have comparable voltage gains, the common-emitter arrangement is the more useful type; in fact, it is the basic BJT configuration.

5-3 ALPHA OF BJTs

From the inception of the bipolar transistor, the symbol α has been used to designate the effect of the current of one electrode upon that of another and, in the consideration so far, we have designated the effect of i_E upon i_C as α_{fb} and that of i_B upon i_C as α_{fe}.* Since it is perhaps the most important parameter of the BJT, alpha deserves further consideration.

5-3.1 Factors Affecting Alpha

The common-base alpha, defined as

$$\alpha_{fb} = -\left.\frac{di_C}{di_E}\right|_{v_{CB}\text{ fixed}} \tag{5-5}$$

may be expressed as

$$\alpha_{fb} = \gamma\beta\delta \tag{5-26}$$

* Although α_{fb} and α_{fe} are IRE-approved symbols, their use is not recommended by the IEEE, the preferred symbols being $-h_{21b}$ or $-h_{fb}$ for α_{fb} and h_{21e} or h_{fe} for α_{fe} [see "Standard Letter Symbols for Semiconductor Devices," *IEEE No. 255* (Dec. 1963), The Institute of Electrical and Electronics Engineers, Inc., New York, New York]. Because the practice is so prevalent, we shall continue the use of the α designation in our consideration of BJT properties, but the h designation will be used in network considerations.

where γ is the emitter efficiency, β is the base-transport factor, and δ is the collector efficiency. The terms *emitter efficiency* and *base-transport factor* are introduced in Sec. 5-2.2 [see Eqs. (5-2) and (5-3)] and *collector efficiency* is alluded to in the common-emitter collector-breakdown discussion of Sec. 5-2.3.

As implied in Sec. 5-2, γ is an indication of the effectiveness of the emitter in injecting minority carriers into the base, β indicates the effectiveness of the base in transporting these injected minority carriers from the emitter junction to the collector junction, and δ indicates the effect on collector current of minority carriers reaching the collector junction by way of the base. On the basis of the α_{fb} definition of Eq. (5-5), γ, β, and δ of an *npn* BJT may be defined as

$$\gamma = \left.\frac{di_{ne}}{di_E}\right|_{v_{CB}\text{ fixed}} \cong \left.\frac{i_{ne}}{i_E}\right|_{v_{CB}\text{ fixed}} \tag{5-27}$$

$$\beta = \left.\frac{di_{nc}}{di_{ne}}\right|_{v_{CB}\text{ fixed}} \cong \left.\frac{i_{nc}}{i_{ne}}\right|_{v_{CB}\text{ fixed}} \tag{5-28}$$

and

$$\delta = \left.\frac{di_C}{di_{nc}}\right|_{v_{CB}\text{ fixed}} \cong \left.\frac{i_C}{i_{nc}}\right|_{v_{CB}\text{ fixed}} \tag{5-29}$$

where

i_{ne} = the electron current through the emitter junction.
i_E = the total emitter-junction current.
i_{nc} = the electron current through the collector junction.
i_C = the total collector-junction current.

For a *pnp* BJT, hole currents i_{pe} and i_{pc} replace electron currents i_{ne} and i_{nc} in these definitions.

The approximate relationships of Eqs. (5-27) through (5-29) are based on the assumption that current changes are proportional to current magnitudes. They are the basis of the derivations of Sec. 5-3.2 and 5-3.3 for γ and β in terms of bulk properties of the emitter and base regions. Additional factors affecting γ, β, and δ are d-c bias levels, temperature, and operating frequency.

5-3.2 Emitter Efficiency

The approximate minority-carrier distributions in the emitter and base regions of an *npn* BJT biased for active-mode operation are shown by Fig. 5-24. The emitter contact is considered to be several diffusion lengths away from the emitter junction, and negligible recombination in the base is implied by the linear electron-density plot for the base region.

The electron-current component i_{ne} of emitter current i_E is given by

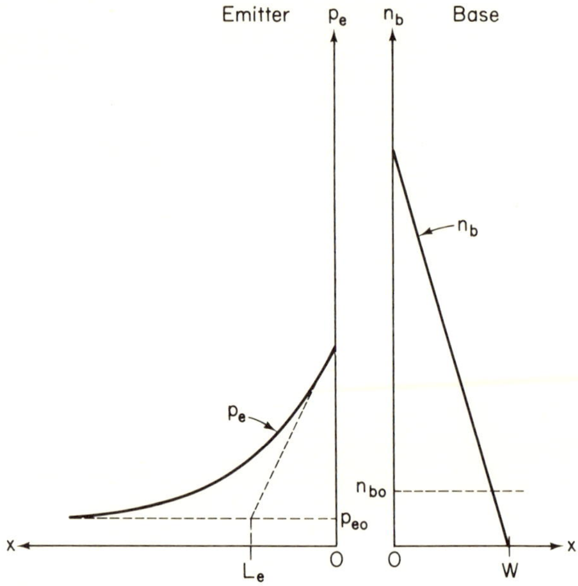

Fig. 5-24. Minority-carrier concentrations in the emitter- and base-regions of an *npn* BJT biased for active-mode operation.

$$i_{ne} = -eAD_b \left.\frac{dn_b}{dx}\right|_{x=0} \tag{5-30}$$

since i_{ne} is a diffusion current. The electron-density gradient in the base is seen to be approximately $-n_b(0)/W$, where

$$n_b(0) = n_{bo}e^{ev_{BE}/kT}$$

Therefore, Eq. (5-30) becomes

$$i_{ne} \cong \frac{eAD_b n_{bo}}{W} e^{ev_{BE}/kT} \tag{5-31}$$

The hole-current component i_{pe} of i_E is given by

$$i_{pe} = -eAD_e \left.\frac{dp_e}{dx}\right|_{x=0} \tag{5-32}$$

since i_{pe} is also a diffusion current. The hole-density gradient in the emitter at the junction edge is seen to be $[p_e(0) - p_{eo}]/L_e$, where

$$p_e(0) = p_{eo}e^{ev_{BE}/kT}$$

Therefore, Eq. (5-32) becomes

$$i_{pe} = -\frac{eAD_e p_{eo}}{L_e}(e^{ev_{BE}/kT} - 1) \cong -\frac{eAD_e p_{eo}}{L_e} e^{ev_{BE}/kT} \tag{5-33}$$

since the exponential term is usually large compared to unity.

Since the conventional emitter current i_E is the difference between the hole- and electron-current components, Eqs. (5-31) and (5-33) give us

$$i_E \cong -\left(\frac{eAD_e p_{eo}}{L_e} + \frac{eAD_b n_{bo}}{W}\right)e^{ev_{BE}/kT} \tag{5-34}$$

From Eqs. (5-31) and (5-34) and the approximate emitter-efficiency definition of Eq. (5-27), we see that

$$\gamma \cong \frac{(eAD_b n_{bo})/W}{(eAD_b n_{bo})/W + (eAD_e p_{eo})/L_e} = \frac{1}{1 + (D_e p_{eo} W / D_b n_{bo} L_e)} \tag{5-35}$$

For a *pnp* BJT, the emitter efficiency is given by

$$\gamma \cong \frac{1}{1 + (D_e n_{eo} W / D_b p_{bo} L_e} \tag{5-36}$$

In the use of these emitter-efficiency equations, it should be recognized that D_e and L_e are minority-carrier properties of the emitter region, D_b is the base-region minority-carrier diffusion constant, and all carrier densities are equilibrium minority values.

We see from Eqs. (5-35) and (5-36) that the smaller the ratios D_e/D_b, W/L_e, and (emitter minority-carrier density)/(base minority-carrier density), the higher the emitter efficiency. The term D_e/D_b expresses the relative ease with which minority carriers diffuse in the emitter and base regions, respectively. If this term is less than unity, as for an *npn* BJT, base minority-carrier flow is favored and γ is increased. The ratio W/L_e relates the minority-carrier concentration gradients at the junction. The smaller W is, the greater the base minority-carrier flow; the larger L_e is, the smaller the flow of minority carriers in the emitter. The emitter-base equilibrium minority-carrier density ratio relates the minority-carrier concentrations at the junction, which also relates the concentration gradients. As mentioned earlier, the emitter is normally heavily doped compared to the base in order that this ratio be very small, causing γ to be very slightly less than unity.

A somewhat more convenient expression for γ may be obtained by recognizing that diffusion constants are proportional to carrier mobilities and that minority-carrier concentrations are inversely proportional to majority concentrations [see Eqs. (1-34), (1-35), and (1-102)]. Thus Eq. (5-35) for the *npn* BJT may be written as

$$\gamma \cong \frac{1}{1 + (\mu_e p_b W / \mu_b n_e L_e)} \tag{5-37}$$

If we take the liberty of assuming that the base majority-hole mobility is equal to emitter minority-hole mobility μ_e, Eq. (5-37) becomes

$$\gamma \cong \frac{1}{1 + (\sigma_b W / \sigma_e L_e)} \tag{5-38}$$

where σ_b and σ_e are conductivities of the base and emitter regions.

5-3.3 Base-Transport Factor

To determine base-transport factor, we must obtain an expression for minority-carrier current in the base as a function of distance from the emitter junction. Since this is a diffusion-current expression, its determination requires a minority-carrier-concentration expression that includes the effect of recombination in the base. The plot of such an expression for an *npn* BJT is somewhat as shown by Fig. 5-25.* The decreasing slope of the $n_b(x)$ plot with increasing distance from the emitter junction is an indication of carrier recombination and its effect on electron diffusion current.

In obtaining an $n_b(x)$ expression for an *npn* BJT, we begin with the diffusion equation

$$\frac{dn_b}{dt} = D_b \frac{d^2 n_b}{dx^2} - \frac{n_b}{\tau_b} \tag{5-39}$$

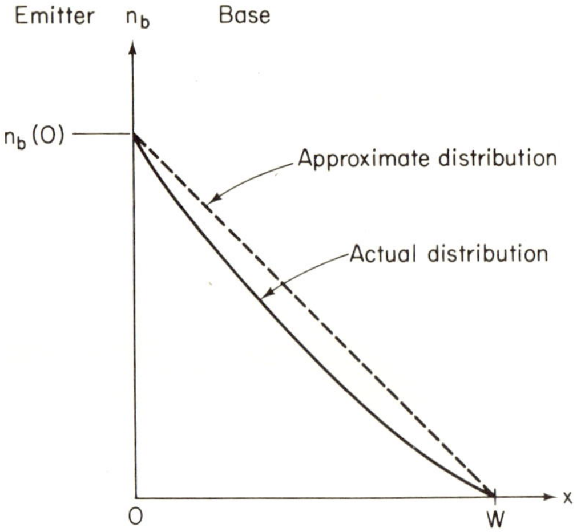

Fig. 5-25. Electron-density, n_b, distribution in the base of an *npn* BJT. Carrier recombination in the base is indicated by decreasing slope of the $n_b(x)$ plot with increasing distance from the emitter junction, the change in slope being greatly exaggerated.

* The deviation from linearity of the $n_b(x)$ plot of Fig. 5-25 is greatly exaggerated, since

$$\left| \frac{dn_b}{dx} \right|_{x=W}$$

is very slightly less than

$$\left| \frac{dn_b}{dx} \right|_{x=0}$$

for typical BJTs.

For low-frequency steady-state conditions, $dn_b/dt = 0$. Therefore,

$$D_b \frac{d^2 n_b}{dx^2} - \frac{n_b}{\tau_b} = 0$$

and

$$\frac{d^2 n_b}{dx^2} - \frac{n_b}{D_b \tau_b} = 0 \tag{5-40}$$

Since $L_b^2 = D_b \tau_b$ [see Eq. (1-60)], Eq. (5-40) becomes

$$\frac{d^2 n_b}{dx^2} - \frac{n_b}{L_b^2} = 0 \tag{5-41}$$

which has a solution of the form

$$n_b(x) = k_1 e^{-x/L_b} + k_2 e^{x/L_b} \tag{5-42}$$

The constants k_1 and k_2 may be determined from the boundary conditions

$$n_b(W) = 0 \tag{5-43}$$

and

$$\left. \frac{dn_b}{dx} \right|_{x=0} = -\frac{i_{nb}(0)}{AeD_b} \tag{5-44}$$

where $i_{nb}(0)$ is the electron current in the base at the emitter edge.* Applying these boundary conditions, we may show that

$$n_b(x) = \frac{-i_{nb}(0)L_b}{AeD_b(e^{2W/L_b} + 1)}[-e^{(2W-x)/L_b} + e^{x/L_b}] \tag{5-45}$$

Since

$$i_{nb}(x) = -AeD_b \frac{dn_b(x)}{dx} \tag{5-46}$$

we see from Eqs. (5-45) and (5-46) that

$$i_{nb}(x) = \frac{i_{nb}(0)}{(e^{2W/L_b} + 1)}[e^{(2W-x)/L_b} + e^{x/L_b}] \tag{5-47}$$

From Eq. (5-47) and the approximate base-transport factor definition of Eq. (5-28), we obtain

$$\beta \cong \frac{i_{nc}}{i_{ne}} = \frac{i_{nb}(W)}{i_{nb}(0)} = \frac{2e^{W/L_b}}{e^{2W/L_b} + 1} = \frac{2}{(e^{W/L_b} + e^{-W/L_b})} \tag{5-48}$$

Equation (5-48) may be simplified by expanding the exponential terms into a series of the form

$$e^x \cong 1 + x + \frac{x^2}{2} + \cdots$$

* Current $i_{nb}(0)$ is also i_{ne}, the electron-current component of the emitter current.

and recognizing that

$$e^x \cong 1 + x + \frac{x^2}{2} \quad \text{for } x \ll 1$$

Since $W \ll L_b$, we can write

$$e^{W/L_b} \cong 1 + \frac{W}{L_b} + \frac{W^2}{2L_b^2} \tag{5-49}$$

and

$$e^{-W/L_b} \cong 1 - \frac{W}{L_b} + \frac{W^2}{2L_b^2} \tag{5-50}$$

From Eqs. (5-48) to (5-50), we obtain

$$\beta \cong \frac{2}{2 + (W^2/L_b^2)} = \frac{1}{1 + \frac{1}{2}(W/L_b)^2} \tag{5-51}$$

verifying our earlier statement that β is very slightly less than unity if the base width is very small compared to the minority-carrier diffusion length of the base region.

5-3.4 Collector Efficiency

The mechanism of collector multiplication has been discussed in Sec. 5-2 in connection with collector breakdown. It will be recalled that this involves the generation of secondary carriers within the collector space-charge region as a result of the ionization of atoms by collision with accelerated primary carriers. These secondary carriers enhance the collector current, causing it to be greater than the primary current i_{nc} of carriers impinging on the collector junction, and the collector efficiency δ is greater than unity. Although no quantitative theory for this action exists, the effect is described fairly well for many BJTs by the empirical relationship

$$\delta \cong \frac{1}{1 - (v_{CB}/V_A)^n} \tag{5-52}$$

where V_A is the collector-base junction avalanche-breakdown voltage and the factor n is about 3. From this expression, we see that δ is essentially unity for $v_{CB} \ll V_A$ but rises above unity for v_{CB} values well below V_A. Since γ and β are usually very slightly below unity, δ may cause α_{fb} to exceed unity over an appreciable range of collector voltage.

There is another collector-multiplication mechanism that is virtually independent of collector voltage. It is recalled that part of the collector cutoff current I_{CBO} is the result of minority carriers thermally generated in the collector region. These carriers reach the collector junction by diffusion, and only those generated within a few diffusion lengths of the junction affect I_{CBO}. However, for collector currents appreciably greater than I_{CBO}, the small electric field produced in the collector bulk region by IR drop accelerates minority carriers toward the junction. This is a second-order effect, but it enhances i_C significantly and causes δ to rise above unity.

5-3.5 Bias Dependence of Alpha

It was originally supposed that the common-base alpha (α_{fb}) of a BJT should be virtually independent of d-c operating conditions, this conclusion being based on the assumption that the concentration of minority carriers injected into the base is small compared to the majority-carrier density. In reality, the base minority-carrier density becomes significant at moderate current levels and may be of the order of the majority-carrier density at high currents, causing alpha to be current dependent. In addition, small changes in α_{fb} that might appear insignificant are of primary importance because of their large effects on α_{fe}, as seen from the expression

$$\alpha_{fe} = \frac{\alpha_{fb}}{1 - \alpha_{fb}}$$

by recognizing that α_{fb} is usually slightly less than unity.

Effects of collector voltage. The increase in α_{fb} with collector voltage as a result of the effect of collector multiplication on collector efficiency δ has been considered in Sec. 5-3.4. In addition to δ, both emitter efficiency γ and base-transport factor β increase with collector voltage as a result of base-width modulation. As pointed out in Sec. 5-2, base width W decreases with increasing collector voltage, and this results in an increased minority-carrier-concentration gradient in the base, increasing the electron component of the emitter current.* Since the hole component of i_E is unaffected, γ is increased. Decreasing W also decreases electron storage in the base, decreasing recombination and increasing β. The effects of W on γ and β are given quantitatively by Eqs. (5-38) and (5-51).

Another effect of collector voltage on alpha occurs at very low voltages for which it is no longer possible to assume zero electron density at the collector junction. As v_{CB} is decreased to zero and made negative, increased electron density at the collector-junction edge of the base decreases both γ and β. This effect is referred to as *collector injection* and is the predominant factor affecting the common-emitter collector characteristic for v_{CE} values below the knees of the characteristic curves. The effects of collector voltage on α_{fb} are summarized in the sketch of Fig. 5-26.

Because changes in α_{fb} with operating conditions are so small, factors affecting α_{fb} are usually investigated through their effects on either α_{fe} or the quantity $(1 - \alpha_{fb})$. The continuous increase of α_{fe} with collector voltage may be seen from the common-emitter collector characteristics of Figs. 5-17 and 5-18, being indicated by the increasing vertical separation of the

* The *npn* transistor continues to be the basis for discussion.

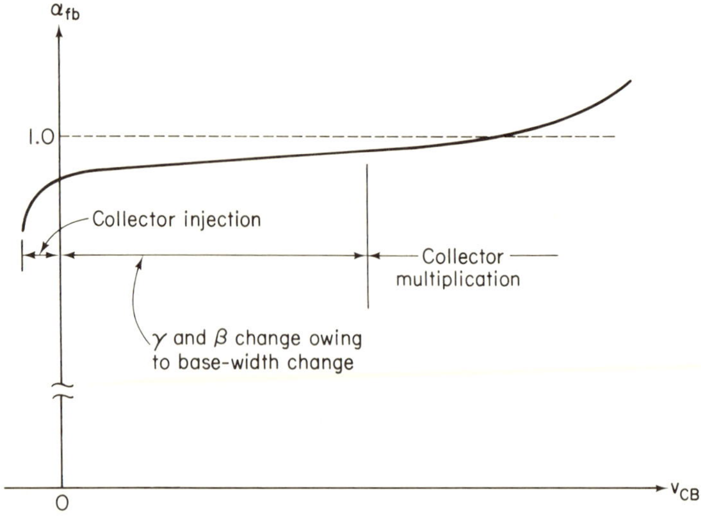

Fig. 5-26. The variation of common-base alpha with collector voltage and the principal factors affecting each region of the $\alpha_{fb}(v_{CE})$ plot.

characteristic curves as v_{CE} increases. Plots of α_{fe} and $(1 - \alpha_{fb})$ versus collector voltage are shown by Fig. 5-27 for a typical *npn* BJT. Careful experimental investigation shows close agreement between the common-emitter collector-breakdown voltage of a particular BJT and the v_{CB} value for which $(1 - \alpha_{fb})$ is zero.

Effects of current level. The increase in α_{fb} with collector current caused by the effect on collector efficiency δ of *IR* drop in the collector region has been considered in Sec. 5-3.4. Clearly, this effect is more pronounced in grown and diffused BJTs, with their characteristically low collector-region conductivities, than in high-collector-conductivity alloy BJTs.

In addition to δ, both emitter efficiency γ and base-transport factor β are dependent upon current level, owing to the effect of current on *majority-carrier* concentration in the base. As pointed out in Sec. 4-5, diode operation is affected by the increases in neutral-region majority-carrier concentrations that occur with forward bias, and it is logical to expect an effect on transistor operation.

Figure 5-28 shows the distribution of each type of charge present in the base region of an *npn* BJT. In addition to minority-electron concentration previously shown in base-region charge-distribution plots, we show acceptor-ion concentration, majority-hole concentration produced by acceptor atoms, and majority-hole concentration resulting from electrons injected into the base. In the absence of injected electrons (that is, without emitter bias), the density of negative acceptor ions and that of positive majority-holes are

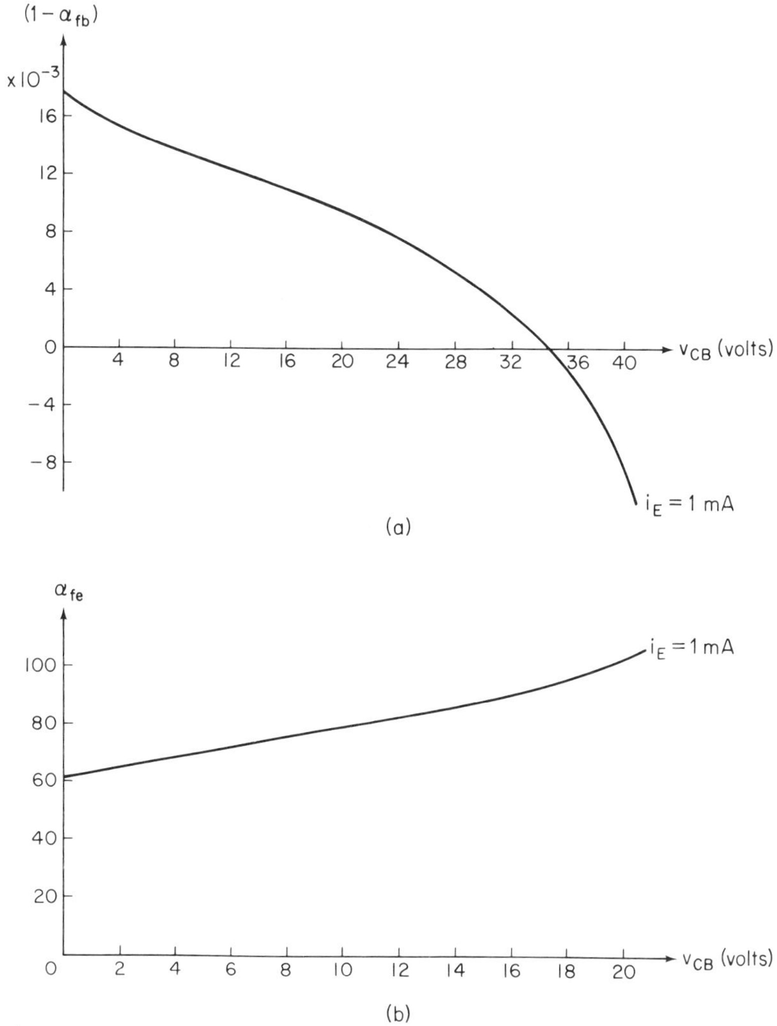

Fig. 5-27. Effect of collector voltage on: (a) $(1 - \alpha_{fb})$ and (b) α_{fe} of a typical (type 2N634A) *npn* germanium BJT. (Note the smaller voltage scale in (b).)

equal at every point in the base and electrical neutrality (zero electric field) exists. When electrons are injected into the base as a result of emitter bias, the familiar linear electron distribution exists with electron density ranging from $n_b(0)$ at the emitter edge of the base to zero at the collector edge.* This

* For purposes of illustration, Fig. 5-28 represents injected minority-carrier density $n_b(0)$ as larger than the equilibrium majority-carrier density. This is not the usual situation, although it may be true at high current levels.

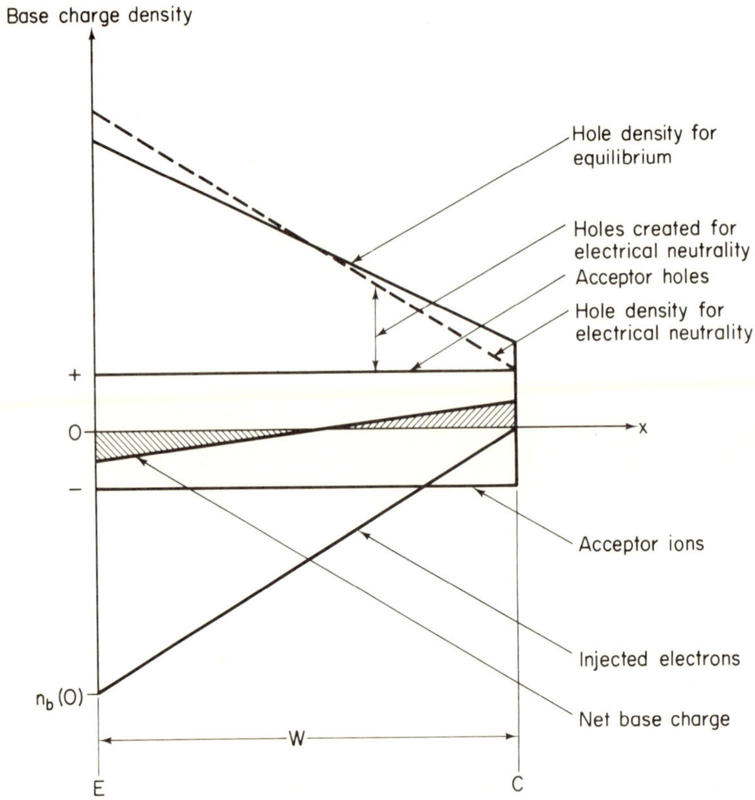

Fig. 5-28. Distribution of each type of charge present in the base of an *npn* BJT based for active-mode operation. The net charge in the base established an electric field that enhances electron flow across the base from emitter *E* to collector *C*.

large increase in negative charge in the base causes an increase in hole density, holes being created by electrons flowing out the base contact to the external circuit. If electrical neutrality were to exist throughout the base, the increase in hole density at each point would equal the increase in electron density, and the hole distribution would be as shown by the electrical-neutrality plot of Fig. 5-28. Such a distribution results in a hole-concentration gradient that produces a diffusion flow of holes toward the collector edge of the base, reducing the density at the emitter edge. Unlike electrons, holes arriving at the junction by diffusion are unable to leave the base, being repelled by the collector-space-charge field, and the hole density increases at the junction edge. This increase, and a corresponding decrease in hole density at the emitter edge of the base, establishes an electric field in the base that causes a drift flow of holes from the collector edge of the base to the emitter edge. When equilibrium conditions are established, this drift flow of holes equals the diffusion flow and the net hole current across the base is zero. The hole distribution in the base for equilibrium and the net charge distribution that results are also shown in Fig. 5-28.

The net charge in the base of the *npn* BJT is seen to cause the collector edge of the base to be positive with respect to the emitter edge. Thus the electric field in the base accelerates injected electrons across the base to the collector junction. This drift field is quite negligible at low current levels, but when the injected-electron concentration becomes comparable with the hole concentration, the field significantly enhances electron flow to the collector. This reduces the charge stored in the base (per unit current), decreasing recombination and increasing the base-transport factor.

Another factor causing alpha to increase with current in the low-level region is emitter-junction leakage current. This component of emitter current causes no injection of carriers into the base and may be a significant component of i_E at low levels, causing the emitter efficiency to be low. This effect and the change in base-transport factor combine to cause a pronounced increase in alpha with current, at moderately low levels, as may be seen from the common-emitter collector characteristics of Figs. 5-17 and 5-18, in which α_{fe} increase with current is indicated by the larger vertical separation of the constant-base-current curves in the higher i_C regions. The α_{fe} versus i_E plots of Fig. 5-29 for several typical BJTs also show α_{fe} increasing with current, but they further show this to be a low-level phenomenon with alpha

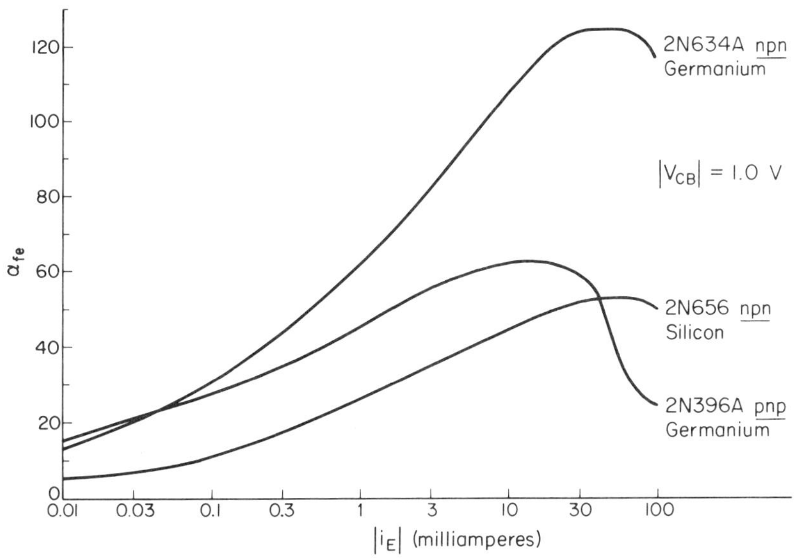

Fig. 5-29. Effect of current level on BJT alpha.

decreasing for currents in the high-level region. The lower alpha at the higher current levels is the result of a lower emitter efficiency with higher majority-carrier density in the base, as indicated by Eq. (5-37). At high-current levels, the density of injected minority carriers is large enough to cause a significant increase in majority-carrier density, which increases the current of carriers injected into the emitter. This increase is more pronounced for *pnp* BJTs, as seen by comparing *npn* and *pnp* α_{fe} plots of Fig. 5-29, since the emitter diffusion constant is greater for minority electrons than for minority holes. This effect is sometimes called *alpha crowding*, since it causes the collector characteristic curves to be more closely spaced in the high-current region.

5-3.6 Temperature Dependence of Alpha

If one examines the common-emitter collector characteristic at elevated temperatures, as may be done qualitatively by extending operation into higher current and voltage levels, an upward displacement of the constant-i_B curves is observed. This is as expected, since I_{CBO} increases with temperature, increasing I_{CEO} which acts as a pedestal for the remainder of the characteristic. However, closer inspection shows a change in the vertical separation of the characteristic curves at elevated temperature; that is, α_{fe} is temperature dependent.

The direction of the change in α_{fe} with temperature turns out to be opposite that which we might expect from our consideration thus far. From the Einstein relationship [Eq. (1-34)], the base diffusion constant D_b is

$$D_b = \frac{kT\mu_b}{e}$$

where μ_b varies as $T^{-1.66}$ for electrons and $T^{-2.33}$ for holes, as given in Table 1-2. Therefore, D_b decreases as temperature increases, and this suggests a decrease in alpha. However, a pronounced increased in α_{fe} with temperature is observed, as shown by the data of Table 5-1, and this must arise from in-

Table 5-1 Temperature dependence of alpha

	2N395 *pnp* Ge Alloy	2N634A *npn* Ge Alloy	2N292 *npn* Ge Grown
T Deg C	α_{fe} at -5 V, -1 mA	α_{fe} at 5 V, 1 mA	α_{fe} at 5 V, 1 mA
25	110	98	37
−78	46	43	20
−196	23	20	8

creased base minority-carrier lifetime τ_b. Since most of the recombination of injected carriers takes place at recombination centers, as pointed out in Sec.

1-3.3, the increased thermal motion of the carriers at higher temperature is thought to make their capture by these centers less probable. Thus τ_b increases with temperature, increasing the base-transport factor.

Another factor causing α_{fe} to increase with temperature is the increase in collector efficiency caused by increased field (*IR* drop) in the "neutral" collector region, owing to increased bulk resistance. However, this effect cannot be significant in alloy BJTs with their characteristically high collector-region conductivities.

5-4 THE DRIFT-FIELD BJT

So far, our treatment of the physical electronics of the BJT has been based on a device model having a homogeneous base region. Commercial BJTs having such base regions are those produced by either the alloy technology or the double-doped-grown technology, the latter being obsolete. These devices are often referred to as *diffusion* BJTs,* being so designated because of the mechanism by which injected minority carriers are transported across the base. BJTs produced by either the rate-grown, the grown-diffused, or the vapor-diffusion technology have base regions of nonuniform, or graded, impurity concentration. Such devices have "built-in" electric fields in their base regions, since, as we shall see, an electric field is established in a semiconductor when an impurity-concentration gradient exists. If the gradient in the base is such that the electric field causes a drift flow of injected minority carriers across the base to the collector which significantly adds to the diffusion flow, the device is a *drift* or *drift-field* BJT.

Drift transport of injected carriers. The impurity profile of a single-diffused *pnp* BJT is somewhat as shown by Fig. 5-30(a), in which N_d represents the donor-atom density and N_a is the acceptor-atom density. The emitter junction is abrupt, having been produced by alloying, and the collector junction is graded, owing to the gradual decrease in the density of diffused donor atoms. Transistors having such base-impurity profiles are known as *graded-base* BJTs.

Figure 5-30(b) shows the distributions of positive donor ions and majority electrons in the base region of the BJT of Fig. 5-30(a) when it is biased for active-mode operation. In these plots, we assume that donor-atom ionization is complete and that the current level is too low for injected minority carriers to have a significant effect on the majority-carrier distribution. The electron distribution that would exist for electrical neutrality is shown, each region of

* The *diffusion BJT* should not be confused with the *diffused BJT*, which is so designated because of the technology of its fabrication.

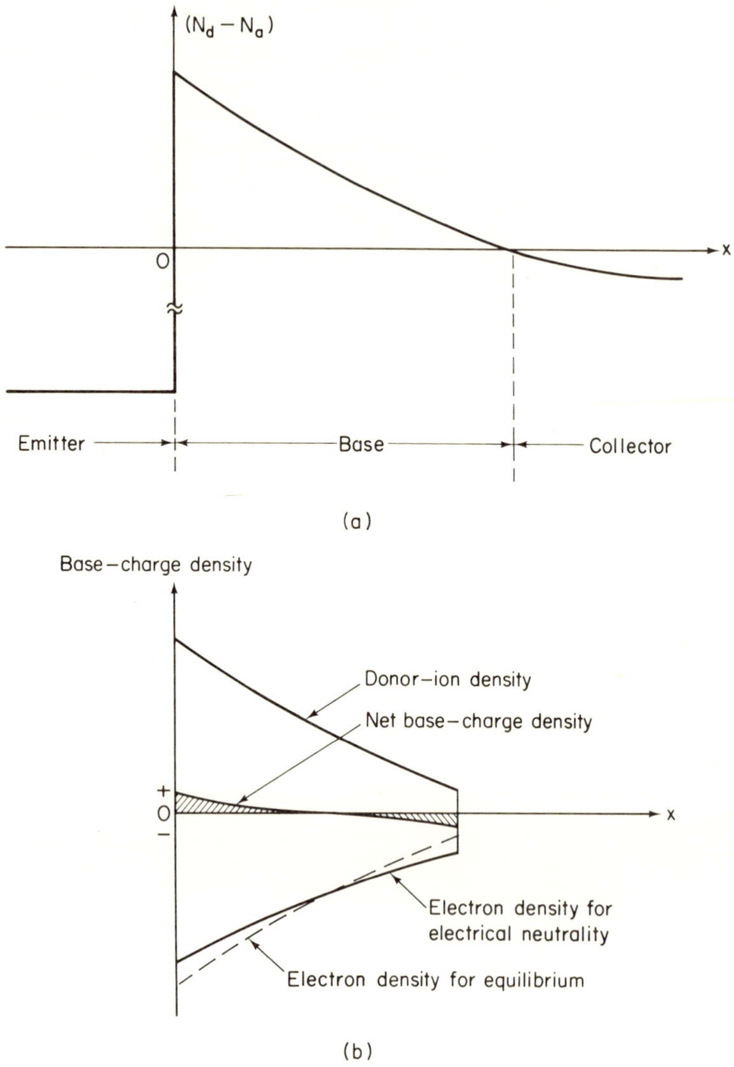

Fig. 5-30. (a) Impurity distribution in a *pnp* drift-field BJT. (b) Charge distributions in the base of the BJT of (a) when biased for active-mode operation.

the base having an electron density equal to the donor-ion density. However, such an electron distribution results in an electron-concentration gradient that produces a diffusion flow of electrons toward the collector edge of the base, reducing the electron density at the emitter edge and increasing it at the collector edge. This change in electron distribution sets up an electric field that causes a drift flow of electrons from the collector edge of the base to the emitter edge, the drift current being equal to the diffusion current when equilibrium conditions are established. The electron distribution in the base for equilibrium and the resulting net charge distribution are also shown in Fig. 5-30(b). The electric field established by the net charge produces a drift flow of injected holes across the base to the collector.

Equating the majority-carrier currents in the base through use of Eqs. (1-10) and (1-32), we obtain

$$Ae\mu_n E n_b = -AeD_n \frac{dn_b}{dx} \tag{5-53}$$

from which

$$E = -\frac{D_n}{\mu_n n_b}\frac{dn_b}{dx} \tag{5-54}$$

From the Einstein relationship [Eq. (1-35)],

$$\mu_n = \frac{eD_n}{kT} \tag{5-55}$$

and Eq. (5-54) becomes

$$E = -\frac{kT}{en_b}\frac{dn_b}{dx} \tag{5-56}$$

Since the impurity concentration resulting from vapor diffusion decreases exponentially with distance below the surface, the net impurity density N in the base of the transistor of Fig. 5-39 is given by

$$N = N_d - N_a = N(0)e^{-x/\mathscr{L}} \tag{5-57}$$

If 100 percent donor-atom ionization is assumed, the electron density in the base is

$$n_b = n_b(0)e^{-x/\mathscr{L}} \tag{5-58}$$

From Eqs. (5-56) and (5-57), we obtain

$$E = \frac{kT}{e\mathscr{L}} \tag{5-59}$$

indicating that a uniform electric field exists in the base. Since kT/e is about 0.025 V at room temperature and $\mathscr{L}$ is of the order of the base-width, which is of order of 10^{-4} cm in typical drift transistors, we see that E is of the order of 250 V/cm.

5-5 THE DOUBLE-DIFFUSED BJT

The fabrication processes for double-diffused BJTs are described in Sec. 5-1, and the structure of an *npn* device is shown in Fig. 5-2. Double-diffused *pnp* BJTs have similar structures, except that each region has a

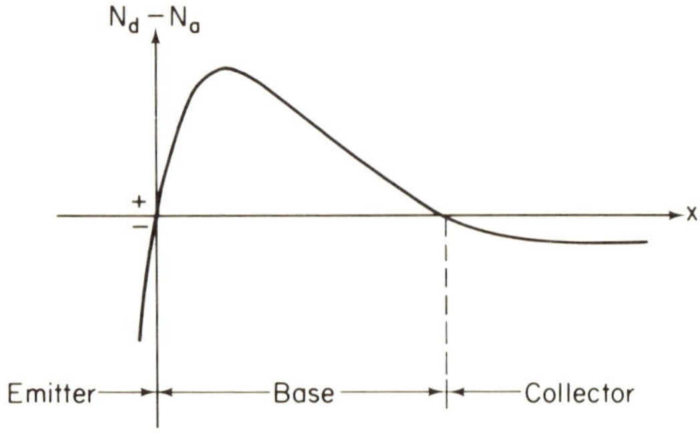

Fig. 5-31. Impurity distribution in a double-diffused *pnp* BJT.

conductivity type opposite that of the *npn* device. Of the new BJT types introduced in recent years, almost all are double-diffused devices.

Since the emitter region of the double-diffused BJT is formed by a second diffusion, the impurity concentration decreases with distance from the surface, resulting in a structure having an impurity profile somewhat like that shown for a *pnp* BJT in Fig. 5-31. The emitter-base junction is seen to be graded, and this causes the internal behavior of the BJT to differ from that of the single-diffused device. The principal difference is that the flow across the base of holes injected at the emitter is initially opposed by the built-in field near the emitter. Throughout the rest of the base region, the hole flow is aided by the built-in field. In the typical double-diffused BJT, the net effect of the field is small, and the transport of minority carriers through the base is much as though the base impurity concentration were uniform and the carriers were transported by diffusion only.

5-6 COMPARISON OF ALLOY AND DIFFUSED BJTs

Although the diffused BJT is used in most new designs, the use of the alloy BJT continues, and it is of interest to compare the characteristics of the two. These characteristics differ in a number of respects that are of primary importance in their application.

An important difference between diffused and alloy BJTs is that the doping of the base region of the diffused device is much heavier than that of the collector region. Because of this, most of the collector space-charge region of a diffused BJT is on the collector side of the metallurgical junction. This is in direct contrast with the alloy device with its heavily doped collector region. Therefore, the reach-through (punch-through) voltage of a diffused

BJT is higher than that of an alloy device having the same base width. Since high-frequency BJTs require thin base regions, as will be shown in Sec. 6-4, high frequency and high collector voltage (and therefore high power) are incompatible in the alloy device. The upper frequency limit of alloy BJTs is of the order of 10 MHz, but diffused transistors capable of operation at 1000 MHz or higher are in existence.

In the interest of emitter efficiency, the doping levels of the emitter regions of both alloy and diffused BJTs must be high compared to that of the base. With both regions of the emitter-base junction having relatively high conductivity, the diffused BJT is characterized by a low emitter-base reverse-breakdown voltage. This is in direct contrast with the alloy device, which has comparable emitter and collector reverse-breakdown voltages, owing to its low-conductivity base region. Emitter reverse-breakdown voltage is an important specification of BJTs used in switching applications, and diffused BJTs commonly require the use of series-connected diodes in order to prevent excessive reverse emitter-base current.

The difference in the doping levels of the collector regions of alloy and diffused BJTs results in an enormous difference in the inverse alphas of the two types of devices. The inverse alpha is the alpha of a device with the normal collector region acting as an emitter and the normal emitter region serving as a collector. The inverse alpha of a diffused BJT is quite low, since the low-conductivity collector region is quite ineffective as an emitter. This is in direct contrast with the alloy device, which has comparable forward and inverse emitter efficiencies, owing to the high-conductivity collector region. The difference in the forward and inverse alphas of an alloy BJT is due largely to the difference in the geometries of the emitter and collector regions, the collector usually having the larger area in order to be more effective. If the emitter and collector areas are made equal, the result is a bilateral BJT—a device having equal forward and inverse alphas. This bilateral property is useful in certain applications.

The essential differences between alloy and diffused BJTs are summarized in Table 5-2.

Table 5-2 Comparison of alloy and diffused BJTs

Characteristic	*Alloy*	*Diffused*
Collector conductivity	High	Low
Base conductivity	Low	High
Collector depletion-region location	Base	Collector
Base reach-through voltage (high-frequency transistor)	Low	High
Compatibility of high frequency and high power	Poor	Good
Emitter reverse-breakdown voltage	High	Low
Inverse alpha	High	Low

PROBLEMS

5-1. A germanium BJT has an I_{CBO} of 5 μA and an α_{fb} of 0.98. What values of I_{CEO} and α_{fe} might be expected?

5-2. The common-emitter characteristic of a silicon BJT is shown in Fig. P5-2. Determine α_{fe} and α_{fb} for the operating point $v_{CE} = 3$ V, $i_B = 3$ μA.

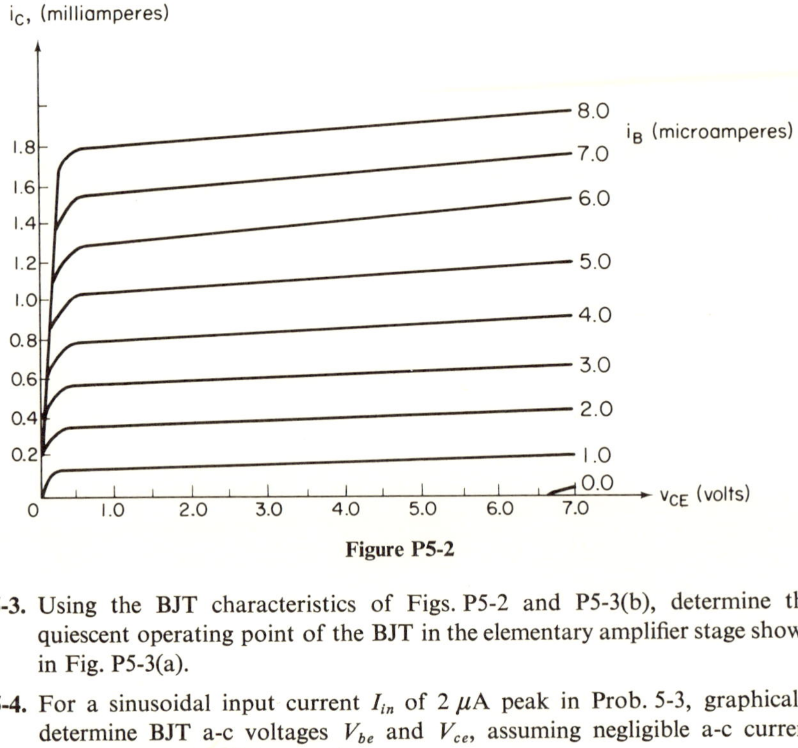

Figure P5-2

5-3. Using the BJT characteristics of Figs. P5-2 and P5-3(b), determine the quiescent operating point of the BJT in the elementary amplifier stage shown in Fig. P5-3(a).

5-4. For a sinusoidal input current I_{in} of 2 μA peak in Prob. 5-3, graphically determine BJT a-c voltages V_{be} and V_{ce}, assuming negligible a-c current through the 1.1 MΩ base resistor.

5-5. Base resistor R_B in the circuit of Fig. P5-5 is adjusted for a d-c base current I_B of 4 μA. Using the collector characteristics of Fig. P5-2, determine:
(a) The quiescent operating point of the BJT.
(b) A-C output voltage V_2 for a sinusoidal input current I_{in} of 3 μA peak at frequencies such that $1/\omega C_2 \ll 2\text{k}\Omega$, assuming negligible a-c current through R_B.
(c) If I_{in} is increased until clipping of the V_2 wave begins, what is the magnitude of V_2?

5-6. An *npn* germanium BJT has a base-region thickness of 1.0 mil. If the basc-

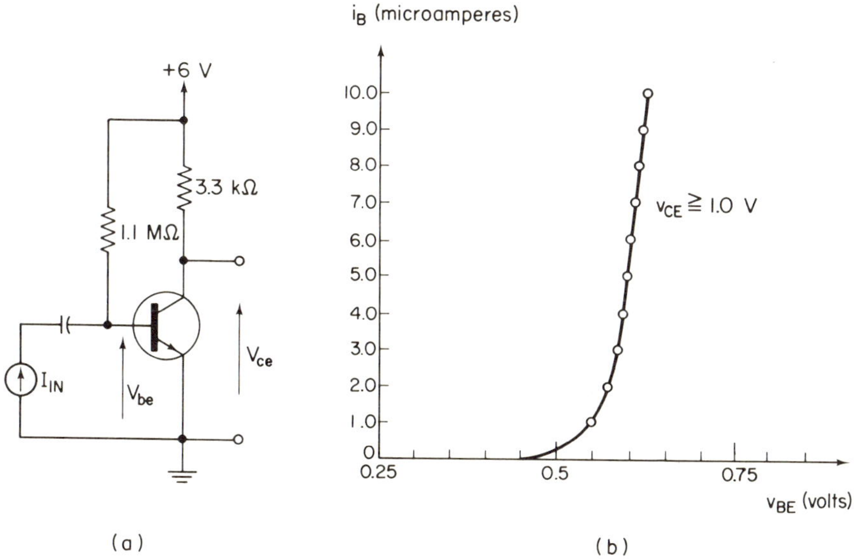

Figure P5-3

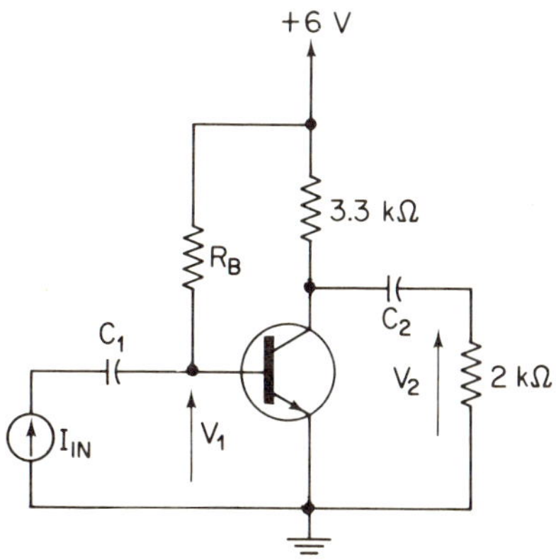

Figure P5-5

region minority-carrier lifetime is 20 μs, what base-transport factor might be expected?

5-7. The base region of the BJT of Prob. 5-6 has a resistivity of 5 ohm-cm. Holes injected in the *n*-type emitter region have a 10-μs lifetime. What emitter efficiency might be expected for an emitter resistivity of 5 ohm-cm?

5-8. Repeat Prob. 5-7 for an emitter resistivity of 0.50 ohm-cm.

5-9. The BJT of Probs. 5-7 and 5-8 has a collector-base breakdown voltage

BV_{CBO} of 50 V. What collector-emitter breakdown voltage BV_{CEO} might be expected?

5-10. In an *n*-type semiconductor, donor-atom density N_d variation with distance x from the surface is given by

$$N_d(x) = 10^{15}e^{-10^4 x} \text{ atoms/cm}^3$$

for x expressed in centimeters. Determine the electric field in the semiconductor for a temperature of 100°C.

5-11. The acceptor-atom density in a semiconductor segment decreases uniformly from 10^{16} atoms/cm^3 to 10^{14} atoms/cm^3 in a distance of 0.01 cm. Determine the electric field at each end of the segment for a temperature of 300°K.

6

BJT Equivalent-Circuit Models and Parameters

Although the BJT is basically a nonlinear device, its current-voltage characteristics are approximately linear if the operating range is adequately restricted. This condition suggests linear representation of the device for small-signal (incremental-signal) operation. BJT circuits may be analyzed either by equivalent-circuit methods or by graphical methods, such as that suggested in Sec. 5-2.

6-1 TWO-PORT NETWORK REPRESENTATIONS

For small-signal operation, the BJT may be considered a three-terminal linear network having two independent terminal currents and two independent terminal-pair voltages. These variables, which are superimposed on d-c bias currents and voltages, are defined in Fig. 6-1. They are related by two linear equations that can be written in six different ways, there being six ways of choosing two of the four variables as the independent ones. We now consider three ways of describing the network in terms of these variables.

***z* parameters.** Choosing input current I_1 and output current I_2 as independent variables, we can write

$$V_1 = z_{11}I_1 + z_{12}I_2 \tag{6-1}$$

Fig. 6-1. Definition of the terminal variables of a two-port network.

and

$$V_2 = z_{21}I_1 + z_{22}I_2 \tag{6-2}$$

The impedances of Eqs. (6-1) and (6-2) are defined as

$$z_{11} = \left.\frac{V_1}{I_1}\right|_{I_2=0}$$

= *open-circuit input impedance*

= input impedance with output terminals open-circuited

$$z_{12} = \left.\frac{V_1}{I_2}\right|_{I_1=0}$$

= *open-circuit reverse transfer impedance*

= voltage developed between the open-circuited input terminals per unit current applied at the output terminals

$$z_{21} = \left.\frac{V_2}{I_1}\right|_{I_2=0}$$

= *open-circuit forward transfer impedance*

= voltage developed between the open-circuited output terminals per unit current applied at the input terminals

$$z_{22} = \left.\frac{V_2}{I_2}\right|_{I_1=0}$$

= *open-circuit output impedance*

= output impedance with input terminals open-circuited

From these definitions, we see that z parameters are open-circuit impedances. The z-parameter representation of a two-port network is shown in Fig. 6-2.

***y* parameters.** If we choose input voltage V_1 and output voltage V_2 as independent variables, the network equations are

$$I_1 = y_{11}V_1 + y_{12}V_2 \tag{6-3}$$

and

$$I_2 = y_{21}V_1 + y_{22}V_2 \tag{6-4}$$

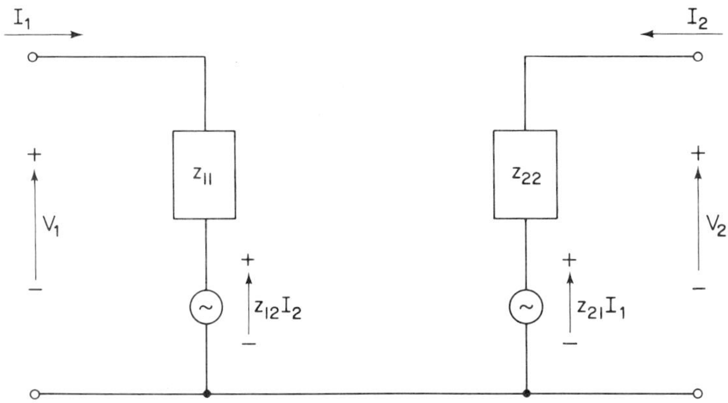

Fig. 6-2. z-parameter representation of a linear two-port network.

where

$$y_{11} = \left.\frac{I_1}{V_1}\right|_{V_2=0}$$

= *short-circuit input admittance*

= input admittance with output terminals short-circuited

$$y_{12} = \left.\frac{I_1}{V_2}\right|_{V_1=0}$$

= *short-circuit reverse transfer admittance*

= current developed in the short-circuited input terminals per unit voltage applied at the output terminals

$$y_{21} = \left.\frac{I_2}{V_1}\right|_{V_2=0}$$

= *short-circuit forward transfer admittance*

= current developed in the short-circuited output terminals per unit voltage applied at the input terminals

$$y_{22} = \left.\frac{I_2}{V_2}\right|_{V_1=0}$$

= *short-circuit output admittance*

= output admittance with input terminals short-circuited

From these definitions, we see that y parameters are short-circuit admittances. The y-parameter representation of a two-port network is shown in Fig. 6-3.

***h* parameters.** Neither the z-parameter representation nor the y-parameter representation lends itself particularly well to BJTs because of the difficulty in measuring certain parameters of each representation. The open-circuit

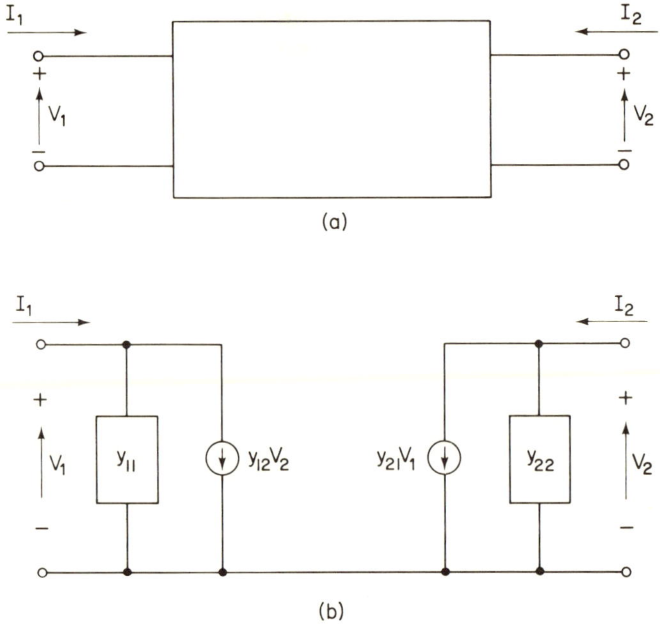

Fig. 6-3. (a) Definition of terminal variables for a two-port network. (b) *y*-parameter representatation of the linear two-port network of (a)

and short-circuit conditions required in the parameter measurements must be alternating-current conditions with d-c bias conditions maintained. Since it is quite difficult to obtain an a-c open in the high-impedance back-biased collector circuit or to obtain an a-c short of the low-impedance forward-biased emitter-base circuit, a representation consisting of parameters measured with the collector short-circuited and the emitter-base open-circuited is preferred. This is obtained by choosing input current I_1 and output voltage V_2 as independent variables. The network equations for this choice of variables are

$$V_1 = h_{11}I_1 + h_{12}V_2 \tag{6-5}$$

and

$$I_2 = h_{21}I_1 + h_{22}V_2 \tag{6-6}$$

The parameters of Eqs. (6-5) and (6-6) are known as hybrid parameters, since they are a combination of open-circuit and short-circuit parameters. The use of lower-case letters in the designation of network parameters identifies them as small-signal quantities.

6-2 BJT *h*-PARAMETER REPRESENTATIONS

The terminal currents and terminal-pair voltages of the two-port-network representation of a BJT operated in the common-emitter configuration are defined in Fig. 6-4(a). The *h*-parameter equations describing the network may be written as

$$V_{be} = h_{ie}I_b + h_{re}V_{ce} \tag{6-7}$$

and

$$I_c = h_{fe}I_b + h_{oe}V_{ce} \tag{6-8}$$

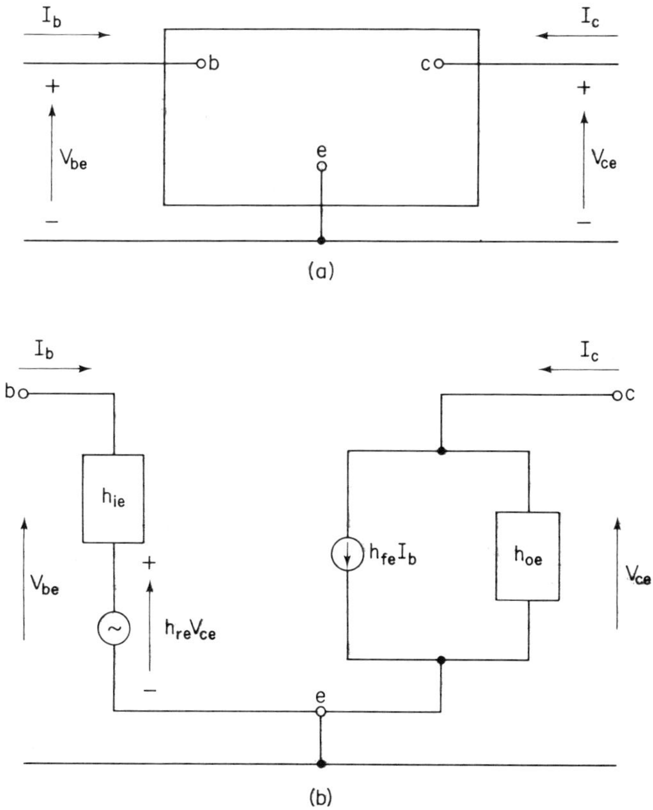

Fig. 6-4. (a) Definition of terminal variables for a two-port network representation of a BJT in the common-emitter configuration. (b) Common-emitter *h*-parameter representation of the linear network of (a).

These equations are Eqs. (6-5) and (6-6) modified through the use of BJT notation. In this notation, the first subscript of a network-parameter symbol indicates whether it is an input parameter, an output parameter, a forward parameter (i.e., from input toward output), or a reverse parameter (i.e., from output toward input). The second subscript of the parameter symbol indicates the BJT connection being used by referring to the terminal that is common to the input and output loops. In Eqs. (6-7) and (6-8), the letter *e* indicates the common-emitter connection. Similarly, h_{ib} is a common-base parameter and h_{ic} is a common-collector parameter. The same subscript system is applied to z and y parameters; that is, z_{ie} is the common-emitter open-circuit input impedance, y_{oe} is the common-emitter short-circuit output admittance, etc. Additional IEEE-approved symbols for the BJT parameters of Eqs. (6-7) and (6-8) are h_{11e}, h_{12e}, h_{21e}, and h_{22e}. Although this subscript system has much to commend it, we make little use of it because it has not been adopted by device manufacturers.

From Eqs. (6-7) and (6-8), we see that the common-emitter h parameters are defined as

$$h_{ie} = \left.\frac{V_{be}}{I_b}\right|_{V_{ce}=0}$$

= *common-emitter short-circuit input impedance*

= impedance measured at the base-emitter terminals with the collector-emitter terminals short-circuited

$$h_{re} = \left.\frac{V_{be}}{V_{ce}}\right|_{I_b=0}$$

= *common-emitter open-circuit reverse voltage-transfer ratio*

= voltage developed at the open circuited base-emitter terminals per unit voltage applied at the collector-emitter terminals

$$h_{fe} = \left.\frac{I_c}{I_b}\right|_{V_{ce}=0}$$

= *common-emitter short-circuit forward current-transfer ratio*

= current developed in the short-circuited collector-emitter terminals per unit current applied at the base-emitter terminals

$$h_{oe} = \left.\frac{I_c}{V_{ce}}\right|_{I_b=0}$$

= *common-emitter open-circuit output admittance*

= admittance measured at the collector-emitter terminals with the base-emitter terminals open-circuited

From these definitions, we see that each hybrid parameter has a different dimension, h_{ie} being an impedance, h_{re} a voltage ratio, h_{fe} a current ratio, and h_{oe} an admittance. An unfortunate feature of the h-parameter symbols is that the letter h gives no clue as the nature of the parameter. The h-parameter representation of a BJT in the common-emitter configuration is shown in Fig. 6-4(b).

For BJTs operated at low frequencies, the a-c voltages and currents of the h-parameter definitions are either in phase or of opposite phase. Therefore, these voltages and currents may be represented by differential changes in total electrode voltage and current, permitting the definition of low-frequency h parameters as slopes of state characteristic curves. Thus, *for low frequencies*,*

$$h_{ie} = \frac{V_{be}}{I_b}\bigg|_{V_{ce}=0} = \frac{dv_{BE}}{di_B}\bigg|_{v_{CE}\text{ fixed}} \tag{6-9}$$

$$h_{re} = \frac{V_{be}}{V_{ce}}\bigg|_{I_b=0} = \frac{dv_{BE}}{dv_{CE}}\bigg|_{i_B\text{ fixed}} \tag{6-10}$$

$$h_{fe} = \frac{I_c}{I_b}\bigg|_{V_{ce}=0} = \frac{di_C}{di_B}\bigg|_{v_{CE}\text{ fixed}} \tag{6-11}$$

$$h_{oe} = \frac{I_c}{V_{ce}}\bigg|_{I_b=0} = \frac{di_C}{dv_{CE}}\bigg|_{i_B\text{ fixed}} \tag{6-12}$$

From Eqs. (5-21) and (6-11), we see that the common-emitter short-circuit forward current-transfer h_{fe} and the common-emitter alpha α_{fe} are identical parameters. That is,

$$h_{fe} \equiv \alpha_{fe} \tag{6-13}$$

Parameters h_{ie}, h_{fe}, and h_{oe} of typical BJTs lend themselves readily to determination from static characteristics, but h_{re} is less easily obtained, owing to its small magnitude.

Common-base and common-collector h parameters. Regardless of the transistor configuration, the BJT may be represented between its collector, emitter, and base terminals by the small-signal equivalent circuit of Fig. 6-4(b). However, in the common-base configuration it is advantageous to take the emitter current, rather than the base current, as an independent variable, since input current is supplied to the emitter. Also, the input and output voltages are taken with respect to the base, as indicated in Fig. 6-5(a). Therefore, for the common-base configuration the two-port h-parameter network equations are

$$V_{eb} = h_{ib}I_e + h_{rb}V_{cb} \tag{6-14}$$

* By low frequencies, we mean the frequency range over which the h parameters may be considered to be real (rather than complex) quantities that are independent of frequency. That is, h_{ie} is a resistance, h_{oe} is a conductance, and h_{fe} and h_{re} are ratios having zero phase angles.

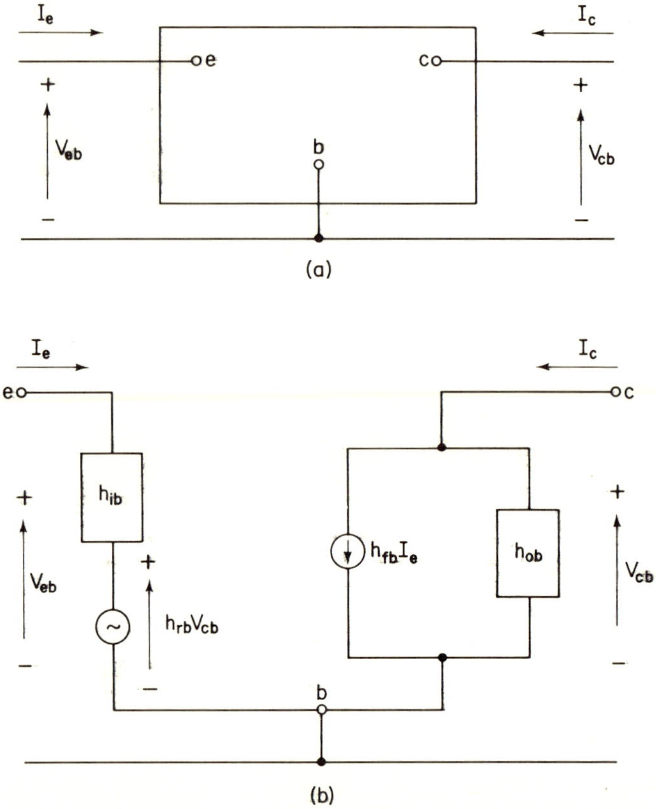

Fig. 6-5. (a) Definition of terminal variables for a two-port network representation of a BJT in the common-base configuration. (b) Common-base *h*-parameter representation of the linear network of (a).

and

$$I_c = h_{fb}I_e + h_{ob}V_{cb} \tag{6-15}$$

and the common-base *h*-parameter definitions are

$$h_{ib} \equiv \frac{V_{eb}}{I_e}\bigg|_{V_{cb}=0} = \frac{dv_{EB}}{di_E}\bigg|_{v_{CB}\text{ fixed}} \quad \text{at low frequencies} \tag{6-16}$$

$$h_{rb} \equiv \frac{V_{eb}}{V_{cb}}\bigg|_{I_e=0} = \frac{dv_{EB}}{dv_{CB}}\bigg|_{i_E\text{ fixed}} \quad \text{at low frequencies} \tag{6-17}$$

$$h_{fb} \equiv \frac{I_c}{I_e}\bigg|_{V_{cb}=0} = \frac{di_C}{di_E}\bigg|_{v_{CB}\text{ fixed}} \quad \text{at low frequencies} \tag{6-18}$$

$$h_{ob} \equiv \frac{I_c}{V_{cb}}\bigg|_{I_e=0} = \frac{di_C}{dv_{CB}}\bigg|_{i_E\text{ fixed}} \quad \text{at low frequencies} \tag{6-19}$$

Unlike its common-emitter counterpart, the common-base short-circuit forward current-transfer ratio h_{fb} is negative at low frequencies, since emitter current i_E and collector current i_C are of opposite sign. From Eqs. (5-5) and (6-18), we see that $-h_{fb}$ and the common-base alpha α_{fb} are identical parameters. That is,

$$h_{fb} \equiv -\alpha_{fb} \tag{6-20}$$

The common-base h-parameter representation of a BJT is shown in Fig. 6-5(b).

Although the common-emitter h-parameter representation of a BJT is quite convenient for considering operation in the common-collector configuration, input current being supplied to the base as in the common-emitter configuration, the common-collector representation is sometimes desired. The two-port network currents and voltages for the common-collector configuration are defined in Fig. 6-6, and the h-parameter equations are

$$V_{bc} = h_{ic}I_b + h_{rc}V_{ec} \tag{6-21}$$

and

$$I_e = h_{fc}I_b + h_{oc}V_{ec} \tag{6-22}$$

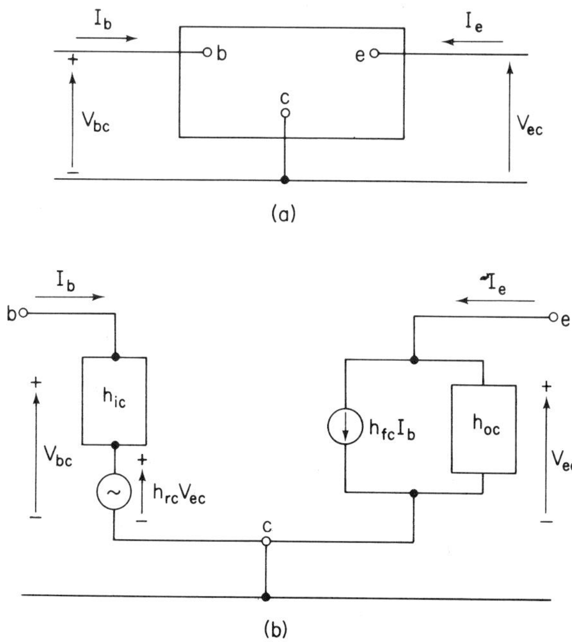

Fig. 6-6. (a) Definition of terminal variables for a two-port representation of a BJT in the common-collector configuration. (b) Common-collector h-parameter representation of the linear network of (a).

and the common-collector h-parameter definitions are

$$h_{ic} = \frac{V_{bc}}{I_b}\bigg|_{V_{ec}=0} \tag{6-23}$$

$$h_{rc} = \frac{V_{bc}}{V_{ec}}\bigg|_{I_b=0} \tag{6-24}$$

$$h_{fc} = \frac{I_e}{I_b}\bigg|_{V_{ec}=0} \tag{6-25}$$

$$h_{oc} = \frac{I_e}{V_{ec}}\bigg|_{I_b=0} \tag{6-26}$$

The common-collector h-parameter representation of a BJT is shown in Fig. 6-6(b).

The orders of magnitude of low-frequency BJT h parameters are shown by Table 6-1, which lists the three sets of h parameters of a typical type 2N1382 *pnp* germanium BJT. Since each parameter set defines the BJT, the parameters of the other two sets can be determined if those of one set are known. Of the parameters listed in Table 6-1, h_{ib}, h_{ob}, h_{rb}, and h_{fe} are specified by the manufacturer, and the others are calculated by using the conversion equations derived in Sec. 6-3.

Table 6-1 Hybrid parameters of a typical type 2N1382 BJT at $i_E = -1$ mA, $v_C = -5$ V, $f = 1$ kHz

Common-base h Parameters	
$h_{ib} = 30$ ohms	$h_{rb} = 5 \times 10^{-4}$
$h_{fb} = -0.988$	$h_{ob} = 0.5\ \mu$mho
Common-emitter h Parameters	
$h_{ie} = 2400$ ohms	$h_{re} = 7 \times 10^{-4}$
$h_{fe} = 80$	$h_{oe} = 40\ \mu$mhos
Common-collector h Parameters	
$h_{ic} = 2400$ ohms	$h_{rc} = 1.00$
$h_{fc} = -81$	$h_{oc} = 40\ \mu$mhos

6-3 TWO-PORT NETWORK-PARAMETER CONVERSION EQUATIONS

It was pointed out in Sec. 6-2 that knowledge of one h-parameter representation of a BJT permits determination of the other two representations. However, the general situation is that any set of BJT parameters, be they z, y, or h parameters for any of the three configurations, defines the BJT as a small-signal device at a particular frequency and provides information for determining all other parameter sets. Therefore, equations for converting from one parameter set to another prove to be useful, particularly for determining certain parameters that do not lend themselves readily to direct measurement.

h-parameter conversion equations. As an example in converting from one parameter set to another and to illustrate the use of h-parameter representations, we now derive the equations for converting from common-base h parameters to common-emitter h parameters. We see from the definitions of Eqs. (6-9) and (6-11) that common-emitter parameters h_{ie} and h_{fe} may be determined from the common-base representation of Fig. 6-5(b) by short-circuiting the collector-emitter terminals, thus reducing V_{ce} to zero, and solving for I_b and I_c in terms of applied voltage V_{be}. This arrangement of the common-base representation is shown in Fig. 6-7.

From Fig. 6-7, we see that

$$I_e = -(I_b + I_c) \tag{6-27}$$

$$V_{be} = -h_{rb}V_{cb} - h_{ib}I_e = h_{rb}V_{be} - h_{ib}I_e \tag{6-28}$$

since

$$V_{cb} = -V_{be}$$

and

$$V_{be} = -\frac{I_c - h_{fb}I_e}{h_{ob}} \tag{6-29}$$

From Eqs. (6-27) and (6-28) we obtain

$$V_{be} = \frac{h_{ib}(I_b + I_c)}{1 - h_{rb}} \cong h_{ib}(I_b + I_c) \tag{6-30}$$

since $h_{rb} \ll 1$ for typical BJTs at low frequencies, as seen from Table 6-1. Substituting for I_e in Eq. (6-29) gives

$$V_{be} = -\frac{I_c + h_{fb}(I_b + I_c)}{h_{ob}} \tag{6-31}$$

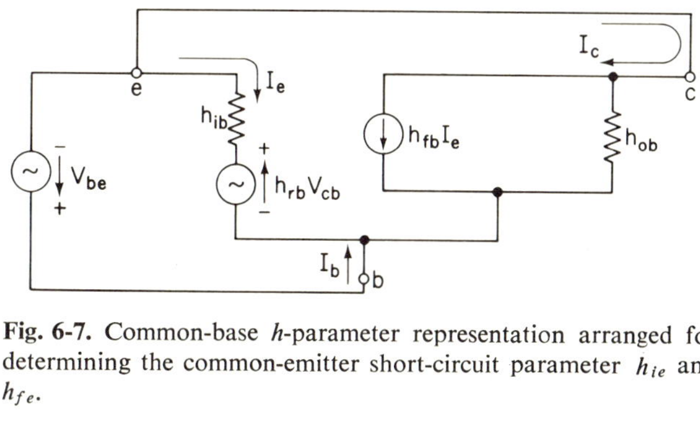

Fig. 6-7. Common-base h-parameter representation arranged for determining the common-emitter short-circuit parameter h_{ie} and h_{fe}.

Solving Eqs. (6-30) and (6-31) for I_b yields

$$I_b \cong V_{be}\frac{[h_{ib}h_{ob} + (h_{fb} + 1)]}{h_{ib}} \tag{6-32}$$

from which

$$h_{ie} = \frac{V_{be}}{I_b} \cong \frac{h_{ib}}{h_{ib}h_{ob} + (h_{fb} + 1)} \tag{6-33}$$

For typical BJTs,

$$h_{ib}h_{ob} \ll (h_{fb} + 1)$$

Therefore,

$$h_{ie} \cong \frac{h_{ib}}{h_{fb} + 1} = \frac{h_{ib}}{1 - \alpha_{fb}} \tag{6-34}$$

Solving Eqs. (6-30) and (6-31) for I_c yields

$$I_c \cong -\frac{V_{be}(h_{fb} + h_{ib}h_{ob})}{h_{ib}} \tag{6-35}$$

From Eqs. (6-32) and (6-35),

$$h_{fe} = \frac{I_c}{I_b} \cong \frac{-(h_{fb} + h_{ib}h_{ob})/h_{ib}}{h_{ib}h_{ob} + (h_{fb} + 1)/h_{ib}} \tag{6-36}$$

Since $h_{ib}h_{ob} \ll (h_{fb} + 1) \ll h_{fb}$ for typical BJTs, Eq. (6-36) may be written

$$h_{fe} \cong -\frac{h_{fb}}{h_{fb} + 1} = \frac{\alpha_{fb}}{1 - \alpha_{fb}} \tag{6-37}$$

From the definitions of Eqs. (6-10) and (6-12), we see that the common-emitter open-circuit parameters h_{re} and h_{oe} may be determined by open-circuiting the base-emitter terminals of the common-base representation, thus reducing I_b to zero, and solving for I_c and V_{be} in terms of applied voltage V_{ce}. This arrangement is shown by Fig. 6-8.

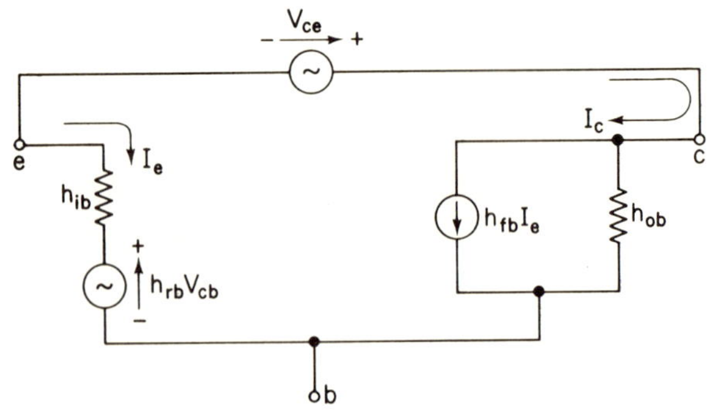

Fig. 6-8. Common-base *h*-parameter representation arranged for determining the common-emitter open-circuit parameters h_{re} and h_{oe}.

From Fig. 6-8, we see that

$$I_c = -I_e \tag{6-38}$$

$$V_{ce} = \frac{I_c - h_{fb}I_e}{h_{ob}} - h_{rb}V_{cb} - h_{ib}I_e \tag{6-39}$$

and

$$V_{cb} = \frac{I_c - h_{fb}I_e}{h_{ob}} \tag{6-40}$$

Substituting in Eq. (6-39) for I_e and V_{cb}, we obtain

$$V_{ce} = \frac{I_c + h_{fb}I_c}{h_{ob}} - \frac{h_{rb}(I_c + h_{fb}I_c)}{h_{ob}} + h_{ib}I_c \tag{6-41}$$

from which

$$V_{ce} = I_c\left[\frac{(h_{fb} + 1)(1 - h_{rb})}{h_{ob}} + h_{ib}\right] \tag{6-42}$$

For typical BJTs,

$$h_{rb} \ll 1 \quad \text{and} \quad \frac{h_{fb} + 1}{h_{ob}} \gg h_{ib}$$

Thus, from Eq. (6-42),

$$I_c \cong \frac{V_{ce}h_{ob}}{h_{fb} + 1} \equiv \frac{V_{ce}h_{ob}}{1 - \alpha_{fb}} \tag{6-43}$$

and

$$h_{oe} = \frac{I_c}{V_{ce}} \cong \frac{h_{ob}}{h_{fb} + 1} \equiv \frac{h_{ob}}{1 - \alpha_{fb}} \tag{6-44}$$

From Fig. 6-8, we see that

$$V_{be} = -h_{rb}V_{cb} - h_{ib}I_e \tag{6-45}$$

Substituting for I_e and V_{cb}, as given by Eqs. (6-38) and (6-40), we obtain

$$V_{be} = \frac{-h_{rb}I_c(h_{fb} + 1)}{h_{ob}} + h_{ib}I_c \tag{6-46}$$

from which

$$V_{be} = \left[\frac{h_{ib}h_{ob} - h_{rb}(h_{fb} + 1)}{h_{ob}}\right]I_c \tag{6-47}$$

Substituting for I_c as given by Eq. (6-43), we obtain

$$V_{be} \cong \frac{V_{ce}h_{ob}}{(h_{fb} + 1)}\left[\frac{h_{ib}h_{ob} - h_{rb}(h_{fb} + 1)}{h_{ob}}\right] \tag{6-48}$$

from which

$$h_{re} = \frac{V_{be}}{V_{ce}} \cong \frac{h_{ib}h_{ob}}{(h_{fb} + 1)} - h_{rb} \equiv \frac{h_{ib}h_{ob}}{1 - \alpha_{fb}} - h_{rb} \tag{6-49}$$

Equations for converting from common-emitter h parameters to common-base and common-collector parameters may be obtained from the common-emitter equivalent representation.* These equations, along with the conversion equations already derived, are given in Table 6-2.

Table 6-2 Approximate h-parameter conversion formulas

↓ *To* / *From* →	*Common Emitter*	*Common Base*	*Common Collector*
h_{ie}		$\frac{h_{ib}}{1+h_{fb}}$	h_{ic}
h_{re}		$\frac{h_{ib}h_{ob}}{1+h_{fb}} - h_{rb}$	$1 - h_{rc}$
h_{fe}		$-\frac{h_{fb}}{1+h_{fb}}$	$-(1 + h_{fc})$
h_{oe}		$\frac{h_{ob}}{1+h_{fb}}$	h_{oc}
h_{ib}	$\frac{h_{ie}}{1+h_{fe}}$		$-\frac{h_{ic}}{h_{fc}}$
h_{rb}	$\frac{h_{ie}h_{oe}}{1+h_{fe}} - h_{re}$		$h_{rc} - 1 - \frac{h_{ic}h_{oc}}{h_{fc}}$
h_{fb}	$-\frac{h_{fe}}{1+h_{fe}}$		$-\frac{1+h_{fc}}{h_{fc}}$
h_{ob}	$\frac{h_{oe}}{1+h_{fe}}$		$-\frac{h_{oc}}{h_{fc}}$
h_{ic}	h_{ie}	$\frac{h_{ib}}{1+h_{fb}}$	
h_{rc}	$1 - h_{re}$	1	
h_{fc}	$-(1 + h_{fe})$	$-\frac{1}{1+h_{fb}}$	
h_{oc}	h_{oe}	$\frac{h_{ob}}{1+h_{fb}}$	

h-, y-, and z-parameter conversion equations. Equations for conversion between sets of h, y, and z parameters may be derived in the same manner as those for conversion between sets of h parameters. These equations are given in Table 6-3; standard network notation is used for the h parameters. A single set of equations takes care of all three BJT configurations; that is, the basic equation for converting y_{11e} to z_{11e} is the same as for y_{11b} to z_{11b}, etc.

* Common-base parameters may also be obtained from common-emitter parameters by solving Eqs. (6-34), (6-37), (6-44), and (6-49) for h_{ib}, h_{rb}, h_{fb}, and h_{ob}.

Table 6-3 Formulas for conversion between *h*, *y*, and *z* parameters

↓ To \ From →	z	y	h
z	$z_{11}\quad z_{12}$ $z_{21}\quad z_{22}$	$\frac{y_{22}}{\Delta y}\quad -\frac{y_{12}}{\Delta y}$ $-\frac{y_{21}}{\Delta y}\quad \frac{y_{11}}{\Delta y}$	$\frac{\Delta h}{h_{22}}\quad \frac{h_{12}}{h_{22}}$ $-\frac{h_{21}}{h_{22}}\quad \frac{1}{h_{22}}$
y	$\frac{z_{22}}{\Delta z}\quad -\frac{z_{12}}{\Delta z}$ $-\frac{z_{21}}{\Delta z}\quad \frac{z_{11}}{\Delta z}$	$y_{11}\quad y_{12}$ $y_{21}\quad y_{22}$	$\frac{1}{h_{11}}\quad -\frac{h_{12}}{h_{11}}$ $\frac{h_{21}}{h_{11}}\quad \frac{\Delta h}{h_{11}}$
h	$\frac{\Delta z}{z_{22}}\quad \frac{z_{12}}{z_{22}}$ $-\frac{z_{21}}{z_{22}}\quad \frac{1}{z_{22}}$	$\frac{1}{y_{11}}\quad -\frac{y_{12}}{y_{11}}$ $\frac{y_{21}}{y_{11}}\quad \frac{\Delta y}{y_{11}}$	$h_{11}\quad h_{12}$ $h_{21}\quad h_{22}$

Note: Δ represents the determinant (that is, $\Delta z = z_{11}z_{22} - z_{21}z_{12}$, $\Delta y = y_{11}y_{22} - y_{21}y_{12}$, and $\Delta h = h_{11}h_{22} - h_{21}h_{12}$).

6-4 HIGH-FREQUENCY *h*-PARAMETER REPRESENTATION

The two-port network models considered so far can be profitably used for narrow-band considerations and at low frequencies where the parameters may be considered to be independent of frequency. They have distinct disadvantages in general use, however, since each *h*, *y*, or *z* parameter is a complex number that varies with frequency. Clearly, for broad-band use, the ideal BJT equivalent-network representation would have no frequency-dependent parameters. Such a representation is nonexistent, but it may be approached. One such approach is by modification of the *h*-parameter representations.

***h*-parameter frequency dependence.** From our device consideration so far, junction capacitances seem the most likely factors affecting BJT frequency response. These are the collector-base transition capacitance* and the diffusion capacitance of the forward-biased emitter-base junction. Therefore, a first step in eliminating frequency-dependent equivalent-circuit elements might be the representation of Fig. 6-9, in which h_{ib} and h_{ob} are shown as low-frequency parameters shunted by capacitances. In this representation,

* As explained later, the collector-base capacitance includes diffusion capacitance and depends upon operating-point current as well as voltage.

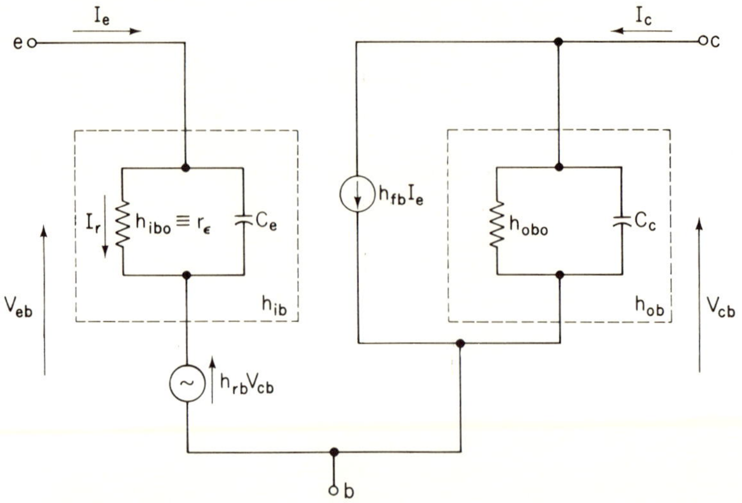

Fig. 6-9. BJT common-base h-parameter representation in which h_{ib} is represented as the low-frequency value h_{ibo} shunted by emitter-junction capacitance C_e and h_{ob} is the low-frequency value h_{obo} shunted by collector-junction capacitance C_c. Parameter h_{ibo} is also defined as the dynamic emitter-junction resistance r_ϵ, the symbol r_ϵ being used to distinguish the parameter from the T equivalent-circuit parameter r_e.

the subscript o added to parameter symbols identifies them as low-frequency parameters, and C_c and C_e are collector-junction and emitter-junction capacitances.

Perhaps the most important BJT common-base h parameter is the short-circuit forward current-transfer ratio h_{fb} (or $-\alpha_{fb}$). Therefore, the frequency dependence of h_{fb} must be taken into consideration if the h-parameter representation of Fig. 6-9 is to be utilized. Fortunately, a general expression for h_{fb} as a function of frequency exists, greatly simplifying the problem.

By inspection of Fig. 6-9, we see that a component of a-c emitter current I_e flows through emitter-junction capacitance C_e. This is the current associated with the change in base-charge storage that occurs during an emitter-voltage change. Since this charge is stored, rather than transmitted through the base to the collector, the current associated with it is not responsible for a-c collector current. Instead, the collector current $h_{fb}I_e$ of Fig. 6-9 is largely the result of emitter-current component I_r, permitting us to write

$$h_{fb}I_e \cong h_{fbo}I_r \qquad (6\text{-}50)$$

where h_{fbo} is the low-frequency value of h_{fb}. From Fig. 6-9, we see that

$$I_r = \frac{I_e/j\omega C_e}{r_\epsilon + 1/j\omega C_e} = \frac{I_e}{1 + j\omega r_\epsilon C_e} \qquad (6\text{-}51)$$

From Eqs. (6-50) and (6-51), we obtain

$$h_{fb} \cong \frac{h_{fbo}}{1 + j\omega r_\epsilon C_e} \tag{6-52}$$

The frequency at which $h_{fb} = 0.707\, h_{fbo}$ is known as the *h_{fb}-cutoff frequency f_{hfb}* or the *common-base alpha-cutoff frequency $f_{\alpha b}$*. From Eq. (6-52) we see that $\omega r_\epsilon C_e = 1$ at f_{hfb}. Therefore,

$$\omega_{hfb} = \frac{1}{r_\epsilon C_e} \tag{6-53}$$

and

$$h_{fb} \cong \frac{h_{fbo}}{1 + j\omega/\omega_{hfb}} \tag{6-54}$$

indicating that the h_{fb}-frequency characteristic is defined by the low-frequency value of h_{fb} and the cutoff frequency. These two quantities may be measured for a particular "low-frequency" BJT, and their values are specified by the manufacturer, manufacturer's specifications usually being in the form of limits and typical values.*

The common-base h-parameter equivalent circuit, with the two most important frequency-dependent circuit elements included, is shown by Fig. 6-10. Since the circuit input is often supplied by a constant-current source, or one that is nearly constant, the frequency dependence of h_{ib} and h_{rb} is of less concern than that of the other two h parameters. Collector capacitance C_c ranges from a few picofarads for low-power BJTs to several hundred picofarads for power BJTs. This is a *medium-frequency* equivalent circuit

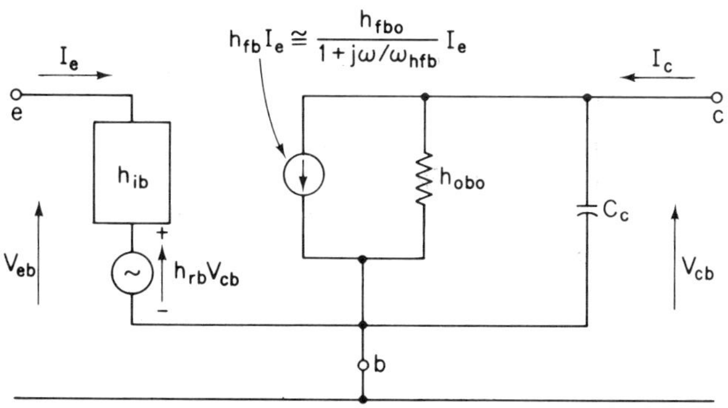

Fig. 6-10. Common-base h-parameter equivalent circuit showing the two most important frequency-dependent elements. Collector capacitance C_c ranges from a few picofarads for low-power BJTs to several hundred picofarads for power BJTs.

* A "low-frequency" BJT might have an f_{hfb} of the order of 10 MHz or less.

useful at frequencies well below f_{hfb}. For frequencies of the order of f_{hfb}, a different equivalent circuit is preferred.

It is often convenient to express the magnitude of h_{fb} in decibels as 20 times the logarithm of h_{fb} expressed as a current ratio; that is,

$$h_{fb} = 20 \log |h_{fb}| \text{ dB} \tag{6-55}$$

A plot of the approximate h_{fb} versus frequency characteristic of Eq. (6-54) is shown by Fig. 6-11. Since the magnitude of h_{fb} is less than unity, h_{fb} is negative when expressed in decibels. Although a BJT is not usually operated at frequencies appreciably greater than f_{hfb}, the plot of Fig. 6-11 includes this region in order to show the 6 dB per octave (or 20 dB per decade) change in h_{fb} with frequency when ω/ω_{hfb} is large compared to unity. This asymptote of the plot is seen to reach h_{fbo} at the corner frequency f_{hfb} for which h_{fb} is 3 dB less than h_{fbo}.

Emitter-junction diffusion capacitance. The emitter-junction capacitance C_e of Fig. 6-9 and Eq. (6-52) consists of a small transition capacitance in

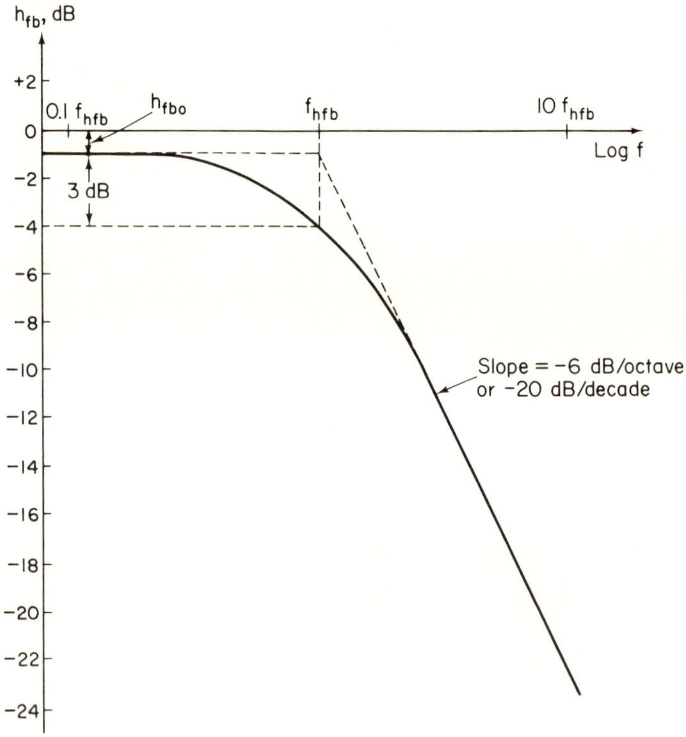

Fig. 6-11. Plot of the approximate magnitude of the common-base short-circuit forward current-transfer ratio h_{fb} versus frequency as given by Eq. (6-54).

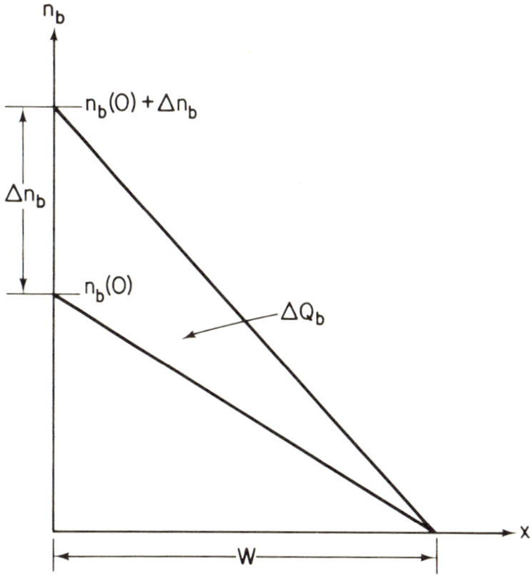

Fig. 6-12. Minority-carrier density in the base region of an *npn* BJT for two values of emitter-base voltage. The change ΔQ_b in stored charge with emitter-voltage change is responsible for the emitter diffusion capacitance.

parallel with a large diffusion capacitance. An approximate value of C_e may be found by considering the charge stored in the base region. Figure 6-12 shows the minority-carrier distribution in the base of an *npn* BJT for two values of emitter-base voltage, $n_b(0)$ being the electron density at the junction edge for the initial value of v_{BE} and $n_b(0) + \Delta n_b$ is the density for $v_{BE} + \Delta v_{BE}$. The change Δi_E in emitter current caused by the change in emitter voltage is seen to be

$$\Delta i_E = \frac{AeD_b \, \Delta n_b}{W} \tag{6-56}$$

where W is the base width, and the change ΔQ_b in stored base charge is given by

$$\Delta Q_b = \frac{Ae \, \Delta n_b W}{2} \tag{6-57}$$

Since C_e approximately equals the emitter diffusion capacitance,

$$C_e \cong \frac{\Delta Q_b}{\Delta v_{BE}} = \frac{\Delta Q_b}{\Delta i_E r_\epsilon} \tag{6-58}$$

where r_ϵ is the dynamic resistance of the emitter junction. Combining Eqs. (6-56), (6-57), and (6-58), we obtain

$$C_e \cong \frac{W^2}{2D_b r_\epsilon} \tag{6-59}$$

This expression is similar to Eq. (2-86) or Eq. (2-87), depending on the BJT type, the diffusion length being replaced by the base width because W is so much shorter than the diffusion length. Also, only one component of emitter current has a significant effect on C_e, since the emitter efficiency is very nearly unity.

Alpha-cutoff frequency. Combining Eqs. (6-53) and (6-59), we see that the common-base alpha-cutoff frequency $f_{\alpha b}$ or f_{hfb} is given by

$$f_{\alpha b} = \frac{D_b}{\pi W^2} \tag{6-60}$$

indicating the necessity of a thin base in high-frequency BJTs. We also see from Eq. (6-60) that, for the same base width, *npn* BJTs have higher alpha-cutoff frequencies than those of the *pnp* type, and silicon BJTs are inherently lower frequency devices than germanium BJTs.

The effects of temperature and operating point on alpha-cutoff frequency may also be seen from Eq. (6-60). An increase in temperature decreases $f_{\alpha b}$, because of the effect on minority-carrier diffusion constant D_b, and d-c operating conditions affect $f_{\alpha b}$ through their effect on base width W. An increase in collector voltage V_{CB} increases $f_{\alpha b}$ because of base-width modulation, the effect being quite pronounced in alloy BJTs whose collector-junction depletion regions lie almost entirely in the base. For a particular value of alpha-cutoff frequency, *npn* BJTs are less sensitive to $f_{\alpha b}$ variation with V_{CB} than *pnp* BJTs and germanium BJTs are less affected than silicon BJTs, thinner base-width devices being more sensitive to collector voltage. Diffused BJTs having collector depletion regions lying almost entirely in the collector region, as a result of their high-conductivity base regions, have $f_{\alpha b}$ values relatively unaffected by collector voltage.

Equation (6-60) suggests that $f_{\alpha b}$ is independent of d-c current level, but experimental results show otherwise. As d-c emitter current I_E is increased, the acceleration of minority carriers through the base by the electric field resulting from increased majority-carrier density reduces the per-unit-current charge stored in the base. This reduction in per-unit-current emitter diffusion capacitance increases the alpha-cutoff frequency.

Common-emitter frequency characteristics. Since a small change in h_{fb} has a large effect on the common-emitter short-circuit forward current-transfer ratio h_{fe}, h_{fe} change with frequency occurs at much lower frequencies than those having a significant effect on h_{fb}. Using the h_{fb}-to-h_{fe} conversion equation

$$h_{fe} = \frac{-h_{fb}}{1 + h_{fb}} \tag{6-37}$$

in conjunction with the $h_{fb}(\omega)$ relationship enables us to investigate h_{fe}

variation with frequency. Substituting in Eq. (6-37) for h_{fb}, as given by Eq. (6-54), we obtain

$$h_{fe} \cong \left(\frac{-h_{fbo}}{1 + \dfrac{j\omega}{\omega_{hfb}}}\right)\left(\frac{1}{1 + \dfrac{h_{fbo}}{1 + j\omega/\omega_{hfb}}}\right) \tag{6-61}$$

which reduces to

$$h_{fe} \cong \frac{-h_{fbo}}{1 + h_{fbo} + j\omega/\omega_{hfb}} \tag{6-62}$$

Since the low-frequency value of h_{fe} is given by

$$h_{feo} = \frac{-h_{fbo}}{1 + h_{fbo}} \tag{6-63}$$

we see from Eq. (6-62) that $h_{fe} = 0.707h_{feo}$ when

$$\frac{\omega}{\omega_{hfb}} \cong 1 + h_{fbo} \tag{6-64}$$

The frequency at which this occurs is known as the *h_{fe}-cutoff frequency f_{hfe}* or the *common-emitter alpha-cutoff* frequency $f_{\alpha e}$. Thus, from Eq. (6-64),

$$f_{hfe} \equiv f_{\alpha e} \cong f_{\alpha b}(1 + h_{fbo}) \equiv f_{\alpha b}(1 - \alpha_{fbo}) \tag{6-65}$$

indicating that $f_{\alpha e}$ is much less than $f_{\alpha b}$ since α_{fbo} is slightly less than unity.

If we divide both numerator and denominator of Eq. (6-62) by the quantity $(1 + h_{fbo})$, we obtain

$$h_{fe} \cong \frac{-h_{fbo}/(1 - h_{fbo})}{1 + j[\omega/\omega_{hfb}(1 + h_{fbo})]} \cong \frac{h_{feo}}{1 + j\omega/\omega_{hfe}} \tag{6-66}$$

indicating that h_{fe} variation with frequency is similar to that of h_{fb}. A plot of the approximate h_{fe} versus frequency characteristic of Eq. (6-66) is shown by Fig. 6-13 for a BJT having an h_{feo} of 15.8 or 24 dB. The frequency for unity h_{fe} (or for $h_{fe} = 0$ dB) is known as the *transition frequency f_T*. For transistors having f_T values of the order of tens of megacycles per second or less, f_T and $f_{\alpha b}$ are of the same order of magnitude. For higher frequency transistors, $f_{\alpha b}$ has little significance, often being unmeasurable because of capacitive coupling between emitter and collector leads, and f_T is usually specified.

For frequencies large compared to $f_{\alpha e}$, we see from Eq. (6-66) that

$$|h_{fe}| \cong \frac{h_{feo}}{\omega/\omega_{hfe}} \tag{6-67}$$

and that

$$f|h_{fe}| \cong f_{\alpha e}h_{feo} \cong f_T \tag{6-68}$$

The approximately constant $f|h_{fe}|$ product *for high frequencies* is a useful relationship.

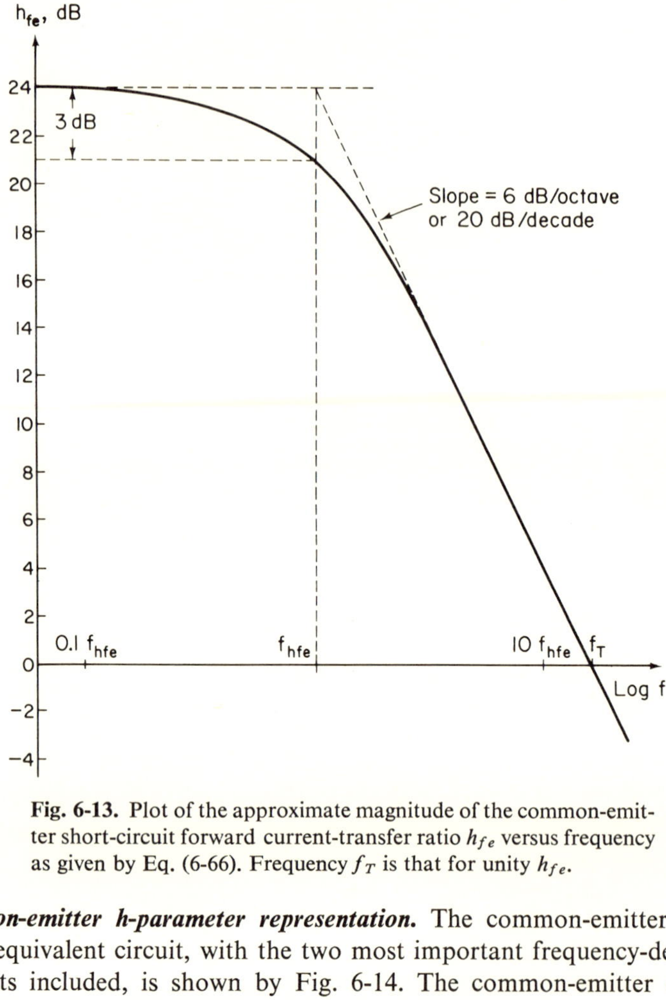

Fig. 6-13. Plot of the approximate magnitude of the common-emitter short-circuit forward current-transfer ratio h_{fe} versus frequency as given by Eq. (6-66). Frequency f_T is that for unity h_{fe}.

Common-emitter h-parameter representation. The common-emitter *h*-parameter equivalent circuit, with the two most important frequency-dependent elements included, is shown by Fig. 6-14. The common-emitter collector capacitance C_{ce} is obtained from the common-base value C_c through use of the h_{ob}-to-h_{oe} conversion equation

$$h_{oe} \cong \frac{h_{oe}}{1 + h_{fb}} \tag{6-44}$$

Thus the capacitative component ωC_{ce} of h_{oe} is given by

$$\omega C_{ce} \cong \frac{\omega C_c}{1 + h_{fb}}$$

and

$$C_{ce} \cong \frac{C_c}{1 + h_{fb}} = C_c(1 + h_{fe}) \cong C_c h_{fe} \tag{6-69}$$

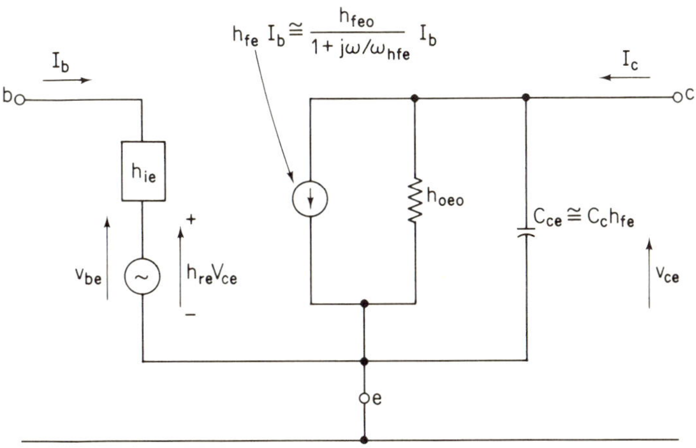

Fig. 6-14. Common-emitter *h*-parameter equivalent circuit showing the two most important frequency-dependent elements.

It is of interest to compare $f_{\alpha b}$ and $f_{\alpha e}$ of the typical 2N1382 BJT whose low-frequency h parameters are listed in Table 6-1. The minimum $f_{\alpha b}$ specification at this operating point is 2.0 MHz. From Eq. (6-65), we see that

$$f_{\alpha e} = f_{\alpha b}(1 + h_{fb}) = 2 \times 10^6(1 - 0.988) = 24 \text{ kHz}$$

indicating that $f_{\alpha e}$ is in the upper audio range. Although the BJT is usable at frequencies up to f_T, the common-emitter h-parameter equivalent circuit of Fig. 6-14 is most useful for frequencies of the order of $f_{\alpha e}$ and below. Even at these frequencies, variation of h_{ie} and h_{re} with frequency may affect circuit behavior unless input current I_b is supplied by a current source that is essentially constant.

6-5 THE HYBRID-π EQUIVALENT CIRCUIT

If we look further into representing h parameters as combinations of elements of resistance and capacitance, the result is the hybrid-π model shown by Fig. 6-15. This model is widely used because it describes the small-signal behavior of a BJT for all frequencies at which the device has significant gain and because it lends insight into relationships between the physical structure of the BJT and the equivalent-circuit elements. Nevertheless, the model fails to meet the specifications of an ideal equivalent circuit for broadband use, owing to the frequency dependence of base resistance $r_{bb'}$. Fortunately, two $r_{bb'}$ values usually suffice, one for low frequencies and a smaller value for high-frequency operation. Although the hybrid-π model appears

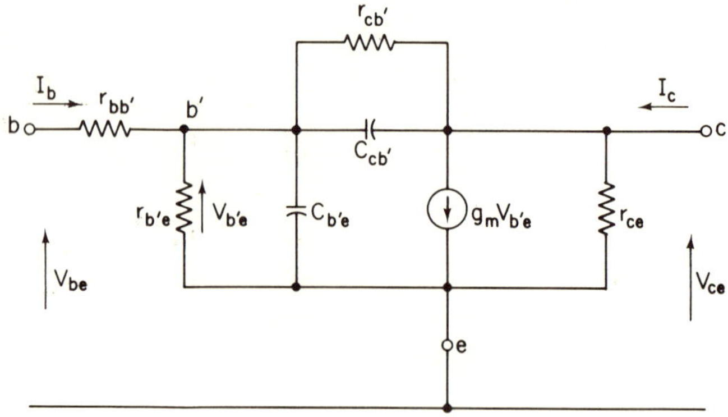

Fig. 6-15. The hybrid-π equivalent circuit. This circuit is good for frequencies up to the highest useful frequency. Although it appears formidable, it may be simplified for particular frequency ranges.

quite formidable compared to two-port models, it can be simplified for particular frequency ranges. Let us now consider the physical significance of each circuit element of the hydrid-π model.

Base resistance $r_{bb'}$ is the resistance that accounts for voltage drops in the base region caused by the transverse flow of majority carriers. Although these voltage drops are modeled in Fig. 6-15 by a series resistance in the base lead, in reality it should be represented by a lossy *RC* transmission line with series resistance in the base and shunt resistance and capacitance to the emitter. This accounts for $r_{bb'}$ variation with frequency.

Capacitance $C_{b'e}$ and *resistance* $r_{b'e}$ are the dynamic emitter capacitance C_e and resistance r_ϵ of Fig. 6-9, except that the magnitude of the resistance must be changed to give the correct low-frequency value of $V_{b'e}$ for a current of I_b, the major part of I_e being considered to flow through equivalent generator $g_m V_{b'e}$. Therefore, the base-emitter resistance $r_{b'e}$ is given by

$$r_{b'e} \cong \frac{r_\epsilon}{(1 - \alpha_{fbo})} \tag{6-70}$$

The base-emitter capacitance is unaffected by this mechanism, since it relates $V_{b'e}$ to stored base charge rather than to current.

Capacitance $C_{cb'}$ is the familiar collector-base transition capacitance C_{TC} plus a diffusion capacitance C_d resulting from base-width modulation. That is,

$$C_{cb'} = C_{TC} + C_d \tag{6-71}$$

When the load impedance is such that V_{ce} is large compared to V_{be} (that is, when voltage gain occurs), the base width changes with signal. The effect of collector voltage on base width is shown by Fig. 5-9, and the collector diffusion capacitance C_d is the capacitance associated with change in stored base charge caused by a change in collector voltage. That is,

$$C_d = \frac{\Delta Q_b}{\Delta v_{CB}} \tag{6-72}$$

and C_d is seen to be dependent on d-c collector current. C_d is usually a minor part of $C_{cb'}$.

The change in base-charge storage resulting from a change in collector voltage also causes a change in the recombination component of base current. This base-current change is accounted for in the hybrid-π model of Fig. 6-15 by *collector-base resistance* $r_{cb'}$.

As pointed out earlier and shown by Fig. 5-10, collector current changes with collector voltage because of the change in slope of the minority-carrier distribution in the base. This collector current change is accounted for in the model by *collector-emitter resistance* r_{ce}. Since collector-current change is the major part of the emitter-current change that results from a change in collector voltage, r_{ce} is much smaller than the collector-base resistance $r_{cb'}$. Except for *equivalent generator* $g_m V_{b'e}$, we have now considered the significance of each element of the hybrid-π model. Generator $g_m V_{b'e}$ is the current generator of the y-parameter model of Fig. 6-3, except for modification so that is controlled by the *internal* base-emitter voltage rather than by the external base voltage. The frequency dependence of current generator $g_m V_{b'e}$ is determined by that of voltage $V_{b'e}$, g_m being independent of frequency. The presence of a single generator in the hybrid-π model is an advantage over two-port models, all of which have two active elements.

6-5.1 Determination of Hybrid-π Parameters

The hybrid-π parameter values at a given operating point can be determined from a combination of h-and y-parameter measurements taken at different frequencies, the intent being to make each measurement at a frequency where it is unaffected by more than one or two parameters. Hybrid-π parameters are not usually specified by the manufacturer, but BJT specification sheets often provide sufficient information for their determination.

Determination of g_m. This is the one parameter that need not be measured, since it can be determined from physical relationships. For emitter currents large compared to the emitter-junction saturation current I_s, we obtain from the diode equation the *npn* transistor relationship

$$i_E \cong -I_s e^{ev_{B'E}/kT} \tag{6-73}$$

where $v_{B'E}$ is the internal base-emitter voltage (that is, $v_{B'E}$ is the voltage applied to the base-emitter junction). From Eq. (6-73), we obtain

$$\frac{di_E}{dv_{B'E}} \cong -\frac{e}{kT} I_s e^{ev_{B'E}/kT} \cong \frac{e}{kT} i_E \tag{6-74}$$

Assuming the quiescent collector current $|I_C|$ equal to the quiescent emitter current, we see from Eq. (6-74) that

$$g_m = \frac{di_C}{dv_{B'E}} \cong \frac{e}{kT} |I_C| \tag{6-75}$$

For a BJT having an α_{fb} value significantly less than unity, Eq. (6-75) becomes

$$g_m \cong \alpha_{fbo} \frac{e}{kT} |I_E| \tag{6-76}$$

and at room temperature

$$g_m \cong \alpha_{fbo} \frac{|I_E|}{25} \text{ mhos} \tag{6-77}$$

for I_C in milliamperes.

Determination of $r_{b'e}$. We see from the hybrid-π model of Fig. 6-15 that the low-frequency short-circuit forward current gain h_{feo} is given by

$$h_{feo} = \frac{g_m V_{b'e}}{I_b} \cong \frac{g_m V_{b'e}}{V_{b'e}/r_{b'e}} \tag{6-78}$$

since resistance $r_{cb'}$ is very large compared to $r_{b'e}$. Therefore, the base-emitter resistance $r_{b'e}$ is given by

$$r_{b'e} \cong \frac{h_{feo}}{g_m} \tag{6-79}$$

Determination of $r_{bb'}$. Again assuming $r_{cb'}$ large compared to $r_{b'e}$, we see that the low-frequency short-circuit input impedance h_{ieo} of the hybrid-π model is given by

$$h_{ieo} = r_{bb'} + r_{b'e} \tag{6-80}$$

Therefore, the low-frequency value of base resistance $r_{bb'}$ is given by

$$r_{bb'} = h_{ieo} - r_{b'e} \tag{6-81}$$

The high-frequency value of $r_{bb'}$ may be determined by measuring h_{ie} as a function of frequency. For frequencies such that $1/\omega C_{b'e} \ll r_{b'e}$ and $1/\omega C_{cb'} \ll r_{cb'}$, the real component of h_{ie} is essentially independent of frequency and equal to $r_{bb'}$, as seen from Fig. 6-15. Either the high-frequency value of $r_{bb'}$ and/or the value of the $r_{bb'}C_{cb'}$ product are often specified by the manufacturer.

Determination of $r_{cb'}$. We see from the hybrid-π model of Fig. 6-15 that the low-frequency open-circuit reverse voltage-transfer ratio h_{re} is given by

$$h_{reo} = \frac{r_{b'e}}{r_{b'e} + r_{cb'}} \cong \frac{r_{b'e}}{r_{cb'}} \tag{6-82}$$

Therefore, the collector-base resistance $r_{cb'}$ is given by

$$r_{cb'} \cong \frac{r_{b'e}}{h_{reo}} \tag{6-83}$$

Determination of r_{ce}. We see from Fig. 6-15 that an expression for the low-frequency open-circuit output admittance h_{oeo} of the hybrid-π model includes the terms r_{ce}, $r_{cb'}$, $r_{b'e}$, and g_m, the latter being included because voltage $V_{b'e}$ is established when h_{oeo} is determined. Except for r_{ce}, each of these terms is now known. Therefore, the h_{oeo} expression can be solved for collector-emitter resistance r_{ce}, resulting in

$$r_{ce} = \frac{r_{cb'} + r_{b'e}}{(r_{cb'} - r_{b'e})h_{oe} - (1 + g_m r_{b'e})} \cong \frac{r_{cb'}}{r_{cb'}h_{oe} - g_m r_{b'e}} \tag{6-84}$$

since $r_{cb'} \gg r_{b'e}$ and $g_m r_{b'e} \gg 1$

Determination of $C_{cb'}$. The value of the feedback capacitor $C_{cb'}$ can be determined by measuring capacitance C_{ob} between the collector and base leads of the transistor with the emitter open-circuited to alternating current. From the hybrid-π model, we see that

$$C_{ob} = C_{cb'} + C'_{cb} + C'_{ce} \tag{6-85}$$

where C'_{cb} and C'_{ce} are stray interelectrode capacitances. Therefore, $C_{cb'}$ is given by

$$C_{cb'} = C_{ob} - C'_{cb} - C'_{ce} \tag{6-86}$$

Stray capacitances C'_{cb} and C'_{ce} may be determined from high-frequency measurements of y_{re} and y_{oe} at frequencies such that

$$y_{re} = j\omega C'_{cb} \tag{6-87}$$

and

$$y_{oe} \cong j\omega(C'_{cb} + C'_{ce} + C_{cb'}) \tag{6-88}$$

Capacitance $C_{cb'}$ is sometimes specified by the manufacturer, but more often the value of C_{ob} is given.

Determination of $C_{b'e}$. As pointed out earlier, the frequency dependence of equivalent generator $g_m V_{b'e}$ is determined by that of $V_{b'e}$, $V_{b'e}$ varying with frequency because of base-emitter capacitance $C_{b'e}$. Therefore, $C_{b'e}$ can be determined by measuring the short-circuit forward current gain h_{fe} as a function of frequency. We see from the hybrid-π model that, at medium frequencies where $1/\omega C_{cb'} \ll r_{cb'}$, with the collector-emitter terminals short-circuited $V_{b'e}$ is given by

$$V_{b'e} = \frac{I_b}{1/r_{b'e} + j\omega(C_{b'e} + C_{cb'})} \tag{6-89}$$

Thus the short-circuit forward current gain h_{fe} is

$$h_{fe} = \frac{g_m V_{b'e}}{I_b} = \frac{g_m}{1/r_{b'e} + j\omega(C_{b'e} + C_{cb'})} \tag{6-90}$$

and we see that the h_{fe}-*cutoff frequency* f_{hfe} (the frequency at which $|h_{fe}|$ is 0.707 of its low-frequency value) is given by

$$f_{hfe} = \frac{1}{2\pi r_{b'e}(C_{b'e} + C_{cb'})} \tag{6-91}$$

Since $r_{b'e}$ and $C_{cb'}$ are known, $C_{b'e}$ can be obtained by using Eq. (6-91) if f_{hfe} is determined.

At frequencies well above f_{hfe}, say $3 f_{hfe}$ and above, $r_{b'e}$ has negligible effect on the magnitude of h_{fe} and Eq. (6-90) becomes

$$|h_{fe}|_{f \gg f_{hfe}} \cong \frac{g_m}{\omega(C_{b'e} + C_{cb'})} \tag{6-92}$$

Solving this expression for $C_{b'e}$, we obtain

$$C_{b'e} \cong \frac{g_m}{\omega\, |h_{fe}|_{f \gg f_{hfe}}} - C_{cb'} \tag{6-93}$$

which becomes

$$C_{b'e} \cong \frac{g_m}{\omega_T} - C_{cb'} \tag{6-94}$$

for the transition frequency f_T at which $|h_{fe}| = 1$. Equations (6-93) and (6-94) may also be written as

$$C_{b'e} \cong \frac{g_m}{\omega_{hfe} h_{feo}} - C_{cb'} \tag{6-95}$$

since we have shown that

$$\omega\, |h_{fe}|_{f \gg f_{hfe}} \cong \omega_{hfe} h_{feo} \cong \omega_T \tag{6-96}$$

Either f_{hfe}, or f_T, or the value of $|h_{fe}|$ at a frequency very large compared to f_{hfe} is usually specified by the manufacturer.

As an illustration of the methods for obtaining values of hybrid-π parameters and to provide an indication of their orders of magnitude, let us determine the parameters of the typical type 2N1382 whose h parameters are given in Table 6-1 on p. 250.

From Eq. (6-77), we see that at room temperature

$$g_m \cong \alpha_{fbo} \left| \frac{I_C}{25} \right| \cong 0.988 \left(\frac{1}{25} \right) \cong 0.04 \text{ mho}$$

Using Eq. (6-79), we find that the value of $r_{b'e}$ is

$$r_{b'e} = \frac{h_{feo}}{g_m} = \frac{80}{0.04} = 2000 \text{ ohms}$$

and from Eq. (6-81)

$$r_{bb'} = h_{ieo} - r_{b'e} = 2400 - 2000 = 400 \text{ ohms}$$

This is the low-frequency value of $r_{bb'}$. It so happens that the manufacturer provides no information that enables us to obtain the high-frequency value of $r_{bb'}$.

From Eq. (6-83), we find

$$r_{cb'} \cong \frac{r_{b'e}}{h_{reo}} = \frac{2000}{7 \times 10^{-4}} = 3 \times 10^6 = 3 \text{ megohms}$$

and from Eq. (6-84),

$$r_{ce} \cong \frac{r_{cb'}}{r_{cb'}h_{oe} - g_m r_{b'e}} = \frac{3 \times 10^6}{3 \times 10^6 \times 40 \times 10^{-6} - 0.04 \times 2 \times 10^3}$$
$$= \frac{3 \times 10^6}{120 - 80} = 75{,}000 \text{ ohms}$$

The type 2N1382 specification sheet provides no information that enables us to determine $C_{cb'}$. Usually C_{ob} is given and, for a low-power germanium BJT such as the 2N1382, might be of the order of 10 pF. We shall consider this to be the "worst-case value" of $C_{cb'}$.

The minimum value of $f_{\alpha b}$ of the 2N1382 transistor is specified as 2 MHz. Assuming $f_T = f_{\alpha b}$, we obtain from Eq. (6-94)

$$C_{b'e} + C_{cb'} \cong \frac{g_m}{\omega_T} = \frac{0.04}{6.28 \times 2 \times 10^6} = 0.32 \times 10^{-8} = 3200 \text{ pF}$$

and $C_{b'e} \cong 3200$ pF. The same value of $C_{b'e}$ is obtained by using Eq. (6-91) and considering f_{hfe} to be $f_{\alpha b}/h_{feo}$. We see from this calculation that lack of knowledge of the magnitude of $C_{cb'}$ does not preclude determination of $C_{b'e}$ of a low-frequency BJT. For a high frequency device, the $C_{cb'}$ term in Eqs. (6-91) through (6-95) may be significant.

The hybrid-π parameter values of the type 2N1382 BJT are listed in Table 6-4.

Table 6-4 Hybrid-π parameters of a typical type 2N1382 BJT at $I_E = -1$ mA, $V_C = -5$ V

$g_m = 0.04$ mho	$r_{ce} = 75{,}000$ ohms
$r_{b'e} = 2000$ ohms	$C_{cb'} = 10$ pF*
$r_{bb'} = 400$ ohms	$C_{b'e} = 3{,}200$ pF
$r_{cb'} = 3$ megohms	

* Estimated value.

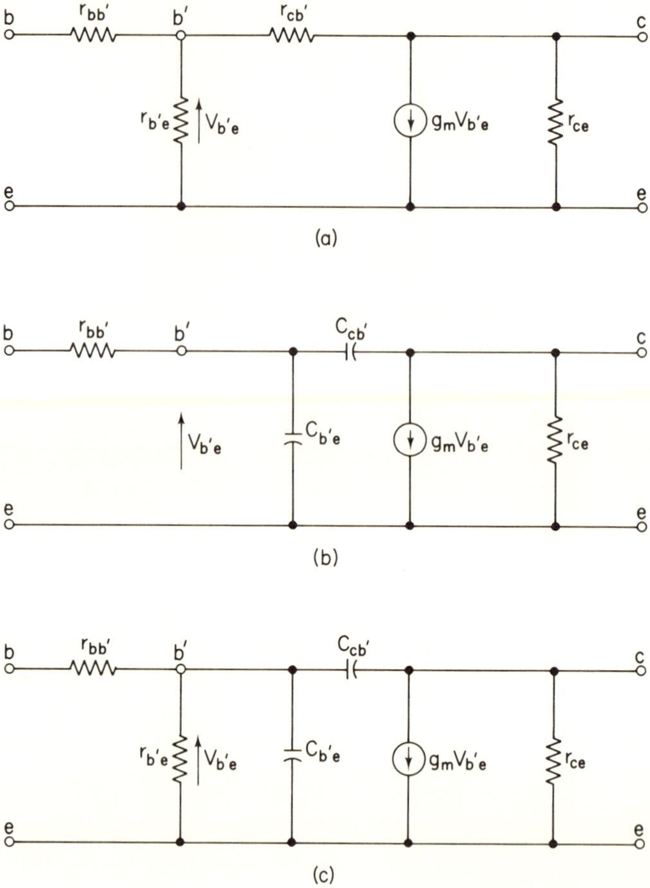

Fig. 6-16. Simplified hybrid-π models. (a) Model for use at low frequencies, where $f \ll 1/(2\pi r_{cb'} C_{cb'})$. (b) Model for use at high frequencies, where $f \gg 1/(2\pi r_{b'e} C_{b'e}) = f_{\alpha e}$. (c) Model for use in the middle- and high-frequency range.

6-5.2 Simplified Hybrid-π Circuits

At first glance, the complete hybrid-π model of Fig. 6-15 may appear somewhat complicated for use in circuit analysis. However, it is not usually necessary to use this "all-frequency" model, since some elements of the model have negligible effect over certain frequency ranges. For example, at "low-frequencies" capacitances $C_{cb'}$ and $C_{b'e}$ act as open circuits resulting in the model of Fig. 6-16(a). In order to utilize this model, we must determine the low-frequency region.

The relative importance of elements $r_{cb'}$ and $C_{cb'}$ may be determined by finding the frequency at which the reactance of $C_{cb'}$ equals the resistance of $r_{cb'}$. This is given by

$$\frac{1}{2\pi r_{cb'}C_{cb'}} = \frac{1}{(2\pi)(3 \times 10^6)(10 \times 10^{-12})} \cong 5.4 \text{ kHz} \qquad (6\text{-}97)$$

for the BJT of Table 6-4. Therefore, at frequencies much lower than this, resistance $r_{cb'}$ is dominant compared with $C_{cb'}$, and $C_{cb'}$ can be neglected. Conversely, at frequencies much higher than that given by Eq. (6-97) capacitor $C_{cb'}$ becomes dominant, and $r_{cb'}$ can be neglected.

In a similar way, the relative importance of elements $r_{b'e}$ and $C_{b'e}$ may be determined. At frequencies much lower than

$$\frac{1}{2\pi r_{b'e}C_{b'e}} = \frac{1}{2\pi(2000)(3200)(10^{-12})} \cong 25 \text{ kHz} \qquad (6\text{-}98)$$

for the BJT of Table 6-4, the resistance $r_{b'e}$ is dominant over $C_{b'e}$, and at frequencies much higher than this, $C_{b'e}$ is dominant over $r_{b'e}$. Since the frequency of Eq. (6-97) is lower than that of Eq. (6-98), Eq. (6-97) is utilized to determine the upper-frequency limit of the low-frequency model.*

The dominance of $C_{b'e}$ over $r_{b'e}$ at frequencies much higher than that given by Eq. (6-98) and the fact that $r_{cb'}$ is negligible compared to $C_{cb'}$ at these frequencies suggest the "high-frequency" model of Fig. 6-16(b). The frequency of Eq. (6-98) is seen to be the common-emitter alpha-cutoff frequency $f_{\alpha e}$ or f_{hfe}.

For many BJTs, the frequency of Eq. (6-98) is sufficiently large compared to that of Eq. (6-97) that a "middle-frequency" range exists. In this range, resistor $r_{cb'}$ and capacitor $C_{b'e}$ can be neglected, resulting in a middle-frequency model. Because of the limited usefulness of this model, it is more common to work with the model of Fig. 6-16(c), in which both $C_{cb'}$ and $C_{b'e}$ are left in and only $r_{cb'}$ is neglected. This model is seen to be applicable over both the middle- and high-frequency ranges.

Equivalent input circuit. Circuit calculations with the hybrid-π model can often be simplified by assuming the feedback impedance $(r_{cb'} \| C_{cb'})$ to be large compared to the parallel combination of resistance r_{ce} and the collector load impedance.† That is, current through the feedback impedance is considered to be negligible compared to the source current $g_m V_{b'e}$. This assumption enables us to develop an equivalent input circuit that permits rapid calculation of voltage $V_{b'e}$ which controls the dependent current source $g_m V_{b'e}$.

* For frequencies less than one-third that of Eq. (6-97), the error in the magnitude of the feedback impedance (parallel impedance of $r_{cb'}$ and $C_{cb'}$) is less than 5 percent if $C_{cb'}$ is neglected.

† Two vertical lines indicate "in parallel with." Hence "$r_{cb'} \| C_{cb'}$" means $r_{cb'}$ in parallel with $C_{cb'}$.

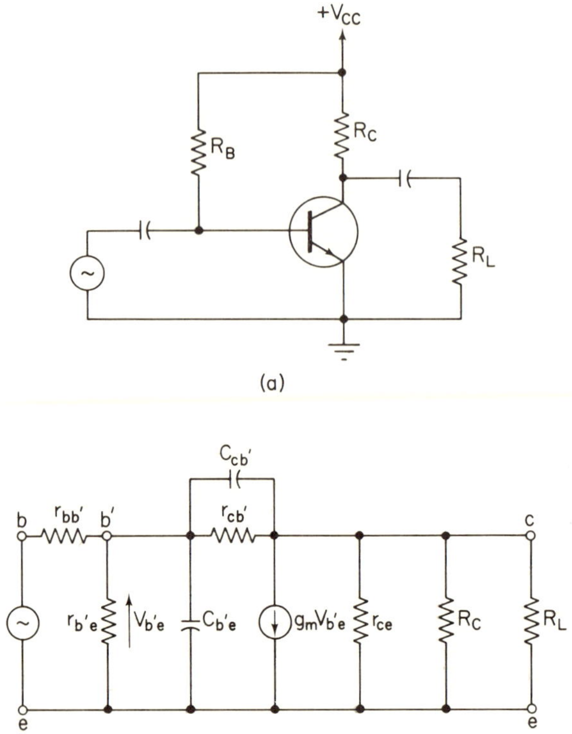

Fig. 6-17. (a) Elementary common-emitter amplifier stage. (b) Equivalent circuit for the amplifier state of (a), using the hybrid-π model of the BJT and assuming negligible signal current through R_B and negligible signal voltages across the coupling capacitors.

As an illustration of the use of simplified hybrid-π models, let us consider the elementary common-emitter amplifier stage of Fig. 6-17(a). The small-signal equivalent circuit of the stage, if we use the hybrid-π model of the BJT and assume negligible signal current through R_B and negligible signal voltages across the coupling capacitors, is shown by Fig. 6-17(b). For low frequencies, this equivalent circuit may be simplified to that of Fig. 6-18(a) by using the low-frequency BJT model and combining parallel resistances r_{ce}, R_C, and R_L into a single resistance R_{ec}.

If we assume current I_f through feedback resistor $r_{cb'}$ to be negligible compared to generator current $g_m V_{b'e}$, voltage V_{ec} of Fig. 6-18(a) is given by

$$V_{ec} = g_m V_{b'e} R_{ec} \tag{6-99}$$

The voltage $V_{b'c}$ across $r_{cb'}$ is seen to be

$$V_{b'c} = V_{b'e} + V_{ec} = V_{b'e} + g_m V_{b'e} R_{ec} \tag{6-100}$$

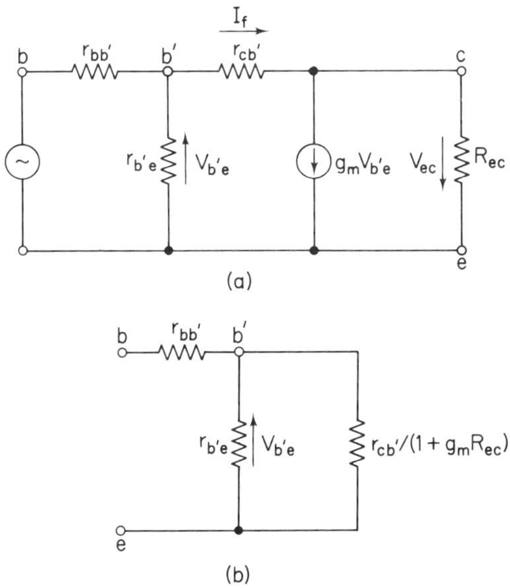

Fig. 6-18. (a) Amplifier equivalent circuit of Fig. 6-17(b) simplified for low frequencies by assuming BJT-model capacitances to be open circuits and representing r_{ce}, R_C, and R_L by R_{ec}. (b) Low-frequency equivalent input circuit of the circuit of (a).

from which

$$I_f = \frac{V_{b'c}}{r_{cb'}} = \frac{V_{b'e}(1 + g_m R_{ec})}{r_{cb'}} \tag{6-101}$$

Therefore, the network to the right of $r_{b'e}$ of Fig. 6-18(a) can be represented by the resistance $r_{cb'}/(1 + g_m R_{ec})$, resulting in the equivalent input circuit of Fig. 6-18(b). Using this input circuit, we can readily calculate voltage $V_{b'e}$, in terms of base-input voltage or current, and determine the output voltage and current.

For middle and high frequencies, the equivalent circuit of Fig. 6-18(b) may be simplified to that of Fig. 6-19(a). From analysis similar to that leading to Eq. (6-101), it may be seen that feedback current I_f is given by

$$I_f = j\omega C_{cb'}(1 + g_m R_{ec})V_{b'e} \tag{6-102}$$

Therefore, the network to the right of $C_{b'e}$ of Fig. 6-19(a) can be represented by the capacitance $(1 + g_m R_{ec})C_{cb'}$, resulting in the equivalent input circuit of Fig. 6-19(b).

An important restriction applying to the input circuits of Figs. 6-18(b) and 6-19(b) is that the BJT load must be a pure resistance. In the general case, a BJT-load impedance Z_L results in an equivalent emitter-collector impedance Z_{ec}, and an admittance Y_{eq}, given by

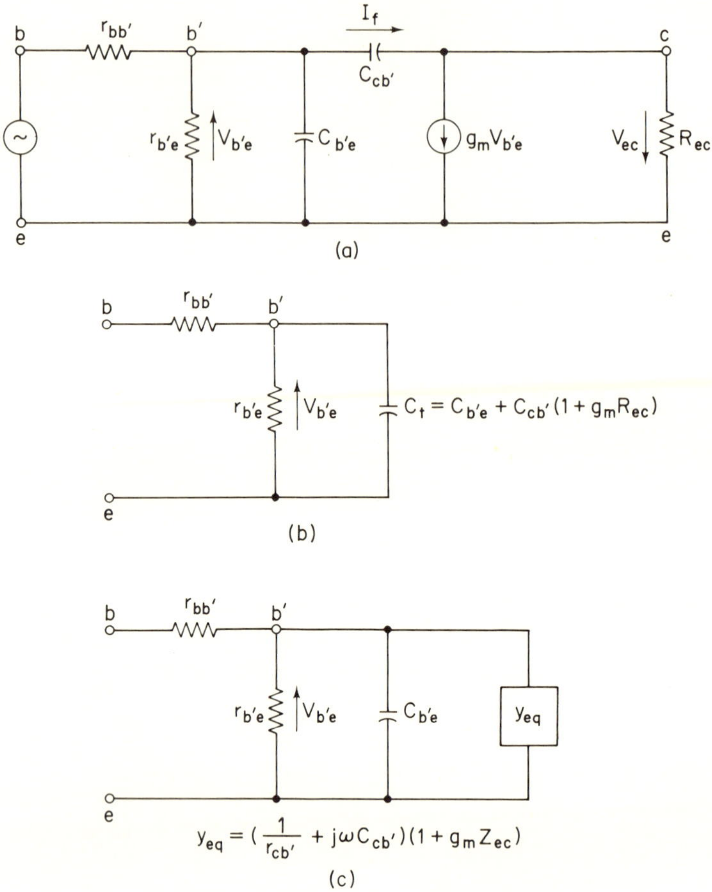

Fig. 6-19. (a) Middle- and high-frequency equivalent circuit of the amplifier of Fig. 6-17(a). (b) Equivalent input circuit of the circuit of (a). (c) All-frequency equivalent input circuit of a common-emitter amplifier stage for a load impedance Z_L. (If R_L in Fig. 6-18(b) is replaced by Z_L, Z_{ec} is the parallel combination of r_{ce}, R_C, and Z_L.)

$$Y_{eq} = (1 + g_m Z_{ec})\left(\frac{1}{r_{cb'}} + j\omega C_{cb'}\right) \tag{6-103}$$

is reflected back to the input circuit. The all-frequency equivalent input circuit for any BJT load impedance is shown in Fig. 6-19(c).

6-6 EQUIVALENT-CIRCUIT PARAMETERS

Although the hybrid-π model is perhaps the most useful small-sginal BJT representation, hybrid-π parameters cannot be measured directly;

instead, they are determined from two-port network parameters. Of the several sets of two-port transistor parameters, h parameters lend themselves most readily to measurement, owing to the ease with which the low-impedance base-emitter circuit can be open-circuited and that with which the high-impedance collector circuit can be short-circuited. For this reason, h parameters are usually specified by manufacturers and are the basis of our discussion of the measurement of parameters and their dependence on operating conditions.

6-6.1 Measurement of *h* Parameters

Transistor-parameter values at low frequencies can be determined either from static characteristics or by dynamic measurement. From the definitions of Eqs. (6-9) through (6-12), we see that each common-emitter h parameter is the slope of a static characteristic curve. This suggests the need for four characteristic-curve families in order to determine an h-parameter set from static characteristics. However, the four h parameters can be determined from two characteristic curve families by utilizing the separations of the curves. The fixed-v_{CE} curves of i_B versus v_{BE} of the common-emitter base characteristic and the fixed-i_B curves of i_C versus v_{CE} of the common-emitter collector characteristic lend themselves readily to the determination of h parameters.

From the definitions of the open-circuit output admittance h_{oe} and the short-circuit forward current-transfer ratio h_{fe}, given as

$$h_{oe} = \left.\frac{di_C}{dv_{CE}}\right|_{i_B \text{ fixed}} \tag{6-12}$$

and

$$h_{fe} = \left.\frac{di_C}{di_B}\right|_{v_{CE} \text{ fixed}} \tag{6-11}$$

we see that h_{oe} is the slope of the fixed-i_B collector-characteristic curves, such as those in Fig. 5-18 and that the vertical separation of the curves is proportional to h_{fe}. We also see from Eqs. (6-9) and (6-10), which define the short-circuit input impedance h_{ie} and the open-circuit reverse voltage-transfer ratio h_{re} as

$$h_{ie} = \left.\frac{dv_{BE}}{di_B}\right|_{v_{CE} \text{ fixed}} \tag{6-9}$$

and

$$h_{re} = \left.\frac{dv_{BE}}{dv_{CE}}\right|_{i_B \text{ fixed}} \tag{6-10}$$

that h_{ie} is the reciprocal of the slope of the fixed-v_{CE} base-characteristic curves, such as those of Fig. 5-22, and that the horizontal separation of the curves is proportional to h_{re}. In general, h_{re} is too small for accurate static determination at v_{CE} values greater than a few tenths of a volt, as may be seen from Fig. 5-22, and all four parameters can be determined more accurately by dynamic methods.

Measurement of h_{ie} and h_{fe}. The short-circuit parameters h_{ie} and h_{fe} may be measured by applying a signal to the base and measuring the alternating current in the base and in the short-circuited collector circuit, as shown by Fig. 6-20. The d-c operating point is adjusted to the desired value by adjusting base resistor R_B and supply voltage V_{CC}. A separate base-supply voltage V_{BB} is shown for greater flexibility, but it may be eliminated by connecting R_B to the positive V_{CC} terminal. The input and output currents are determined from the a-c voltages across resistances R_1 and R_2. It is preferable that R_1 be much larger than h_{ie} and R_2 must be small enough to be considered a short circuit. Making R_2 adjustable facilitates verification that the collector is short-circuited, since R_2 may be considered a short circuit if the effects of doubling it are less than the uncertainty of the measurements.* Because of the low value of R_2, it is essential to verify that C_2 adequately bypasses the collector bias-supply circuit. This may be done by comparing the a-c voltage across C_2 with V_3. With the BJT at the desired d-c operating points, a-c voltages V_1, V_2, and V_3 are measured and used in conjunction with R_1, R_B, and R_2 to determine h_{fe} and h_{ie}. If R_1 is very large compared to h_{ie} (that is, if $V_1 \gg V_2$) and if R_B is so large that signal current through it is negligible, we see that

$$h_{fe} \simeq \frac{V_3 R_1}{V_1 R_2} \tag{6-104}$$

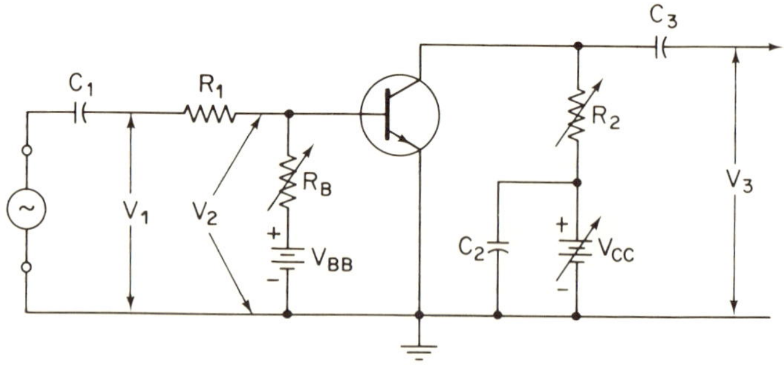

Fig. 6-20. Circuit for measurement of short-circuit parameters h_{fe} and h_{ie}. R_2 must be small enough to effectively short the collector-emitter circuit for alternating current.

* IEEE Standards on Transistor Testing.

and

$$h_{ie} \cong \frac{V_2 R_1}{V_1} \tag{6-105}$$

As a practical set of conditions for low-power transistors, R_2 might be 100 ohms, R_1 might be 100,000 ohms, and input voltage V_1 might be 1.0 V. The reading of V_3 in millivolts would then approximately equal h_{fe}, and V_2 in millivolts would approximately equal h_{ie} in hundreds of ohms.

Since h_{fe} and h_{ie} are small-signal parameters, it is essential that the measurements just described be small signal in nature. This may be verified by doubling a-c input voltage V_1 and determining that the effects on the measured values of h_{fe} and h_{ie} are less than the uncertainty of the measurements.*

Measurement of $(1 + h_{fb})$. We may measure common-base short-circuit parameters h_{fb} and h_{ib} using the arrangement of Fig. 6-20 by operating the BJT in the common-base configuration. However, there is little point in measuring the h_{fb} of typical BJTs by this method because of the difficulty in distinguishing between the magnitude of h_{fb} and unity. Instead, it is preferable to measure the quantity $(1 + h_{fb})$, or $(1 - \alpha_{fb})$, by an arrangement such as that of Fig. 6-21, in which voltage V_3 is an indication of alternating current $(1 + h_{fb})I_e$. Values of external base resistance R_2 that effectively short-circuit the collector-base circuit in this arrangement can be much larger than if R_2 were in the collector, permitting the measurement of small values of $(1 + h_{fb})$. Both positive and negative values of $(1 + h_{fb})$ may be measured, the relative phase of voltages V_3 and V_1 indicating the sign of the quantity. The quantity $(1 + h_{fb})$ is, of course, positive for low values of collector voltage V_{CB}, decreasing to zero and becoming negative with increase in V_{CB}. It is of interest to compare the value of V_{CB} for which $(1 + h_{fb})$ equals zero ($\alpha_{fb} = 1$) with the collector-emitter breakdown voltage for comparable

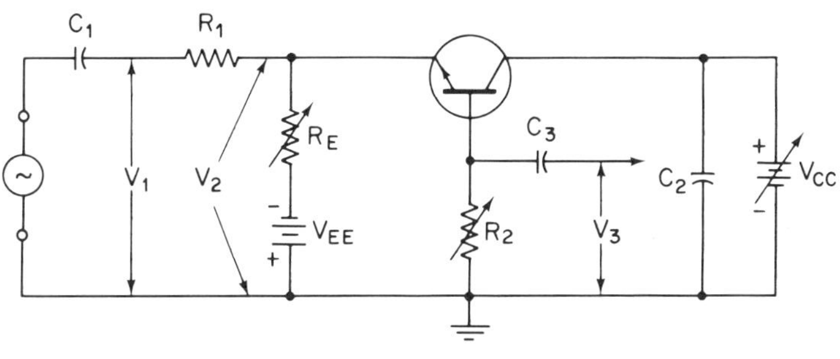

Fig. 6-21. Circuit for measurement of short-circuit parameters $(1 + h_{fb})$ and h_{ib}. R_2 must be small enough to short the collector-base circuit for alternating current.

* IEEE Standards on Transistor Testing.

values of electrode current. By so doing, one can verify that common-emitter collector breakdown may be considered an "alpha-equal-unity phenomenon."* If it is desired, h_{ib} can be determined by measuring a-c voltage V_2 with R_2 set to zero.

Measurement of h_{oe} and h_{re}. The open-circuit parameters h_{oe} and h_{re} may be measured by using the circuit arrangement of Fig. 6-22. The d-c operating point is controlled by adjusting R_B, R_C, V_{CC}, and V_{BB}, maintaining R_C large enough to avoid overloading the signal generator and using an R_B value large enough effectively to open-circuit the base to alternating current. R_B may be considered an open circuit if the effects of halving it are less than the uncertainty of the measurements.† Resistor R_1 is for the purpose of current measurement in the h_{oe} determination, perhaps being of the order of 100 ohms for low-power BJTs, and is shorted for the measurement of h_{re}.

With switch S closed and the BJT at the desired operating point, a-c voltages V_1 and V_3 are measured and, by definition,

$$h_{re} = \frac{V_3}{V_1} \tag{6-106}$$

A sensitive voltmeter is required to measure V_3, since V_3 is likely to be of the order of tenths of a millivolt for small-signal measurements. As in measuring h_{fe} and h_{ie}, the criterion for small-signal measurements is that the effect of doubling input voltage V_1 should be less than the uncertainty of the measurements. It is seen that the peak value of V_1 must not exceed the d-c collector

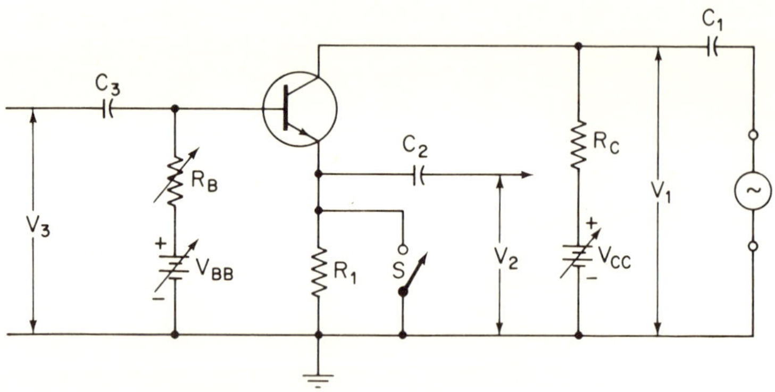

Fig. 6-22. Circuit for measurement of open-circuit parameters h_{oe} and h_{re}. R_B must be large enough to effectively open the transistor base-emitter circuit to alternating current.

* It is, of course, possible that the upper limit of v_{CE} is determined by base reach-through (punch-through).

† IEEE Standards on Transistor Testing.

voltage V_{CE}; otherwise, the collector junction is forward biased during part of the cycle, resulting in gross nonlinearity.

In the determination of h_{oe}, S is opened, the d-c operating point readjusted if necessary, and a-c voltage V_2 measured. From the definition of h_{oe}, it is seen that

$$h_{oe} \cong \frac{V_2}{R_1 V_1} \tag{6-107}$$

and for $V_1 = 1.0$ V and $R_1 = 100$ ohms, V_2 in millivolts approximately equals h_{oe} in tens of micromhos.*

Circuits such as those of Figs. 6-20 through 6-22 are utilized in commercial transistor test sets† so arranged that the parameters are either read directly from an appropriately calibrated a-c voltmeter or represented as ratios of a-c voltages.

Bridge measurement of low-frequency parameters. All four low-frequency common-emitter h parameters may be measured by bridge methods that express them as ratios of either a-c voltages or resistances. In general, bridge methods are more accurate than the methods described so far.

Since h_{ie} and $1/h_{oe}$ are resistances at low frequencies, both may be measured by means of an a-c Wheatstone bridge. In the measurement of h_{ie}, the arrangement of Fig. 6-20 may be used by connecting the bridge to the source terminals. Resistance R_1 is, of course, set to zero, so that the resistance measured is the parallel combination of h_{ie} and R_B, and it may be necessary to increase the magnitude of blocking capacitor C_1. Using an a-c resistance bridge in conjunction with the arrangement of Fig. 6-22 in the measurement of h_{oe} is usually unsatisfactory, since R_C is likely to be much smaller than $1/h_{oe}$. This difficulty may be alleviated by incorporating the collector-emitter circuit as one arm of the bridge, collector bias being obtained by connecting a d-c supply in series with either the null indicator or the a-c source. Commercial bridges usually have a provision for measuring a quantity in the presence of bias.

The parameters h_{ie}, h_{re}, and $1/h_{oe}$ are expressed as ratios of a-c voltages in instruments such as the General Radio Type 1661-B Vacuum-Tube Bridge. For the measurement of h_{ie} and $1/h_{oe}$, the Type 1661-B Bridge is arranged for resistance (r_p) measurement. The voltage-ratio (μ) arrangement of the bridge is used for measuring h_{re}; it is recognized that h_{re} is a reverse parameter. An advantage of the Type 1661-B Bridge is the ease with which electrode bias can be provided.

* Equation (6-107) is approximate as a result of the assumption that V_2 is very small compared to V_1. This is an excellent assumption when R_1 is properly chosen.

† Baird Atomic Corporation Types GP-4 and KP-3 are typical instruments of this type.

A bridge arrangement for measuring h_{fb} (or α_{fb}) is shown by Fig. 6-23. The a-c voltmeter V is a null indicator, and balance is obtained by adjusting the potentiometer. With the bridge in balance ($V = 0$), the collector-base circuit is short-circuited to alternating current, and the a-c currents are as indicated. For conditions of balance, we see that

$$(I_e + h_{fb}I_e)R_1 + h_{fb}I_eR_2 = 0 \tag{6-108}$$

from which

$$h_{fb} = -\frac{R_1}{R_1 + R_2} = -\frac{R_1}{R_T} \tag{6-109}$$

Since R_T is fixed, h_{fb} is a linear function of the angular position of the potentiometer, which can be calibrated for direct reading of h_{fb}. This circuit is capable of high accuracy and can be used for measuring h_{fb} magnitudes quite close to unity. It is utilized in commercial instruments such as the Boonton Radio Corporation Type 275-A Transistor Test Set. Electrode bias can be provided in the circuit of Fig. 6-23 by connecting high-impedance supplies between the BJT terminals, high impedances being required in order to avoid disturbing a-c conditions, or by inserting bias supplies in series with the emitter and collector electrodes.

By modifying the h_{fb} bridge, as shown by Fig. 6-24, h_{fe} can be measured. From Eq. (6-109), we see that for conditions of balance,

$$h_{fb} = -\frac{R_T}{R_T + R_3} \tag{6-110}$$

Since h_{fe} may be expressed in terms of h_{fb} as

$$h_{fe} = -\frac{h_{fb}}{1 + h_{fb}} \tag{6-37}$$

substituting for h_{fb} yields

$$h_{fe} = \frac{R_T}{R_3} \tag{6-111}$$

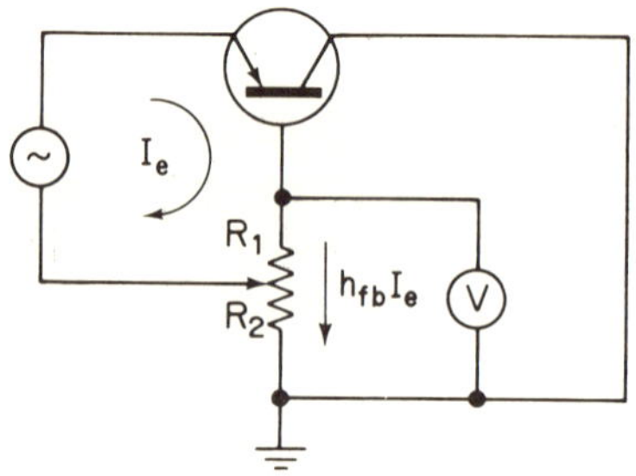

Fig. 6-23. Bridge arrangement for the measurement of h_{fb} or α_{fb}. For balance, $h_{fb} = -R_1/(R_1 + R_2)$.

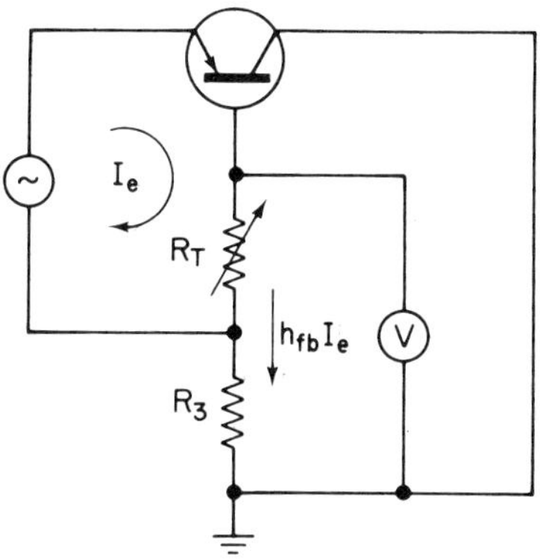

Fig. 6-24. Bridge arrangement for the measurement of h_{fe} or α_{fe}. For balance, $h_{fe} = R_T/R_3$.

indicating that h_{fe} is a linear function of R_T. Thus, R_T can be calibrated for direct reading of h_{fe}.

In addition to h_{fb} and h_{fe}, h_{ib} may be measured by using the Type 275-A Transistor Test Set, the basic arrangement being that of Fig. 6-25(a). The equivalent circuit for this arrangement is shown in Fig. 6-25(b), from which we see that, for conditions of balance,

$$h_{ib} = \frac{R_s R_1}{R_2} \tag{6-112}$$

Therefore, the potentiometer can be calibrated for direct reading of h_{ib}.

Measurement of high-frequency parameters. The circuits of Figs. 6-20 through 6-23 may be used to measure the magnitudes of transistor h parameters as a function of frequency, the upper frequency limit of such measurements being determined by stray capacitances, by resistors becoming reactive, etc. In carefully constructed circuits, such as those in commericial test sets, this upper limit is of the order of 1 MHz. Clearly, higher-frequency methods are required.

Commercial instruments that may be used for high-frequency parameter measurements are three-terminal bridges, such as the Wayne Kerr Type B601 R-F Bridge (15 kHz to 5 MHz) and Type B801 V-H-F Admittance Bridge (1 to 100 MHz), the General Radio Type 1607-A Transfer Function and Immittance Bridge (20 to 1500 MHz), the Boonton Radio Type 250A RX Meter (0.5 to 250 MHz), and the Rhode and Schwarz Diagraph (30 to 2400 MHz). The Wayne Kerr Bridges may be used to measure y-parameter sets, the three-terminal arrangement permitting the measurement of trans-admittances; the Type 1607-A Bridge may be used to measure h-, y-, and z-parameter sets, since it is capable of measuring transadmittances, trans-

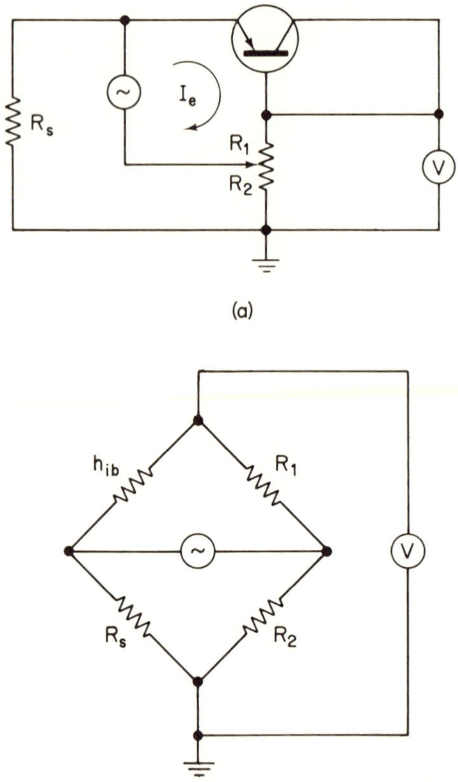

Fig. 6-25. (a) Bridge arrangement for the measurement of h_{ib}. (b) Equivalent circuit for the arrangement of (a).

impedances, transfer-current ratios, and transfer-voltage ratios in addition to open-circuit and short-circuit impedances and admittances; the RX Meter may be used to measure y_{ib}, y_{ie}, and y_o (y_{ob} and y_{oe} being identical); and the Diagraph may be used to obtain y parameters.

In general, the measurement of complex two-port BJT parameters over a wide frequency range is a tedious process. Therefore, it is appropriate to review the principal reason for making parameter measurements in order to minimize the number that need be made. The usual reason is, of course, to evaluate the parameters of a model used to represent the BJT. Except for narrow-band considerations, for which an h, y, or z set at the center frequency might be used, the hybrid-π model is usually preferred. From Sec. 6-5.1, we know that the hybrid-π parameter values can be determined from the low-frequency h parameters, the collector-base capacitance $C_{cb'}$ and a high-frequency measurement of the magnitude of h_{fe}. In addition, the high-frequency value of base-resistance $r_{bb'}$ may be required. Since we have considered methods for measuring low-frequency h parameters in some detail, let us now direct our attention to the measurement of $C_{cb'}$ and the high-frequency values of h_{fe} and $r_{bb'}$.

Determination of collector-base capacitance $C_{cb'}$. As pointed out in Sec. 6-5.1, $C_{cb'}$ can be determined from the common-base output capacitance C_{ob}. Capacitance C_{ob} can be determined from h_{ob} at frequencies such that h_{ob} is greater than about three times the low-frequency value. At these frequencies

$$h_{ob} \cong \omega C_{ob} \tag{6-113}$$

From Eq. (6-113), it may be seen that when h_{ob} is measured at a frequency of 159 kHz, C_{ob} in picofarads is equal to h_{ob} in micromhos, providing Eq. (6-113) applies at 159 kHz. In addition to obtaining C_{ob}, by using one of the several methods already described for determining h_{ob}, C_{ob} may be measured directly by using capacitance bridges, Q meters, etc.

Determination of high-frequency h_{fe}. For low-frequency BJTs, the circuit arrangement of Fig. 6-20 permits measurement of the magnitude of h_{fe} at a frequency sufficiently high for the determination of hybrid-π parameters. The frequency range of this basic arrangement can be extended considerably by constructing two such circuits as nearly identical as possible, so that they have the same frequency characteristics, and using one as a reference. The reference circuit has a short between the base and collector terminals of the BJT socket, as shown in the basic arrangement of Fig. 6-26. A 50-ohm attenuator, and connections are made using 50-ohm co-axial cables. Resistance R_1 should be large compared to 50 ohms and to h_{ie} of the BJT under test. In the measurement procedure, a current I_1 is supplied to the reference-circuit input and output voltage V_1 is noted, V_1 being proportional to I_1. The same current (I_1) is then applied to the attenuator input and the attenuator adjusted until test-circuit output voltage V_2 is equal to V_1. Voltage V_2 is seen to be proportional to the current $h_{fe}AI_1$, where A is the attenuation. Thus, when $V_2 = V_1$,

$$KI_1 = Kh_{fe}AI_1 \tag{6-114}$$

and

$$h_{fe} = \frac{1}{A} \tag{6-115}$$

That is, the magnitude of h_{fe} expressed in decibels is equal to the attenuator setting in decibels. If a continuously adjustable attenuator is not available, it is necessary to set the attenuator and adjust the frequency for $V_1 = V_2$.

The method of Fig. 6-26 may be adapted for measuring the magnitude of h_{fb} as a function of frequency by using appropriate reference and test circuits.

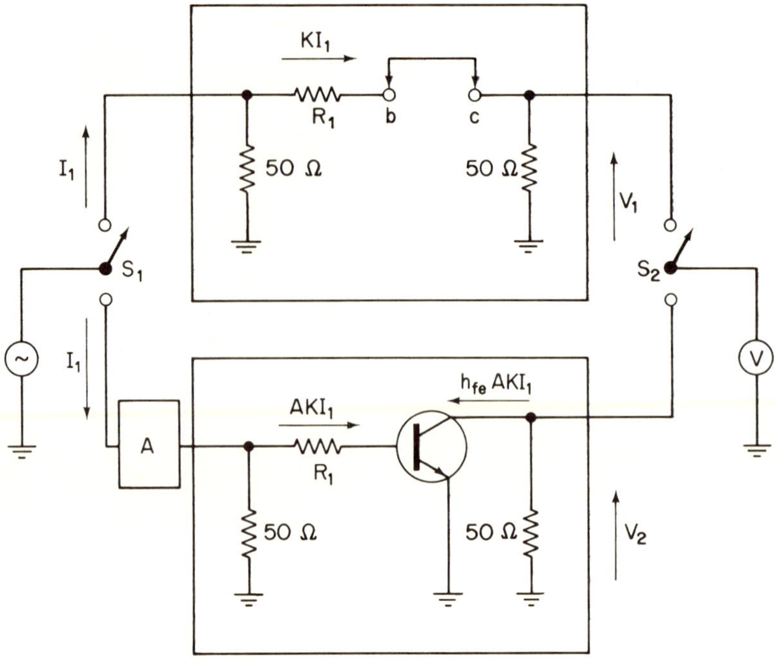

Fig. 6-26. Dual-channel method for measuring the magnitude of h_{fe} as a function of frequency. If the two channels have the same frequency response, h_{fe} expressed in dB is equal to the attenuation of A in dB when A is adjusted so that $V_1 = V_2$.

In this arrangement, the reference-circuit short is between the emitter and collector terminals, and the attenuator is placed in the reference-circuit input, since h_{fb} is less than unity.

The complex value of h_{fe} may be measured directly, by using an instrument such as the Type 1607-A Transfer Function and Immittance Bridge, or may be determined from the forward transadmittance y_{fe} and the input admittance y_{ie}, being given by

$$h_{fe} = \frac{y_{fe}}{y_{ie}} \tag{6-116}$$

as indicated in Table 6-3 on p. 255. Admittance parameters y_{fe} and y_{ie} may be measured by using three-terminal bridges such as the Wayne Kerr Type B601 or Type B801.

Determination of high-frequency $r_{bb'}$. As shown in Sec. 6-5.1, the high-frequency value of base resistance $r_{bb'}$ is the real value of h_{ie} or $1/y_{ie}$ at high frequencies. Thus $r_{bb'}$ can be obtained by using the Type 1607-A Transfer Function and Immittance Bridge to measure h_{ie} or y_{ie}, or from y_{ie} measured

Type 2N3117
Small-signal characteristics (f = 1 kHz)

Symbol	Characteristic	Min.	Typ.	Max.	Units	Test Conditions	
h_{ie}	Input resistance	10	18	24	kohms	$I_C = 1.0$ mA	$V_{CE} = 5.0$ V
h_{oe}	Output conductance		15	40	μmho	$I_C = 1.0$ mA	$V_{CE} = 5.0$ V
h_{re}	Voltage feedback ratio		425	800	$\times 10^{-6}$	$I_C = 1.0$ mA	$V_{CE} = 5.0$ V
h_{fe}	Small-signal current gain	400	620	900		$I_C = 1.0$ mA	$V_{CE} = 5.0$ V

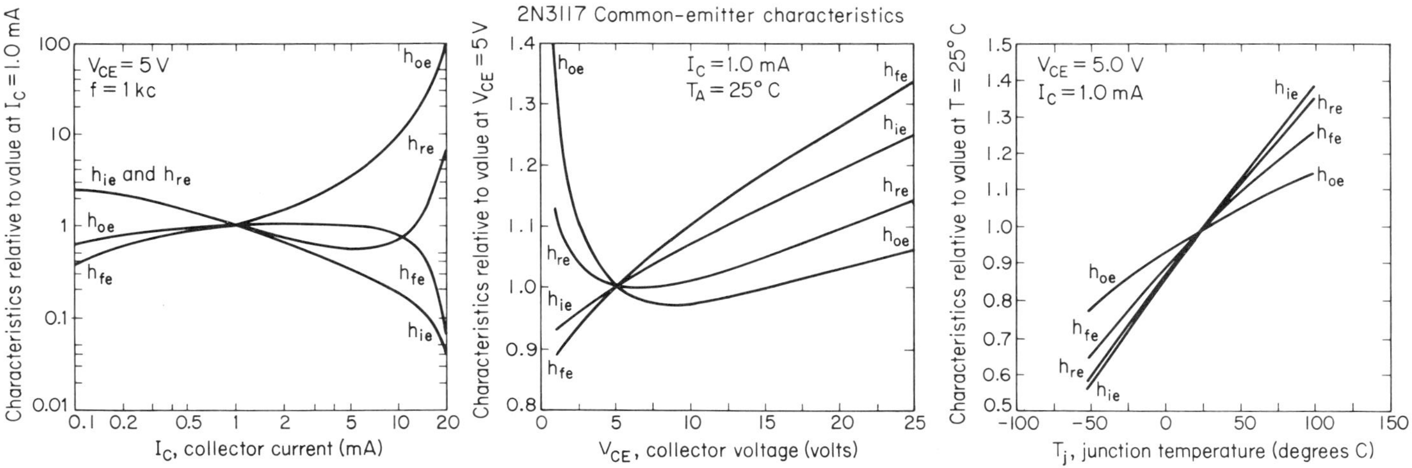

Fig. 6-27. The spread of common-emitter *h* parameters for a typical type of silicon BJT and the dependence of these parameters on d-c operating conditions and temperature. (Courtesy Fairchild Semiconductor Corp.)

by using either the Type 250A RX Meter or a high-frequency admittance bridge such as the Wayne Kerr Type B801.

6-6.2 Parameter Dependence Upon Operating Conditions

The effects of d-c operating point, temperature, and frequency on the common-emitter h parameters of typical BJTs are shown by Figs. 6-27 and 6-28. All four parameters are seen to be sensitive to operating conditions, particularly to d-c current and frequency. We shall briefly consider factors affecting h_{ie}, h_{oe}, and h_{re}, the dependence of h_{fe} upon operating conditions having been considered in Secs. 5-3 and 6-4.

Factors affecting h_{ie}. At low frequencies, the common-base short-circuit input impedance h_{ib} is largely the dynamic resistance of the emitter-base junction. By assuming emitter current I_E large compared to emitter-base junction saturation current I_s, we obtain from Eq. (2-95)

$$h_{ib} \cong \frac{kT}{eI_E} \cong \frac{25}{I_E} \text{ ohms} \tag{6-117}$$

at room temperature for I_E expressed in milliamperes. From Eq. (6-34), the common-emitter short-circuit input impedance h_{ie} is

$$h_{ie} = \frac{h_{ib}}{1 + h_{fb}} \cong h_{fe}h_{ib} \tag{6-118}$$

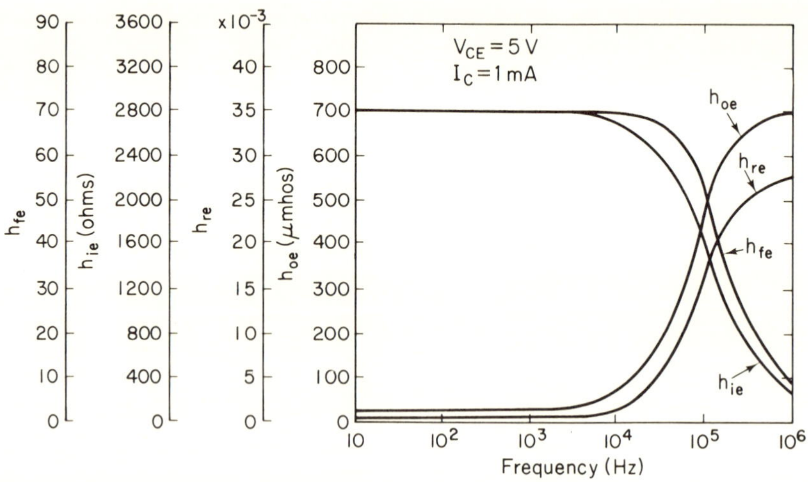

Fig. 6-28. The effects of frequency on the magnitudes of the common-emitter h parameters of a typical 2N1302 germanium BJT.

from which

$$h_{ie} \cong \frac{h_{fe}kT}{eI_E} \cong \frac{25}{I_E} h_{fe} \text{ ohms} \tag{6-119}$$

at room temperature for I_E in milliamperes. From this expression, we see that h_{ie} varies inversely with I_C, since $I_C \cong I_E$, and directly with absolute temperature, and that h_{ie} increases with V_{CE} because of the increase of h_{fe} with collector voltage. This is in agreement with the h_{ie} plots of Fig. 6-27. The decrease in h_{ie} with frequency, shown by Fig. 6-28, is the effect of the base-emitter junction capacitance.

Factors Affecting h_{oe}. At low frequencies, the open-circuit output admittance h_{oe} is an indication of the effect of collector voltage v_{CE} on collector current i_C, v_{CE} affecting i_C through the mechanism of base- width modulation. As base width decreases with increase in v_{CE}, a larger base minority-carrier density is required for fixed base current i_B. This results in a larger carrier-density gradient in the base with a corresponding increase in i_C; the larger the i_B value, the greater the increase in i_C with v_{CE}. Since i_C is proportional to i_B, h_{oe} varies directly with i_C, as shown by Fig. 6-27.

Figure 6-27 shows that h_{oe} is dependent upon collector voltage, decreasing with increase in v_{CE} at low values but increasing with v_{CE} at higher voltages. At low v_{CE} values, minority-carrier injection into the base by the collector reduces i_C, causing i_C to be quite dependent upon v_{CE}. This effect decreases with increase in v_{CE}, decreasing h_{oe}. However, when v_{CE} becomes large enough to significantly decrease the base width, the effect on h_{oe} changes direction, since the effect of v_{CE} on i_C increases as base width decreases.

The increase in h_{oe} with temperature can be attributed to the increase in h_{fe}, since

$$h_{oe} \cong \frac{h_{ob}}{1 + h_{fb}} \cong h_{fe}h_{ob} \tag{6-44}$$

and to the increase in base-region minority-carrier density with temperature.

The increase in h_{oe} with frequency, shown by Fig. 6-28, is the effect of the collector-base junction capacitance. For frequencies such that the magnitude of h_{oe} is three or more times the low-frequency value,

$$h_{oe} \cong \omega C_{ce} \cong \omega h_{fe} C_c \tag{6-120}$$

as obtained from Eq. (6-69).

Factors affecting h_{re}. From the hybrid-π equivalent circuit, it is seen that the low-frequency value of the open-circuit reverse-voltage ratio h_{re} is the ratio $r_{b'e}/r_{cb'}$ or the product $r_{b'e}g_{cb'}$. Resistance $r_{b'e}$ is proportional to the dynamic resistance of the emitter-base junction, and $g_{cb'}$ varies with d-c conditions much like h_{oe}. Since $r_{b'e}$ and $g_{cb'}$ vary oppositely with collector current, h_{re} is not highly dependent upon i_C, as shown by Fig. 6-27. However, the variation of h_{re} with collector voltage is much the same as that of h_{oe}, $g_{cb'}$ being significantly affected by v_{CE}. Therefore, h_{re} decreases with increase in v_{CE} at low values and increases with v_{CE} at higher voltages. Since both $r_{b'e}$ and h_{oe} increase with temperature, h_{re} increases with temperature. From the middle-and high-frequency hybrid-π model of Fig. 6-16(c), it is seen that h_{re} increases with frequency, approaching the value $C_{cb'}/(C_{cb'} + C_{b'e})$ at high frequencies.

Collector-base capacitance $C_{cb'}$. As pointed out in Sec. 6-4, the collector-base capacitance is the junction transition capacitance plus a diffusion capacitance. Thus $C_{cb'}$ decreases with increase in collector voltage, owing to the change in transition capacitance, as shown in Fig. 6-29. Since collector diffusion capacitance is associated with the change in base stored-charge caused by a change in collector voltage, owing to base-width modulation, and base charge increases with collector current, $C_{cb'}$ increases with collector current as shown in Fig. 6-29.

I_{CBO} dependence upon temperature. Although not an equivalent-circuit parameter, the collector-base junction reverse current I_{CBO} and its variation with

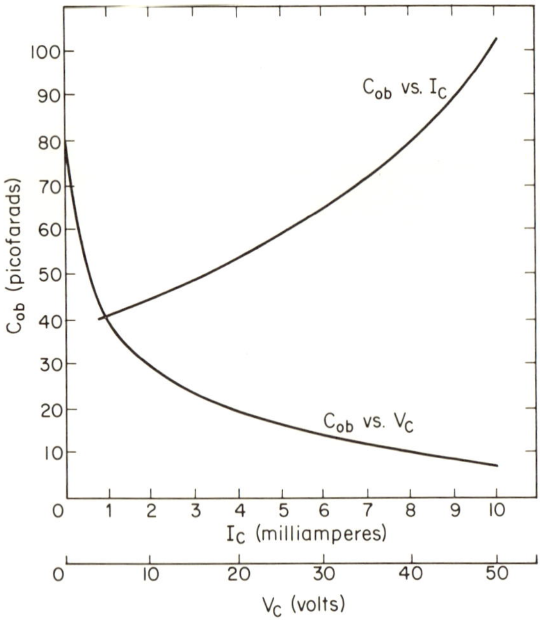

Fig. 6-29. The effects of operating point on the collector capacitance C_{ob} of a typical alloy-junction germanium BJT.

temperature have an important role in BJT circuit design. I_{CBO} varies exponentially with temperature, thus following a straight line on a semilogarithmic plot, as shown by Fig. 6-30 for both a germanium BJT and a silicon BJT. Therefore, when the maximum I_{CBO} value of a BJT type is specified at each of two junction temperatures, the maximum value for any other temperature can be readily determined.

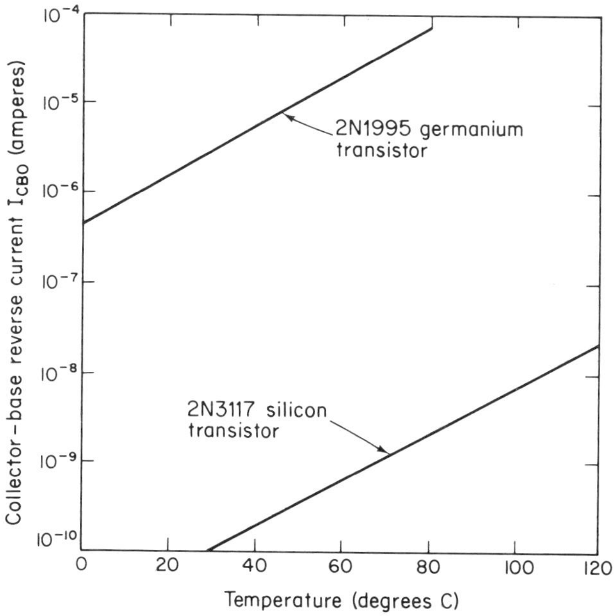

Fig. 6-30. Variation of I_{CBO} with temperature for a typical 2N1995 germanium BJT at $V_C = 20$ V and a typical 2N3117 silicon BJT at $V_C = 45$ V. (Courtesy Texas Instruments, Inc. and Fairchild Semiconductor Corp.)

PROBLEMS

6-1. Determine the approximate value of each common-emitter h parameter of the BJT for which the following date are obtained:

i_C	v_{CE}	i_B	v_{BE}
7 mA	10 V	100 μA	0.150 V
10	20	100	0.160
16	30	200	0.170

6-2. Determine h_{ie}, h_{fe}, and h_{oe} of the BJT of Fig. P5-2 at the operating point of Prob. 5-3.

6-3. Using the h-parameter equivalent-circuit model of the BJT, determine the a-c voltages of Probs. 5-4 and 5-5, assuming $h_{re}V_{ce}$ to have negligible effect.

6-4. Using the common-emitter h-parameter model of a BJT operating in the common-collector configuration, derive expressions for common-collector h parameters in terms of common-emitter parameters (see Table 6-2).

6-5. Repeat Prob. 6-4 for the common-base h parameters expressed in terms of common-emitter h parameters.

6-6. Using the expressions derived in Probs. 6-4 and 6-5, check the common-collector and common base h-parameter values of Table 6-1.

6-7. Using the common-emitter h-parameter representation of a BJT, derive expressions for common-emitter y parameters in terms of common-emitter h-parameters (see Table 6-3).

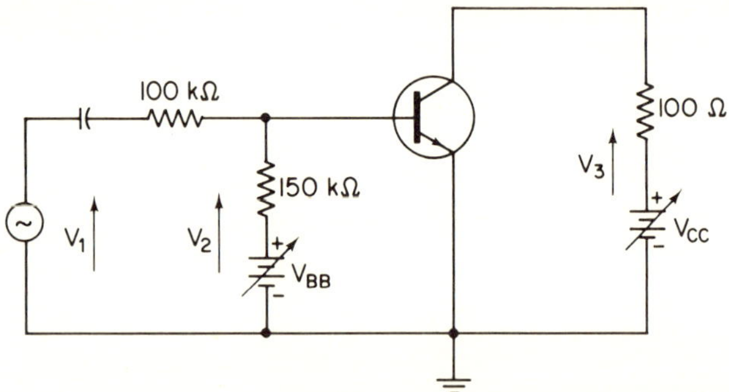

$V_1 = 1.0$ V, $V_2 = 12$ mV, $V_3 = 40$ mV

(a)

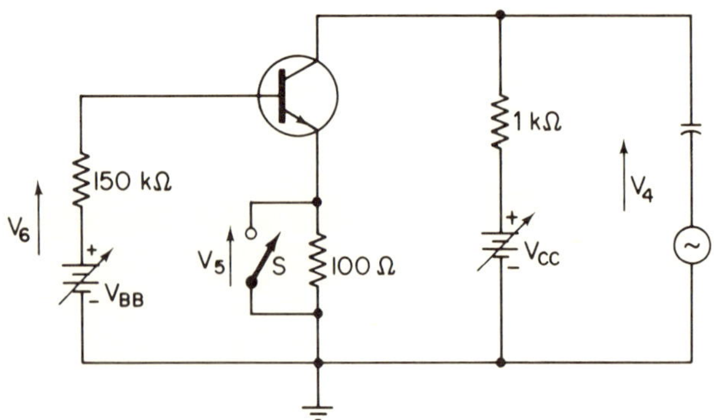

$V_4 = 1.0$ V, $V_5 = 2.0$ mV, $V_6 = 0.22$ mV (S closed)

(b)

Figure P6-10

6-8. Determine y_{ie} and y_{fe} of a BJT having low-frequency h_{fb} and h_{ib} values of -0.965 and 12 Ω, respectively.

6-9. Bridge measurements at a frequency of 50 MHz on the BJT of Prob. 6-8 yield the following results: y_{ie} of $G = 6$ mmhos in shunt with $C = 28$ pF; y_{fe} of 52 mmhos in shunt with $C = -72$ pF. Determine the 50-MHz value of h_{fe}.

6-10. In using the circuits of Fig. P6-10 to measure low-frequency h parameters of a BJT, V_{CC} and V_{BB} are adjusted for the specified operating point, and a-c voltage readings obtained are as indicated. Assuming the 100-Ω resistor to be an a-c short circuit and the 150-kΩ resistor to be an a-c open circuit, determine the common-emitter h parameters.

6-11. Determine the errors resulting from the assumptions of Prob. 6-10 by using the parameters of Prob. 6-10 to determine voltages V_2, V_3, V_3, and V_4 for the specified values of V_1 and V_4.

6-12. In the circuit of Fig. P6-12, the BJT may be considered to be operating with the collector-base terminals short-circuited to alternating current and R_E may be considered to be an a-c open circuit. Using the low-frequency voltage readings indicated, determine h_{ib} and h_{fb}.

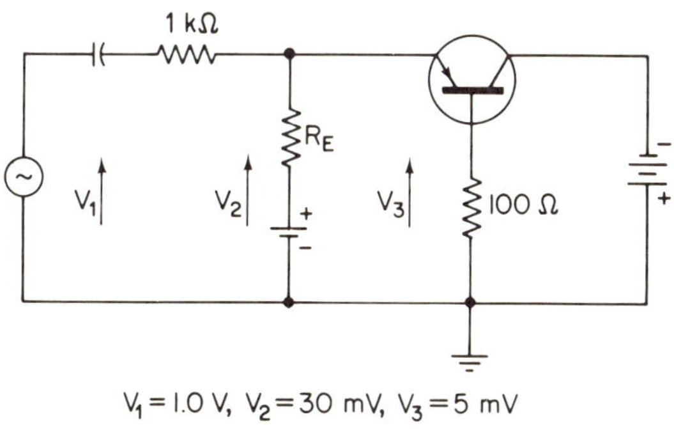

Figure P6-12

6-13. A *pnp* alloy-junction germanium BJT has a 25-mil diameter circular emitter and a base thickness of 0.5 mil. Determine the diffusion capacitance that might be expected for an emitter current of 10 mA.

6-14. An alloy-junction BJT has a base width of 0.5 mil. What common-base alpha-cutoff frequency f_{hfb} might be expected if the BJT type is (a) *pnp* silicon, (b) *npn* silicon, (c) *pnp* germanium, and (d) *npn* germanium?

6-15. A BJT has a low-frequency h_{fb} of -0.98 and an h_{fb}-cutoff frequency of 10 MHz. What h_{fb} magnitude might be expected at 5 MHz?

6-16. What h_{fe}-cutoff frequency might be expected for the BJT of Prob. 6-15? What transition frequency f_T might be expected?

6-17. A BJT has a low-frequency h_{fe} of 100 and an h_{fe} magnitude of 10 at a frequency of 20 MHz. What value might be expected for each of the following: (a) h_{fe} magnitude at 40 MHz, (b) h_{fe}-cutoff frequency, and (c) transition frequency f_T?

6-18. The following measurements were made on a type 2N3638 diffused silicon *pnp* BJT at $V_C = -5$ V, $I_E = 1.0$ mA: $f_{hfe} = 0.76$ MHz, $C_{ob} = 7$ pF, and, at low frequencies, $h_{ib} = 27\ \Omega$, $h_{fe} = 61$, $h_{re} = 3.2 \times 10^{-4}$, $h_{oe} = 32\ \mu$mhos. Assuming $C_{cb'} = C_{ob}$, determine the hybrid-π equivalent-circuit parameter values.

6-19. Using the hybrid-π parameters of Table 6-4 and assuming negligible a-c current through R_B and negligible a-c voltages across the coupling capacitors in the amplifier stage of Fig. P6-19, determine the magnitudes of input impedance V_1/I_1, voltage ratios V_2/V_1 and V_2/V_g, and current gain I_2/I_1 at low frequencies and at a frequency of 100 kHz.

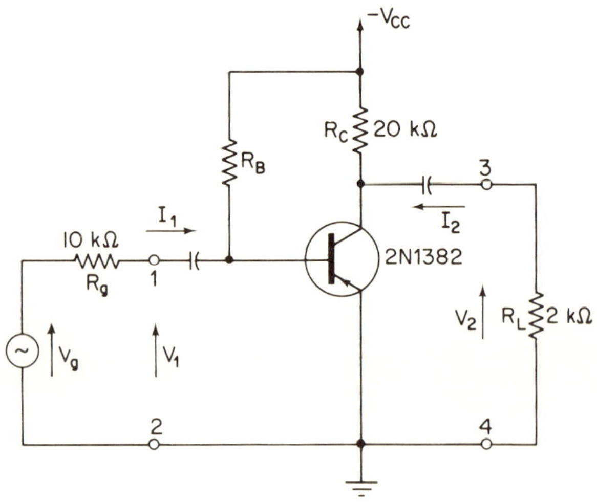

Figure P6-19

6-20. Compare the power developed in the 2-kΩ load resistor of Prob. 6-19 with that for R_L connected to source terminals 1-2. Express this power difference in decibels.

6-21. The maximum I_{CBO} specifications of the type 2N1132 silicon BJT are 1 μA at a junction temperature of 25°C and 100 μA at 150°C. What maximum I_{CBO} might be expected at a junction termperature of 100°C?

7

Basic BJT Amplifier Stages

The use of the BJT in elementary common-base and common-emitter amplifier circuits and graphical analysis of these circuits have been considered in Sec. 5-2. In this chapter, the small-signal performance of basic amplifier configurations is studied by using the equivalent-circuit models presented in Chapter 6. Items of particular interest are current, voltage, and power gains and input and output resistances. The discussion of this chapter will be restricted to low-frequency linear operation of the BJT.

7-1 THE THREE CONFIGURATIONS

As already implied, BJT amplifier stages are classified according to the transistor terminal that is common to the input and output circuits. Thus, in addition to the *common-base* and *common-emitter* circuits already mentioned, we have the *common-collector* configuration. The latter is commonly known as an *emitter follower*, since the potential of the emitter follows that of the base when changes occur.

Common-base amplifier. A common-base amplifier stage using a *pnp* BJT is shown in Fig. 7-1. A disadvantage of this configuration is the requirement of separate emitter and collector bias supplies, d-c emitter current

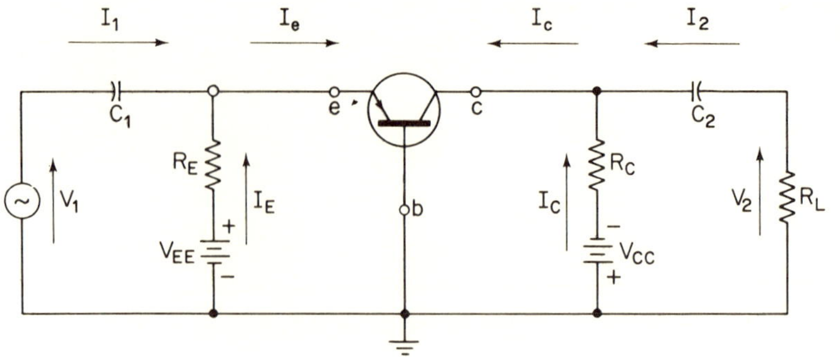

Fig. 7-1. Common-base amplifier configuration. Input-signal current I_1 results in emitter-signal current I_e being superimposed upon emitter-bias current I_E, causing collector-signal current I_c to be superimposed upon collector-bias current I_C. Amplifier current gain I_2/I_1 is less than unity, but voltage gain V_2/V_1 exceeds unity and power gain occurs if R_L is large compared to the input impedance of the stage.

I_E and d-c collector current I_C being of opposite polarity. Emitter current I_E is given by

$$I_E \simeq \frac{V_{EE}}{R_E} \tag{7-1}$$

since R_E is usually large compared to the emitter-base resistance of the BJT in order that most of a-c input current I_1 flow into the BJT. A-C emitter current I_e is superimposed upon I_E and is less than I_1 because of the diversion effect of R_E. As a result of I_e, a-c collector current I_c is generated and superimposed upon I_C. Output current I_2 is less than I_c because of the diversion effect of collector resistor R_C, indicating that R_C should be large compared to load resistance R_L in order that most of I_c flow through R_L. Output current I_2 is seen to be less than input current I_1, because I_c is less than I_e, owing to BJT parameter h_{fb} (or α_{fb}) having a magnitude less than unity, and because of the diversion effects of R_E, h_{ob}, and R_C. Thus the current gain A_I of the stage, defined as

$$A_I = \frac{I_2}{I_1} \tag{7-2}$$

is less than unity, and R_L must exceed the input impedance of the stage in order for power gain to occur. That is, the common-base stage can have power gain only through the existence of voltage gain. Voltage gain A_V is defined as

$$A_V = \frac{V_2}{V_1} \tag{7-3}$$

and power gain A_p is the product of the current and voltage gains; that is,

$$A_p = \frac{V_2 I_2}{V_1 I_1} = A_V A_I \tag{7-4}$$

Common-collector amplifier. Two arrangements of the common-collector or emitter-follower amplifier stage are shown in Fig. 7-2. If the d-c supply has negligible impedance to alternating current, a-c conditions are the same

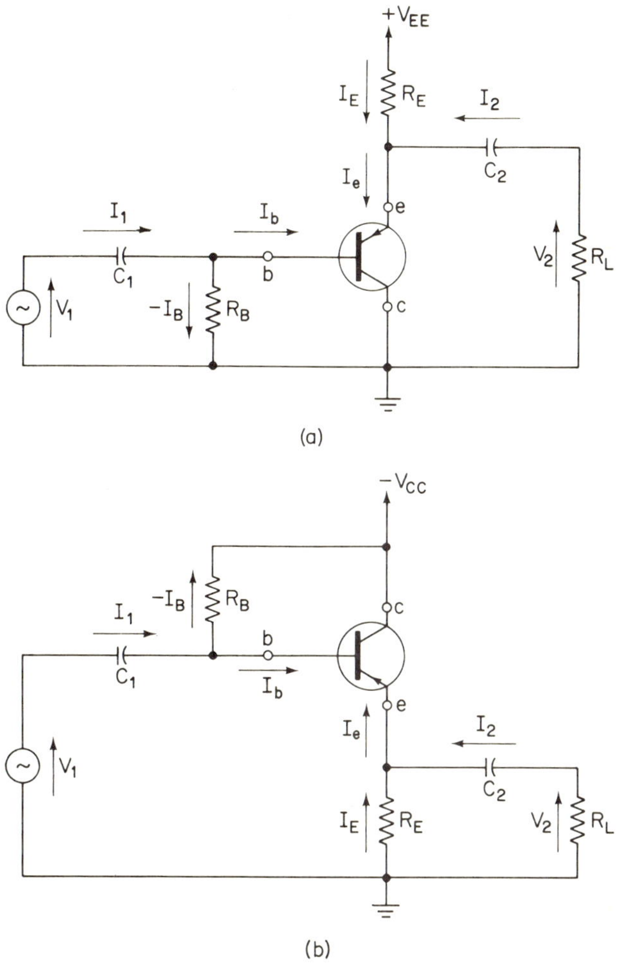

Fig. 7-2. Common-collector (or emitter-follower) amplifier arrangements. For negligible d-c supply impedance to alternating current, identical a-c conditions exist in the two arrangements. Amplifier voltage gain V_2/V_1 is less than unity, but current gain I_2/I_1 may exceed unity, resulting in power gain.

in both arrangements, the collector being at a-c ground potential in Fig. 7-2(b) and literally grounded in Fig. 7-2(a). Unlike the common-base stage, a single d-c supply provides electrode bias, base-bias current I_B being given by

$$I_B = \frac{V_{CB}}{R_B} \cong \frac{V_{CE}}{R_B} \tag{7-5}$$

The dependence of I_B upon collector voltage V_{CE} rather than upon d-c supply voltage provides an element of bias stabilization, since an increase in emitter-bias current I_E caused by an increase in temperature or by a change in BJT characteristics results in a decrease in I_B.

Input-signal current I_1 results in a-c base current I_b being superimposed upon d-c base current I_B and generating a-c emitter current I_e, which may be large compared to I_1, since BJT parameter h_{fc} is large compared to unity. Thus, unlike the common-base stage, the common-collector stage is capable of current gain. However, output voltage V_2 is less than V_1, owing to signal voltage V_{be} being in phase with V_2, and load impedance R_L must be less than the input impedance of the stage in order for power gain to occur. Thus, the emitter follower can have power gain only through the existence of current gain. As in the common-base stage, coupling resistance R_E should be large compared to load resistance R_L in order that most of a-c emitter current I_e flow through R_L.

Common-emitter amplifier. A common-emitter amplifier stage is shown in Fig. 7-3. Like the emitter-follower stage, a single d-c supply provides electrode bias, base-bias current I_B being given by

$$I_B \cong \frac{V_{CC}}{R_B} \tag{7-6}$$

Unlike the emitter-follower stages of Fig. 7-2, this stage has fixed base-bias current in that I_B is independent of the BJT characteristics. As we shall see, this is an impractical situation. Practical bias arrangements for common-emitter stages are considered in Chapter 8.

As in the emitter-follower stage, input current I_1 results in a-c base current I_b being superimposed upon d-c base current I_B, producing a-c collector current I_c, which may be large compared to I_1 since BJT parameter h_{fe} is large compared to unity. With coupling resistance R_C large compared to load resistance R_L, in order that load current I_2 constitute most of I_c, the common-emitter stage is capable of current gain. In fact, common-emitter and emitter-follower stages have comparable current gains, since the magnitudes of common-emitter parameters h_{fe} and h_{oe} are almost identical to those of common-collector parameters h_{fc} and h_{oc}.

Unlike the emitter-follower stage, the common-emitter stage is capable of voltage gain, the voltage gain being comparable to that of the common-

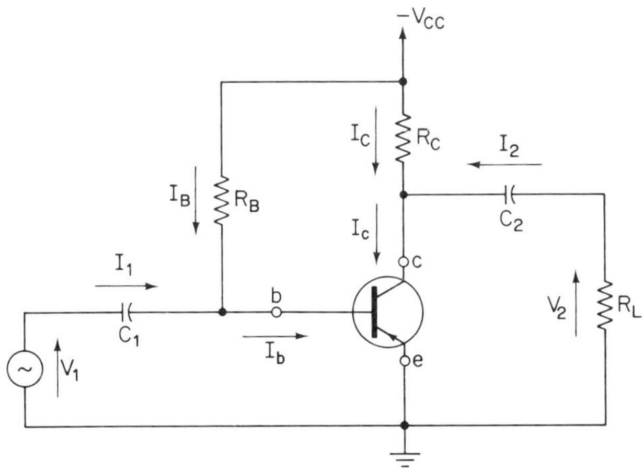

Fig. 7-3. A common-emitter amplifier configuration. Electrode biases are provided by a single d-c supply, base-bias current I_B being approximately V_{CC}/R_B and collector bias current I_C being $h_{FE}I_B$ (providing $I_C R_C$ is significantly less than V_{CC}).

base stage, as will be shown in Sec. 7-2. Since it combines the voltage-gain capability of the common-base stage with the current-gain capability of the emitter-follower stage, the common-emitter stage is by far the most widely used amplifier configuration.

7-2 COMPARISON OF THE BASIC CONFIGURATIONS

The use of the h-parameter transistor model provides a convenient means for comparing low-frequency characteristics of the three basic amplifier configurations, since expressions for current, voltage, and power gains and for input and output resistances are independent of the configuration. Once these general expressions are determined, the appropriate h-parameter set can be used to determine the characteristics of a particular configuration.

Stage gains and input resistance. The general small-signal h-parameter representation of the amplifier stages of Figs. 7-1, 7-2, and 7-3 is shown in Fig. 7-4; coupling capacitors C_1 and C_2 are assumed to have negligible reactances. Resistance R_g' is the equivalent source resistance, in that it includes the input-coupling resistances, and R_L' is the equivalent load resistance, which includes the output-coupling resistance. Thus R_g' is the parallel combination of R_g and R_E for the common-base configuration and of R_g and R_B for the other two configurations. Likewise, R_L' is the parallel com-

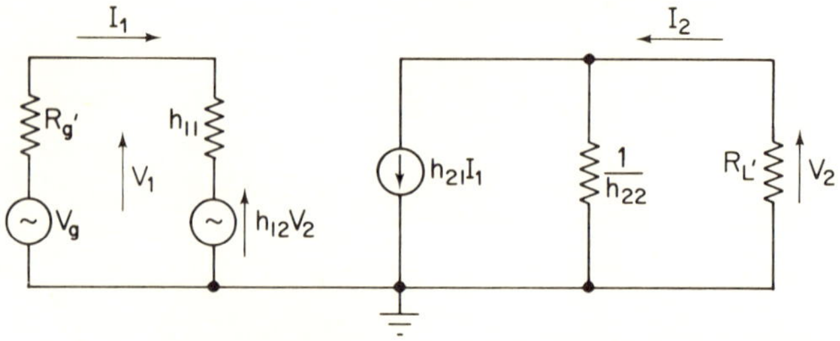

Fig. 7-4. General small-signal low-frequency h-parameter representation of the basic amplifier stages of Figs. 7-1, 7-2, and 7-3. R'_g and R'_L are equivalent input and load resistances, respectively, which include the input- and output-coupling resistances. Coupling capacitors are considered to have negligible reactances.

bination of R_L and R_E for the emitter follower and of R_L and R_C for the common-base and common-emitter configurations.

For the circuit of Fig. 7-4, we see that

$$V_1 = h_{11}I_1 + h_{12}V_2 \tag{7-7}$$

$$I_2 = h_{21}I_1 + h_{22}V_2 \tag{7-8}$$

and

$$V_2 = -I_2R'_L \tag{7-9}$$

Solving these equations for I_1, I_2, and V_2, we may show that current gain A_I is

$$A_I = \frac{I_2}{I_1} = \frac{h_{21}}{1 + h_{22}R'_L} \tag{7-10}$$

that voltage gain A_V is

$$A_V = \frac{V_2}{V_1} = \frac{-h_{21}R'_L}{h_{11} + R'_L\Delta^h} \tag{7-11}$$

where Δ^h is the determinant of Eqs. (7-7) and (7-8), being given by

$$\Delta^h = h_{11}h_{22} - h_{21}h_{12} \tag{7-12}$$

and that input resistance R_{in} is given by

$$R_{\text{in}} = \frac{V_1}{I_1} = \frac{h_{11} + R'_L\Delta^h}{1 + h_{22}R'_L} \tag{7-13}$$

From Eqs. (7-10) and (7-11), we see that power gain A_p is given by

$$A_p = |A_I||A_V| = \frac{h_{21}^2R'_L}{(1 + h_{22}R'_L)(h_{11} + R'_L\Delta^h)} \tag{7-14}$$

The negative sign that appears in Eq. (7-11) is a characteristic of the current and voltage directions assumed in the circuit of Fig. 7-4. For the common-base and common-collector configurations, h_{21} is a negative quantity, and the voltage gain is a positive number. For the common-emitter configuration, the voltage gain is a negative number, owing to h_{21} being positive, indicating that the input and output voltages are of opposite phase when considered with respect to the common terminal. Since power is delivered to the load irrespective of the signs of the voltage and current gains, a negative sign in the power-gain expression of Eq. (7-14) is avoided by defining A_p in terms of absolute values of A_V and A_I.

The voltage-gain equation may be applied directly to the amplifier stages of Figs. 7-2 through 7-3, but the current-gain expression requires correction for signal-current loss in the coupling resistors, and the input resistance should be corrected for the shunting effect of the input-coupling resistance. For example, the corrected A_I and R_{in} expressions for the common-emitter stage of Fig. 7-3 are

$$A_I = \frac{R_B R_C h_{21}}{(R_B + R_{\text{in}})(R_C + R_L)(1 + h_{22}R'_L)} \tag{7-15}$$

and

$$R_{\text{in}} = \frac{R_B(h_{11} + R'_L\,\Delta^h)}{R_B(1 + h_{22}R'_L) + (h_{11} + R'_L\,\Delta^h)} \tag{7-16}$$

Equation (7-15) is seen to approach Eq. (7-10) for $R_B \gg R_{\text{in}}$ and $R_C \gg R_L$, and Eq. (7-16) becomes Eq. (7-13) for values of R_B such that $R_B(1 + h_{22}R'_L) \gg (h_{11} + R'_L\,\Delta^h)$.

Stage output resistance. In order to obtain a general expression for amplifier-stage output resistance R_O in terms of low-frequency h parameters, we replace R'_L in the equivalent circuit of Fig. 7-4 by a signal source and obtain the V_2/I_2 ratio for $V_g = 0$. The equivalent circuit for this arrangement is shown in Fig. 7-5. For this circuit, we see that

$$0 = (h_{11} + R'_g)I_1 + h_{12}V_2 \tag{7-17}$$

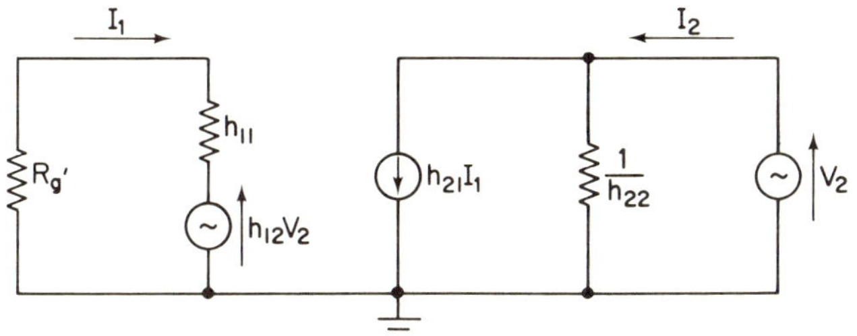

Fig. 7-5. General low-frequency h-parameter equivalent circuit for determining the output resistance of the basic amplifier stages of Figs. 7-1, 7-2, and 7-3.

and

$$I_2 = h_{21}I_1 + h_{22}V_2 \tag{7-18}$$

Solving these equations for V_2 yields

$$V_2 = \frac{(h_{11} + R'_g)I_2}{\Delta} \tag{7-19}$$

where Δ is the determinant of the equations, being given by

$$\Delta = (h_{11} + R'_g)h_{22} - h_{21}h_{12} \tag{7-20}$$

From Eqs. (7-12) and (7-20), we see that

$$\Delta = \Delta^h + R'_g h_{22} \tag{7-21}$$

From Eqs. (7-19) and (7-21), we obtain

$$R_O = \frac{V_2}{I_2} = \frac{h_{11} + R'_g}{\Delta^h + R'_g h_{22}} \tag{7-22}$$

In determining the output resistance of the stages of Figs. 7-1, 7-2, and 7-3, R_L is replaced by a signal source, and the output resistance is seen to be the parallel combination of the output coupling resistance (R_C or R_E) and the resistance given by Eq. (7-22).

Typical h parameters. In comparing the characteristics of the three basic amplifier configurations, we shall use illustrative examples that consider a typical BJT having the low-frequency parameters of Table 7-1. These are seen to be the parameters of Table 6-1 on p. 250, with the addition of the Δ^h entries and the use of general network symbols.

Table 7-1 Hybrid parameters of a typical 2N1382 BJT at $I_E = -1$ mA, $V_C = -5$ V, $f = 1$ kHz

Parameter	*Configuration*		
	COMMON-BASE	COMMON-EMITTER	COMMON-COLLECTOR
h_{11}	30 ohms	2400 ohms	2400 ohms
h_{12}	5×10^{-4}	7×10^{-4}	1.00
h_{21}	-0.988	80	-81
h_{22}	0.5×10^{-6} mho	40×10^{-6} mho	40×10^{-6} mho
Δ^h	5.09×10^{-4}	400×10^{-4}	81.1

Current-gain comparison. From Eq. (7-10), we see that current gain A_I is a function of load resistance, varying from

$$A_I = 0 \qquad \text{for} \quad R'_L = \infty \tag{7-23}$$

to

$$A_I = h_{21} \qquad \text{for} \quad R'_L = 0 \tag{7-24}$$

Using the h-parameter values of Table 7-1, plots of the absolute value of A_I

versus load resistance would be as shown in Fig. 7-6. As pointed out earlier, the common-emitter and common-collector current gains are essentially equal in magnitude and large compared to those of the common-base configuration, which have an upper limit of unity. From the signs of parameter h_{21}, we see that the current gains of the common-base and common-collector configurations are negative, indicating that input current and output current are of opposite phase.

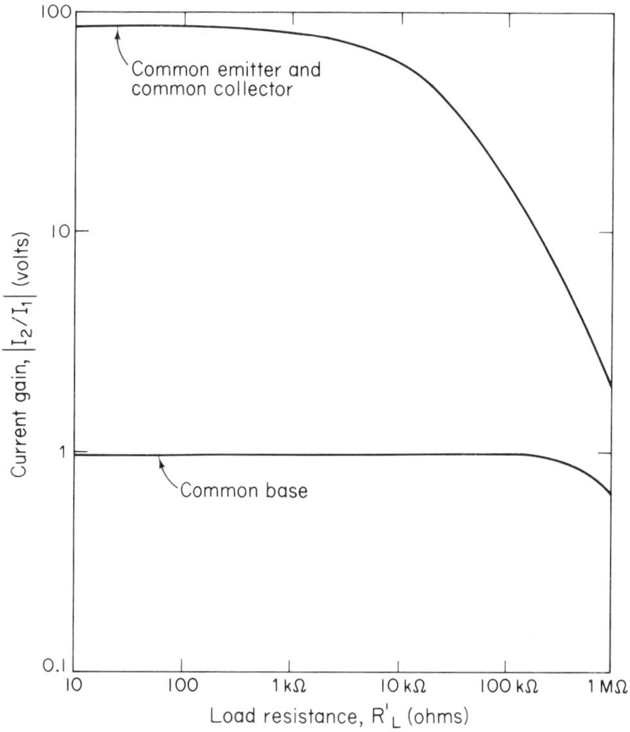

Fig. 7-6. Current-gain magnitude $|A_I|$ versus load resistance for the three basic amplifier configurations using a BJT having the characteristics of Table 7-1.

Voltage-gain comparison. Equation (7-11) shows that voltage gain A_V is also a function of load resistance, varying from

$$A_V = 0 \quad \text{for} \quad R'_L = 0 \tag{7-25}$$

to

$$A_V = \frac{h_{21}}{\Delta^h} \quad \text{for} \quad R'_L = \infty \tag{7-26}$$

Plots of the absolute value of A_V versus load resistance for the BJT parameters of Table 7-1 are shown in Fig. 7-7. The common-base and common-emitter configurations are seen to have equal voltage gains, both being large compared to that of the common-collector configuration, which has an upper limit of unity. As pointed out earlier, the algebraic sign of the voltage gain is an indication of the relative phase of the input and output voltages. Thus, the common-base and common-collector input and output voltages are of the same phase with respect to the common terminal, but a phase reversal occurs in the common-emitter stage.

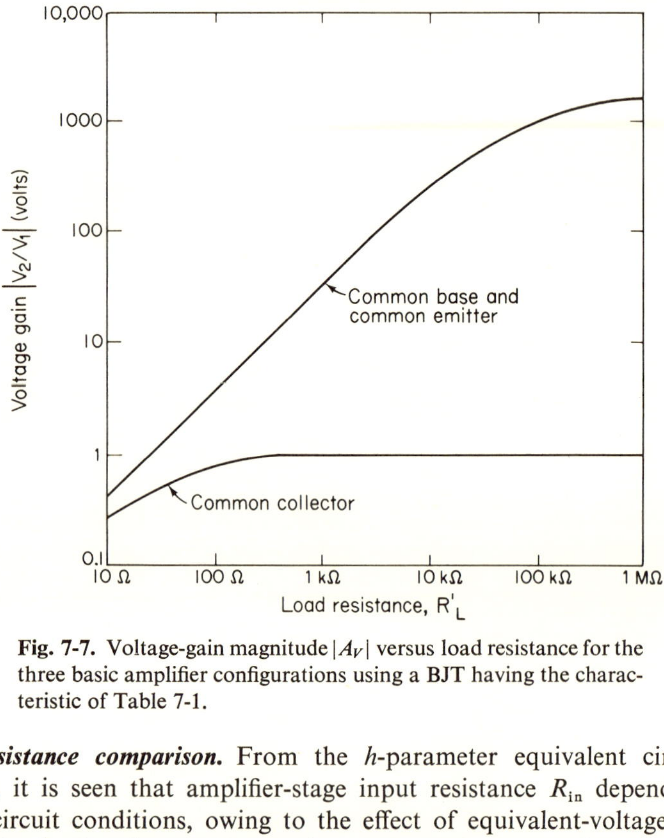

Fig. 7-7. Voltage-gain magnitude $|A_V|$ versus load resistance for the three basic amplifier configurations using a BJT having the characteristic of Table 7-1.

Input-resistance comparison. From the h-parameter equivalent circuit of Fig. 7-4, it is seen that amplifier-stage input resistance R_{in} depends upon output-circuit conditions, owing to the effect of equivalent-voltage source $h_{12}V_2$. For load resistance R_L equal to zero, V_2 is zero and R_{in} is seen to be h_{11}. As R_L is increased, R_{in} either increases or decreases, depending upon the relative phase relationship between the input and output voltages. Thus, R_{in} increases with R_L for the common-base and common-collector configurations, but decreases with increase in R_L for the common-emitter stage. Plots of R_{in} versus load resistance fot the BJT parameters of Table 7-1, obtained through use of Eq. (7-13), are shown in Fig. 7-8. It is seen that the input resistance of the common-collector (or emitter-follower) stage *may be* several orders of magnitude larger than that of either of the other

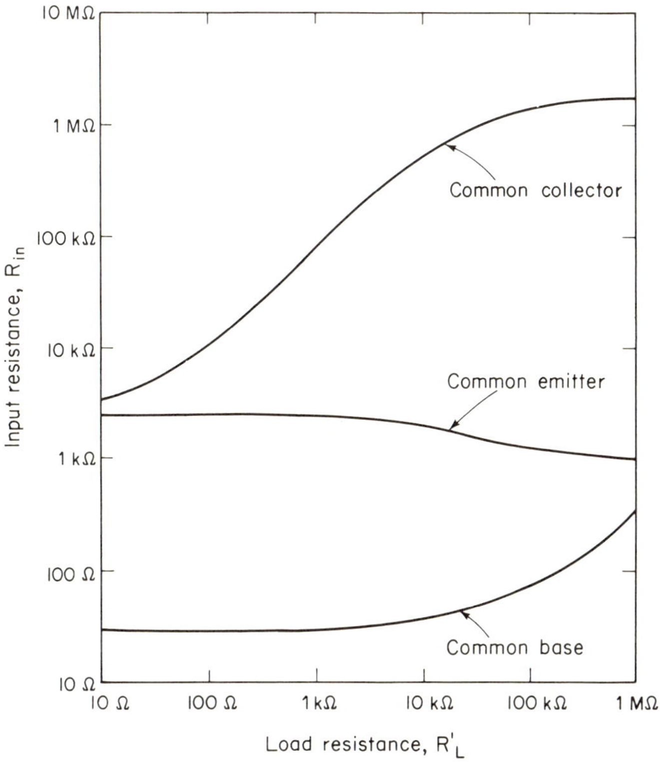

Fig. 7-8. Input resistance R_{in} versus load resistance for the three basic amplifier stages using a BJT having the characteristics of Table 7-1.

two configurations and is *highly dependent upon load resistance*. If a stage having high input impedance is required, the emitter follower is a likely candidate.

Output-resistance comparison. It may be seen from the h-parameter equivalent circuit of Fig. 7-5 that amplifier-stage output resistance R_O depends upon input conditions, owing to the effect of equivalent current source $h_{21}I_1$. For source resistance R'_g equal to infinity, I_1 is zero and R_O is seen to be $1/h_{22}$. As R_g is decreased, R_O either increases or decreases, depending upon the relative phase of $h_{21}I_1$ and I_2. Since positive V_2 causes I_1 to be negative, R_O increases with decrease in R_g for the common-emitter stage but decreases with R_g for the other two configurations, owing to h_{21} being negative. Plots of R_O versus load resistance for the transistor parameters of Table 7-1, obtained through use of Eq. (7-22), are shown in Fig. 7-9. It is seen that the output resistance of the common-collector (or emitter-follower) stage *may be* several orders of magnitude smaller than that of either of the other two configurations and is *highly dependent upon source resistance*.

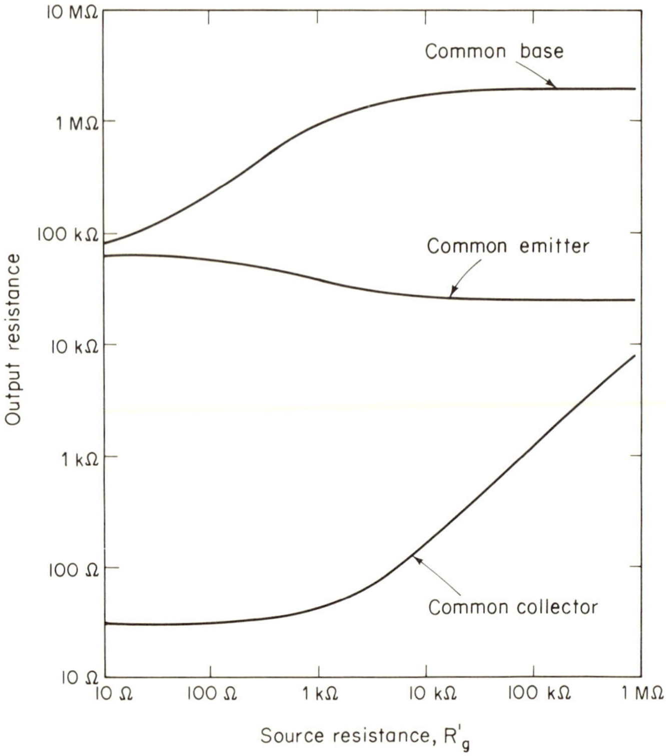

Fig. 7-9. Output resistance R_0 versus source resistance for the three basic amplifier stages using a BJT having the characteristics of Table 7-1.

Thus, the emitter follower is a likely candidate for applications requiring a low output impedance.

Power-gain comparison. From the effects of load resistance on current and voltage gains, it is clear that power gain is a function of R_L. Since A_V is zero for $R_L = 0$ and A_I is zero for $R_L = \infty$, the power gain is zero for both limits of R_L and is maximum for some intermediate value. The value of R_L yielding maximum power gain may be obtained from Eq. (7-14) and is found to be

$$R'_L = \sqrt{\frac{h_{11}}{h_{22}\Delta^h}} \tag{7-27}$$

Using this expression, we find from Eq. (7-14) that the maximum value of power gain from a BJT is

$$A_{p,\max} = \frac{h_{21}^2}{(\sqrt{\Delta^h} + \sqrt{h_{11}h_{22}})^2} \tag{7-28}$$

Plots of power gain versus load resistance for the BJT parameters of Table 7-1, obtained through use of Eq. (7-14), are shown in Fig. 7-10. It is seen that, for any given load resistance, the power gain for the common-emitter stage is greater than that for either of the other two configurations. This is as expected, since the common-emitter configuration is the only one having both current-gain and voltage-gain capability. As pointed out in Sec. 7-1, this feature of the common-emitter stage is responsible for its widespread usage.

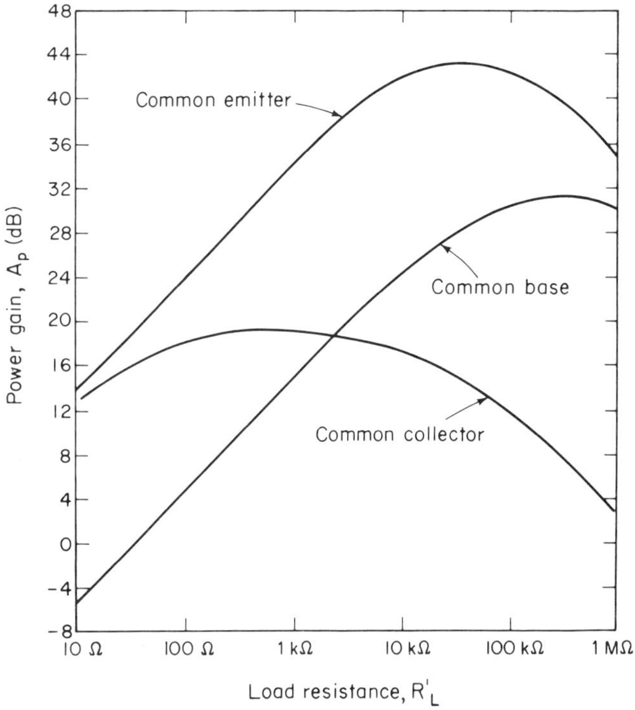

Fig. 7-10. Power gain A_p in decibels versus load resistance for the three basic amplifier configurations using a BJT having the characteristics of Table 7-1.

7-3 STAGES IN CASCADE; ITERATIVE GAIN

Since a single amplifier stage usually fails to provide adequate gain, amplifiers ordinarily consist of stages in cascade. In the most desirable cascade arrangement for broad-band operation, the input impedance of the second stage is the load impedance of the first stage. That is, except for provisions for maintaining bias conditions, the stages are directly connected, there being no impedance-transforming networks between stages. When identical stages are so operated, the arrangement is known as an *iterative connection*. Thus for iterative operation, the load impedance of an amplifier stage is equal to its own input impedance, and the source impedance of the stage is equal to its own output impedance. The current, voltage, and power gains of such a stage are known as *iterative gains*. Since the load impedance of a common-base stage must exceed the input impedance if the stage is to have voltage gain, and current gain occurs in the common-collector stage only for load impedances less than the input impedance, only the common-emitter configuration has iterative gains. Therefore, common-emitter stages are operable in cascade without the use of impedance-transforming coupling networks, but the other configurations cannot be so operated and are rarely used in cascade.* Since high input impedance and low output impedance are often desired in an amplifier, the emitter follower may be used as the first stage of a cascade chain in order to obtain high input impedance, and as the last stage of the chain in the interest of low output impedance. Should either low input impedance or high output impedance be required (a less likely occurrence), the common-base configuration can be used as a terminal stage.

7-4 POWER GAIN

Power gain is probably the most important characteristic of a BJT stage. As we have defined power gain, it is an indication of the ability of a BJT to amplify a given amount of power applied to its input terminals. However, the value of R_L that results in maximum power gain is not necessarily the R_L that results in maximum load power for a particular source, owing to the effect of R_L on BJT input resistance. That is, it may be possible to obtain greater load power by using an R_L value that results in an increase in the transfer of power from the source to the BJT—an increase that more than compensates for the reduction in BJT power gain.

In order to obtain maximum load power from a stage operating with a particular source, R_L should be the value given by Eq. (7-27) for maximum power gain, and the input resistance of the stage should match the source resistance R_g. For stages in cascade, conditions for optimum power gain usually require impedance-transforming networks between stages and are

* Two or more emitter-follower stages are sometimes directly cascaded in order to obtain high input impedance.

not ordinarily met, owing to the advantages of direct coupling.* In order to provide an indication, for a particular circuit, of how well the source and load are matched, use is made of a gain characteristic called *transducer gain* G_T. By definition, transducer gain is the ratio of the power developed in the load to the power available from the source. Recognizing that the source in the circuit of Fig. 7-4 supplies maximum power to the transistor when $R_{in} = R_g$, we see that the power available from the source is $V_g^2/4R_g$. Thus

$$G_T = \frac{I_2^2 R_L'}{V_g^2/4R_g'} = \left(\frac{I_2}{I_1}\right)^2 \frac{4R_L' R_g'}{(R_g' + R_{in})^2} \tag{7-29}$$

Substituting for I_2/I_1 and R_{in} as given by Eqs. (7-10) and (7-13), we find that Eq. (7-29) becomes

$$G_T = \frac{4h_{21}^2 R_g' R_L'}{R_g' R_L' h_{22} + R_L' \Delta^h + R_g' + h_{11}} \tag{7-30}$$

Since the maximum value that exists for transducer gain is the maximum power gain given by Eq. (7-28), as may be shown by substituting expressions for R_L' and R_g' for optimum power transfer, a comparison of transducer gain [given by Eq. (7-30)] with the maximum power gain [given by Eq. (7-28)] gives an indication of the improvement that is possible through impedance matching.

PROBLEMS

7-1. For the BJT of Table 7-1 operated in common-base, common-emitter, and common-collector configurations, determine the limit values of input resistance R_{in} as load resistance R_L' is varied in the circuit of Fig. 7-4.

7-2. Repeat Prob. 7-1 for output resistance R_o as source resistance R_g' is varied.

7-3. Determine the load resistance R_L' that results in maximum power gain for each configuration of Prob. 7-1.

7-4. In the emitter follower circuit of Fig. 7-2, $R_E = 20\ \text{k}\Omega$, $R_L = 2\ \text{k}\Omega$, the a-c source has a resistance $R_g = 10\ \text{k}\Omega$, and R_B, C_1, and C_2 may be considered large enough to have negligible effects. Using Eqs. (7-10) through (7-22), determine A_I, A_V, R_{in}, R_O, and V_2/V_g (where V_g is the open-circuit source voltage) for the BJT of Table 7-2.

7-5. Determine the emitter-follower characteristics of Prob. 7-4, using the common-emitter *h*-parameter model of the BJT.

* Except in large-signal amplifiers, it is usually preferable to use direct coupling with additional stages to compensate for the lack of optimum power-transfer conditions.

8

BJT Biasing Networks

In general, the biasing arrangements of the amplifier circuits considered so far render them impractical because of changes in d-c operating point that occur as a result of changes in operating temperature. These circuits also require bias adjustments when transistors are changed, owing to the spread in characteristics of BJTs bearing the same type number. In the design of an amplifier stage, a collector operating point is selected on the basis of the largest expected signal level, and, ideally, the bias arrangement used should result in the maintenance of this operating point as different BJTs of the same type are used and as operating temperature fluctuates. Practical bias arrangements usually involve the use of d-c degeneration for operating-point stabilization.

8-1 BIAS-STABILIZER ANALYSIS AND DESIGN

Although a number of bias-stabilization methods exist, the arrangement of Fig. 8-1 for the common-emitter stage is the most widely used and will receive our major attention. As we shall see, this method may also be used to stabilize common-collector stages. In Fig. 8-1, stabilization occurs as a result of using a "stiff source" for the base-to-ground voltage. Resistances R_2 and R_3 are chosen so that the d-c bleeder current (current I_2 through R_2)

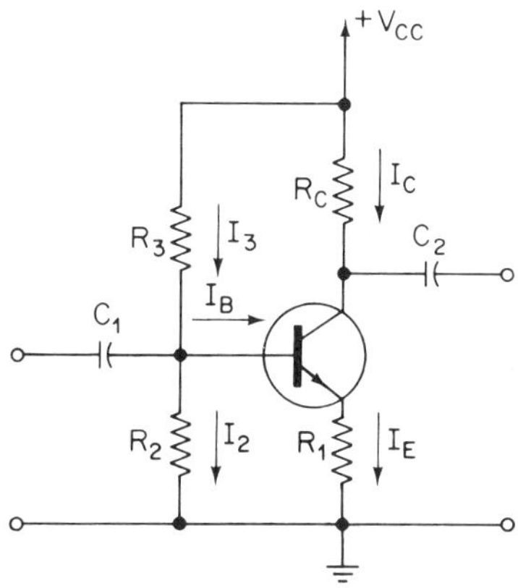

Fig. 8-1. Bias-stabilized common-emitter stage.

is large compared to d-c current I_B, causing the base-to-ground voltage to be essentially independent of I_B. When a change in d-c emitter current I_E occurs, either because of a change in temperature or because of the use of a BJT having a different h_{FE} value ($h_{FE} = I_C/I_B$), a change in d-c voltage $I_E R_1$ across R_1 results. This change in $I_E R_1$ causes a change in d-c base-emitter voltage V_{BE}, changing I_B in the direction that minimizes the change in I_E. Thus, the larger I_E, the smaller I_B, and the circuit is quite capable of providing negative I_B values so that the BJT can be operated at I_C values well below I_{CEO}.

The effect on I_E of a change in BJT characteristics is referred to in terms of the current stability factor S_I, which may be defined as

$$S_I = \frac{dI_E}{dI_{CBO}} \tag{8-1}$$

From this definition, we see that the ideal value of S_I is zero. For this condition, a change in I_{CBO} would affect only I_B and I_C, with all of the I_{CBO} change occurring in the base. Achieving zero stability factor in the circuit of Fig. 8-1 is impractical, and typical S_I values range upward from unity.

In order to obtain an S_I expression for the circuit of Fig. 8-1, we need to consider additional currents. For a d-c emitter current I_E, we recall that d-c collector current I_C is given by

$$I_C \cong I_{CBO} + \alpha_{fb} |I_E| \tag{8-2}$$

D-C base current I_B is then given by

$$I_B \cong (1 - \alpha_{fb}) I_E - I_{CBO} \tag{8-3}$$

since the direction of base-current component I_{CBO} is opposite that of the component due to carrier recombination in the base and injection into the emitter.

If we assume d-c base-emitter voltage V_{BE} to be very small compared to voltage $I_E R_1$, the voltage across R_2 approximately equals $I_E R_1$. Therefore, d-c current I_2 is given by

$$I_2 \cong \frac{I_E R_1}{R_2} \tag{8-4}$$

and d-c current I_3 is seen to be

$$I_3 = I_B + I_2 = \frac{I_E R_1}{R_2} + (1 - \alpha_{fb})I_E - I_{CBO} \tag{8-5}$$

Since supply voltage V_{CC} is the sum of the voltages across R_2 and R_3, we see from Eqs. (8-4) and (8-5) that

$$V_{CC} \cong I_E R_1 + R_3\left[\frac{I_E R_1}{R_2} + (1 - \alpha_{fb})I_E - I_{CBO}\right] \tag{8-6}$$

from which

$$I_E \cong \frac{V_{CC} + R_3 I_{CBO}}{R_1 + R_3(1 - \alpha_{fb}) + (R_1 R_3/R_2)} \tag{8-7}$$

From Eq. (8-7) we see that

$$S_I = \frac{dI_E}{dI_{CBO}} \cong \frac{R_3}{R_1 + R_3(1 - \alpha_{fb}) + (R_1 R_3/R_2)} \tag{8-8}$$

Dividing the numerator and denominator of Eq. (8-8) by $R_1 R_3$ gives

$$S_I = \frac{1/R_1}{(1/R_3) + (1 - \alpha_{fb})/R_1 + (1/R_2)} = \frac{G_1}{G_3 + (1 - \alpha_{fb})G_1 + G_2} \tag{8-9}$$

Since α_{fb} is very close to unity and G_1 is of the same order of magnitude as G_2 and G_3, $(1 - \alpha_{fb})G_1$ may be considered negligible compared to G_2 and G_3. Thus, from Eq. (8-9)

$$S_I \cong \frac{G_1}{G_2 + G_3} \tag{8-10}$$

From this expression, we see the futility of obtaining ideal stabilization, since an S_I value of zero requires that G_1 be zero or that either G_2 or G_3 be infinite. The former requires an infinite value of V_{CC}, and the latter would short-circuit the input terminals of the stage. Bias stabilization is obtained at the cost of decreased efficiency, because of d-c power dissipated by resistors R_1, R_2, and R_3, and decreased gain owing to part of the input current being diverted to ground by R_2 and R_3. Values of S_I appreciably less than unity are usually considered too costly to be practical.

Bias-stabilizer design. We now develop a stabilizer-design procedure based on a set of approximations that are reasonably accurate for high-h_{fe} BJT's operated at d-c collector currents large compared to I_{CBO}. These approximations are

1. $I_E \cong I_C$, since α_{fb} is very close to unity.
2. $V_{BE} \ll I_E R_1$, because of the choice of R_1.
3. $I_B \ll I_2$, because of the high h_{fe} and the choice of R_2 and R_3.
4. $I_2 \cong I_3 = I_R \cong I_E R_1 / R_2$, because of approximations (2) and (3).

On the basis ot these approximations, d-c conditions in the amplifier circuit are as shown in Fig. 8-2.

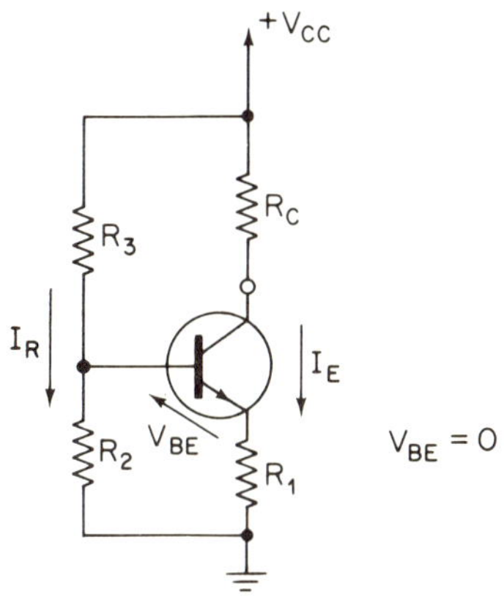

Fig. 8-2. Approximate d-c conditions in the bias-stabilizer circuit of Fig. 8-1.

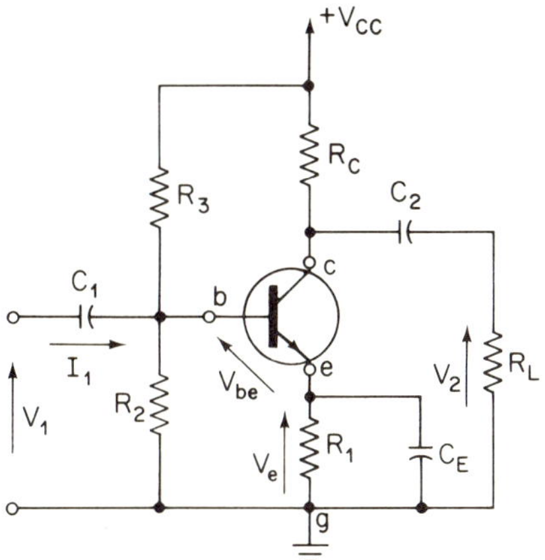

Fig. 8-3. Practical bias-stabilized common-emitter stage. Emitter bypass capacitor C_E is required to reduce a-c degeneration.

An amplifier-stage design usually begins with a choice of quiescent operating point values V_{CE} and I_C based on the largest expected a-c signals V_{ce} and I_c. A value of R_1 is then selected such that

$$I_E R_1 \gg V_{BE} \tag{8-11}$$

Since V_{BE} is of the order of tenths of volts, $I_E R_1$ should be of the order of volts. Thus R_1 is of the order of megohms if I_E is of the order of microamperes, of the order of kilohms if I_E is of the order of milliamperes, of the order of ohms if I_E is of the order of amperes, etc.

After the selection of R_1, R_C may be determined if V_{CC} is specified; otherwise, a value of R_C is selected, such that R_C is large compared to the expected load impedance, and V_{CC} is determined. Clearly, the choice of R_C must result in a reasonable value of V_{CC}.

Current stability factor S_I is now selected. This choice is determined by the maximum I_E change that can occur without the quiescent operating point becoming unsatisfactory for the largest expected I_c and V_{ce} values, and by the change in I_{CBO} for the maximum expected temperature change. The I_{CBO} versus temperature characteristic may be provided by the manufacturer or the I_{CBO} change can be predicted. Thus, by definition,

$$S_I = \frac{\Delta I_E}{\Delta I_{CBO}} \tag{8-12}$$

From Fig. 8-2, we see that

$$V_{CC} = I_R(R_2 + R_3) \tag{8-13}$$

from which

$$R_3 = \frac{V_{CC} - I_R R_2}{I_R} \tag{8-14}$$

Since $I_R R_2 = I_E R_1$, Eq. (8-14) becomes

$$R_3 = \frac{V_{CC} - I_E R_1}{I_E R_1/R_2} \tag{8-15}$$

$$R_3 = \frac{R_1(V_{CC}/R_1 - I_E)}{I_E R_1/R_2} = \frac{G_1 V_{CC} - I_E}{I_E G_2} \tag{8-16}$$

Simultaneous solution of Eqs. (8-10) and (8-16) yields

$$R_3 = \frac{S_I V_{CC}}{I_E} \tag{8-17}$$

After R_3 has been determined, R_2 can be found by using the d-c conditions of Fig. 8-2. If an expression for R_2 is desired, it can be obtained from Eq. (8-10). Thus

$$G_2 = \frac{G_1 - S_I G_3}{S_I} \tag{8-18}$$

from which

$$R_2 = \frac{R_1 R_3 S_I}{R_3 - S_I R_1} \tag{8-19}$$

We may now summarize the design procedure for the bias-stabilized stage of Fig. 8-1 as follows:

1. Select I_E and V_{CE} on the basis of I_c and V_{ce} (for very small values of I_c and V_{ce}, this choice may be based on the transistor characteristics).
2. Select R_1 so that $I_E R_1 \gg V_{BE}$.
3. Select R_C so that $R_C \gg Z_L$, or choose a V_{CC} value.
4. Determine V_{CC} or R_C, depending on the option exercised in step 3.
5. Select S_I so that change in I_E with temperature is not excessive.
6. Determine R_3 from $R_3 = S_I V_{CC}/I_E$.
7. Determine R_2.

It is of interest that, with the possible exception of step 1, BJT characteristics are not considered in the design of the stage.

Alpha stability factor S_α. Except for assuming h_{fe} (or α_{fe}) to be large compared to unity, BJT parameter alpha has been neglected in the design procedure. The dependence of bias current I_E on alpha is indicated by Eq. (8-7), which may be used to obtain an expression for alpha stability factor S_α, where

$$S_\alpha \equiv \frac{dI_E}{d\alpha} \tag{8-20}$$

However, such an expression is of limited usefulness, since S_α is highly dependent on alpha and cannot be readily used to predict I_E variation for large changes in alpha. Therefore, the final step in the design procedure might be to use Eq. (8-7) to determine I_E boundaries for the α_{fb} limits specified by the manufacturer for the particular BJT type. If the I_E boundaries are excessive, the design procedure should be repeated using a smaller S_I value. When silicon BJTs are used over a moderate temperature range, I_{CBO} may be too small to have significant effect on I_E; for this condition, I_E change with alpha is the basis for S_I selection in step 5 of the design procedure. If the effect of I_{CBO} is negligible, I_E is less than the design value when α_{fb} is less than unity, as may be seen from Eq. (8-7),* and S_I should be selected so that I_E is adequate for the lower limit of α_{fb}.

a-c conditions. In practice, emitter resistor R_1 in the bias-stabilized circuit of Fig. 8-1 must be bypassed in order to reduce *a-c degeneration*. Thus the practical bias-stabilized stage is as shown by Fig. 8-3, the presence of bypass capacitor C_E having no effect on d-c conditions. If *a-c* emitter voltage V_e is appreciable compared to a-c base voltage V_{be}, a larger input voltage V_1 is required for a given output voltage V_2. That is, an impedance in the emitter circuit decreases the voltage gain. In addition, the increased input voltage results in a larger fraction of input current I_1 being diverted to ground by resistors R_2 and R_3, reducing the current gain. If, as a rough approximation, BJT parameters h_{re} and h_{oe} are considered to be negligible, it is seen from the *h*-parameter representation that the base-to-ground input impedance Z_{bg} of the BJT is given by

$$Z_{bg} \cong h_{ie} + (h_{fe} + 1)Z_e \tag{8-21}$$

where Z_e is the impedance of the parallel R_1C_E combination. Typically, over the operating frequency range,

$$Z_e \cong \frac{1}{j\omega C_E} \tag{8-22}$$

the reactance of C_E being much smaller than R_1, and h_{ie} and $(h_{fe} + 1)Z_e$ add in quadrature to give Z_{bg}. Therefore, if the magnitude of Z_{bg} is to be unaffected by Z_e (so that the voltage- and current-gain magnitudes are unaffected)

* Since Eq. (8-7) is approximate, I_E is less than the design value, even though $\alpha_{fb} = 1$.

$$\frac{(h_{fe}+1)}{\omega C_E} < \frac{h_{ie}}{10} \tag{8-23}$$

and

$$C_E > \frac{10(h_{fe}+1)}{\omega h_{ie}} \tag{8-24}$$

at the lowest frequency of operation.*

If a BJT having the parameters given by Table 7-1 on p. 300 is to be used in a bias-stabilized stage operating at frequencies as low as 100 Hz, we see from Eq. (8-24) that

$$C_E > \frac{(10)(31)}{(2\pi)(100)(2400)} = 0.54 \times 10^{-3} = 540\ \mu\text{F}$$

Because of the large capacitance required for essentially ideal emitter bypassing at this frequency, C_E may be selected so that it determines the lower limit of the operating frequency. If C_1 of Fig. 8-3 is considered to have negligible reactance and h_{re} is negligible, application of Thévenin's theorem at the base-ground terminals results in the equivalent input circuit of Fig. 8-4, as suggested by Eq. (8-21). Equivalent source resistance R'_g is the parallel

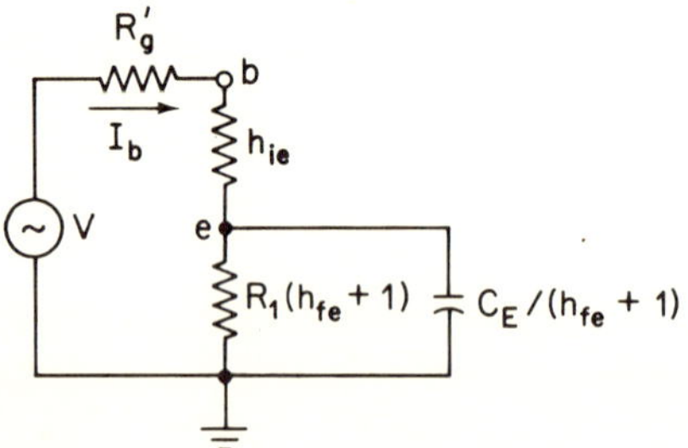

Fig. 8-4. Equivalent input circuit of the common-emitter stage of Fig. 8-3 assuming C_1 to have negligible reactance.

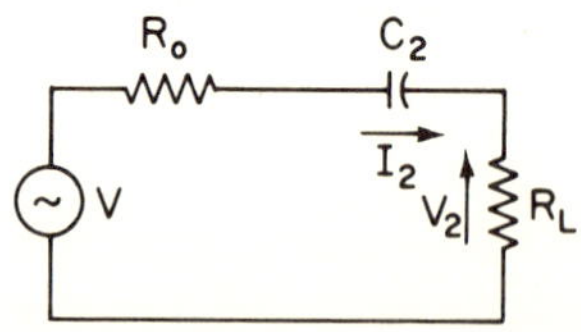

Fig. 8-5. Equivalent output circuit of the common-emitter stage of Fig. 8-3.

combination of R_2, R_3, and the source resistance. If the reactance of C_E is small enough (compared to R_1) that R_1 may be considered an open circuit to alternating current, it is seen that the magnitude of a-c base current I_b in Fig. 8-4 is reduced by the factor $\sqrt{2}$ (or by 3 dB) below the value for ideal bypassing when

$$\frac{(h_{fe}+1)}{\omega C_E} = R'_g + h_{ie} \tag{8-25}$$

* It is quite common to consider the sum of two components adding in quadrature to be the larger component for ratios of three or higher, since the error in the approximation (5 percent or less) is comparable to the uncertainty of typical component values. However, a ratio of ten is suggested in Eqs. (8-23) and (8-24), since Z_e is not entirely reactive.

reducing the output voltage by the same factor and the output power by the factor 2. Therefore, if C_1 and C_2 of Fig. 8-1 have negligible effect on frequency response, the lower half-power frequency of the stage may be obtained from Eq. (8-25). By comparing Eqs. (8-23) and (8-25), we see that C_E is at least an order of magnitude smaller if it is to determine the lower half-power frequency of the stage rather than provide essentially ideal bypassing at this frequency.

Coupling-capacitor requirement. If ideal emitter bypassing occurs, the network to the left of coupling capacitor C_2 in Fig. 8-3 may be represented by a Thévenin equivalent generator, as shown by Fig. 8-5. Output resistance R_o is given by Eq. (7-22) and is usually appreciably larger than load resistance R_L, which may be the input resistance of the following stage. From Fig. 8-5, it is seen that the magnitude of output current I_2 and that of output voltage V_2 are reduced by the factor $\sqrt{2}$ (or by 3 dB) below the value for negligible C_2 reactance when

$$\frac{1}{\omega C_2} = (R_o + R_L) \tag{8-26}$$

Therefore, Eq. (8-26) may be used to determine the lower half-power frequency of the stage or to determine C_2 if the half-power frequency is specified. If C_2 is to have negligible effect on the magnitude of I_2 and V_2,

$$\frac{1}{\omega C_2} < \frac{(R_o + R_L)}{10} \tag{8-27}$$

and

$$C_2 > \frac{10}{\omega(R_o + R_L)} \tag{8-28}$$

at the lowest frequency of operation. Thus, coupling capacitor C_2 is determined by Eq. (8-28), if C_E determines the lower half-power frequency of the stage, or from Eq. (8-26) if C_2 is to determine the half-power frequency.

8-2 BIAS-STABILIZED EMITTER FOLLOWER

The stabilization method of Fig. 8-2 may also be used for the emitter-follower stage, as shown by Fig. 8-6. The design procedure is the same as for the common-emitter stage, except that the absence of a collector resistor results in emitter resistor R_E being either determined by supply voltage V_{CC} or selected so that it is large compared to the expected load resistance.

If the emitter-follower stage is to be used to obtain high input resistance R_{in}, the circuit of Fig. 8-6 proves disappointing, since the upper limit of

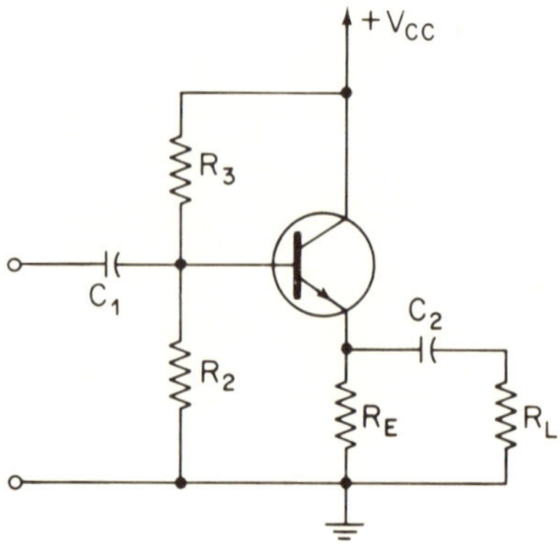

Fig. 8-6. Bias-stabilized emitter-follower stage. For large values of R_L the input impedance of the stage may consist largely of the parallel combination of bias-stabilizing resistances R_2 and R_3.

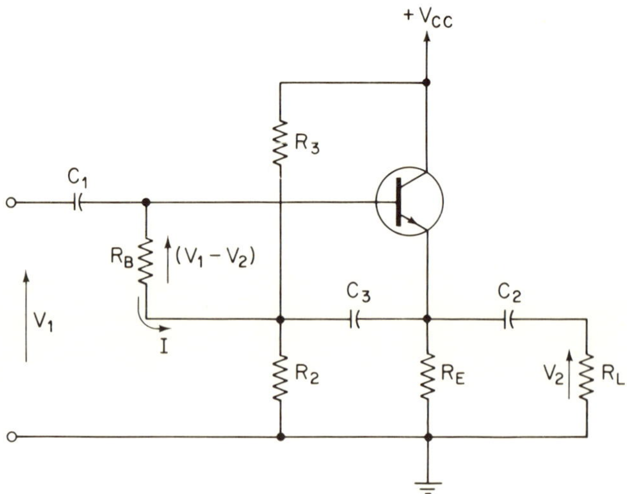

Fig. 8-7. Bootstrap scheme for reducing the shunting effect of bias-stabilizing resistors of an emitter-follower stage. Capacitor C_3 couples R_2 and R_3 to the output terminals, for a-c conditions, so that the voltage across R_B is (V_1-V_2). This significantly reduces diversion current I if R_B is comparable to the resistance of the parallel combination of R_2 and R_3.

R_{in} is the parallel resistance of bias-stabilizing resistors R_2 and R_3. The shunting effect of the biasing resistors may be reduced by use of the *bootstrap* arrangement of Fig. 8-7. In this arrangement, C_3 (which has no effect on d-c conditions) couples output voltage V_2 to the R_2-R_3 combination. The signal current I diverted from the transistor base is seen to be given by

$$I = \frac{V_1 - V_2}{R_B} \tag{8-29}$$

and the a-c impedance *of the biasing resistors* is

$$R_{ac} = \frac{V_1}{I} = \frac{V_1 R_B}{V_1 - V_2} = \frac{R_B}{1 - A_V} \tag{8-30}$$

Since voltage gain A_V is quite close to unity and R_B may be of the order of resistance of the R_2-R_3 combination without seriously disturbing d-c conditions, the circuit is an extremely effective impedance changer. The increase in input impedance is obtained at the expense of a slight decrease in voltage gain, since R_2 and R_3 are shunted across the load, and A_V should be calculated by using the total shunt load.

8-3 ADDITIONAL BIAS-STABILIZATION METHODS

In the interest of completeness, some less widely used bias-stabilizing methods will be briefly described.

Current-feedback bias. The base-bias current may be obtained by connecting a resistor between the base and collector, as shown in Fig. 8-8. This is sometimes referred to as *self bias*, since base bias is provided by the collector voltage. The d-c bias current is seen to be given by

$$I_B = \frac{V_C - V_B}{R_F} \cong \frac{V_C}{R_F} \tag{8-31}$$

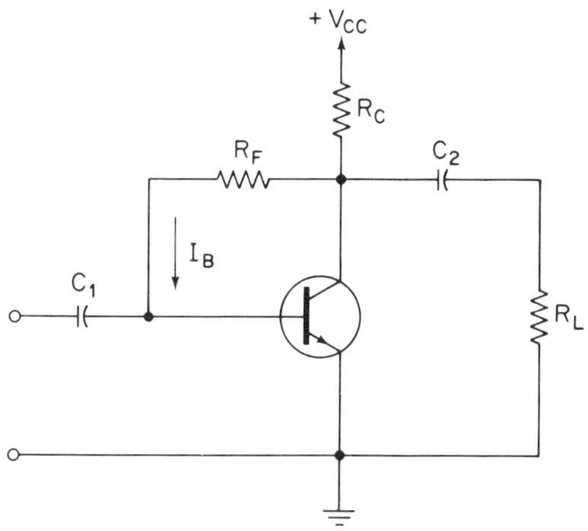

Fig. 8-8. Current-feedback bias stabilization.

and stabilizing action occurs as a result of I_B dependence upon V_C. Unless BJT tolerances are close and the junction temperature range is quite restricted, this circuit provides inadequate stabilization.

Combined current-voltage feedback. The bias-stabilizing methods of Figs. 8-1 and 8-8 are combined in the stabilizing arrangement shown in Fig. 8-9.

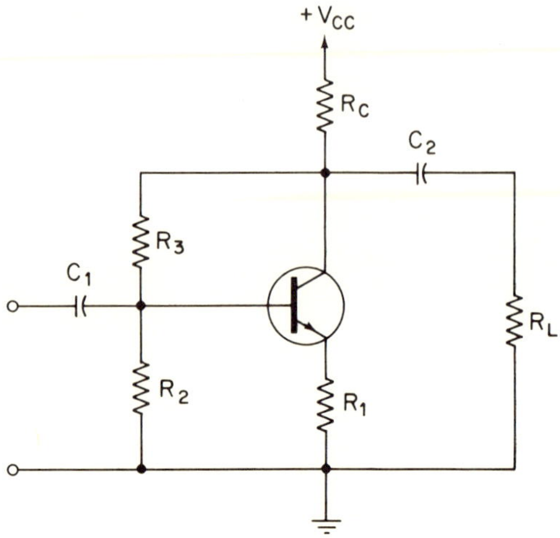

Fig. 8-9. Combined current-voltage feedback bias stabilization. This arrangement combines the voltage feedback of Fig. 8-1 and the current feedback of Fig. 8-8.

In this arrangement, the base-to-ground voltage decreases with an increase in I_C, and bias stabilization is superior to that of the voltage-feedback arrangement of Fig. 8-1. Analysis of the circuit discloses that the current stability factor is given by

$$S_I \cong \frac{G_1}{[G_2 + G_3 + G_3(G_1 + G_2)/G_C]} \tag{8-32}$$

which becomes the stability-factor expression of Eq. (8-10) if current feedback is eliminated by making $R_C = 0$. The design procedure for the circuit is essentially the same as that for the circuit of Fig. 8-1. That is, R_1 is selected, R_3 is approximately $S_I V_{CC}/I_E$, and R_2 and R_C are determined from d-c conditions.

Bias resistors R_F of Fig. 8-8 and R_3 of Fig. 8-9 have degenerative effects on a-c operation of the circuits by providing paths for a-c components of base current 180 deg out of phase with the input current, since the output voltage with respect to ground is 180 deg out of phase with the input voltage.* This reduces the current gain of the stage, but has little effect on the voltage

* From the point of view of branch currents, output voltage V_2 aids input voltage V_1 in diverting input-signal current from the base.

gain. As a result, the input impedance is reduced. The amount of degeneration depends on the magnitude of the output voltage and is relatively small for a low-impedance load, such as the input impedance of a succeeding common-emitter stage. For high load impedances, this degeneration may be eliminated by splitting the feedback resistor into two series resistances and shunting the feedback current from the junction of the two resistances to ground through a capacitor, as shown in Fig. 8-10. If the reactance of C_3 is negligible compared to R_a and R_b, R_a is seen to be shunted across the input terminals and R_b is across the output. These shunting effects reduce the gain somewhat,

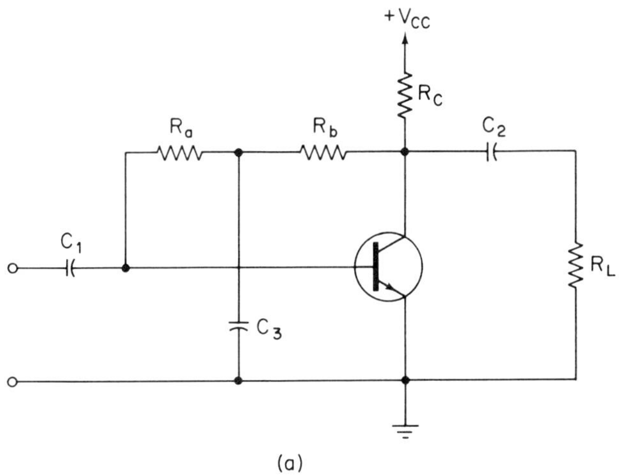

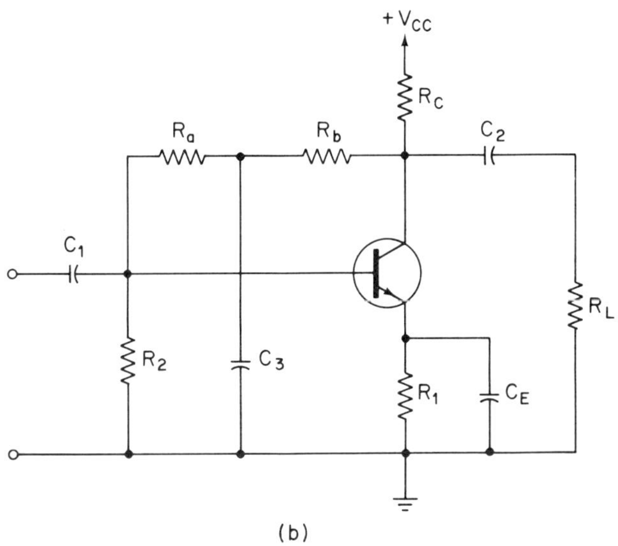

Fig. 8-10. Arrangement of the stabilization circuits of Figs. 8-8 and 8-9 to reduce a-c degeneration.

but with proper circuit design the reduction is small compared to that resulting from the degeneration that might otherwise exist.

Stabilization of transformer-coupled stages. In transformer-coupled power-amplifier stages, isolation from ground of the input signal permits avoiding diversion of input-signal current by the biasing resistors, as shown in Fig. 8-11. Capacitor C bypasses both R_1 and R_2 so that the input voltage is applied to the base-emitter terminals of the BJT.

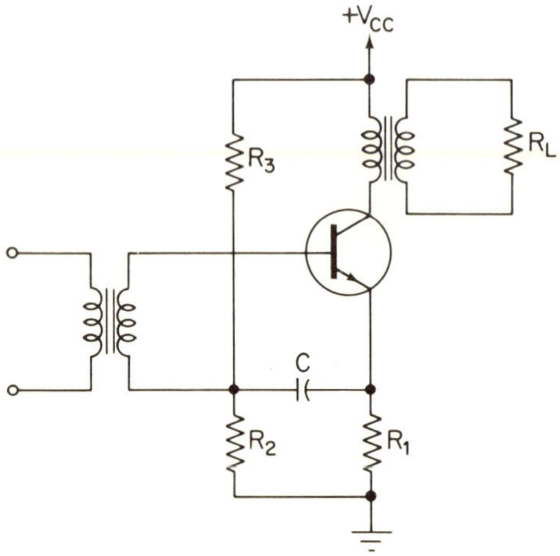

Fig. 8-11. Bias-stabilized transformer-coupled stage. Isolation from ground of the input signal permits avoiding diversion of input-signal current by the biasing resistors.

Nonlinear stabilization. The bias-stabilization methods described so far are referred to as *linear stabilization*, since linear elements are used. Greater stabilization for a given power loss may be obtained by using nonlinear temperature-sensitive elements in the stabilizing network. Several nonlinear stabilization arrangements, using both negative-temperature-coefficient (NTC) and positive-temperature-coefficient (PTC) resistors, are shown in Fig. 8-12. In each arrangement it is seen that a temperature change results in a change in base-emitter voltage that changes the collector-emitter bias in a direction opposite that caused by the change in I_{CBO}.

8-4 JUNCTION TEMPERATURE AND THERMAL STABILITY

The change in I_{CBO} used in the selection of S_I in bias-stabilizer design [see Eq. (8-12)] depends upon the largest expected change in junction tem-

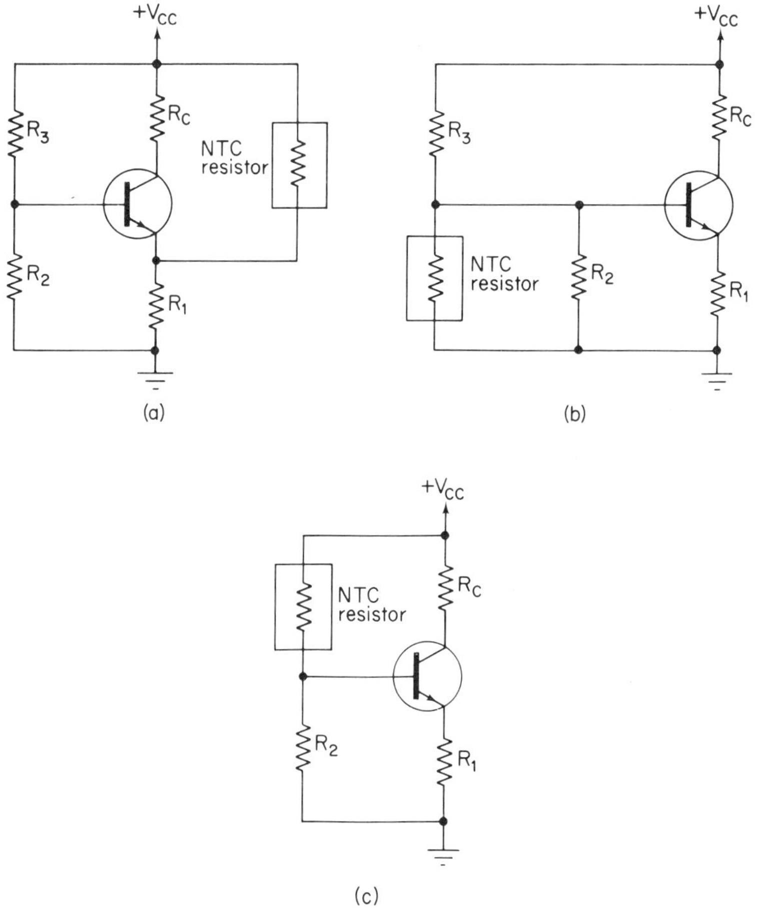

Fig. 8-12. Nonlinear bias-stabilization arrangements using positive- and negative-temperature-coefficient (PTC and NTC) resistance.

perature T_J. Frequently ΔI_{CBO} is considered to be I_{CBO} for the largest expected T_J value, since I_{CBO} is very small at low temperatures, owing to its exponential variation with T_J. T_J depends on ambient temperature T_A and the power P_i dissipated by the transistor under thermal equilibrium. When the BJT is turned on with a d-c power input P_C, T_J rises above T_A and would stabilize at a value such that the device loses heat to its environment at the rate P_C, *if its power input were independent of* T_J. In reality, the current increases with T_J, increasing the power input, which further increases T_J, and so forth. Without proper bias stabilization, the result may be *thermal runaway* and BJT destruction.

***Thermal resistance* θ.** When a BJT operates at thermal equilibrium with a power dissipation P_i, the rise of T_J above T_A is found to be proportional to P_i. That is,

$$T_J = T_A + \theta P_i \tag{8-33}$$

where θ is the thermal resistance of the transistor. Typically, BJT specifications include a maximum power-dissipation rating for a T_A of 25°C and a derating factor for T_A values above 25°C. Clearly, θ is the reciprocal of the derating factor, and the device is derated to zero power dissipation for T_A equal to maximum-rated T_J. If there were no thermally dependent currents, P_i would be equal to the initial power input and Eq. (8-33) would become

$$T_J = T_A + \theta P_C = T_O \tag{8-34}$$

T_O proves to be a convenient reference temperature.

The process whereby T_J is affected by thermally dependent current may be thought of as an electrical-thermal feedback loop, as shown in Fig. 8-13. From Fig. 8-13 we have

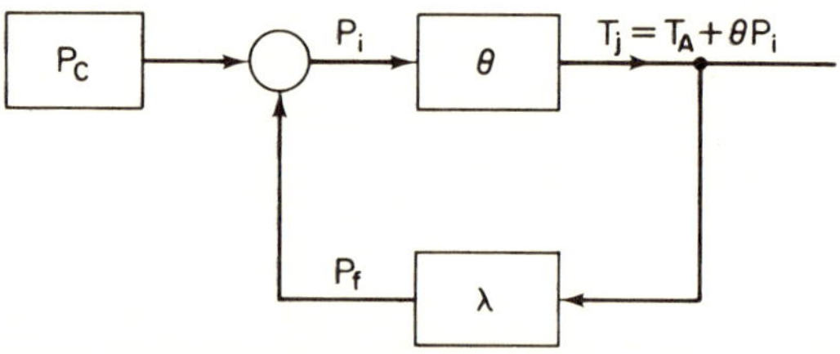

Fig. 8-13. Thermal-electrical feedback-loop representation of a BJT having thermally dependent current.

$$P_i = P_C + P_f \tag{8-35}$$

where P_f is the feedback power given by

$$P_f = \lambda T_J \tag{8-36}$$

From Eqs. (8-33) through (8-36), we obtain

$$\begin{aligned} T_J &= T_A + \theta P_i = T_A + \theta(P_C + P_f) \\ &= T_A + \theta(P_C + \lambda T_J) \end{aligned} \tag{8-37}$$

Solving for T_J yields

$$T_J = \frac{T_A + \theta P_C}{1 - \theta\lambda} = \frac{T_O}{1 - \theta\lambda} \tag{8-38}$$

This expression is of limited use in the determination of T_J, because λ is not necessarily constant, but it clearly shows that thermal runaway is possible and occurs if $\theta\lambda \geq 1.0$.

As a specific example, we now consider a BJT operating at fixed collector-emitter voltage V_{CE}. The power input at thermal equilibrium is given by

$$P_i = V_{CE}I_{E,T_J} \tag{8-39}$$

where I_{E,T_J} is the equilibrium value of emitter current. By definition of S_I,

$$I_{E,T_J} = I_{E,T_A} + S_I\,\Delta I_{CBO} \cong I_{E,T_A} + S_I I_{CBO,T_J} \tag{8-40}$$

where I_{E,T_A} is the emitter current when the BJT is turned on. Since I_{CBO} varies exponentially with T_J, it may be expressed as

$$I_{CBO,T_J} = I_{CBO,T_{\text{ref}}}e^{a(T_J - T_{\text{ref}})} \tag{8-41}$$

where a is a constant depending on the transistor. From Eqs. (8-33) and (8-39) through (8-41), we obtain

$$T_J \cong T_A + \theta V_{CE}I_{E,T_A} + \theta V_{CE}S_I I_{CBO,T_{\text{ref}}}e^{a(T_J - T_{\text{ref}})} \tag{8-42}$$

This expression can be solved for T_J by using successive approximations. If S_I happens to be too large for thermal stability to exist, there is no solution (other than $T_J = \infty$).

An approximate expression for the upper limit of S_I for thermal stability may be obtained by using T_O as the reference temperature in Eq. (8-42). From Eqs. (8-34) and (8-42) we obtain

$$S_I \cong \frac{(T_A - T_O)e^{-a(T_J - T_O)}}{\theta V_{CE}I_{CBO,T_O}} \tag{8-43}$$

Conditions for the maximum value of S_I may be found by obtaining $dS_I/d(T_J - T_O)$ and equating it to zero. Upon doing so, it is found that the maximum occurs when

$$T_J - T_O = \frac{1}{a} \tag{8-44}$$

and that

$$S_{I,\max} \cong \frac{1}{2.718a\theta V_{CE}I_{CBO,T_O}} \tag{8-45}$$

Therefore, thermal runaway occurs unless S_I is less than the value of Eq. (8-45), and solutions to the equilibrium equation [Eq. (8-42)] exist only for such S_I values. From Eq. (8-45), we see that the highest value of S_I depends on the highest expected ambient temperature and, conversely, the value of S_I limits the highest ambient temperature for thermal stability.

It should be recognized that the example just considered represents a "worst-case" situation, since V_{CE} is held constant. In an actual circuit, circuit resistance causes a decrease in V_{CE} as current increases, and the heating is less than in our example. A rather interesting situation exists when the resistance is so large that V_{CE} is one-half the supply voltage, or less, when the BJT is turned on. It may be shown that such a circuit is stable for all values of S_I. This is known as the half-power-supply-voltage principle. In most circuits, the variation of operating point has more stringent requirements than that of simply avoiding thermal runaway, necessitating the use of S_I values significantly lower than that of Eq. (8-45).

8-5 OPERATING-POINT SELECTION

The first step in the design of a bias-stabilized stage is the selection of a quiescent operating point. The d-c collector voltage V_{CE} should be such that the superimposed a-c component V_{ce} causes the instantaneous collector voltage v_{CE} neither to rise to the *sustaining value* nor to fall to the *saturation value*. The sustaining voltage $V_{CE,\text{sus}}$ is the collector voltage in the breakdown region and may be significantly less than the base-open breakdown value BV_{CEO} [see Fig. 5-19(b)]; the saturation voltage $V_{CE,\text{sat}}$ is the collector voltage at the knee of the collector characteristic.

Since maximum collector voltage and minimum collector current occur simultaneously in the amplifier stage, V_{CE} might be midway between the $V_{CE,\text{sus}}$ and the $V_{CE,\text{sat}}$ values in order to permit the largest possible collector-voltage swing. However, this is an inefficient operating condition for smaller a-c collector voltages. Therefore, if the upper limit of the peak value $v_{ce,\max}$ of the a-c collector voltage is known, often being either specified or determined by the a-c collector current and the load resistance, V_{CE} should be chosen so that

$$(V_{CE} - V_{CE,\text{sat}}) > v_{ce,\max} \tag{8-46}$$

Since V_{CE} decreases as temperature increases, the difference between $(V_{CE} - V_{CE,\text{sat}})$ and $v_{ce,\max}$ at the design-center ambient temperature, together with the values of R_C and R_1 (see Figs. 8-1 and 8-6), determines the maximum permissible ΔI_C with temperature.

The upper limit of d-c collector current I_C may now be determined from the collector power-dissipation rating and V_{CE}. The magnitude of I_C is the upper limit of the peak value $i_{c,\max}$ of the a-c component of collector current, and this may be the basis for the choice of I_C. In the interest of efficiency, I_C should not be appreciably larger than the maximum expected value of $i_{c,\max}$ unless using a larger I_C results in a major improvement in transistor characteristics, say in h_{fe}. When the d-c operating point and R_L have been selected, the upper limit of $i_{c,\max}$ is either I_C or $(V_{CE} - V_{CE,\text{sat}})/R'_L$ whichever is smaller, where R'_L is the resistance of the parallel combination of R_C and R_L.

8-6 THE DIFFERENCE AMPLIFIER[1,2]

The difference, or differential, amplifier is introduced at this juncture because it illustrates the use of a second BJT for stabilizing an amplifier stage. The difference amplifier is so named because its output is proportional to the difference between two input signals. Because of its excellent bias stability, it is widely used for the amplification of direct current or of signals having frequencies ranging down to zero.

One of the major problems in the amplification of direct current is the inability of the output to distinguish between the effect of an input signal and that of a change in a transistor characteristic such as I_{CBO} or V_{BE}. In the difference amplifier, these changes for each transistor are balanced by those of another BJT.

The basic circuit of an emitter-coupled difference-amplifier stage is shown in Fig. 8-14. Ideally, it consists of a pair of identical BJTs with the emitter current of the pair supplied by a constant-current source. The requirement of matched BJTs and the need for a second power supply $-V_{EE}$ are disadvantages of the difference amplifier. In order to approach a constant-current source of emitter current, R_E and $-V_{EE}$ should be as large as practical. Approximately matched BJT pairs in a common enclosure are commercially available for this type of application.

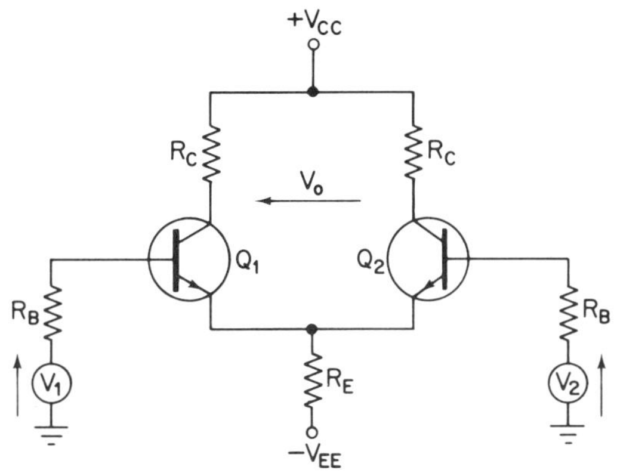

Fig. 8-14. Basic emitter-coupled difference amplifier.

[1] D. W. Slaughter, "The Emitter-Coupled Differential Amplifier," *IRE Trans.*, Vol. CT-3 (1956), pp. 51–53.

[2] R. D. Middlebrook, *Differential Amplifier*. New York: John Wiley and Sons, Inc., 1964.

If the BJTs and drain resistors of Fig. 8-14 are matched and $V_1 = V_2 = 0$ V, output V_o is seen to be zero. If d-c collector currents change because of changes in I_{CBO}, V_{CC}, or $-V_{EE}$, the change in the collector potential of Q_1 equals that of Q_2, and V_o remains zero. Thus, the output of the stage is stabilized against changes in power-supply voltages as well as in I_{CBO}. For silicon BJTs at moderate temperatures, I_{CBO} may be so small that its effect on collector current is unimportant, but V_{BE} changes with temperature (about -2 mV/deg C) can be of major significance. For example, if Q_2 in Fig. 8-14 were open-circuited and V_o taken with respect to ground, the effect on V_o of a 1°C change in the temperature of Q_1 is as though a d-c input V_1 of 2 mV were present. With Q_2 in the circuit, the input loop includes two base-emitter voltages of opposite polarity, and the d-c input current is unaffected by V_{BE} change with temperature. It should be recognized that compensation for a change in V_{BE} of one BJT by that of the other reduces the temperature dependence of V_o even though it is taken with respect to ground.

For idealized conditions (matched components and $R_E = \infty$), the output voltage of the amplifier of Fig. 8-14 can be shown to be given by

$$V_o = -\frac{h_{fe}R_C(V_1 - V_2)}{(h_{ie} + R_B)} \tag{8-47}$$

provided $h_{oe}R_C \ll 1$, indicating that the output is proportional to the difference in the input voltages. In practice, it is found that an output occurs with equal input voltages, and the circuit is said to have *common-mode gain* A_c. For $V_1 = V_2 = V_s$, it may be shown that

$$A_c = \frac{V_o}{V_s} = \frac{2(h_{oe}R_E - h_{fe})R_C}{2R_E(1 + h_{fe}) + (R_B + h_{ie})(2h_{oe}R_C + 1)} \tag{8-48}$$

provided h_{re} is negligible and $h_{oe}R_C \ll 1$. From Eq. (8-48) it is seen that A_c approaches zero as R_E approaches infinity. The common-mode gain should be as small as possible, because its existence causes V_o to depend on $(V_1 + V_2)$ as well as $(V_1 - V_2)$. That is, V_o would not be the same for $V_1 = +1$ mV and $V_2 = -1$ mV as for $V_1 = 11$ mV and $V_2 = 9$ mV, even though $(V_1 - V_2)$ is 2 mV for each arrangement.

In practice, the difference amplifier is often operated with a single input, the other input being zero. Each R_B in Fig. 8-14 includes a source resistance R_g, and the BJTs usually operate with bases returned to ground through equal resistances. That is, when R_g is changed, R_B is adjusted accordingly. Frequently, R_E is replaced by a BJT circuit, as in Fig. 8-15. In this arrangement, R_1, R_2, and R_3 are selected to give the same quiescent conditions for Q_1 and Q_2 as in Fig. 8-14. Because of the degenerative effect of R_3 described in Sec. 8-1, the result is an effective R_E several orders of magnitude greater

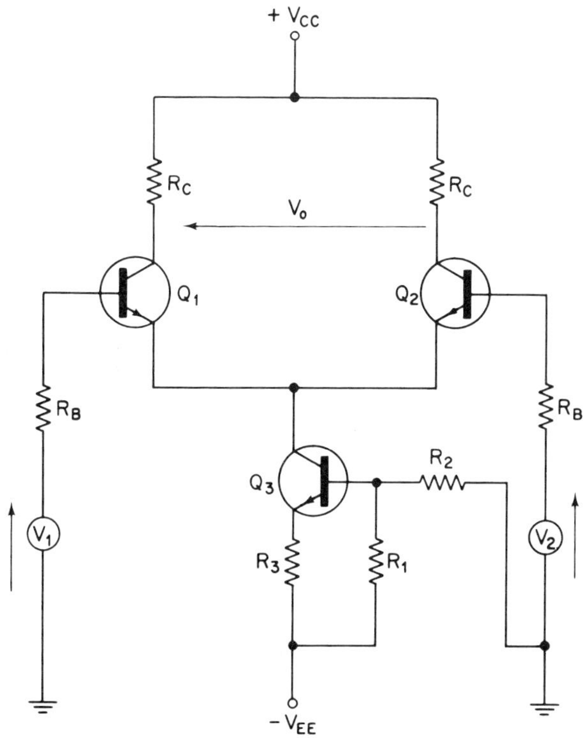

Fig. 8-15. Difference amplifier having a constant-current stage in the emitter circuit.

than R_3. That is, Q_3 acts as a constant source of emitter current for Q_1 and Q_2.

As mentioned earlier, the requirement of matched BJTs and drain resistors is a disadvantage of difference amplifiers using discrete components. Fortunately, this is not a problem when integrated-circuit techniques are employed (Chapter 11), and difference amplifiers are widely used in linear integrated circuits, as pointed out in Sec. 11-12.*

REFERENCES

[1] D. W. Slaughter, "The Emitter-Coupled Differential Amplifier," *IRE Trans.*, Vol. CT-3 (1956), pp. 51–53.

* See Figs 11-43 and 11-50

[2] R. D. Middlebrook, *Differential Amplifiers*. New York: John Wiley and Sons, Inc., 1964.

PROBLEMS

8-1. Determine approximate values of d-c emitter current I_E and d-c collector-emitter voltage V_{CE} in the bias-stabilized circuit of Fig. P8-1.

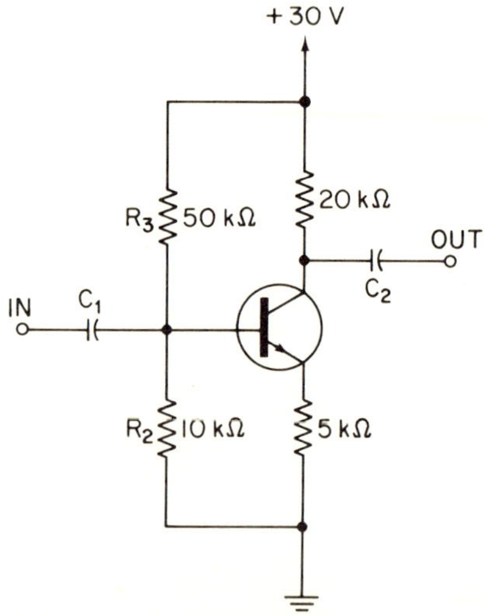

Figure P8-1

8-2. An increase in junction temperature causes I_{CBO} of the BJT of Prob. 8-1 to increase from 1 μA to 50 μA. What value of V_{CE} might be expected at the higher temperature?

8-3. Specify values of R_2 and R_3 that result in the same d-c operating point in Prob. 8-1, but limit the V_{CE} change to 1.0 V for the I_{CBO} change of Prob. 8-2.

8-4. The BJT in the circuit of Fig. P8-4 has an h_{fe} of 80 and an h_{ie} of 2400 Ω. Assuming h_{re}, h_{oe}, and the reactances of C_1 and C_2 to be negligible, determine the input impedance V_1/I_1 and the voltage ratio V_2/V_g for: (a) ideal emitter bypassing (C_3 acting as an emitter-to-ground a-c short circuit) and (b) C_3 removed from the circuit.

8-5. What value of C_3 is required at a frequency of 100 Hz in Prob. 8-4 for: (a) a magnitude of V_2/V_1 essentially the same as for ideal emitter bypassing, (b) a magnitude of V_2/V_g essentially the same as for ideal emitter bypassing, and (c) a V_2/V_g magnitude 3 dB below the value for ideal emitter bypassing?

8-6. Using the BJT parameters of Table 7-1 and assuming the reactances of C_1,

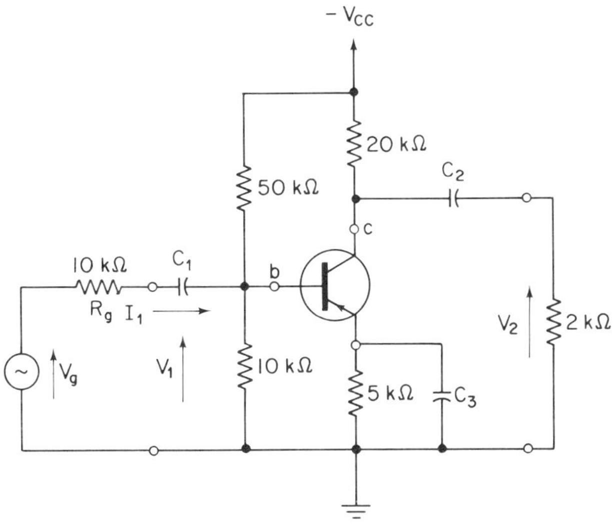

Figure P8-4

C_2, and C_3 to be negligible, determine values of input impedance, output impedance, V_2/V_1, and V_2/V_g for the circuit of Fig. P8-4.

8-7. What value of C_2 is required at a frequency of 100 Hz in Prob. 8-6 for: a magnitude of V_2/V_g essentially the same as though C_2 were an a-c short circuit, and (b) a V_2/V_g magnitude 3 dB below that of (a)?

8-8. Determine how much effect the C_2 value of Prob. 8-7(a) has on the magnitude of V_2/V_g at a frequency of 100 Hz when the C_3 value of Prob. 8-5(c) is used in the circuit of Prob. 8-6.

8-9. Silicon BJTs used in the circuit of Fig. P8-1 have h_{fe} values that range from 25 to 100 at the operating point $V_{CE} = 5$ V, $I_E = -1$ mA. If I_{CBO} has negligible effect on the operating point, what V_{CE} range might be expected?

8-10. For the emitter-follower of Fig. P8-10, determine approximate values of V_{CE}, I_E, S_I, and maximum obtainable undistorted V_2.

8-11. Assuming coupling capacitors and BJT parameter h_{re} to have negligible effects, determine the input impedance, the voltage ratios V_2/V_1 and V_2/V_g, and the output impedance of the circuit of Fig. P8-10 when the BJT of Table 7-1 is used.

8-12. The circuit of Prob. 8-11 is modified as shown in Fig. P8-12. Assuming all capacitors to have negligible effects and the BJT parameters to be those of Prob. 8-11, determine the input impedance, the voltage ratios V_2/V_1 and V_2/V_g, and the output impedance.

8-13. A type 2N1132 silicon BJT has a power-dissipation rating of 0.6 W at an ambient temperature of 25°C and a derating factor of 4 mW/deg C. What is the maximum safe junction temperature?

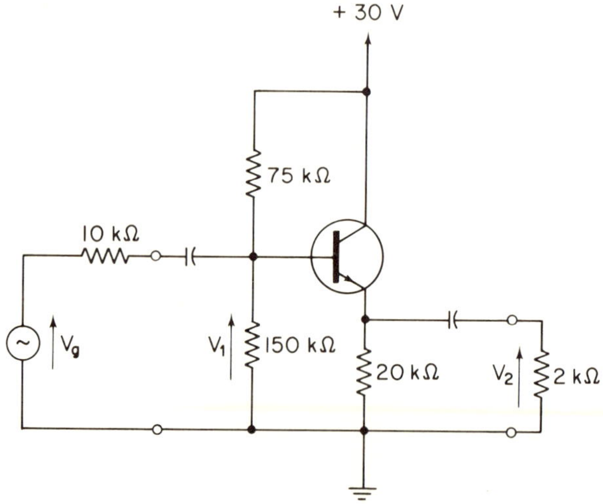

Figure P8-10

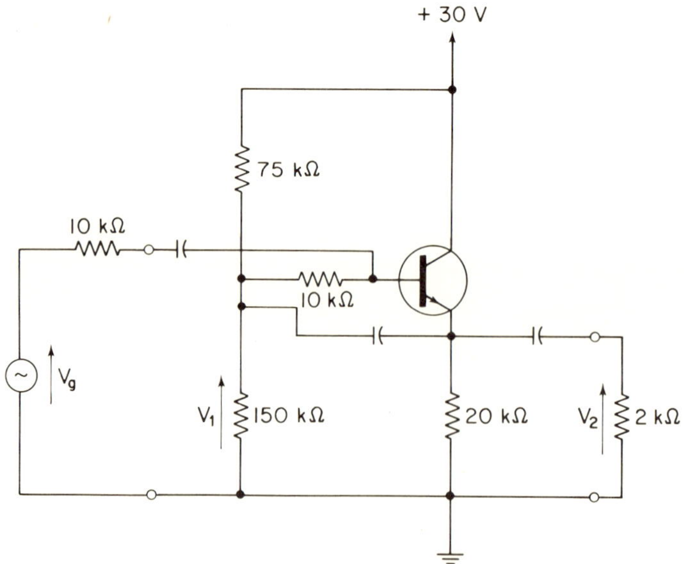

Figure P8-12

8-14. A germanium BJT has a power-dissipation rating of 150 mW at an ambient temperature of 25°C and a junction-temperature rating of 85°C. What is the thermal resistance θ?

8-15. The BJT of Prob. 8-13 is operated at a collector voltage of 10 V and an ambient temperature of 100°C. What is the maximum safe collector current?

8-16. The BJT of Prob. 8-13 has maximum I_{CBO} specifications of 1 μA at 25°C and 100 μA at 150°C. What maximum I_{CBO} might be expected at an ambient temperature of 60°C if $V_{CE} = -10$ V and $I_C = -10$ mA?

8-17. A germanium BJT has an I_{CBO} value of 2 μA at 25°C and a derating factor of 1 mW/°C. At an elevated junction temperature T, I_{CBO} is given by $I_{CBO} = 2 \exp((T - 25)/14)$ μA. If V_{CE} is fixed at 10 V, I_E is 2 mA for a junction temperature of 25°C, and the ambient temperature is 50°C, what current stability factor is required to prevent thermal runaway?

8-18. Using the $T_j(T_o)$ relationship for maximum stability factor, determine the junction temperature for the stability factor of Prob. 8-17 and check the temperature by use of the $T_j(S_I)$ relationship.

8-19. At what temperature will the junction stabilize if a current stability factor of 5.0 is used in the circuit of Prob. 8-17?

8-20. If the BJT in the circuit of Prob. 8-17 has an h_{fe} of 50 and the base current is fixed at the value that results in an I_E of 2 mA at 25°C, what is the maximum ambient temperature without thermal runaway?

8-21. A BJT similar to that of Prob. 8-17 is operated with a collector voltage of 10 V and an emitter current of 3 mA for a junction temperature of 25°C. If the current stability factor is 8.0 and the junction temperature is not to exceed 80°C, what is the maximum permissible ambient temperature?

8-22. Assuming identical BJTs having negligible h_{oe} and h_{re}, derive Eq. (8-48) for the output voltage of the emitter-coupled difference amplifier of Fig. 8-44.

8-23. Assuming identical BJTs having a negligible h_{re} value and an h_{oe} value such that $h_{oe}R_C \ll 1$, derive Eq. (8-48) for the common-mode voltage gain of the difference amplifier of Fig. 8-44.

9

The BJT as a Switch

So far, our consideration of the BJT has emphasized active-mode operation with the emitter-base junction forward biased and the collector-base junction back biased. In this operating mode, the BJT has been considered to be essentially a linear device with limits of the active region being the collector-current *cutoff* and *saturation* regions (see Figs. 5-17 and 5-18). The small current of the cutoff region, owing to zero or negative (for an *npn* BJT) base current, and the small collector-emitter voltage of the saturation region, owing to the forward-biased collector-base junction, suggest use of the BJT as a switch. That is, the cutoff and saturation regions are, respectively, the OFF and ON operating modes of the BJT when it is acting as a switch. BJT switches are widely used in electronic systems.

9-1 SWITCHING MODES

In the normal use of the BJT as a switch, the load is in series with the collector, as shown in the circuit of Fig. 9-1. Collector current i_C and voltage v_{CE} dependence on base current i_B may be determined by drawing a load line on the collector characteristic, as in Fig. 9-1. Points A and B indicate the OFF and ON conditions of the switch as well as the limits of the active region of the BJT in the switching circuit. Although $i_{C,\mathrm{OFF}}$ is represented as

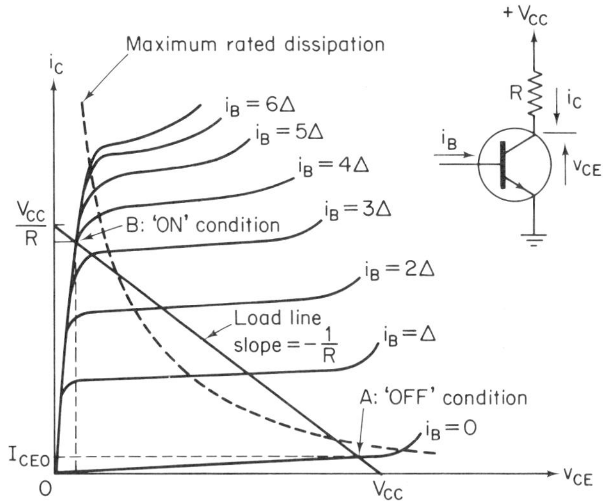

Fig. 9-1. Generalized collector characteristics of an *npn* BJT used as a switch.

I_{CEO} for $v_{CE} = V_{CC}$, we see from Fig. 5-17 that it can be made significantly lower by back-biasing the base-emitter junction. For an ideal switch, $i_{C,\text{OFF}}$ would be zero, as would the ON value of collector voltage v_{CE}. These characteristics can be more nearly achieved in the BJT switch than in any other electronic switch.

The active mode. Active-mode operation of a BJT switch is a transient condition, since the device is operated either ON or OFF and does not dwell for appreciable time in the active region. It is of interest to observe that a considerable portion of the active-region operating path (load line) of Fig. 9-1 is in regions where the power input of the BJT exceeds its dissipation rating. This is not harmful to the BJT, because it is transient operation, and is quite common in switching circuits. That is, the average power dissipated by the BJT is not usually increased appreciably by operating-point excursions into normally forbidden regions. This increase depends on the switching rate, of course, and may be significant for switching rates approaching the maximum capability of the transistor. For example, it may be shown that for $v_{CE,\text{OFF}} = 6$ V and $i_{C,\text{ON}} = 10$ mA, 10 mW of average power is dissipated as a result of switching at the ultimate rate.* Even so, we see that a BJT is capable of switching powers that are large compared to its dissipation rating. This is so because the power dissipated in the OFF and ON modes is also low, owing to i_C being small when v_{CE} is large and vice versa.

* *General Electric Co. Transistor Manual*, 7th ed., p. 142.

The OFF mode. Three types of OFF operation of a BJT are shown by Fig. 9-2. These are (1) base-open or zero-i_B operation, (2) use of a base-emitter

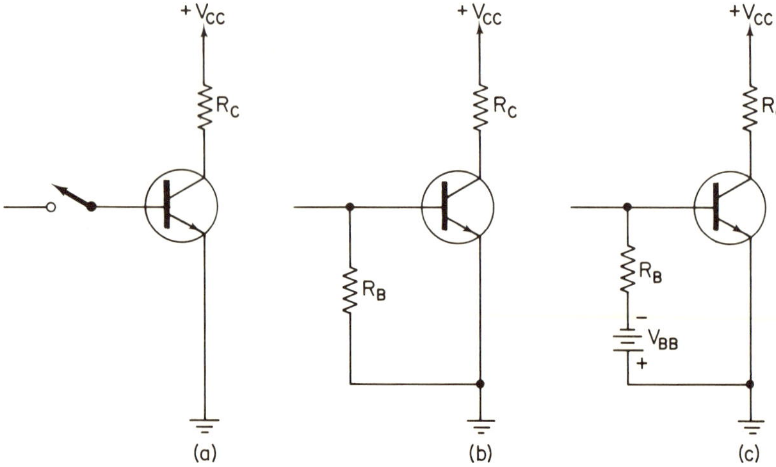

Fig. 9-2. Methods for maintaining a BJT switch in the OFF mode.

resistor, (3) back-biasing the base-emitter junction. With the base open as in Fig. 9-2(a), $i_{C,\mathrm{OFF}}$ flows through the base-emitter junction, forward-biasing the junction and injecting minority carriers into the base. As a result, $i_{C,\mathrm{OFF}}$ is given by

$$i_{C,\mathrm{OFF}} = \frac{I_{CBO}}{1 - \alpha_{fbo}} \cong h_{fe}I_{CBO} \cong h_{FE}I_{CBO} \tag{9-1}$$

That is, $i_{C,\mathrm{OFF}}$ is I_{CEO} as given by Eq. (5-17). From Eq. (9-1), we see that the condition $i_B = 0$ can be considered the result of supplying a base current equal to and opposite I_{CBO}, $i_{C,\mathrm{OFF}}$ being the result of common-emitter amplification of this effective i_B value. For many applications, particularly germanium BJTs operated at elevated temperatures, this value of $i_{C,\mathrm{OFF}}$ is excessive.

The use of base-emitter resistor R_B as in Fig. 9-2(b) results in diversion of part of $i_{C,\mathrm{OFF}}$ from the base-emitter junction. This reduces minority-carrier injection into the base, causing $i_{C,\mathrm{OFF}}$ to be less than that with the base open. Thus

$$I_{CBO} < i_{C,\mathrm{OFF}} < h_{FE}I_{CBO} \tag{9-2}$$

for the arrangement of Fig. 9-2(b), and the smaller the resistance of R_B, the greater its diversion effect and the closer $i_{C,\mathrm{OFF}}$ is to I_{CBO}. Even for $R_B = 0$, $i_{C,\mathrm{OFF}} > I_{CBO}$, since short-circuiting the base-emitter terminals fails to divert all of $i_{C,\mathrm{OFF}}$ from the base-emitter junction. For silicon BJTs at moderate temperatures, the use of moderate R_B values, say of the order of 10 kΩ, results in $i_{C,\mathrm{OFF}}$ values that are acceptably low for most applications. As added advantage in the use of R_B is an increase in the upper limit of $v_{CE,\mathrm{OFF}}$, which is collector-emitter breakdown voltage BV_{CE}. Since collector-emitter breakdown occurs when α_{fbo} becomes unity, as pointed out in Sec. 5-2.3, and α_{fbo} is dependent on i_C, reducing $i_{C,\mathrm{OFF}}$ through the use of R_B increases the breakdown voltage. As R_B approaches zero, BV_{CE} approaches BV_{CBO}, the collector-base breakdown value, which may be several times larger than BV_{CEO}, the value with the base open. For switching BJTs, manufacturers commonly specify BV_{CER}, the common-emitter breakdown voltage for a specific R_B value (say 10 kΩ), rather than BV_{CEO}.

Of the methods shown by Fig. 9-2 for maintaining a BJT switch in the OFF mode, back-biasing the base-emitter junction, as in Fig. 9-2(c), results in the smallest value of $i_{C,\mathrm{OFF}}$. The effect on i_C of this bias may be seen from base-region minority-carrier distribution plots such as those in Fig. 9-3(b)

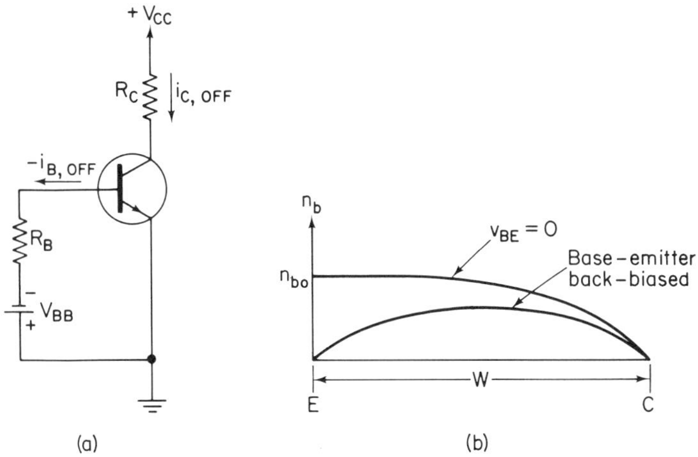

Fig. 9-3. The effect of base-emitter junction back-bias on base minority-carrier density n_b of an *npn* transistor switch operating in the OFF mode. (a) Circuit diagram. (b) n_b distribution.

for an *npn* BJT. For purposes of illustration, a base width W of several diffusion lengths is assumed, and the effect on W of base-bias voltage v_{BE} is neglected. For $v_{BE} = 0$, electron density n_b ranges from the equilibrium value n_{bo} at the emitter edge of the base ($x = 0$) to essentially zero at the collector edge ($x = W$), owing to the back-biased collector-base junction. With appreciable base back-bias, n_b is reduced to zero at the emitter edge of the base with consequent reductions of n_b throughout the base. Of particular interest is the effect of v_{BE} on dn_b/dx at $x = W$, since this concentration gradient is indicative of the electron component of i_C. Since

$$\left.\frac{dn_b}{dx}\right|_{x=W}$$

for $v_{BE} = 0$ is indicative of the electron component of I_{CBO}, we see from Fig. 9-3 that back-biasing the base-emitter junction results in a collector current somewhat less than I_{CBO}. The usual practice in switching-circuit calculations is to *assume* $i_{C,\text{OFF}}$ *to be* I_{CBO}.

Base current $i_{B,\text{OFF}}$ of the transistor in the OFF mode is of interest, since it flows in the base-bias circuit. From the circuit of Fig. 9-3(a), we see that, with appreciable base-emitter back-bias, $-i_{B,\text{OFF}}$ is the sum of $i_{C,\text{OFF}}$ and the current through the base-emitter junction. Since the latter is less than I_{EBO} (the value with $i_C = 0$), owing to collector-base junction back-bias, and $i_{C,\text{OFF}}$ is less than I_{CBO}, which is of the order of I_{EBO}, it is customary to *assume* $-i_{B,\text{OFF}}$ *to be* I_{CBO}.

With base-emitter back-bias, the upper limit of $v_{CE,\text{OFF}}$ is determined by the base-bias voltage and collector-base breakdown voltage BV_{CBO}. From Fig. 9-2(c), we see that

$$v_{CB} = V_{CC} + V_{BB} \tag{9-3}$$

for negligible voltage drops across R_B and R_C. Therefore,

$$v_{CE,\text{OFF}} \leq BV_{CBO} - V_{BB} \tag{9-4}$$

and we see that the upper limit of $v_{CE,\text{OFF}}$ approaches BV_{CBO}, since V_{BB} need be merely large enough to provide a base-emitter back-bias of the order of tenths of a volt. Because of the favorable values of $i_{C,\text{OFF}}$ and $v_{CE,\text{OFF}}$, base back-bias is widely used for OFF-mode operation of germanium-BJT switches and is quite common for silicon BJTs. In addition, as we shall see, base back-bias aids in the turnoff of a BJT that has been operating in the ON mode.

The ON mode. As stated earlier and indicated in Fig. 9-1, the BJT switch operates in the collector-current saturation region when ON. As the operating point moves through the active region from the OFF condition, collector current i_C increases with base current i_B according to the relationship

$$i_C \cong h_{FE} i_B \tag{9-5}$$

h_{FE} being approximately constant in the active region for $i_C \gg i_{C,\text{OFF}}$. Since i_C in the circuit of Fig. 9-1 cannot exceed V_{CC}/R, Eq. (9-5) no longer applies as i_C approaches V_{CC}/R_C, and BJT operation enters the saturation region in which i_C is essentially independent of i_B. That is, the BJT operates in its saturation region when circuit conditions are such that $i_C < h_{FE} i_B$. In this region, both the emitter-base and the collector-base junctions are forward biased, as pointed out in Sec. 5-2.3, and the collector-emitter voltage is usually quite small compared to the forward-bias voltage of either junction.

The state of affairs inside the BJT as operation proceeds through the active region into saturation is shown for the *npn* BJT of Fig. 9-1 by the base-region electron-distribution $n_b(x)$ plots of Fig. 9-4. For purposes of illustration, the effects of junction bias on base width W is neglected. For i_B values of Δ, 2Δ, and 3Δ, electron density $n_b(W)$ at the collector edge of

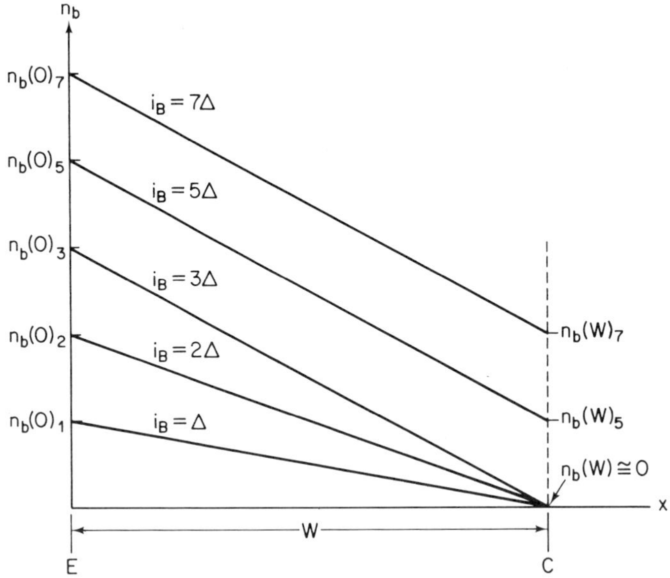

Fig. 9-4. Base-region minority-carrier distribution for the BJT of Fig. 9-1 with active-region base currents of Δ, 2Δ, and 3Δ and saturation-region currents of 5Δ and 7Δ.

the base ($x = W$) is essentially zero, since operation is in the active region and the collector-base junction is back-biased. As i_B is increased, the area under the $n_b(x)$ plot increases because i_B is largely recombination current, which depends on the number of minority carriers stored in the base region. With $n_b(W)$ held at zero, this increase in area results in an increased $n_b(x)$ slope with consequent increase in i_E and i_C. From Fig. 9-1, we see that the BJT is operating at the edge of the saturation region for an i_b value of 3Δ and that further increases in i_B cause little or no increase in i_C. Therefore, the slope of the $n_b(x)$ plot cannot increase significantly above that for $i_B = 3\Delta$, and the necessary increase in area under the plot for further increase in i_B can occur only through increase in $n_b(W)$. Thus the $n_b(x)$ plots for i_B values of 5Δ and 7Δ have essentially the same slope as that for 3Δ, but have $n_b(W)$ values that are positive and proportional to $(i_B - 3\Delta)$. For $n_b(W)$ values greater than the equilibrium value n_{bo}, the collector-base junction is *forward biased*, as we know from Eq. (2-10). Because of the exponential relationship between $n_b(W)$ and v_{CB} and between $n_b(0)$ and v_{BE}, there is little difference between the emitter-base and collector-base junction voltages of a BJT operating in saturation. Therefore, $v_{CE,\mathrm{ON}}$ is largely *IR* drop in the collector and emitter bulk-regions and contacts.

9-2 TRANSIENT RESPONSE CHARACTERISTICS

Now that we have considered the static characteristics of the ON and OFF modes of the transistor switch, we shall consider transitions between these modes.

In order to turn a BJT switch on, charge must be supplied to the base region to neutralize the junction depletion-region charges of the OFF mode and to establish sufficient excess charge in the base region to support the ON-mode collector current. Conversely, to turn a BJT switch off, all excess charge must be removed from the base region and charge must be removed from the depletion regions so that they can support the OFF-mode junction voltages. During turn-on, neutralization of depletion-region charge is responsible for a *time delay* between application of the turn-on signal and appreciable change in collector current, and the time required to establish the excess base charge is the collector-current *rise time*. During turn-off, the time required for removal of the excess base charge responsible for collector-base junction forward-bias is responsible for turn-off delay or *storage time*, and the time required for removal of the remaining excess charge and the depletion-region charge is the collector-current *fall time*. For purposes of illustration of BJT switching characteristics, let us consider the circuit of Fig. 9-5(a).

The BJT of Fig. 9-5(a) is turned OFF and ON by periodically opening and closing switch *S*. With *S* open, the -10-V source is connected to the base through the 100-ohm resistor and base resistor R_B, and the base-emitter voltage is -10 V plus a small base-circuit voltage drop due to $i_{B,\mathrm{OFF}}$. Therefore, the BJT is OFF, and its base-emitter transition capacitance C_E is charged to approximately 10 V. For this condition, the collector-emitter voltage is $+10$ V minus a small collector-circuit voltage drop due to $i_{C,\mathrm{OFF}}$, and the collector-base transition capacitance C_C is charged to approximately 20 V. Electrode voltages for the OFF mode are shown in Fig. 9-5(b), in which the BJT is considered out of the circuit.

Circuit conditions at the instant of closure of switch *S* are shown by Fig. 9-5(c). Since the collector and emitter voltages cannot change immediately, owing to associated capacitances, a voltage of 20 V has been placed across R_B, resulting in a base current i_B of 1.0 mA. Base-current flow discharges C_E and C_C, but the BJT remains OFF as long as v_{BE} is negative. As base

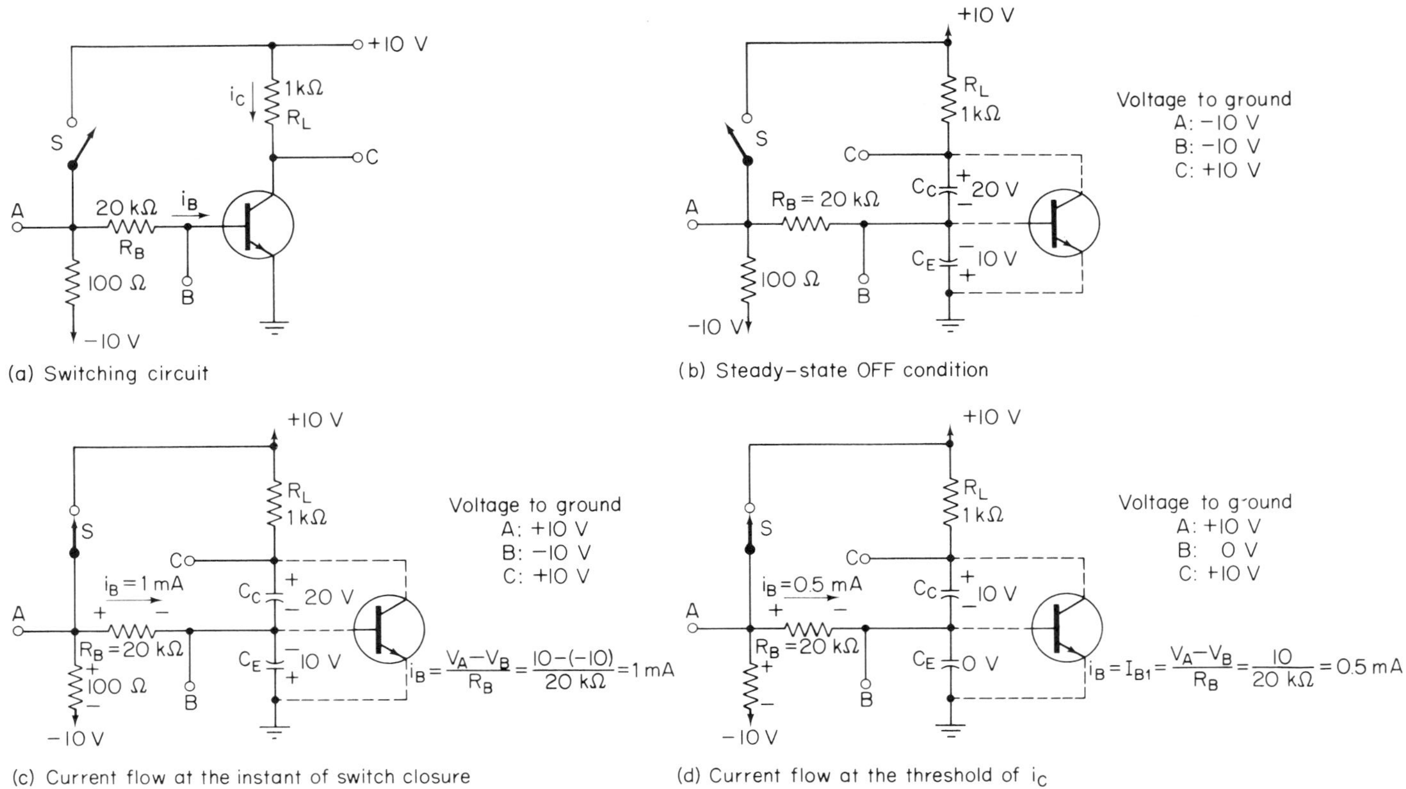

Fig. 9-5. (a) BJT switching circuit. (b) Circuit conditions before closure of *S*. (c) Circuit conditions at the instant of closure of *S*. (d) Circuit conditions at the collector-current threshold.

current continues to flow, v_{BE} falls to zero and becomes positive, forward-biasing the base-emitter junction so that turn-on begins. Circuit conditions at the threshold of turn-on are shown in Fig. 9-5(d). With $v_{BE} = 0$, the voltage across R_B is seen to be 10 V, resulting in an i_B of 0.5 mA. Thus, turn-on is delayed by the time required for a capacitance of $(C_E + C_C)$ to experience a discharge of 10 V by a current that decreases exponentially from 1.0 mA to 0.5 mA. This is likely to be the principal cause of delay time. Another cause is the time required for injected minority carriers to diffuse through the base to the collector. We shall also see that, by definition, delay time includes a small portion of the collector-current rise time. For a particular BJT, delay time is seen to be dependent on bias voltages and base-circuit resistance. It also depends on transistor base width W, owing to base-region diffusion time.

When the base-emitter junction becomes forward biased, v_{BE} is of the order of tenths of a volt, and the voltage across R_B continues to be approximately 10 V. Therefore, i_B during the remainder of turn-on, and thereafter, is approximately 0.5 mA. This current is known as the turn-on base current I_{B1}. With base-emitter forward-bias, the BJT enters the active region, collector-current flow begins, and the collector voltage begins to decrease because of collector resistor R_L. Since, initially, I_{B1} is greater than the rate of recombination of charge in the base, the charge stored in the base increases, increasing the charge gradient, which increases the collector current. This process continues until collector-current saturation occurs, providing $I_{B1} > i_{C,\mathrm{ON}}/h_{FE}$, and the time required to reach the edge of saturation is the collector-current rise time t_r. Since the rate of charge storage in the base is proportional to I_{B1}, t_r varies inversely with I_{B1} or with the extent that the base is overdriven. From the $n_b(x)$ plots of Fig. 9-4, we see that the amount of stored charge required for saturation varies directly with base width W. indicating that rise time is an inverse function of BJT alpha-cutoff frequency [see Eq. (6-60)].

Let us now consider the effect of opening switch S in the circuit of Fig 9-5(a), assuming that S has been closed long enough that equilibrium conditions exist. That is, charge stored in the transistor base is no longer increasing, being such that the recombination rate is equal to I_{B1}. Approximate conditions in the circuit of Fig. 9-5(a) just prior to opening S are shown for a silicon BJT by Fig. 9-6(a). Immediately after S is opened, excess base charge maintains forward bias of the collector-base and emitter-base junctions, and collector-current flow is virtually unaffected by opening S. In order for i_C to fall, v_{CE} must rise, and this cannot occur until the collector-base junction becomes reverse biased. With the emitter-base junction forward biased, virtually all of the -10-V supply is across R_B, resulting in a base current of approximately -0.5 mA, as shown in Fig. 9-6(b). This negative base current is known as the turn-off base current I_{B2}. The flow of I_{B2} and

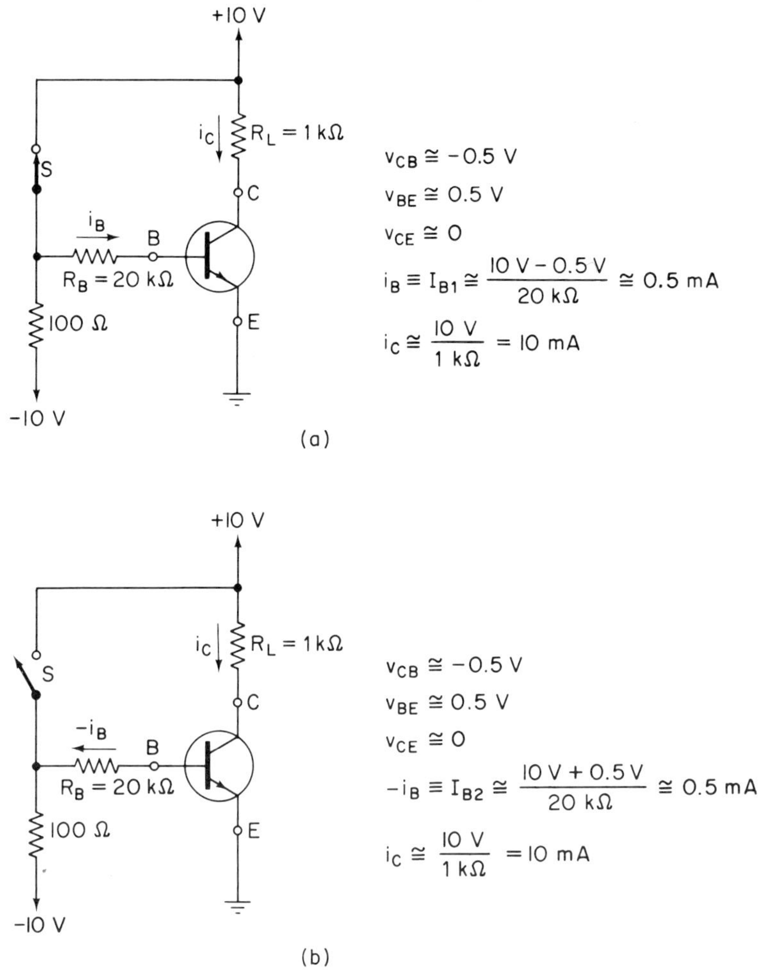

Fig. 9-6. Approximate conditions in the circuit of Fig. 9-5 for a silicon BJT. (a) ON-mode conditions. (b) Conditions immediately after opening switch *S*.

continuation of i_C at the saturation magnitude removes stored charge from the base and, after the turn-off delay or storage time t_s, minority-carrier density $n_b(W)$ at the collector edge falls to essentially zero. This permits the collector-base junction to become reverse biased with consequent decrease in i_C. Clearly, for a particular BJT, storage time varies inversely with I_{B2} and directly with I_{B1}.

Although i_C decreases when the collector-base junction becomes reverse biased, i_C cannot immediately fall to $i_{C,\mathrm{OFF}}$, owing to the excess charge remaining in the base. Thus i_C gradually decreases as base charge is removed by I_{B2} and by continuation of i_C, and the time required for charge removal is the collector-current fall time t_f. Clearly, t_f varies inversely with I_{B2} and directly with the excess charge in the base when the BJT is at the edge of collector-current saturation. This charge varies directly with $i_{C,\mathrm{ON}}$ and inversely with base width W.

Figure 9-7(b) shows current and voltage wave forms of the switching circuit of Fig. 9-7(a) for periodic operation of S. Switching-time definitions are shown by the i_C wave form and are seen to involve the 10 and 90 percent points. The four switching times may be combined into the turn-on and turn-off times, turn-on time being $(t_d + t_r)$ and turn-off time being $(t_s + t_f)$. Clearly, the turn-on and turn-off times limit the operating rate of the BJT switch. Because of this, switching times are specified (by the manufacturer) for BJTs intended for use in switching applications. Other conditions being equal, all four switching times have been shown to decrease with BJT base width C or with increase in alpha-cutoff frequency $f_{\alpha b}$. Thus, as one might expect, higher-frequency BJTs are higher-speed switches. For switching BJTs, $f_{\alpha b}$ or transition frequency f_T may not be specified, owing to switching-time specifications, although this is an important specification if the BJT is used as an amplifier.

9-3 NONREGENERATIVE SWITCHING CIRCUITS

Switching circuits are classified as regenerative or nonregenerative, depending on whether or not an externally applied drive is required to hold the active devices in their desired operating states. In a regenerative circuit, the switching action is initiated by an externally applied pulse; after this, action of the circuit carries the change-of-state to completion. The nonregenerative circuit is analogous to a simple relay that requires a continuous current to keep its contacts closed. The switching circuit that we considered in Sec. 9-2 is seen to be of the nonregenerative type. Nonregenerative circuits are widely used, particularly in logic applications.

Figure 9-8 shows a nonregenerative switching circuit that is basic to many applications. In this circuit, the BJT is turned ON by the input pulse of amplitude V_1 and turns OFF after the end of the pulse. Base-bias supply voltage V_{BB} aids in turning the transistor OFF and holds it OFF until the occurrence of the next pulse. Let us now consider the ON and OFF conditions of the circuit.

When the BJT is in the ON mode, we see from Fig. 9-8 that current i_1 through resistor R_1 is given by

$$i_1 = \frac{V_1 - v_{BE,\mathrm{ON}}}{R_1} \tag{9-6}$$

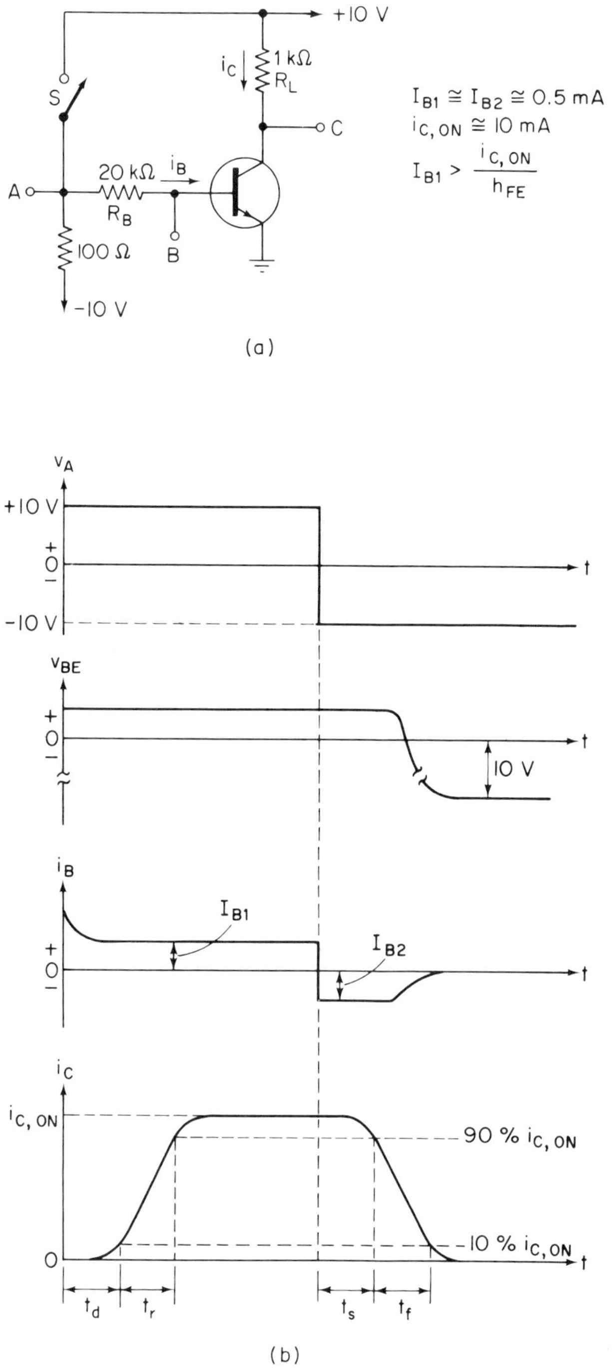

Fig. 9-7. (a) Switching circuit of Fig. 9-5(a). (b) Waveforms for the circuit of (a).

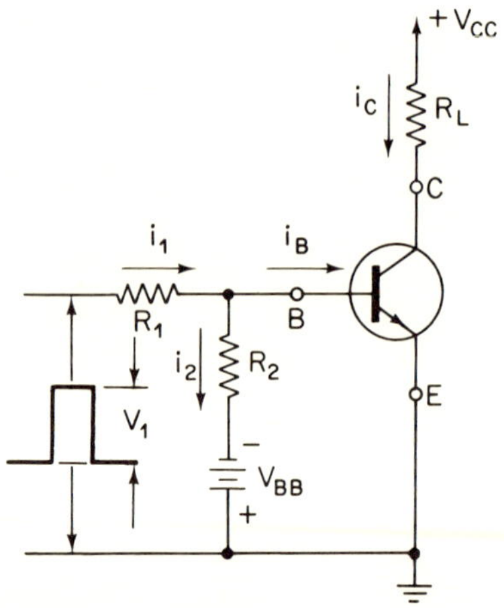

Fig. 9-8. Typical nonregenerative switching circuit.

and current i_2 through resistor R_2 is

$$i_2 = \frac{V_{BB} + v_{BE,\mathrm{ON}}}{R_2} \tag{9-7}$$

Therefore,

$$i_B = i_1 - i_2 \tag{9-8}$$

which is the turn-on base current I_{B1}. If I_{B1} exceeds $V_{CC}/h_{FE}R_L$, the BJT is in the ON mode and

$$i_{C,\mathrm{ON}} \cong \frac{V_{CC}}{R_L} \tag{9-9}$$

From Eqs. (9-6), (9-7), and (9-8), we see that V_{BB}/R_2 should be small compared to V_1/R_1, in order that most of input current i_1 flow into the base of the BJT. Thus, if the magnitudes of V_1 and V_{BB} are of the same order, R_2 might be an order of magnitude larger than R_1.

The investigation of circuit conditions during turn-off, and for the OFF mode, is facilitated by the application of Thévenin's theorem at the base-emitter terminals. If the pulse-source impedance is negligible, the network to the left of terminals B and E, *when the input is off*, may be represented by a voltage $V_{BB}R_1/(R_1 + R_2)$ in series with the parallel combination of R_1 and R_2, as shown by Fig. 9-9. Since the base-emitter voltage is $v_{BE,\mathrm{ON}}$ during turn-off, we see from Fig. 9-9(a) that the turn-off base current I_{B2} is given by

$$I_{B2} = \frac{V_{BB}R_1/(R_1 + R_2) + v_{BE,\mathrm{ON}}}{R_1 \| R_2} \tag{9-10}$$

where $R_1 \| R_2$ is the parallel combination of R_1 and R_2. In order to calculate $v_{BE,\mathrm{OFF}}$, we assume that $-i_{B,\mathrm{OFF}}$ is I_{CBO}, as concluded in Sec. 9-1 and indicated

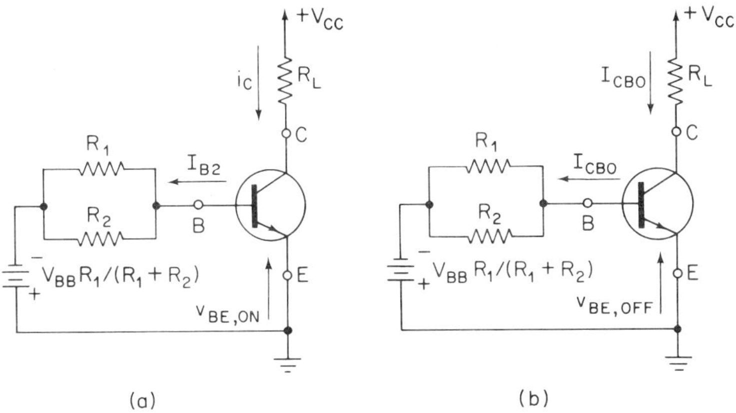

Fig. 9-9. Equivalent representations of the circuit of Fig. 9-8. (a) During turn-off. (b) In the OFF mode, showing approximate values of $-i_{B,OFF}$ and $i_{C,OFF}$.

in Fig. 9-9(b). Thus, using the Thévenin-equivalent circuit of Fig. 9-9(b), we obtain

$$v_{BE,\mathrm{OFF}} = -\frac{V_{BB}R_1}{R_1 + R_2} + I_{CBO}(R_1 \parallel R_2) \tag{9-11}$$

From Eq. (9-11) we see that $v_{BE,\mathrm{OFF}}$ may be positive if the $I_{CBO}(R_1 \parallel R_2)$ is excessive. Since this conditions is undesired (for an *npn* BJT), the circuit should be designed so that it does not occur under normal operating conditions.

Speed-up capacitors. We have seen that, when a BJT is turned ON with constant base current I_{B1} and turned OFF by an abrupt reduction of base drive, a trade-off exists between the turn-on and turn-off times. That is, any rise-time reduction obtained by increasing I_{B1} is at the expense of increased turn-off storage time. Therefore, the ideal base drive might be a pulse so shaped that i_B is very large during the turn-on interval, but falls instantly to the value just necessary for saturation after turn-on and is constant at this value during the ON interval. This condition can be approached *for a particular BJT* by the addition of a capacitor, as shown in Fig. 9-10.

The circuit of Fig. 9-10 is seen to be the same as that of Fig. 9-8, except that capacitor C_s has been added in parallel with R_1. The circuit is designed so that, without C_s, I_{B1} is the value just necessary for saturation. With C_s, an input pulse of amplitude V_1 results (ideally) in an *impulse* of charge supplied to the base as C_s is charged to V_1. If the value of C_s is such that this charge is that required by the base to bring BJT operation to the edge of saturation, the turn-on time would be *zero*. In practice, limitations of the

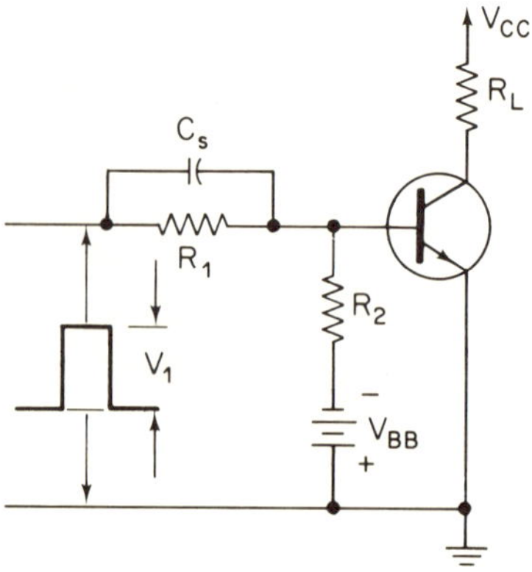

Fig. 9-10. The use of a speed-up capacitor in a switching circuit.

base-drive source and the absence of an ideal base-emitter diode cause operation to be far from ideal, but a significant improvement in switching times can be achieved through the use of C_s. Therefore, C_s is called a *speed-up capacitor*.

Capacitor C_s in the circuit of Fig. 9-10 also assists in BJT turn-off. When the base-drive voltage falls to zero, the voltage on C_s tends to reverse-bias the emitter-base junction. Ideally, the base charge removal necessary for reverse-bias of both junctions occurs in an *impulse*, and the storage and fall times are zero. In practice, significant reductions in turn-off time can be obtained through the use of speed-up capacitors.

As one might expect, there are disadvantages in the use of speed-up capacitors. Clearly, the R_1C_s time constant limits the maximum pulse-repetition rate of the circuit of Fig. 9-10. In logic circuits having more than one input to a common base (see Fig. 9-11), speed-up capacitors reduce the isolation of the inputs. Because of these disadvantages, it is often preferable to use higher-frequency BJTs *without speed-up capacitors* when additional speed is required.

Logic circuits. Figure 9-11 shows several BJT-logic circuits that utilize the basic circuit of Fig. 9-8. The circuit of Fig. 9-11(a) is simply that of Fig. 9-8 and is known as an *inverter* circuit, since there is *no* output voltage when v_A is present and there is an output when there is no v_A. Figure 9-11(b) shows a NOR (not-OR) circuit, from which there is *no* output voltage if either v_A or v_B are present. If the NOR circuit is followed by an inverter circuit, as in Fig. 9-11(c), the result is an OR circuit, since there is an output if either v_A or v_B are present. The circuit if Fig. 9-11(d) is a NAND (not-AND) circuit, since there is *no* output if v_A *and* v_B are present. This is so because both BJTs must be ON for zero output voltage. A voltage v_B causes BJT Q_2 to be ON,

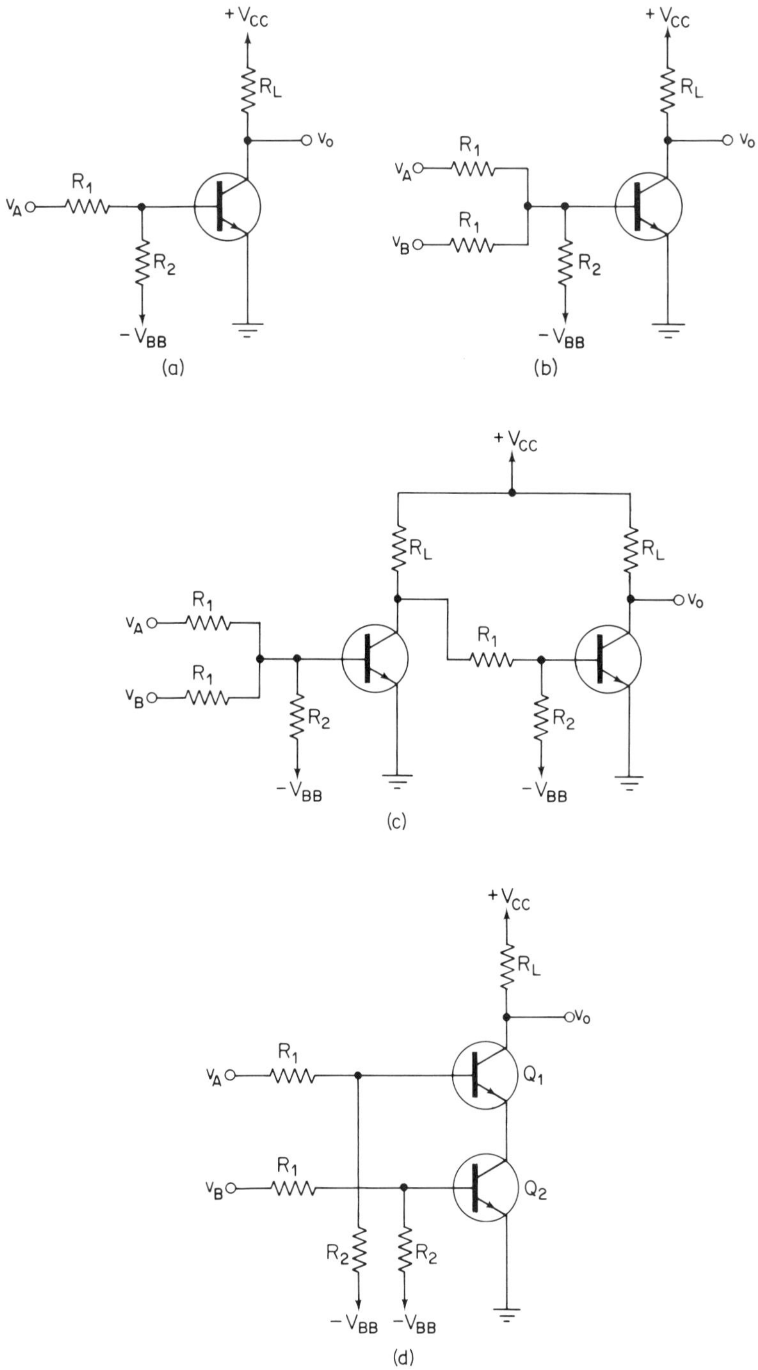

Fig. 9-11. BJT logic circuits. (a) Inverter circuit. (b) NOR circuit. (c) OR circuit. (d) NAND circuit.

in that its collector-emitter voltage is essentially zero, but Q_2 cannot conduct collector current unless Q_1 is turned ON by v_A. Similarly, voltage v_A can supply no base current to turn Q_1 ON unless Q_2 is ON. If the NAND circuit is followed by an inverter, the combination is an AND circuit.

9-4 REGENERATIVE SWITCHING CIRCUITS

Regenerative switching circuits are similar to nonregenerative circuits in that each BJT is either OFF or ON. In general, regenerative circuits have two stable states and are classified on the basis of whether the circuit can remain indefinitely in one or either of these states. If, in the absence of *triggering* pulses, the circuit can remain permanently in either state, it is *bistable*; if it can remain permanently in only one state, it is a *monostable* circuit; if the circuit cannot remain permanently in either state, it is *astable*.

Bistable circuits are commonly referred to as *flip-flops* and astable circuits are called *multivibrators*, owing to the "many frequencies" present in their wave forms. Because of their similarity, all three circuits are often considered to be types of multivibrators, being known as bistable, monostable, and astable multivibrators. This will be our practice as we consider representative multivibrator circuits.

9-4.1 The Bistable Multivibrator

The basic circuit of a bistable multivibrator or flip-flop is shown in Fig. 9-12. It is seen to consist of two inverter circuits in cascade, with the input of the first inverter supplied by the output of the second. The circuit has two stable states in that either BJT may be OFF, as a result of V_{BB}, and the collector-emitter voltage of the OFF unit provides base drive that keeps the other unit ON. It might appear that both BJTs could operate in the active mode with equal collector currents, but this is an unstable condition. In this condition, the slightest disturbance in the conduction of either BJT will drive the circuit toward one of its stable states. The circuit can be triggered from one stable state to the other by momentarily short-circuiting the base-emitter terminals of the ON unit. With its base current removed by the short circuit, the previously ON unit turns OFF, turning ON the other unit.

The preceding discussion presumes that each inverter of the bistable circuit is operating properly, being biased OFF in the absence of base drive and fully ON when driven by the other inverter. This condition may be checked, for a particular circuit, by calculations much like those for the inverter circuit of Fig. 9-8. In making this check, we shall assume one BJT

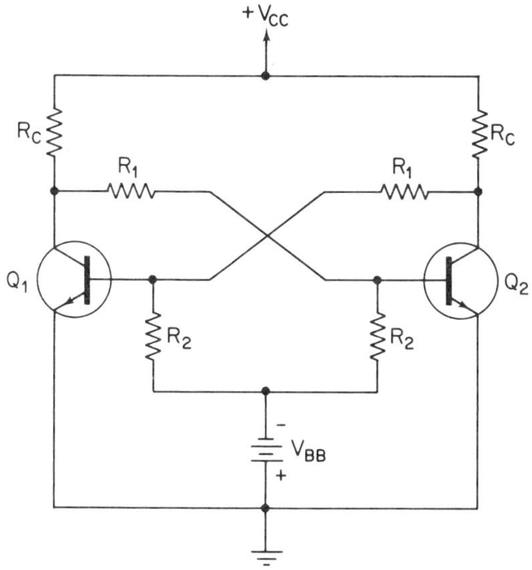

Fig. 9-12. Basic circuit of a bistable multivibrator or flip-flop.

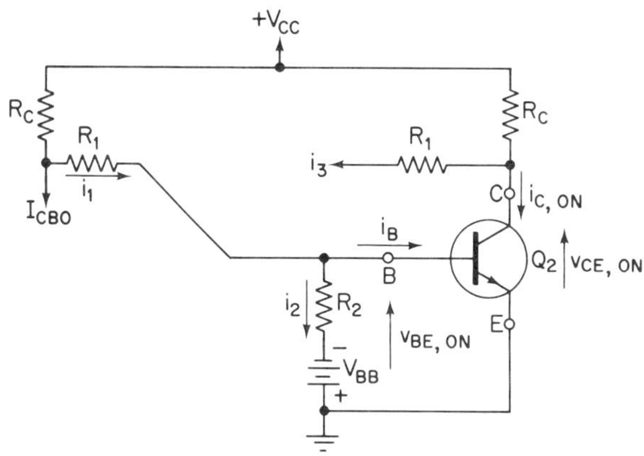

Fig. 9-13. Steady-state conditions for the ON BJT of Fig. 9-12, assuming Q_1 to be OFF and $i_{C,\text{OFF}}$ to be I_{CBO}.

to be OFF and see if this causes the other unit to be ON. If so, we shall then determine whether or not the first BJT is biased OFF.

Figure 9-13 shows ON-unit conditions for the circuit of Fig. 9-12, if BJT Q_1 is assumed to be OFF and $i_{C,\text{OFF}}$ to be I_{CBO}. Current $i_{C,\text{ON}}$ is somewhat less than that of Fig. 9-8, owing to the current i_3, but we shall see that this difference is usually small enough to be neglected. Thus

$$i_{C,\text{ON}} \cong \frac{V_{CC} - v_{CE,\text{ON}}}{R_C} \cong \frac{V_{CC}}{R_C} \tag{9-12}$$

The determination of i_B includes the effects on current i_1 of collector resistor R_C and current I_{CBO} of Q_1, the latter probably being negligible except at elevated temperatures. Neglecting the effect of I_{CBO}, we see that

$$i_B \cong \frac{V_{CC} - v_{BE,\mathrm{ON}}}{R_C + R_1} - \frac{V_{BB} + v_{BE,\mathrm{ON}}}{R_2} \tag{9-13}$$

In order for Q_2 to be ON, its h_{FE} for the $i_{C,\mathrm{ON}}$ value of Eq. (9-12), must be such that $i_B \geq i_{C,\mathrm{ON}}/h_{FE}$.

Figure 9-14 shows OFF-unit conditions for the circuit of Fig. 9-12, if

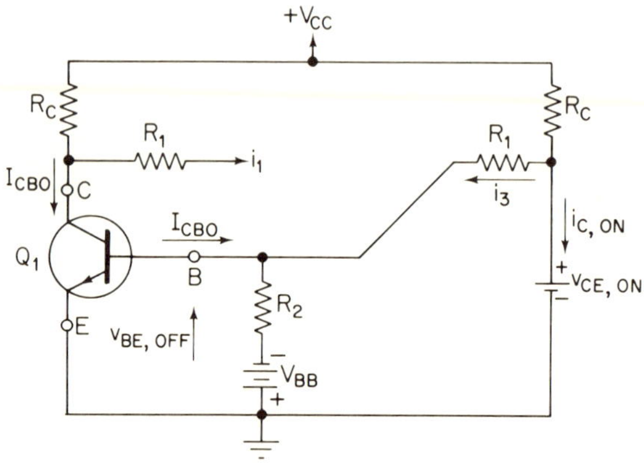

Fig. 9-14. Steady-state conditions for the OFF BJT of Fig. 9-12, assuming Q_2 to be ON and $-i_{B,\mathrm{OFF}}$ to be I_{CBO}.

Q_2 is assumed to be ON and $-i_{B,\mathrm{OFF}}$ to be I_{CBO}. Voltage $v_{BE,\mathrm{OFF}}$ of Q_1 may be calculated by the procedure for Fig. 9-8 [see Fig. 9-9(a) and Eq. (9-11)], except that the effect of $v_{CE,\mathrm{ON}}$ should be included if $v_{CE,\mathrm{ON}}$ is significant compared to V_{BB}. BJT Q_1 is OFF if $v_{BE,\mathrm{OFF}}$ is appreciably negative.

From Fig. 9-14, we see that i_3 is given by

$$i_3 = \frac{V_{BB} - I_{CBO}R_2}{R_1 + R_2} \tag{9-14}$$

if $v_{CE,\mathrm{ON}}$ is negligible compared to V_{BB}. Since $(R_1 + R_2)$ is much larger than R_C in the typical circuit and V_{BB} is usually less than V_{CC}, i_3 has negligible effect on $i_{C,\mathrm{ON}}$ as assumed in deriving Eq. (9-12).

Triggering bistable circuits. In general, a bistable circuit is used to register the occurrence of an event by changing its operating state, the event usually being in the form of a triggering pulse. For example, if a bistable multivibrator is triggered by a regular train of pulses, the output voltage at the collector of either BJT is a square wave having a repetition rate that is half that of the triggering pulses. This is known as *binary* operation.

In order to change the state of a bistable multivibrator, the circuit must be driven beyond the point where the two collector currents are equal. If the collector current of the previously OFF BJT is larger than that of the other unit at the end of the trigger pulse, a change of state occurs. If the trigger pulse does not achieve this condition, the circuit reverts to its previous state at the end of the pulse.

Triggering may be accomplished either by increasing the collector current of the OFF BJT, with consequent decrease in ON-unit current, or by decreasing the collector current of the ON BJT. In general, it is easier to effect a turn-off rather than a turn-on, because of the smaller bias voltages of the ON unit. A triggering arrangement of this type is shown in Fig. 9-15. In

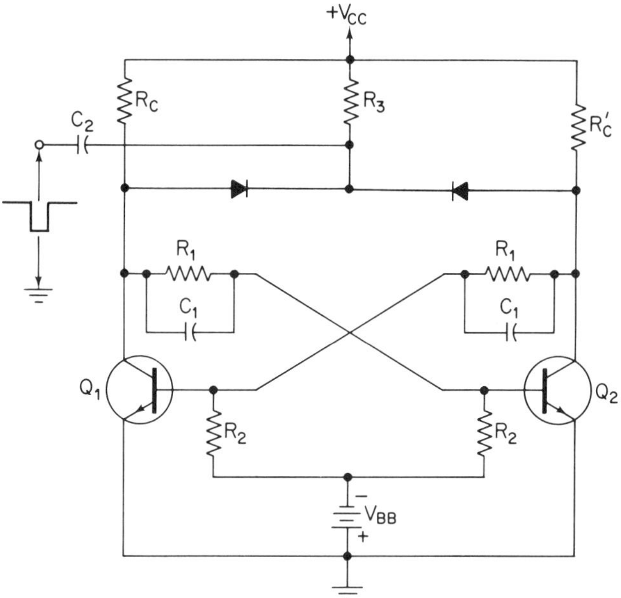

Fig. 9-15. Collector-triggered bistable multivibrator.

this circuit, a negative-going trigger pulse will pass through the diode connected to the OFF BJT, but will be blocked by the other diode. This occurs because the diode connected to the ON BJT is reverse biased by approximately V_{CC} by way of R_3, while the other diode has only slight reverse-bias. After passing through the diode, the trigger pulse is coupled by "speed-up" capacitor C_1 to the base of the ON BJT. If the pulse removes enough charge from the base of the ON unit to cause its collector current to fall below that of the other unit, whose current is increasing as a result of the rising ON-unit collector voltage, the pulse can be removed, and regenerative action of the circuit carries the change-of-state to completion.

The diodes of Fig. 9-15 are called pulse-steering diodes, because they steer the trigger pulses to the base of the ON BJT and block them from the OFF-BJT base, where they would prevent turn-on. The "speed-up" capacitors of inverter circuits are called *commutating capacitors* in the bistable circuit, since they do more than merely speed up the switching action. In fact, they are essential in order to transmit collector-potential changes of a BJT to the base of its companion unit, and *triggering is virtually impossible* without them. Since trigger pulses are applied to the collectors in Fig. 9-15, the arrangement is known as *collector triggering*. Base-triggering and emitter-triggering arrangements also exist, but collector triggering is usually preferred.

The direct-coupled flip-flop. A bistable circuit of particular interest because of its simplicity is shown in Fig. 9-16. It is known as a DCTL (direct-coupled transistor-logic) flip-flop because of its use in direct-coupled logic circuits.*

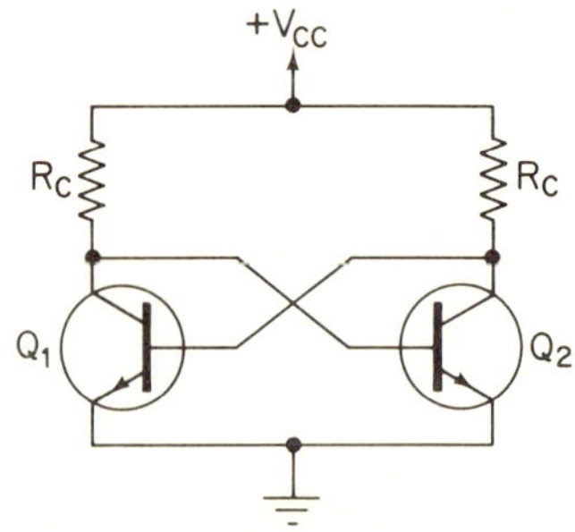

Fig. 9-16. A direct-coupled flip-flop.

The principle basic to its operation is that the collector-emitter voltage $v_{CE,\mathrm{ON}}$ of a BJT operating well into saturation is significantly lower than the base-emitter voltage $v_{BE,\mathrm{ON}}$ required for appreciable collector current. Therefore, if one BJT in Fig. 9-16 is ON, the other unit is essentially OFF because of its low base-emitter voltage. Since either BJT may be ON, the circuit has two stable operating states. The collector voltage $v_{CE,\mathrm{OFF}}$ of the OFF BJT is seen to be $v_{BE,\mathrm{ON}}$ of the ON unit; therefore, the collector potential of each BJT swings between $v_{CE,\mathrm{ON}}$ and $v_{BE,\mathrm{ON}}$ as the unit switches between its ON and OFF modes. This voltage swing is only a few tenths of a volt, but it is large enough for use in DCTL circuitry.

The circuit operation just described can be verified by considering the currents of an ON unit; they are seen to be given by

$$I_B = \frac{V_{CC} - v_{BE}}{R_C} \tag{9-15}$$

* Direct-coupled logic circuits have no resistors, capacitors, or diodes between stages.

and

$$I_C = \frac{V_{CC} - v_{CE}}{R_C} \tag{9-16}$$

If V_{CC} is large relative to v_{BE}, I_B drives the BJT into saturation and $v_{CE} \ll V_{CC}$. Consequently, as a first approximation,

$$I_B \simeq \frac{V_{CC}}{R_C} \simeq I_C \tag{9-17}$$

Since $I_B \gg I_C/h_{FE}$, the ON unit is driven well into saturation.

If values of I_B and I_C more accurate than those of Eq. (9-17) are desired, these first-approximation values can be used to obtain approximate values of $v_{CE,\mathrm{ON}}$ and $v_{BE,\mathrm{ON}}$ from the manufacturer's data. Equations (9-15) and (9-16) can then be used to obtain second-approximation values of I_B and I_C. This procedure can be repeated again, but it is usually unnecessary to do so, since the process converges rapidly.

Like the flip-flop of Fig. 9-12, the circuit can be triggered from one stable state to the other by momentarily short-circuiting the base-emitter terminals of the ON unit. However, it is exceedingly difficult to trigger the circuit by the usual method (Fig. 9-15) of applying a pulse to the OFF-unit collector, since it is tied directly to the ON-unit base. Triggering usually requires the use of an amplifier for each BJT, as shown in Fig. 9-17. Since

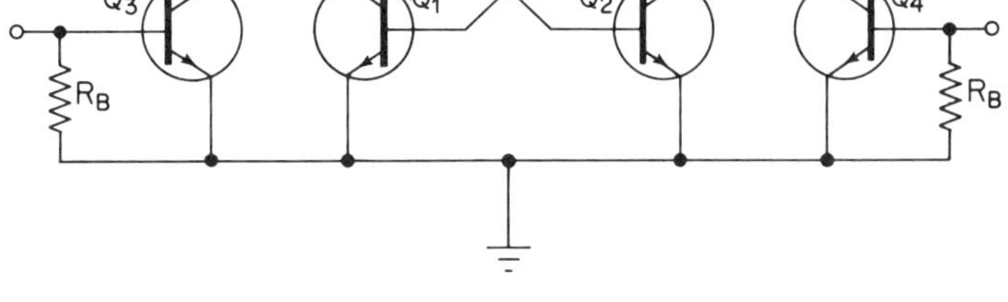

Fig. 9-17. The use of trigger amplifiers in conjunction with a DCTL flip-flop.

BJTs Q_3 and Q_4 have zero base-to-emitter bias, they are essentially OFF and have no effect on the flip-flop unless they receive positive input pulses. If Q_3 receives a positive base pulse when Q_2 is ON, Q_3 is driven into saturation, virtually short-circuiting the base-emitter terminals of Q_2. Therefore, Q_2 turns OFF, and the increase in the base-emitter voltage of Q_1 causes it to turn ON.

The collector-voltage swing of the DCTL flip-flop can be increased by the use of series base resistors, as in Fig. 9-18. These resistors permit $v_{CE,\mathrm{OFF}}$ to exceed $v_{BE,\mathrm{ON}}$, as in the basic flip-flop of Fig. 9-12. Like the OFF BJT in the DCTL flip-flop, the OFF unit in Fig. 9-18 conducts slightly, owing to

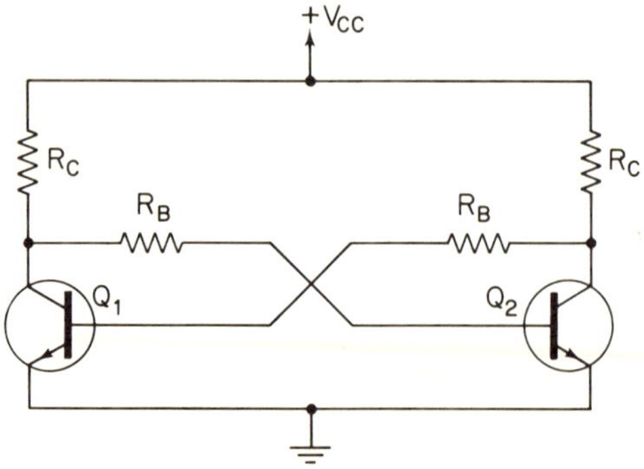

Fig. 9-18. Modified DCTL flip-flop.

forward base-emitter voltage provided by v_{CE} of the ON unit. Although the unit is not cut off, its currents are very small compared with those of the ON BJT, and the preceding analysis of the DCTL flip-flop is essentially correct and applicable to the circuit of Fig. 9-18. From this analysis, it is seen that

$$R_C + R_B < h_{FE}R_C \tag{9-18}$$

for bistable operation in Fig. 9-18.

9-4.2 The Monostable Multivibrator

If one of the equilibrium states of a bistable multivibrator is suppressed, the circuit can be made to exhibit monostable behavior. Monostable circuits are useful for standardizing pulses of random widths and in the generation of pulses of controllable width or time delay.

The basic two-BJT monostable circuit shown in Fig. 9-19 is seen to be similar to the bistable circuit, except that BJT Q_2 is a-c coupled to Q_1 and the base-bias supply (V_{CC}) of Q_2 causes it to tend to be ON. Under equilibrium conditions, Q_2 is ON, since timing resistor R_t is chosen so that

$$\frac{V_{CC}}{R_t} \geq \frac{V_{CC}}{h_{FE}R_C} \tag{9-19}$$

or

$$R_t \leq h_{FE}R_C \tag{9-20}$$

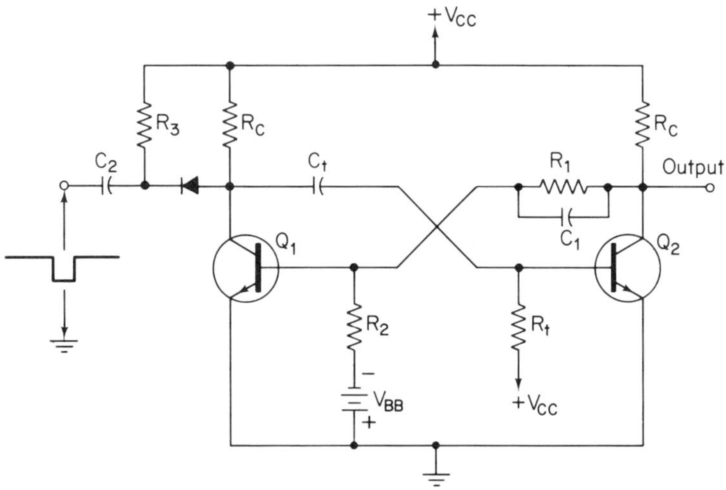

Fig. 9-19. Basic circuit of a monostable (one-shot) multivibrator.

With Q_2 ON, V_{BB} causes Q_1 to be OFF, as in the bistable circuit. This is the equilibrium state of the circuit. In this operating state, timing capacitor C_t is charged to approximately V_{CC}.

Resistor R_1, like the coupling resistor of the bistable circuit, is chosen so that Q_1 is ON if Q_2 happens to be OFF, say as a result of the negative-going trigger pulse. If this pulse succeeds in turning Q_2 OFF, Q_1 then turns ON and remains ON as long as Q_2 is OFF. When Q_1 goes ON, the left-hand terminal of C_t is effectively connceted to ground and the base of Q_2 is driven negative by approximately V_{CC}.* Therefore, Q_2 is held OFF by the charge of C_t. However, this charge decreases as C_t charges toward $+V_{CC}$ through R_t, and Q_2 turns back ON when the voltage across C_t reaches approximately zero. This turns Q_1 OFF, and the circuit has reverted to its equilibrium state. Since circuit activity subsides after a single cycle of operation, following activation by a triggering pulse, the monostable circuit is sometimes called a *one-shot multivibrator.*

The circuit output is usually taken to be the collector-to-emitter voltage of Q_2. This output consists of a pulse having a height of approximately $+V_{CC}$ and a width T given by

$$T \cong R_t C_t \ln 2 = 0.69 R_t C_t \tag{9-21}$$

In the derivation of Eq. (9-21), the following assumptions are made: (1) C_t charges to V_{CC} when Q_1 is OFF. (2) $v_{CE,\text{ON}}$ of Q_1 is negligible. (3) Discharging

* If V_{CC} exceeds the reverse-breakdown voltage of the emitter-base junction, a protective diode should be used (Sec. 9-4.3).

of C_t by $i_{B,\mathrm{OFF}}$ of Q_2 is negligible. (4) Q_2 turns ON when the voltage of C_t reaches zero, as C_t charges through R_t toward $+V_{CC}$.

We may control the width of the output pulse by adjusting either C_t or R_t, or both, recognizing that Eq. (9-20) must be satisfied for monostable operation. If R_t happens to be larger than $h_{FE}R_C$, astable operation (oscillation) occurs. In using the monostable circuit to generate a delayed pulse, the output pulse of the circuit is differentiated. The pulse derived from the trailing edge of the wave form is delayed from the triggering pulse by the controlled time T.

9-4.3 The Astable Multivibrator

If both stable states of a multivibrator circuit are only temporarily stable, the circuit is an astable multivibrator or oscillator. Such a circuit switches periodically between its two stable states, resulting in an output (collector-to-emitter) voltage that resembles a square wave. Square-wave generation constitutes the principal application of astable or free-running multivibrators.

The basic astable circuit shown in Fig. 9-20 is derived by extending the

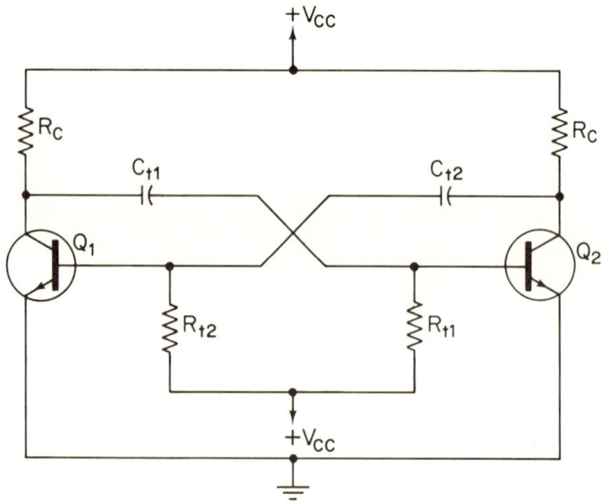

Fig. 9-20. Basic circuit of an astable (free-running) multivibrator.

bistable-circuit modifications that resulted in the monostable circuit. That is, there is no d-c coupling between the BJTs and both units tend to be ON as a result of the base-bias supply (V_{CC}). Like the monostable-circuit timing resistor, resistors R_{t1} and R_{t2} are selected so that both BJTs would be ON in the absence of coupling (timing) capacitors. If V_{CC} is increased slowly from zero to the operating value, both BJTs may be ON, causing the circuit to be inoperative, but an abrupt application of power causes one BJT to turn

ON first, owing to component tolerances. When a BJT turns ON, the capacitor connected to its collector drives the base of the other BJT negative, turning it OFF. As in the monostable circuit, this BJT is held OFF by the capacitor until the capacitor voltage falls to approximately zero. The BJT then turns ON, turning OFF the other unit, and the process is repeated.

Since the OFF conditions of each BJT in the astable circuit are the same as those for the normally ON unit of the monostable circuit, we can use Eq. (9-21) to determine the conduction time of each BJT. Conduction time t_1 of BJT Q_1 is the OFF time of Q_2, which is seen to be given by

$$t_1 \cong R_{t1}C_{t1} \ln 2 = 0.69R_{t1}C_{t1} \tag{9-22}$$

and conduction time t_2 of Q_2 is

$$t_2 \cong R_{t2}C_{t2} \ln 2 = 0.69R_{t2}C_{t2} \tag{9-23}$$

The total cycle time T, or period of oscillation, is then

$$T = t_1 + t_2 \cong 0.69(R_{t1}C_{t1} + R_{t2}C_{t2}) \tag{9-24}$$

In a symmetrical multivibrator, $R_{t1} = R_{t2} = R_t$ and $C_{t1} = C_{t2} = C_t$, Eq. (9-24) becomes

$$T \cong 1.4R_tC_t \tag{9-25}$$

and the oscillating frequency f is

$$f = \frac{1}{T} \cong \frac{1}{1.4R_tC_t} \tag{9-26}$$

Electrode-voltage wave forms for a symmetrical astable multivibrator

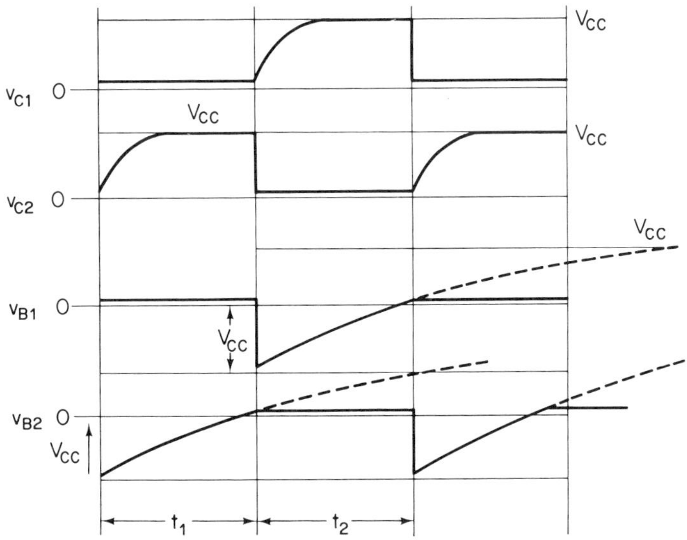

Fig. 9-21. Electrode-voltage waveforms of a symmetrical astable multivibrator.

are somewhat as sketched in Fig. 9-21. Rounding of the leading edge of the collector-voltage wave form is due to the charging time of the capacitor, which depends on the time constant $R_C C_t$. This effect is significant when $R_C C_t$ is comparable to t_1 and t_2. It is important to observe that, like the base of the normally ON BJT of the monostable circuit, each base is driven negative by approximately V_{CC}. If V_{CC} exceeds the reverse-breakdown voltage of the base-emitter junctions, protective diodes should be used in the base-emitter circuits. Such a diode may be in series with either the base or the emitter, the latter usually being preferred to avoid degradation of BJT turn-off time that could occur if the diode turned OFF in advance of the BJT.

PROBLEMS

9-1. The BJT in the circuit of Fig. P9-1 has an h_{fe} of 100 and an I_{CBO} of 10 μA. Determine the approximate values of: (a) V_{CE} with switch S open, and (b) maximum value of resistance R if $V_{CE} \cong 0$ when S is closed.

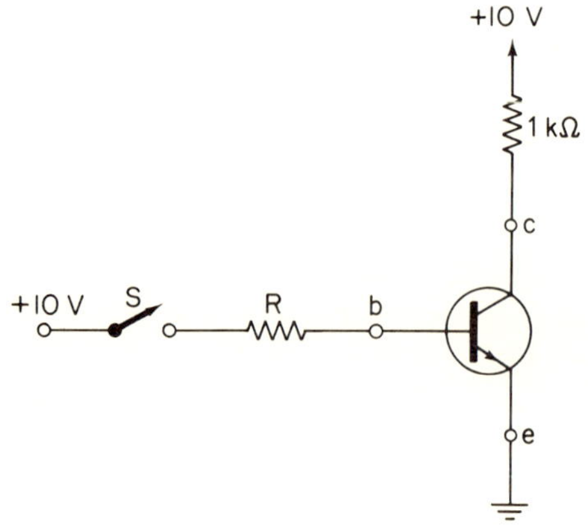

Figure P9-1

9-2. The BJT of Prob. 9-1 is used in the circuit of Fig. P9-2. Determine approximate values of: (a) V_{BE} with switch S open, (b) V_{CE} with S open, (c) maximum value of resistance R if $V_{CE} \cong 0$ when S is closed and $V = +10$ V, and (d) V_{BE} for the R value of (c) with S closed and $V = 0$ V.

9-3. Assuming the input pulse in the circuit of Fig. P9-3 to have zero rise and fall times, determine the approximate values of: (a) initial value of i_B during turn-on, (b) turn-on base current I_{B1}, (c) turn-off base current I_{B2}, and (d) the minimum value of BJT h_{FE} for collector-current saturation.

9-4. The switching circuit of Fig. P9-4 may be used when d-c isolation of the pulse source is desired. Assuming capacitor C large enough that negligible charging

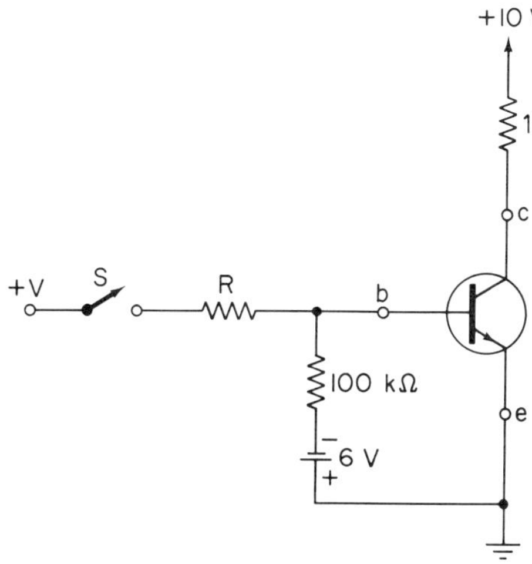

Figure P9-2

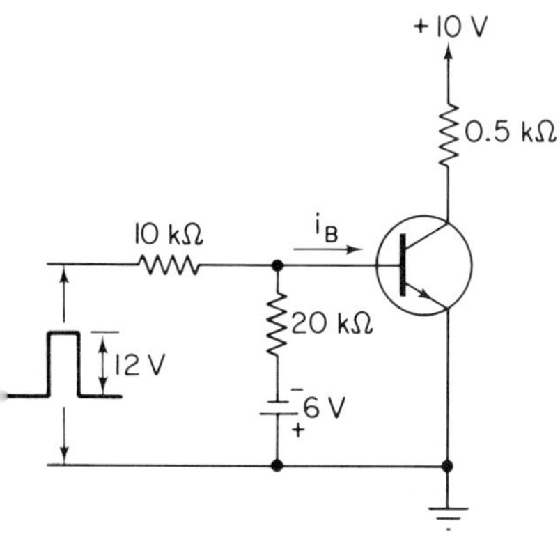

Figure P9-3

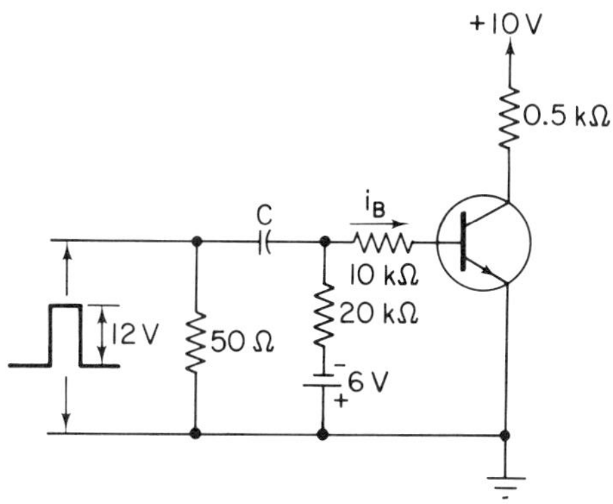

Figure P9-4

and discharging occur during the ON and turn-off times, repeat (a), (b), (c), and (d) of Prob. 9-3.

9-5. Assuming the BJTs of the bistable circuit of Fig. P9-5 to have the maximum-specified I_{CBO} values ($-6\ \mu$A at 25°C and $-120\ \mu$A at 71°C) and the base-emitter and collector-emitter voltages of the ON BJT to be zero, determine, for a junction temperature of 25°C, the minimum value of h_{FE} and the base-emitter voltage of the OFF BJT.

9-6. Repeat Prob. 9-5 for a junction temperature of 71°C.

9-7. Specifications for the BJTs of the bistable circuit of Fig. P9-7 are $V_{BE,\text{ON}} = 0.6$ V, $V_{CE,\text{ON}} = 0.1$ V, $h_{FE} = 20$, and $I_{CBO} = 10\ \mu$A at 150°C. For a junction temperature of 150°C, determine the maximum value of R for saturation of the ON BJT and V_{BE} of the OFF unit when this R value is used.

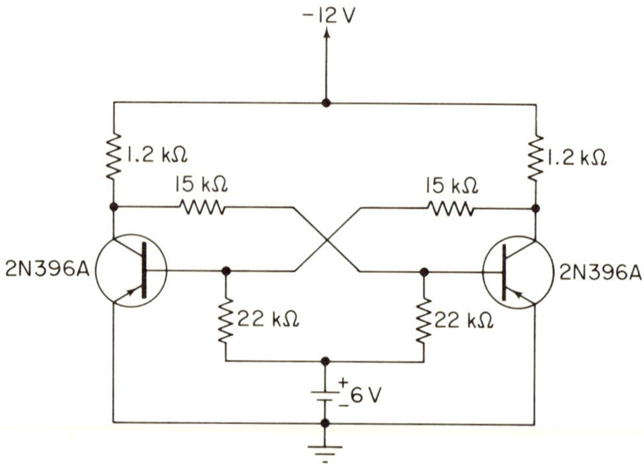

Figure P9-5

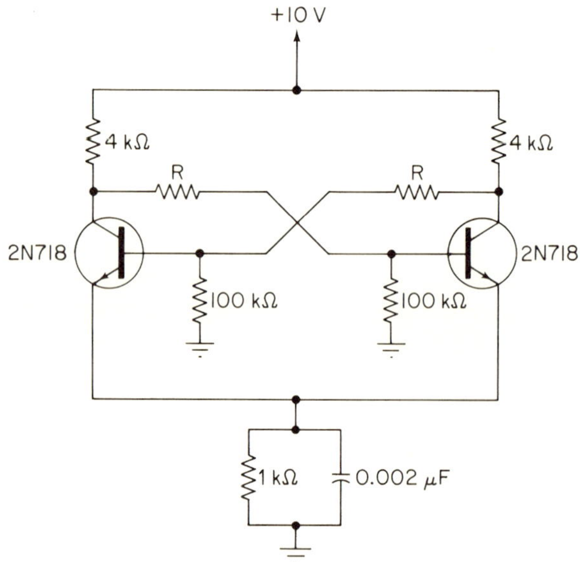

Figure P9-7

9-8. The bistable circuit of Prob. 9-7 is modified to form the monostable circuit of Fig. P9-8. Using the BJT specifications of Prob. 9-7, determine the maximum value of R for monostable operation and the output-pulse duration T that might be expected if $R = 50\ \text{k}\Omega$ and $C = 0.001\ \mu\text{F}$.

9-9. The 2N718 BJTs of Prob. 9-8 have an emitter-base breakdown specification of 5 V minimum. If breakdown occurs at this voltage, how would T compare

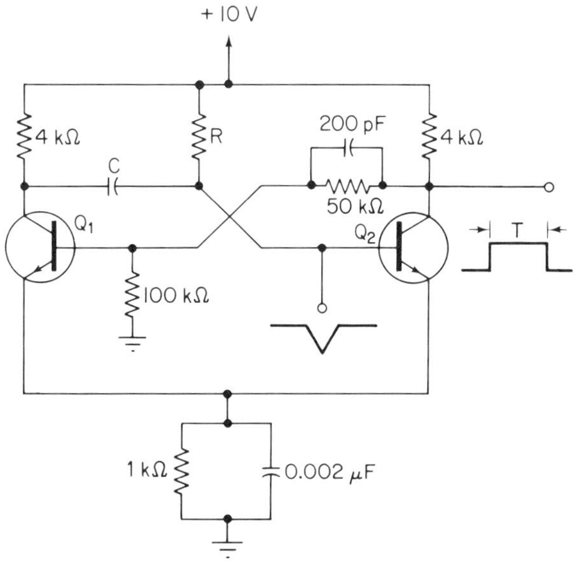

Figure P9-8

with the value predicted in Prob. 8-8? Explain the effect on T of emitter-base breakdown and suggest circuit modifications that would prevent its occurrence.

9-10. The BJTs in the astable multivibrator of Fig. P9-10 have an h_{FE} of 50. If the frequency of oscillation is 1000 Hz, what is the minimum value of C_t that could be used? What BJT emitter-base breakdown-voltage rating is required?

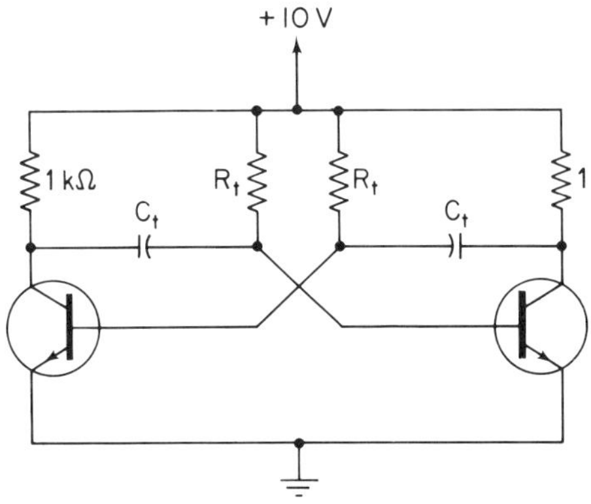

Figure P9-10

10

Field-Effect Transistors (FETs)

The field-effect transistor (FET) is a *pn*-junction device whose operation depends on the control of current by an electric field. It is the dual of the bipolar junction transistor (BJT), being voltage controlled rather than current controlled, and is the solid-state analogue of the vacuum triode. In many applications, it is interchangeable with the vacuum triode and combines the advantages of the triode with many of those of the BJT. Unlike the BJT, its operation depends on the flow of majority carriers only. For this reason, it is often referred to as a *unipolar transistor*.

Although the BJT was the only commercial transistor available for many years, its discovery was soon followed by that of the FET.[1,2,3] Because of the enormous effort directed toward development of the BJT, the FET received little attention for several years, and commercial devices are of relatively recent origin. Basic FET types are the junction type, abbreviated JFET, and the insulated-gate type (IGFET), which is more commonly called the metal-oxide-semiconductor (MOS) transistor or the MOSFET.

[1] O. M. Stuetzer, "A Crystal Amplifier with High Input Impedance," *Proc. IRE*, Vol. **38**, No. 8 (August 1950).

[2] W. Shockley, "A Unipolar Field-Effect Transistor," *Proc. IRE*, Vol. **40**, No. 11 (November 1952).

[3] G. Dacey and I. M. Ross, "Unipolar Field-Effect Transistor," *Proc. IRE*, Vol. **41**, No. 8 (August 1953).

10-1 THE JUNCTION FIELD-EFFECT TRANSISTOR (JFET)[4,5,6,7]

The junction field-effect transistor consists of a segment of semiconductor material whose resistance is modulated by electrical variation of the effective cross-sectional area of a conducting channel in the material. The segment may be of either *n*-type material or *p*-type material, resulting in either an *n-channel* JFET or a *p-channel* JFET.

The basic structure of an *n*-channel JFET and the use of the device in an elementary amplifier circuit are shown in Fig. 10-1. Ohmic contacts are made

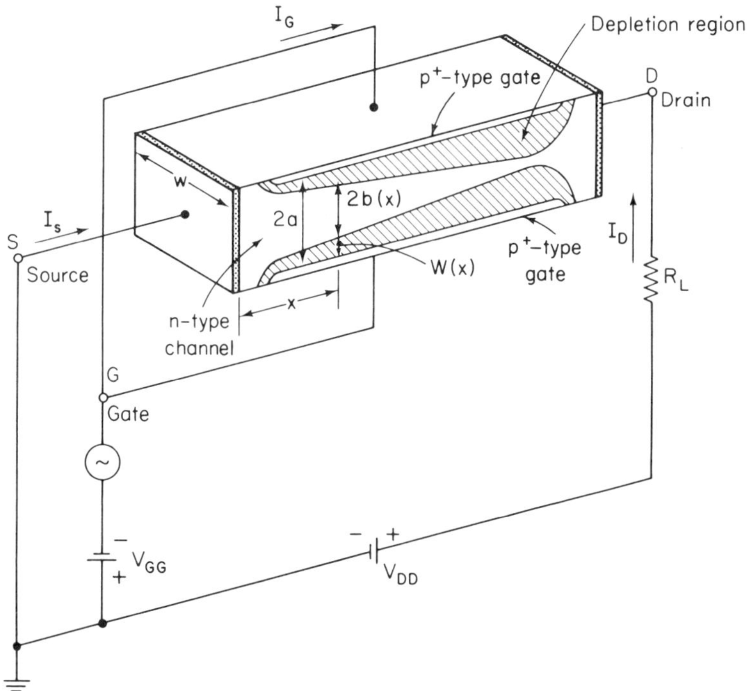

Fig. 10-1. Basic structure and bias-voltage polarities of an *n*-channel JFET operated as an elementary amplifier.

4 L. J. Sevin, Jr., *Field-Effect Transistors.* (New York: McGraw-Hill Book Co., Inc., 1965, pp. 1–50.

5 J. T. Wallmark and H. Johnson, *Field-Effect Transistors.* Englewood Cliffs, N.J.: Prentice Hall, Inc., 1966, pp. 1–15, 155–158.

6 C. D. Todd, *Junction Field-Effect Transistors.* New York: John Wiley & Sons, Inc., 1968, pp. 1–52.

7 W. Golsing, *Field Effect Transistor Applications.* New York: John Wiley & Sons, Inc., 1965, pp. 1–20.

to the two ends of an *n*-type semiconductor bar, and current flows along the length of the bar, owing to the voltage applied between the ends. The left-hand end of the bar is called the *source*, because it is the source of majority carriers, the right-hand end is the *drain*, where carriers are removed, and the two *p*-regions constitute the *gate*, which controls carrier flow. The *p*-regions are formed either by alloying to the *n*-type bar or by vapor diffusion. The source, drain, and gate electrodes are analogous to the cathode, plate, and grid electrodes, respectively, of the vacuum triode, and the polarities of the gate and drain bias voltages are the same as those of the corresponding vacuum-triode bias voltages. For a *p*-channel JFET, the electrode bias-voltage polarities are opposite those of the *n*-channel device, the gate bias being positive and the drain bias negative.

Let us now consider the operation of the JFET and associated circuit of Fig. 10-1. The junctions between the gate regions and the bar are seen to be reverse biased so that the carrier-depletion regions extend into the channel, partly closing it. Channel closing with reverse-bias occurs because the conductivity of the depletion regions is nominally zero, owing to the unavailability of current carriers. Thus the effective height of the channel through which drain-to-source current must flow becomes progressively smaller with increasing reverse-bias. Accordingly, drain current is a function of gate-junction reverse-bias. Therefore, a signal voltage superimposed on gate bias V_{GG}, as in Fig. 10-1, modulates drain current I_D, developing a signal voltage across drain-load resistor R_L that may be much larger than the input value. The required input-signal power is small, being only that required to charge the gate-junction capacitances and that dissipated by the dynamic reverse resistances of the junctions. Low-frequency (d-c) input impedances of the order of 10^4 megohms are typical of commercial silicon JFETs operated at 25°C. Germanium JFETs have much smaller low-frequency input impedances, but their high-frequency input impedances may be larger than those of silicon JFETs, owing to their smaller gate-junction capacitances. The high input impedance of the FET may be an important advantage over the BJT, depending on the application.

Static characteristics. Families of drain current-drain voltage curves, referred to as common-source drain characteristics, for typical *n*-channel and *p*-channel JFETs are shown in Fig. 10-2. Drain characteristics of *p*-channel devices, like collector characteristics of *pnp* BJTs, are customarily shown in the first quadrant; however, it should be noted that drain voltage and current are represented as negative in Fig. 10-2(b), since conventional drain current of the *p*-channel JFET flows from the drain electrode to the external circuit. As expected, the magnitude of I_D varies inversely with reverse-bias gate voltage, and a sufficiently large voltage can reduce I_D to essentially zero by causing the gate depletion region to extend far enough to close off the

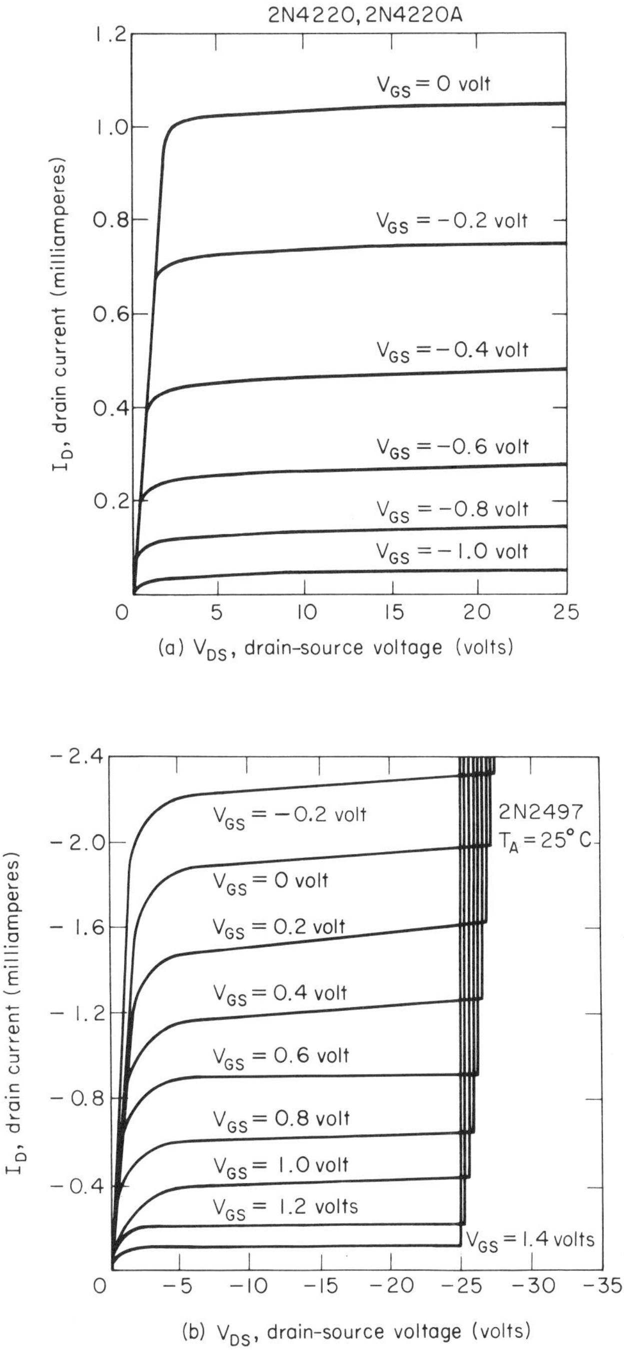

Fig. 10-2. Drain current-drain voltage curves for silicon JFETs. (a) Typical 2N4220 *n*-channel JFET (courtesy Motorola Semiconductor Products, Inc.). (b) Typical 2N2497 *p*-channel JFET (courtesy Texas Instruments, Inc.).

channel completely. This gate voltage is known as the *pinch-off voltage* V_P. For a type 2N4220 JFET operated at a drain voltage of 15 V and a temperature of 25°C, the drain current is specified as 50 μA or less for a gate voltage of -2.5 V and 0.1 nA or less for -4.0 V; that is, the magnitude of V_P should not exceed 4.0 V.

We note in Fig. 10-2 that, if the drain voltage is high enough, each curve approaches a current saturation. This occurs because of the effect of drain voltage on the channel. In Fig. 10-1, the gate carrier-depletion regions are shown thicker near the drain end of the channel than near the source end. This is so because *IR* drop along the channel causes the gate-junction reverse-bias to be higher, and the channel height to be less, at the drain end. For values of drain-source voltage V_{DS} too small to have significant effect on the channel, drain current i_D increases linearly with V_{DS}, but the current begins to level off when V_{DS} becomes large enough to affect channel height. As the channel height decreases near the drain end with increasing V_{DS}, complete pinch-off cannot occur,* but the height becomes so small that the electric field in the region becomes large enough that carrier drift velocity is virtually independent of field strength (see Sec. 1-2). Thus I_D no longer increases appreciably with V_{DS}, and the channel is said to be pinched off. The magnitude of V_{DS} required for this condition, *with gate voltage* V_{GS} *equal to zero*, is that of the pinch-off voltage V_P. Examination of the drain characteristics of Fig. 10-2 shows general agreement between the magnitude of V_{GS} for negligible I_D and that of V_{DS} at the knee of the zero-V_{GS} drain-characteristic curve. The V_{DS} required for "channel pinch-off" is seen to depend on V_{GS}, since pinch-off sets in when the voltage between the gate and the drain end of the channel is equal to V_P; that is, the pinch-off condition exists when

$$|V_{GS} - V_{DS}| \geq |V_P| \tag{10-1}$$

or

$$|V_{DS}| \geq |V_P| - |V_{GS}| \tag{10-2}$$

As $|V_{DS}|$ is increased above the value for channel pinch-off, the voltage drop across the channel is essentially constant and the voltage between the drain terminal and the drain end of the channel is absorbed by the carrier depletion region. The upper limit of V_{DS} is determined by gate-junction breakdown, with breakdown occurring at lower V_{DS} values for larger values of V_{DS}, since the magnitude of the junction voltage is the sum of the two electrode-voltage magnitudes.

The region of the drain characteristic in which I_D is essentially independent of V_{DS} is referred to as the *pinch-off region;* the remainder of the characteristic is the *triode region*. In the triode region, the I_D–V_{DS} curves are essentially

* If complete pinch-off were to occur, the current density would become infinite, which is physically impossible.

linear, passing through the origin ($I_D = 0$, $V_{DS} = 0$), and having slopes ranging from below $(100\ \mathrm{k}\Omega)^{-1}$ to above $(100\ \Omega)^{-1}$, depending on V_{GS} and the operating temperature [see Fig. 10-12(b)]. In the pinch-off region, the characteristic curves are similar to those of the plate characteristic of a vacuum pentode. Unlike the vacuum tube, the JFET has high input impedance for small-signal operation with either zero or small forward gate bias. This is particularly true for silicon JFETs, since a forward-biased silicon junction is virtually an open circuit at room temperature for voltages less than a few tenths of a volt. In the interest of high input impedances at low frequencies, most commercial JFETs are silicon devices.

The pinch-off voltage V_P. An expression for the reverse-bias gate voltage that removes all free charge from the channel of a JFET may be derived for the structure of Fig. 10-1. This analysis was first made by Shockley.*

A cross section of the JFET structure for low drain voltages (that is, $|V_{DS}| \ll |V_{GS}|$) is shown in Fig. 10-3. The carrier-depletion width $(a - b)$ of each *pn* junction is considered to be entirely within the *n*-region, owing to the high-conductivity p^+ gate region, and is uniform throughout the length of the structure for low V_{DS} values. If the junctions are assumed to be abrupt, the depletion-region width of each junction is given by Eq. (2-54), from which we obtain

$$a - b = \left[\frac{2\epsilon(V_d - V_{GS})}{eN_d}\right]^{1/2} \cong \left(\frac{-2\epsilon V_{GS}}{eN_d}\right)^{1/2} \tag{10-3}$$

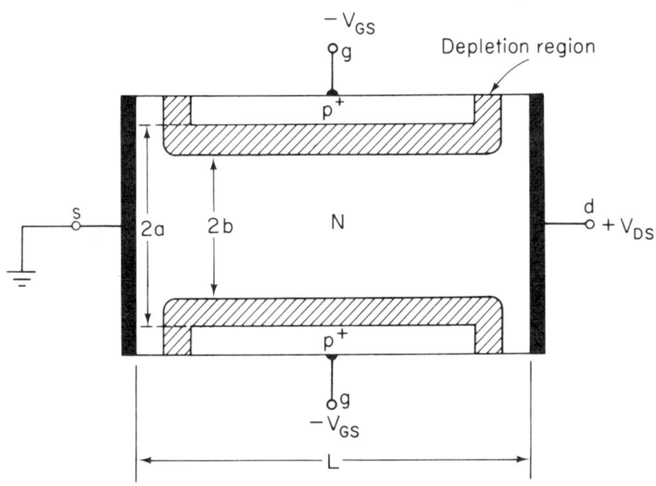

Fig. 10-3. Cross section of the JFET structure of Fig. 10-1 for $|V_{DS}| \ll |V_{GS}|$.

* Reference 2, pp. 1365–1367.

For pinch-off, the channel height $2b$ is zero, and the total depletion-region width is the n-region height $2a$. Since this occurs for $V_{GS} = V_P$, we see from Eq. (10-3) that

$$|V_P| \cong \frac{eN_d}{2\epsilon} a^2 \tag{10-4}$$

From Eqs. (10-3) and (10-4), it may be shown that

$$V_{GS} \cong \left(1 - \frac{b}{a}\right)^2 V_P \tag{10-5}$$

and

$$b \cong a\left[1 - \left(\frac{V_{GS}}{V_P}\right)^{1/2}\right] \tag{10-6}$$

Triode-region I_D–V_{DS} characteristics. For small drain voltages, the resulting small drain current I_D has negligible effect on the channel profile. Under these conditions, the channel cross-sectional area A may be considered uniform throughout the channel length L. Therefore $A = 2bw$, where b is given by Eq. (10-6) and w is the channel width, as indicated in Fig. 10-1. Using Ohm's law, we have

$$I_D = gV_{DS} \tag{10-7}$$

where g is the channel conductance and is given by

$$g = \frac{\sigma A}{L} = \frac{2bw\sigma}{L} \tag{10-8}$$

From Eqs. (10-6) through (10-8), we obtain

$$I_D \cong \frac{2aw\sigma}{L}\left[1 - \left(\frac{V_{GS}}{V_P}\right)^{1/2}\right]V_{DS} \tag{10-9}$$

which indicates that, *for small V_{DS} values*, the JFET behaves like an ohmic resistance whose value is determined by V_{GS}. One of the important applications of the JFET is as a *voltage-controlled resistor*.

The pinch-off region transfer characteristic. When used as an amplifier, the JFET is almost always operated in the pinch-off region, where the magnitude of V_{DS} is comparable to that of V_{GS} and Eq. (10-9) is not applicable. For V_{DS} values large enough to affect depletion-region width, channel height $2b$ varies with distance x from the source as indicated in Fig. 10-1. The depletion-region width at x depends on the voltage between the gate and x, which is given by

$$V = V_{GS} - V_x \tag{10-10}$$

where V_x is the potential of x with respect to the source. Since V_{GS} in Eq. (10-6) is the voltage applied to the gate-channel junction for negligible V_x, we see from Eqs. (10-6) and (10-10) that

$$b(x) \cong a\left[1 - \left(\frac{V_{GS} - V_x}{V_P}\right)^{1/2}\right] \tag{10-11}$$

where $2b(x)$ is the channel height at distance x from the source.

We now apply the preceding analysis [that which led to Eq. (10-9)] to a channel seqment of incremental length Δx having a voltage drop ΔV_x that is very small compared to the gate-channel junction voltage V; this is illustrated by the JFET segment shown in Fig. 10-4, and the analysis is valid,

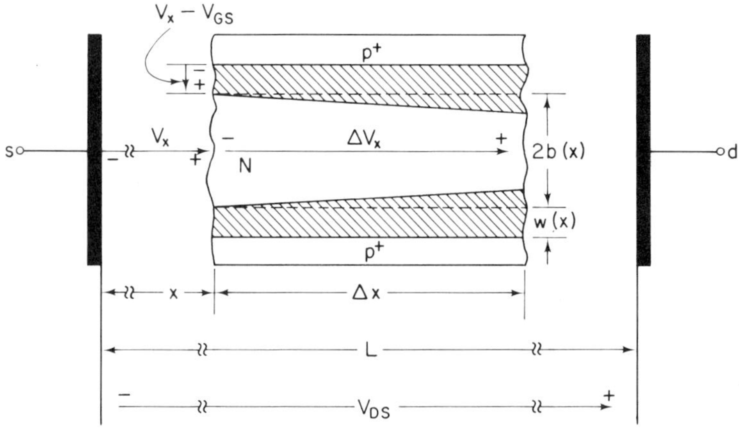

Fig. 10-4. Incremental segment of the JFET of Fig. 10-1. (The change in depletion-region width $w(x)$ with distance x from the source is greatly exaggerated in the segment.)

since the depletion width $w(x)$ may be considered uniform throughout the segment. Applying Eq. (10-8) to the channel segment, the segment conductance g_s is given by

$$g_s = \frac{2b(x)w\sigma}{\Delta x} \tag{10-12}$$

Since the current through the channel segment is the drain current I_D, we see that

$$I_D = g_s \, \Delta V_x \tag{10-13}$$

From Eqs. (10-11) through (10-13), we obtain

$$I_D = 2aw\sigma\left[1 - \left(\frac{V_{GS} - V_x}{V_P}\right)^{1/2}\right]\frac{\Delta V_x}{\Delta x} \tag{10-14}$$

from which

$$I_D \, dx = 2aw\sigma\left[1 - \left(\frac{V_{GS} - V_x}{V_P}\right)^{1/2}\right] dV_x \tag{10-15}$$

An expression for I_D may be obtained by integrating Eq. (10-15) between appropriate limits; that is,

$$\int_0^L I_D\,dx = \int_0^{V_{DS}} 2aw\sigma\left[1 - \left(\frac{V_{GS} - V_x}{V_P}\right)^{1/2}\right] dV_x \qquad (10\text{-}16)$$

Integrating Eq. (10-16) and substituting limits, we obtain

$$I_D L = 2aw\sigma\left[V_{DS} + \frac{2}{3}V_P\left(\frac{V_{GS} - V_{DS}}{V_P}\right)^{3/2} - \frac{2}{3}V_P\left(\frac{V_{GS}}{V_P}\right)^{3/2}\right] \qquad (10\text{-}17)$$

Recognizing that

$$V_{DG} = V_{DS} - V_{GS} \qquad (10\text{-}18)$$

and collecting terms, we may write Eq. (10-17) as follows:

$$I_D = \frac{2aw\sigma}{L}\left\{V_{DG}\left[1 - \frac{2}{3}\left(\frac{-V_{DG}}{V_P}\right)^{1/2}\right] + V_{GS}\left[1 - \frac{2}{3}\left(\frac{V_{GS}}{V_P}\right)^{1/2}\right]\right\} \qquad (10\text{-}19)$$

Since pinch-off occurs when $V_{DG} = -V_P$, resulting in a drain current having the pinch-off or saturation value I_{DS}, we see from Eq. (10-19) that

$$I_{DS} = \frac{2aw\sigma}{L}\left\{\frac{-V_P}{3} + V_{GS}\left[1 - \frac{2}{3}\left(\frac{V_{GS}}{V_P}\right)^{1/2}\right]\right\} \qquad (10\text{-}20)$$

The pinch-off or saturation drain current for $V_{GS} = 0$ is known as I_{DSS}. Thus

$$I_{DSS} = \frac{2aw\sigma}{L}\left(\frac{-V_P}{3}\right) = \frac{1}{3}g_0(-V_P) \qquad (10\text{-}21)$$

indicating that I_{DSS} is one-third the I_D value for $V_{DS} = -V_P$ and the gates floating.

An expression for I_{DS} in terms of I_{DSS} may be obtained from Eqs. (10-20) and (10-21). This is seen to be

$$I_{DS} = I_{DSS}\left\{1 - \frac{3V_{GS}}{V_P}\left[1 - \frac{2}{3}\left(\frac{V_{GS}}{V_P}\right)^{1/2}\right]\right\} = I_{DSS}\left[1 - 3\left(\frac{V_{GS}}{V_P}\right) + 2\left(\frac{V_{GS}}{V_P}\right)^{3/2}\right] \qquad (10\text{-}22)$$

and was originally derived by Shockley.* It is an approximate relationship, since the built-in junction potential (contact potential) V_d was neglected in Eq. (10-3), and results in a transfer characteristic (I_{DS} versus V_{GS}) that is approximately parabolic. The effect of V_d is to cause the transfer characteristic closely to approach the true parabolic relationship

$$I_{DS} = I_{DSS}\left(1 - \frac{V_{GS}}{V_P}\right)^2 \qquad (10\text{-}23)$$

This delightfully simple expression is found to be a good approximation for the transfer characteristic of JFETs no matter how they are manufactured.†

* Reference 2.

† Reference 4, p. 21.

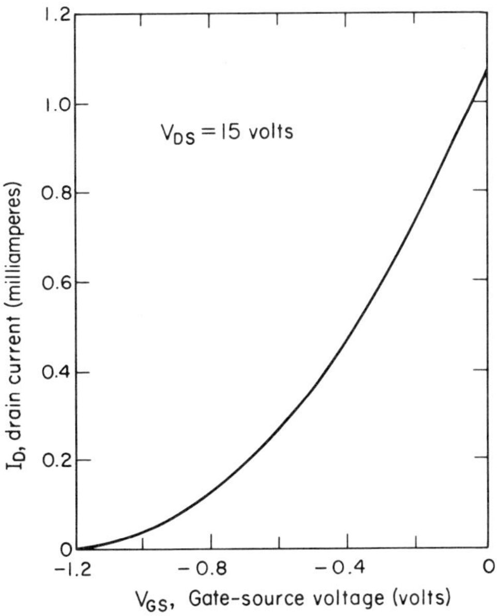

Fig. 10-5. Common-source transfer characteristic of a typical type 2N4220 *n*-channel JFET (courtesy Motorola Semiconductor Products, Inc.).

It may be of interest to check its applicability to the transfer characteristic of Fig. 10-5.

JFET equivalent-circuit model. The elementary amplifier circuit of Fig. 10-1 is reproduced in Fig. 10-6(a), with the JFET represented symbolically. The arrow of the JFET symbol points in the direction of easy flow of conventional gate current and would be reversed, as would the bias-voltage polarities, for a *p*-channel device. Low-frequency small-signal equivalent-circuit representations of the amplifier stage are shown by Figs. 10-6(b) and 10-6(c).* The equivalent-circuit parameters r_d, g_m, and μ are the dynamic (a-c) drain resistance, the gate-to-drain transconductance, and the drain-to-gate amplification factor. Transconductance g_m is also known as the forward transadmittance y_{fs}. These parameters are analogous to vacuum-tube parameters r_p, g_m, and μ and are defined as

$$r_d \equiv \left.\frac{dv_D}{di_D}\right|_{v_G \text{ fixed}} \tag{10-24}$$

$$g_m \equiv \left.\frac{di_D}{dv_G}\right|_{v_D \text{ fixed}} \tag{10-25}$$

and

$$\mu \equiv -\left.\frac{dv_D}{dv_G}\right|_{i_D \text{ fixed}} \tag{10-26}$$

* These FET models are derived from the *y*-parameter model of Fig. 6-3.

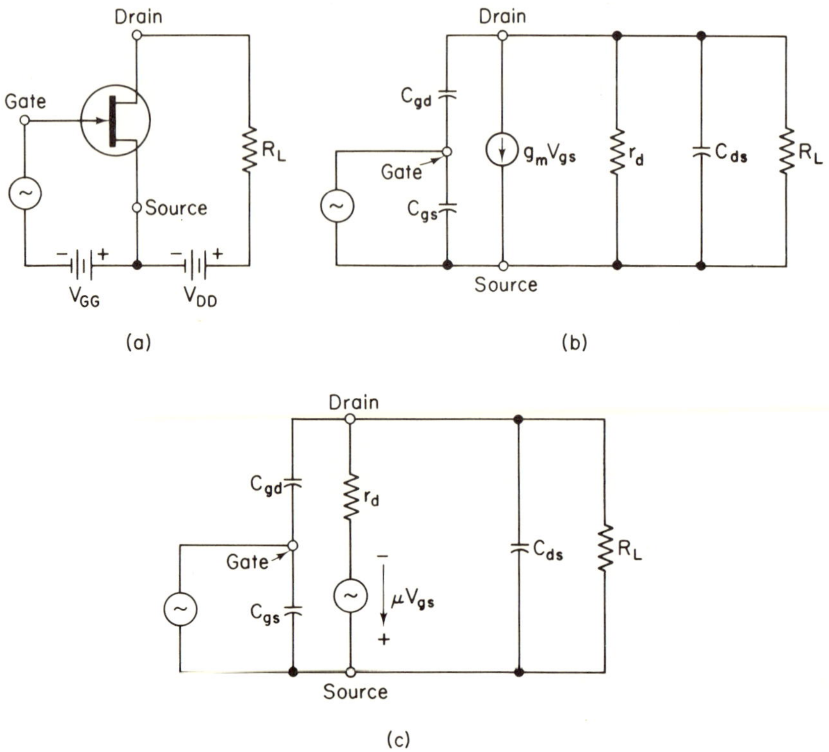

Fig. 10-6. (a) *n*-channel JFET used as an elementary amplifier. (b) and (c) Small-signal low-frequency linear equivalent-circuit representations of the circuit of (a).

where v_D, i_D, and v_G are instantaneous values of drain voltage, drain current, and gate voltage. Values of r_d, g_m, and μ can be obtained from drain-current characteristics, such as those of Fig. 10-2, or they can be measured dynamically. Typical values of g_m and r_d are of the order of magnitude of the g_m and r_p values of high-mu vacuum triodes. For example, at a drain voltage of 15 V and a gate-source voltage of 0 V (at 25°C), the g_m of the type 2N4220 JFET of Fig. 10-2(a) is specified as being not less than 1000 μmhos or more than 4000 μmhos, and r_d is not less than 100 kΩ. Like the g_m of vacuum tubes and BJTs, the g_m of a JFET is highly dependent on its operating point. From Eqs. (10-23) and (10-25), it may be seen that, in the pinch-off region,

$$g_m \cong -\frac{2I_{DSS}}{V_P}\left(1 - \frac{V_{GS}}{V_P}\right) = g_{mo}\left(1 - \frac{V_{GS}}{V_P}\right) \tag{10-27}$$

Other JFET equivalent-circuit parameters shown in Fig. 10-6 are the interelectrode capacitances. The drain-source capacitance C_{ds} is simply the capacitance between connecting lead wires and contacts, but the gate-source and gate-drain capacitances C_{GS} and C_{gd} are the transition capacitances of the gate junctions. Not shown in Fig. 10-6 are junction-leakage resistances shunting C_{gs} and C_{gd}. These resistances have been omitted since they are of the order of 10^4 megohms for typical silicon JFETs at room temperature. Unlike vacuum-type interelectrode capacitances, C_{gs} and C_{gd} of junction-type FETs are bias dependent. In addition, they may be somewhat larger than comparable vacuum-tube capacitances. For example, the parallel combination of C_{gs} and C_{gd}, referred to as the common-source short-circuit input capacitance C_{iss}, of the type 2N4220 JFET of Fig. 10-2 is specified as being not more than 6 pF at a gate-source bias voltage V_{GS} of 0 V and a drain-source bias voltage V_{DS} of 15 V. The comparable capacitance of a type 6C4 miniature vacuum triode is 3.4 pF.

The tetrode JFET. A triode JFET, such as that of Fig. 10-1, has a single gate connection with the two gate regions connected together internally. If external connections to both gates are available, the device is a *tetrode*. The second gate is useful in signal-mixing, frequency-conversion, and automatic-gain-control applications. Although the gain of a triode amplifier stage can be controlled by variation of gate bias, which varies g_m of the JFET, the use of a second gate is preferable in order to avoid detuning by variation of gate-junction capacitance. The effect of gate-two bias on gate-one-to-drain transconductance g_m of three n-channel tetrode JFETs is shown by Fig. 10-7.

The structure of Fig. 10-1 does not lend itself readily to device fabrication by vapor diffusion techniques, owing to the difficulties involved in dif-

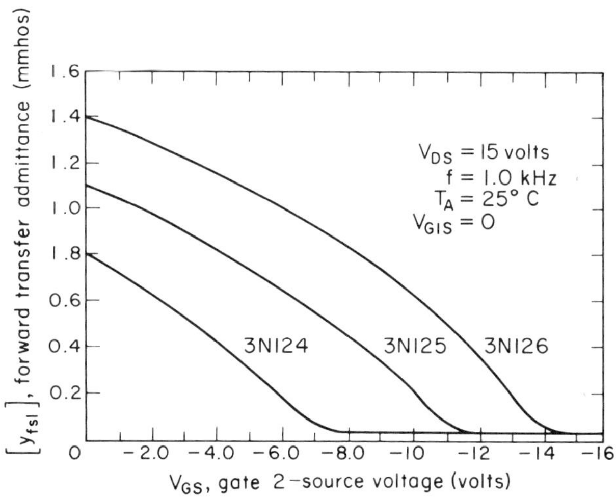

Fig. 10-7. Gate 1-to-drain transconductance g_m versus gate 2 bias voltage for typical n-channel tetrode JFETs (courtesy Motorola Semiconductor Products, Inc.).

fusing impurities into both sides of a semiconductor wafer. Figure 10-8 shows an *n*-channel tetrode JFET structure formed by diffusion from one side only. The device is fabricated by diffusing *p*-type gate 1 in an *n*-type channel that has been epitaxially grown on a *p*-type substrate that serves as gate 2. This type of structure is also used in diffused triode JFETs. Additional JFET structures are shown in Sec. 11-10.1.

The light-sensitive FET (photoFET). The photoFET is a junction-type field-effect transistor so packaged that light may be focused on the gate region. When the gate region is irradiated, the gate-junction reverse-current increases as a result of increased gate-region minority-carrier density. The photoFET is operated with a gate resistor R_G, as shown in Fig. 10-9, so that

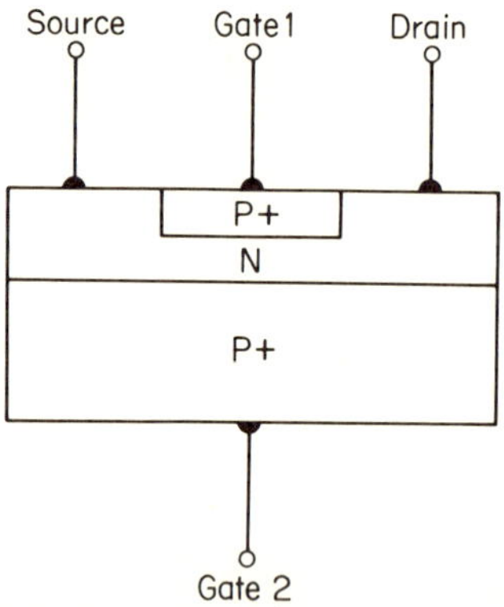

Fig. 10-8. Structure of an *n*-channel tetrode JFET.

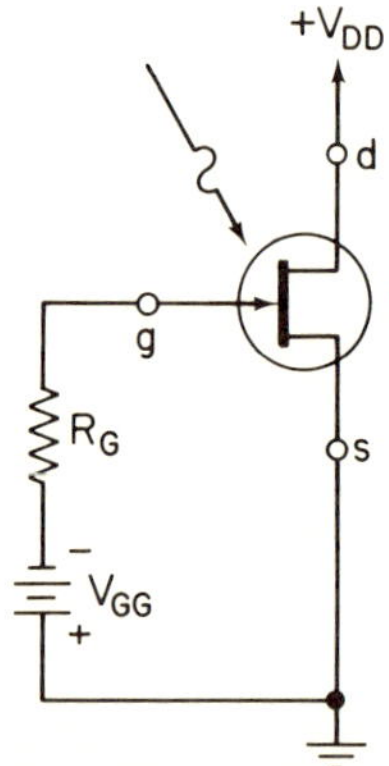

Fig. 10-9. Light-sensitive junction-type FET (photoFET) arranged for drain-current control by gate-region irradiation.

increasing gate current decreases the gate-junction reverse-bias and increases the drain current. The Crystalonics type FF400 *n*-channel silicon photoFET has a gate current change of not less than 15 nA per foot candle of incident light at 2800°K color temperature, and may be used in circuits having R_G values up to 100 MΩ. Clearly, the light-sensitivity of the photoFET varies directly with R_G. Figure 10-10(b) shows the drain current-drain voltage characteristics of a type FF400 photoFET irradiated by a General Electric type LED11 light-emitting diode (LED) in the arrangement of Fig. 10-10(a). The I_D–V_{DS} characteristic for zero gate voltage is included for reference and shows that the photoFET gate-source junction is forward biased for a LED current I_1 of 25 mA. Gate-bias voltage V_{GG} is adjusted for drain-current pinch-off when I_1 is zero.

Temperature effects. An important advantage of the FET over the conventional transistor is its smaller temperature sensitivity. Unlike the col-

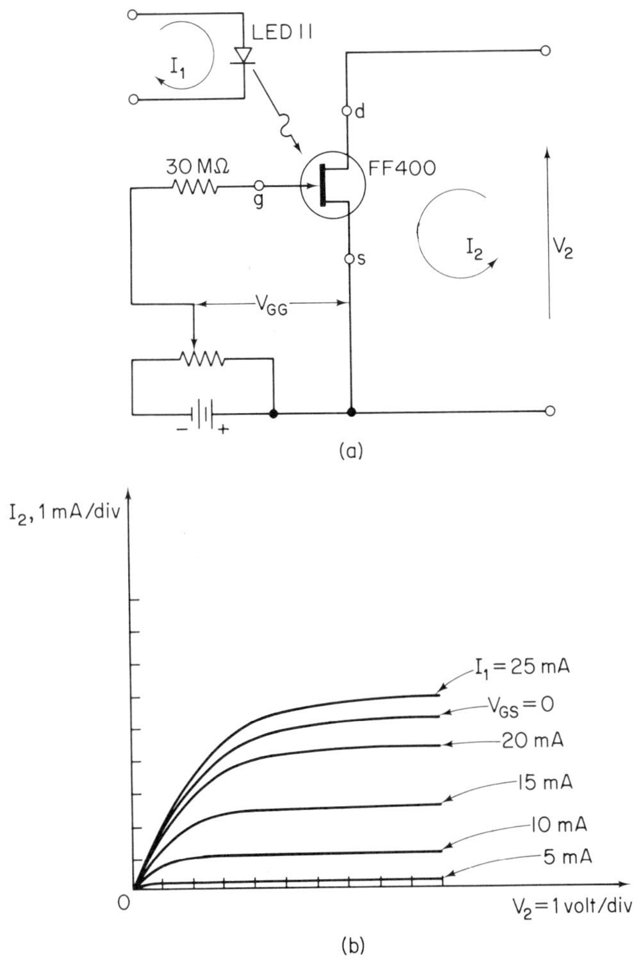

Fig. 10-10. Photon-coupled combination of a General Electric type LED11 light-emitting diode and a crystalonics type FF400 photo-FET. (a) Circuit arrangement. (b) PhotoFET drain characteristics.

lector current of the BJT, the drain current of the FET *usually* decreases with increasing temperature, as shown by the drain characteristics of Fig. 10-11. Therefore, thermal runaway is not encountered in FETs. An increase in temperature decreases the channel conductivity, owing to decreased carrier mobility, but increases the channel height because of the decrease in depletion-region thickness (junction-barrier height) that occurs as a result of Fermi-level shifts with temperatrue. These are opposing effects, as far as drain current is concerned, but Fig. 10-11 suggests that the effect on channel conductivity predominates. However, the transfer characteristics of Fig. 10-12(a) for the same device show that the effect on channel height predominates for

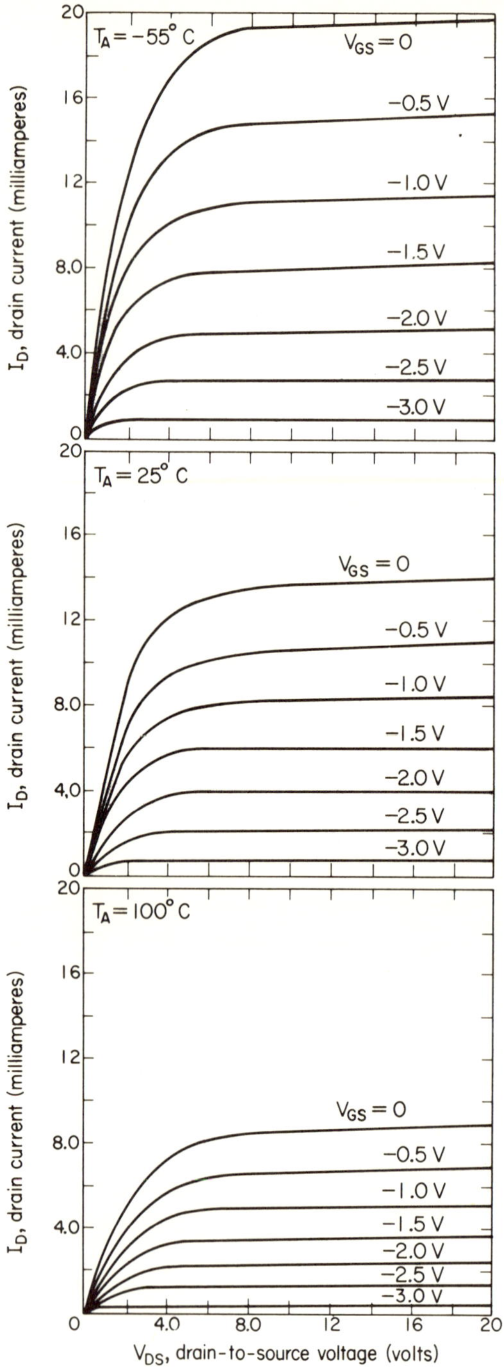

Fig. 10-11. Effect of temperature on the drain characteristic of a typical type 2N5163 *n*-channel silicon junction FET (courtesy Fairchild Semiconductor).

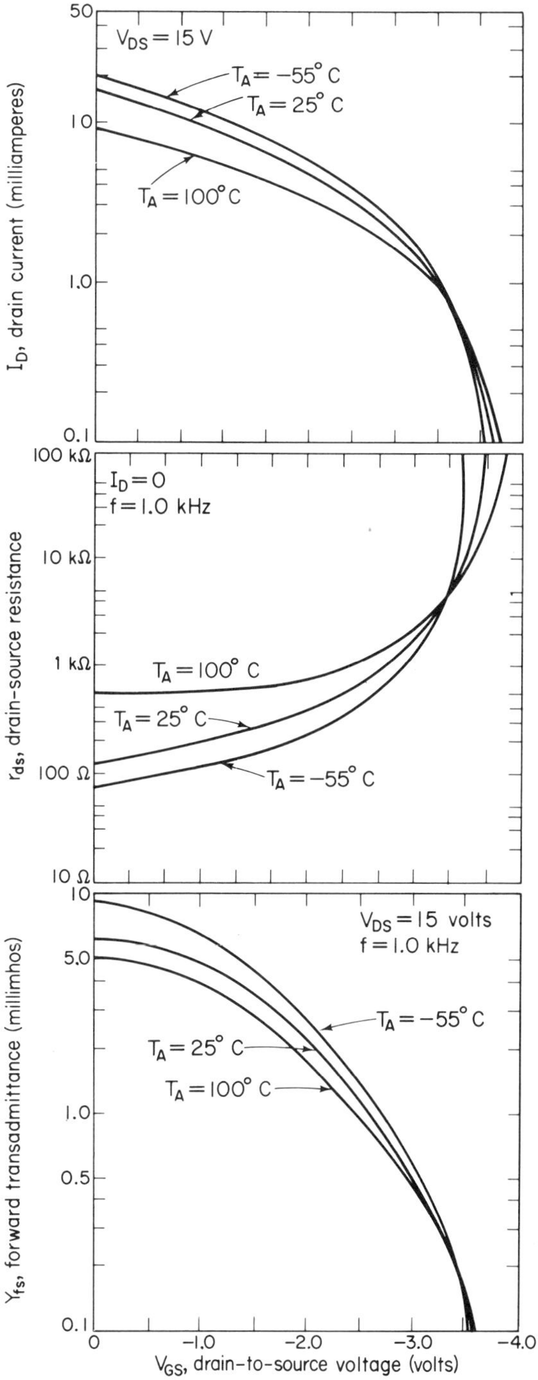

Fig. 10-12. Effect of temperature on (a) drain current I_D, (b) drain-source (channel) resistance r_{ds}, and (c) forward transconductance Y_{fs} of a typical type 2N5163 *n*-channel silicon junction FET (courtesy Fairchild Semiconductor).

gate voltages approaching the value for pinch-off. This is also shown by the drain-source (channel) resistance (r_{ds}) versus gate voltage plots of Fig. 10-12 (b). The effects of temperature on g_m are much the same as those on drain current, as may be seen by comparing Fig. 10-12(c) with Fig. 10-12(a). From Fig. 10-12, we see that the FET may be biased so that temperature has no effect on drain current and forward transconductance.

The effect of temperature on the gate current of a JFET is simply that on the current of a reverse-biased *pn* junction, as seen from Fig. 10-13. Therefore, the low-frequency input resistance of a JFET is highly temperature dependent.

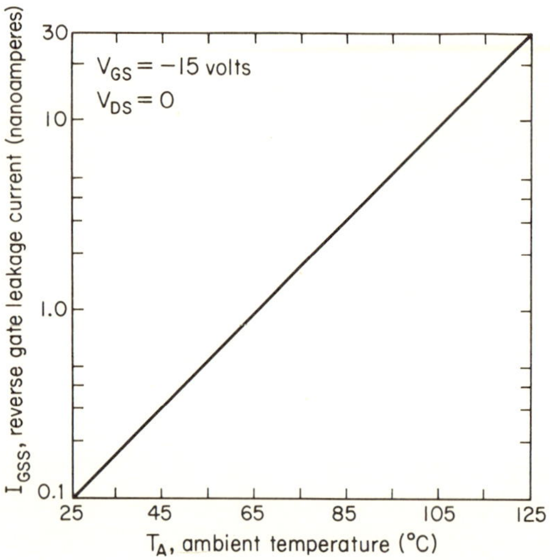

Fig. 10-13. Effect of temperature on the reverse gate current of a typical type 2N5163 *n*-channel silicon JFET (courtesy Fairchild Semiconductor).

Comparison of JFETs with BJTs. Although the JFET has such unique characteristics that it should not be regarded merely as a competitor of the BJT, it is of interest to compare the devices because of their common applications. Higher input impedance and lower temperature sensitivity have already been mentioned as important advantages of JFETs. JFETs are much less noisy, particularly in the low-frequency region, and are generally believed to be less subject to degradation by atomic radiation, since their operation is not dependent on minority-carrier lifetime. The latter difference has been found to be less pronounced, if not nonexistent, in JFETs having low pinch-off voltages, since such devices have low-conductivity channel regions.[8]

[8] B. L. Gregory and F. M. Smits, "A Comparison of Radiation Tolerance of Field-Effect and Bipolar Transistors," *IEEE Trans.*, Vol. **ED 12**, No. 5 (May 1965).

Disadvantages of FETs are their higher ON resistance, when used as switches, and their lower gain-bandwidth product. In these respects, *n*-channel JFETs are superior to *p*-channel devices, owing to the higher mobility of *n*-channel majority carriers. Another disadvantage of the JFET is the requirement of higher-voltage bias supplies, since the knees of JFET drain characteristics occur at voltages higher than those of BJT collector characteristics. This difference between JFETs and BJTs has diminished as JFETs having lower pinch-off voltages have become available.

10-2 THE INSULATED-GATE FIELD-EFFECT TRANSISTOR (MOSFET)[9,10,11,*]

The insulated-gate or metal-oxide-semiconductor FET (MOSFET) uses a metal gate electrode, separated from the semiconductor by a thin layer of silicon-dioxide (glass) insulation, to modulate the resistance of a conducting channel in the semiconductor. It is of interest that an insulated gate field-effect device made from semiconductor material was patented several years before the discovery of the bipolar transistor.† However, no practical devices ensued at that time, owing to the limited knowledge of the physics of surfaces and thin films. Increased understanding of semiconductor properties and the desire to improve upon the JFET led to a revival of interest in insulated-gate devices, with commercial units becoming available in the mid-1960s. Because it lends itself so readily to use in integrated circuits, the MOSFET is proving to be of greater commercial importance than the JFET (see Sec. 11-11).

MOSFETs have two operating modes—the *depletion* mode and the *enhancement* mode. In the depletion mode, a conducting channel exists with zero gate-bias and, as in the JFET, reverse gate-bias depletes the charge carriers of the channel and reduces its conductance. Unlike as in the JFET, however, forward gate-bias enhances the charge carriers and increases channel conductance.‡ Devices that exhibit appreciable channel conductance with

* Reference 4, pp. 24–29; Reference 5, pp. 113–154, 187–214; Reference 6, pp. 14–16; Reference 7, pp. 128–134.

9 S. R. Hofstein and F. P. Heiman, "The Silicon Insulated-Gate Field-Effect Transistor," *Proc. IRE*, Vol. **51**, No. 9 (September 1963).

10 P. Richman, *Characteristics and Operation of MOS Field-Effect Devices*, New York: McGraw-Hill Book Co., Inc., 1967.

11 R. H. Crawford, *MOSFET in Circuit Design*, New York: McGraw-Hill Book Co., Inc., 1967).

† O. Heil, "Improvements in or Relating to Electrical Amplifiers and Other Control Arrangements and Devices," *British Patent* **439,457** (Sept. 26, 1939).

‡ The phenomenon does occur, to a limited degree, in JFETs for small forward gate voltages, as shown in Fig. 10-2(b). For MOSFETs, the effect is of much greater significance.

zero gate-bias are called *depletion-type* or *normally on* MOSFETs; those that have no channel conductance at zero bias are referred to as *enhancement-type* or *normally off* MOSFETs. Since insulated-gate FETs may be either *n*-channel or *p*-channel devices, four types of MOSFETs are obtainable. In 1970 all but the *p*-channel depletion type existed as commercial devices.

The enhancement-type MOSFET. The cross section of a *p*-channel enhancement-type MOSFET is shown in Fig. 10-14. This unit is fabricated with

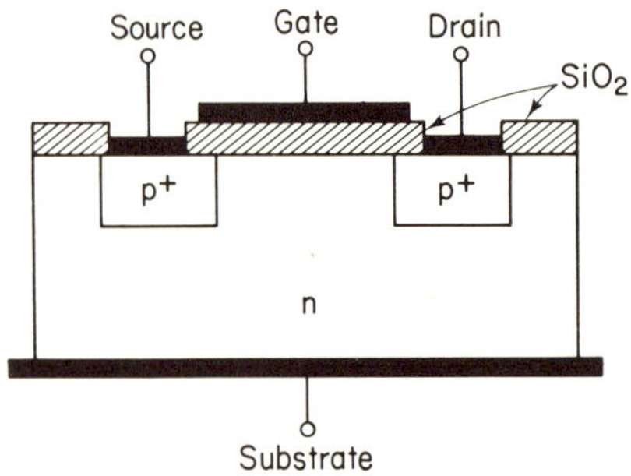

Fig. 10-14. Cross section of a *p*-channel enhancement-type MOSFET.

heavily doped source and drain semiconductor regions of conductivity type opposite that of the semiconductor substrate. Thus series back-to-back diodes exist between the source and drain contacts, and the drain current is essentially zero for zero gate-bias, regardless of the polarity of the drain-to-source voltage; that is, for all practical purposes I_{DSS} of the enhancement-type MOSFET is zero. When the gate of the *p*-channel device is made slightly negative with respect to the substrate, electrons are repelled from the substrate under the gate and a depletion layer is formed between the source and drain regions. If the gate becomes sufficiently negative, free holes are attracted into the depletion layer, and a *p*-type channel or "inversion layer" is formed at the oxide-substrate interface. Since the channel extends from source to drain, source and drain diode action ceases at the substrate surface, and drain current can now flow. The gate voltage required to form the channel is called the threshold voltage V_{th}. An increase in negative gate voltage above the threshold value induces more free holes into the channel, enhancing the drain current.

The common-source drain characteristics of a typical type 2N4120 enhancement type *p*-channel MOSFET are shown in Fig. 10-15. We see that a gate voltage of the order of -4 V is required in order to permit appreciable drain current. This is the threshold voltage and is specified for a particular drain current. For example, the magnitude of the gate-source voltage of the 2N4120 should not exceed -6 V or be less than -3 V for a drain current

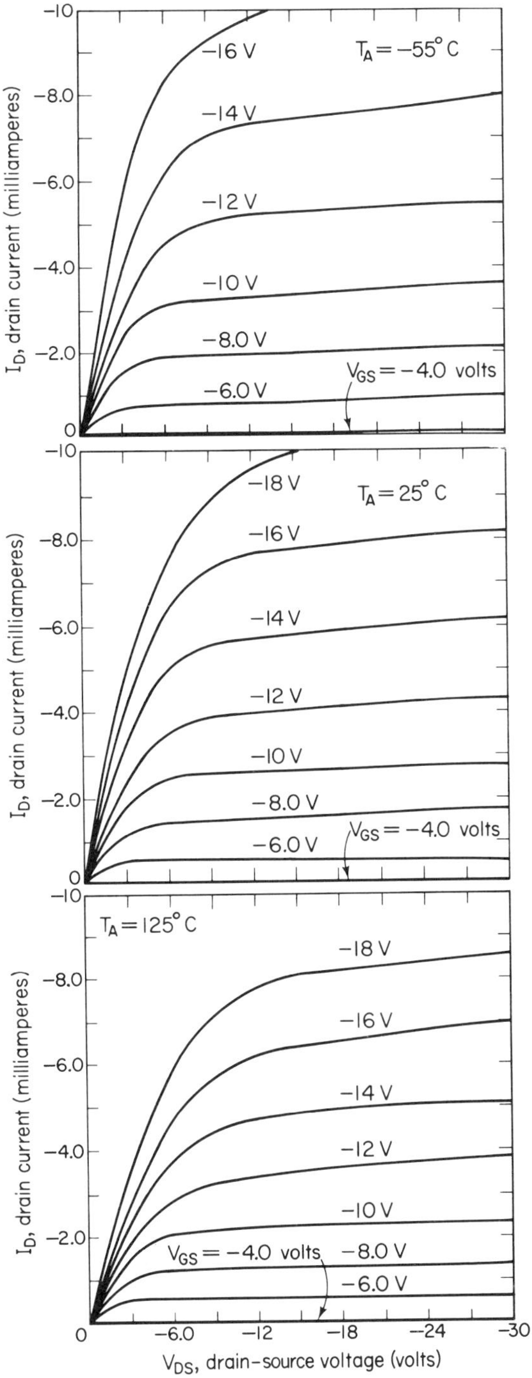

Fig. 10-15. Common-source drain characteristics of a typical type 2N4120 enhancement-type *p*-channel silicon MOSFET. The gate and drain voltages are seen to have the same polarity and to be of the same order of magnitude (courtesy Fairchild Semiconductor).

of $-10\ \mu A$ at a temperature of 25°C. The I_{DSS} specification of the 2N4120, for a drain voltage of -15 V, is that the magnitude of I_{DSS} should not exceed 0.5 nA at 25°C or 1.0 μA at 150°C.

The drain current-drain voltage curves for the *p*-channel enhancement-type MOSFET are seen to resemble those of the JFET, with I_D varying linearly with V_{DS} for small V_{DS} values and leveling off or saturating as V_{DS} increases negatively. As in the JFET, the channel is essentially unaffected by small values of V_{DS}, but is pinched off near the drain end for large negative values. Since V_{th} is the voltage required between the gate and the drain end of the channel in order to maintain the channel, pinch-off sets in when $|V_{GD}|$ falls to $|V_{th}|$ as $|V_{DS}|$ is increased. That is, the channel is pinched off when

$$|V_{DS}| \geq |V_{GS}| - |V_{th}| \tag{10-28}$$

This expression may be verified by examining the characteristic curves of Fig. 10-15. When $-V_{DS}$ is increased above the value for pinch-off, the voltage drop across the channel remains essentially constant at the pinch-off value, and the increase in $-V_{DS}$ is absorbed by the back-biased drain-substrate junction.

An important feature of the enhancement-type MOSFET is that input and output bias voltages have the same polarity and are of the same order of magnitude. Because of this feature, biasing is simplified and the device lends itself particularly well to direct coupling of amplifier stages. This may be seen in the amplifier circuit of Fig. 10-16.

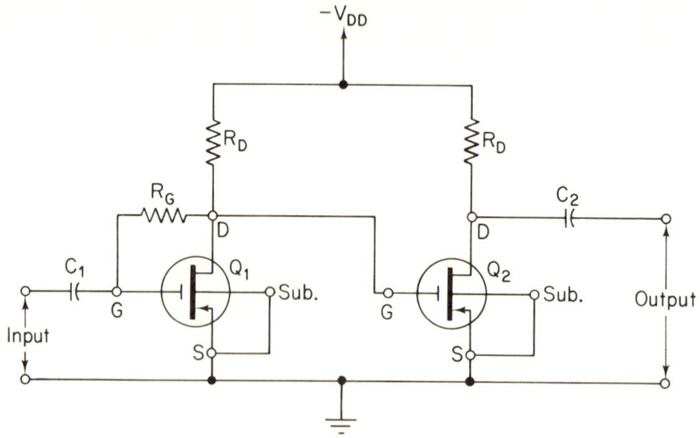

Fig. 10-16. Enhancement-type *p*-channel MOSFET amplifier circuit. Direct coupling between the stages is permissible, since gate- and drain-bias voltages of enhancement-mode transistors have the same polarity and may be of the same magnitude.

The insulated gate results in an input resistance several orders of magnitude higher than that of the JFET, typical input resistances being of the order of 10^{15} ohms. Therefore, the drain-bias voltage of MOSFET Q_1 in Fig. 10-16 is unaffected by the gate of Q_2 and the gate and drain bias voltages of Q_1 are equal even though the resistance of R_G may be of the order of 10^3 megohms or higher. Another advantage of the MOSFET is that the interelectrode capacitances are independent of bias voltage. In addition, they may be smaller than the capacitances of the JFET.

In the amplifier circuit of Fig. 10-16, coupling capacitors C_1 and C_2 prevent disturbance of electrode-bias conditions by the signal-input or load circuits. The source electrode of the MOSFET symbol is identified by an arrow that indicates the channel type, the arrow showing the direction of conventional source current when the device is on. Symmetrical location of the gate with respect to the source and drain leads indicates an enhancement-type device, the gate being displaced toward the source in the symbol for a depletion-type MOSFET (see Fig. 10-20).* The substrate of each MOSFET is shown connected to the source in Fig. 10-16. This is the usual arrangement for enhancement-type devices.

n-channel enhancement-type MOSFETs have n^+ source and drain regions on a p-type substrate, and an n-type surface channel is established when the gate is made sufficiently positive with respect to the substrate. Early n-channel devices had rather large I_{DSS} values, since an inversion (n-type) layer existed in the silicon at the oxide-substrate interface with zero gate voltage. For this reason, most early enhancement-type MOSFETs were p-channel devices. However, n-channel MOSFETs having characteristics comparable to those of p-channel devices are now available. For example, the n-channel type 2N4351 MOSFET is complementary† with the type 2N4352 p-channel device, which has characteristics like those of the type 2N4120 unit of Fig. 10-15. Since it requires no gate bias for the OFF mode, the enhancement-type (normally off) MOSFET lends itself readily to switching applications, and circuits using complementary devices are especially attractive.

The depletion-type MOSFET. MOSFETs of the depletion type have source, drain, and channel regions of the same semiconductor type in order to permit substantial drain current with zero gate-bias. The cross section of an n-channel depletion-type MOSFET is shown in Fig. 10-17. Since the channel is formed on the surface of a p-type substrate, the substrate forms pn junctions with the heavily doped source and drain regions. These junctions act

* The symbols of Figs. 10-16 and 10-20 are not universally used, since FET symbols are yet to be standardized.

† Complementary devices have similar characteristics, but electrode currents and voltages are of opposite polaraties.

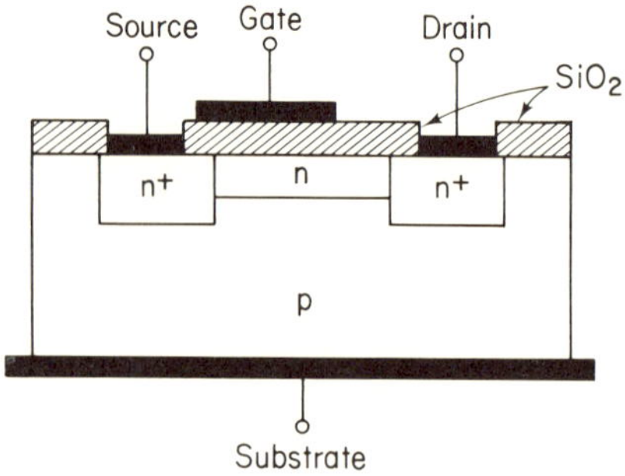

Fig. 10-17. Cross section of an *n*-channel depletion-type MOSFET. Unlike the enhancement-type device, the gate electrode need not overlap the source, and drain, permitting smaller gate-to-drain capacitances.

as series back-to-back diodes in parallel with the channel. When the gate is made positive with respect to the substrate, additional electrons are attracted into the channel from the n^+ source region, increasing channel conductance; that is, enhancement-mode operation occurs with positive gate voltage. If the gate voltage is negative, free electrons are repelled from the channel. This decreases the channel conductance and forms a carrier-depletion region at the oxide-silicon interface, constricting the channel as in a JFET. If the gate becomes sufficiently negative, this depletion region extends completely through the channel and joins the depletion region of the *pn* junction on the other side. For this condition, drain-to-source conduction ceases; the required gate voltage is the pinch-off voltage V_P. The manufacturer usually indicates the V_P value at which I_D is reduced to some specified value. For example, I_D of the type 3N139 *n*-channel depletion-type MOSFET, whose drain characteristics are shown in Fig. 10-18, is specified (at 25°C) as 50 μA or less for a V_{GS} of -6 V and V_{DS} of 15 V.

The drain characteristics of the *n*-channel depletion-type MOSFET are seen to be much like those of the *n*-channel JFET, except that both positive and negative gate voltages may be used without gate-current flow. Depletion-type MOSFET and JFET I_D–V_{DS} curves have the same shapes for much the same reasons. For MOSFET V_{DS} values significant relative to the magnitude of V_P, the depletion layer at the oxide-substrate interface expands at the drain end of the channel, as shown in Fig. 10-19, and channel pinch-off may occur. When V_{DS} is increased above the value required for channel pinch-off, the increase is absorbed by the space-charge region at the drain end of the channel, as in the enhancement-type MOSFET.

The p-channel depletion-type MOSFET. In principle, the foregoing discussion is applicable to a *p*-channel depletion-type MOSFET. However,

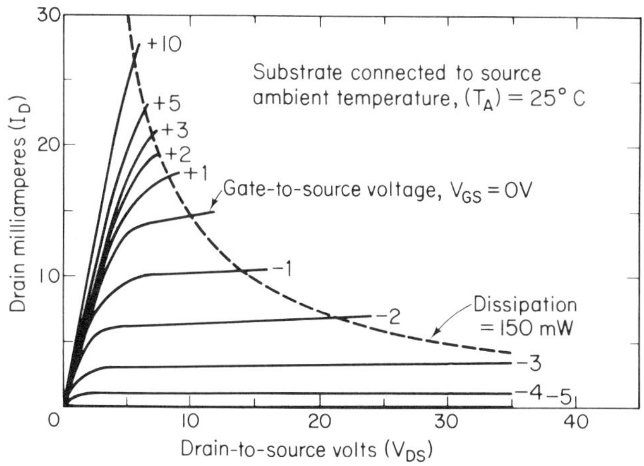

Fig. 10-18. Drain characteristics of a typical type 3N139 depletion type *n*-channel MOSFET (courtesy Radio Corporation of America.)

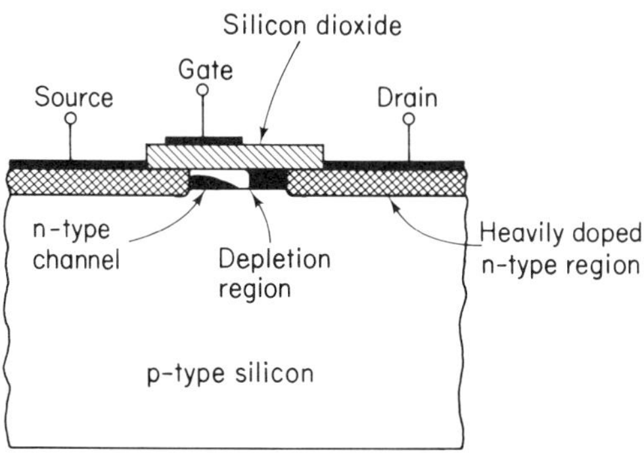

Fig. 10-19. Cross section of an *n*-channel depletion-type MOSFET operating at a drain-to-source voltage near the value for channel pinch-off.

no such device has yet appeared on the market, owing to the difficulty in producing the required *p*-channel at the oxide-substrate interface. This is the problem that has been responsible for high-I_{DSS}, *n*-channel, enhancement-type MOSFETs. As mentioned earlier, the substrate surface tends to be *n*-type whether the substrate itself is *p*- or *n*-type material.

An advantage of the depletion-type MOSFET over the enhancement type is that the metal gate need not overlap both the source and the drain (see Fig. 10-19). This offset geometry results in a much smaller gate-to-drain (feedback) capacitance—a capacitance that has an important effect on circuit behavior and should be as small as possible. With offset geometry, a portion of the channel resistance is unmodulated by the gate. Ideally, this bulk or contact resistance should be in series with the drain, since the effect on amplifier voltage gain of resistance in series with the source is as though a much larger resistance were in the drain-source circuit, as pointed out in Sec. 10-4.

Because of the insulated gate, the depletion-type MOSFET may be operated with zero gate bias, as in the elementary amplifier stage shown in Fig. 10-20. More commonly, however, the MOSFET is operated with reverse gate bias. This results in a smaller value of g_m, as may be seen from Fig. 10-21, but a larger voltage gain can be obtained (through use of a larger

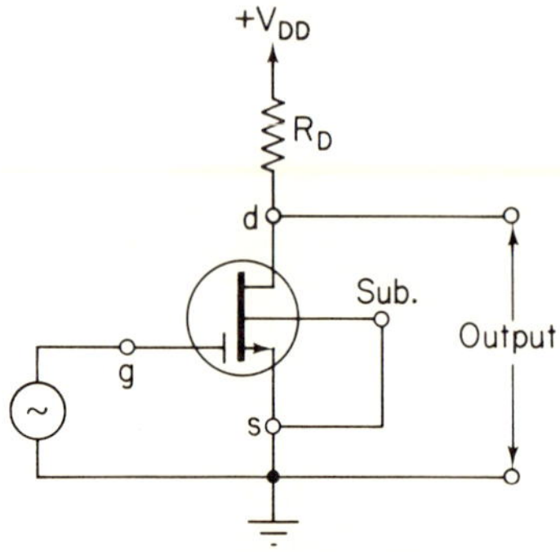

Fig. 10-20. Depletion-type *n*-channel MOSFET amplifier stage operated with zero gate bias. Displacement of the gate electrode of the transistor symbol toward the source electrode indicates that the MOS transistor is of the depletion type.

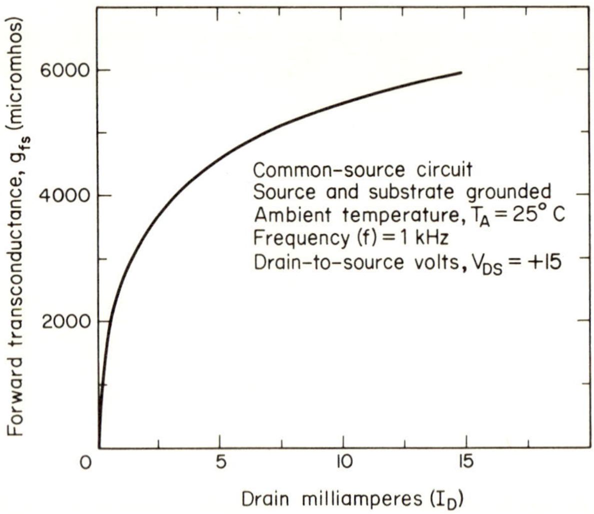

Fig. 10-21. Forward transconductance versus drain current for the type 3N139 depletion-type MOSFET of Fig. 10-18 (courtesy Radio Corporation of America).

drain-load resistance) for a given drain-supply voltage V_{DD}. Also, a problem arises in the circuit of Fig. 10-20 if there is no d-c path through the input circuit. In this case, the MOSFET operates in the enhancement mode, owing to forward gate bias from the internal voltage divider formed by the gate-to-source and gate-to-drain leakage resistances, and the circuit is quite critical. The substrate is shown internally connected to the case in Fig. 10-20. This is the normal arrangement for MOSFETs, since the substrate is usually maintained at a-c ground potential.

Substrate bias. The *p*-type substrate of the depletion-type MOSFET forms a junction with the source-drain contacts and the channel, as shown in Fig. 10-17. If the substrate is left unconnected, its potential floats slightly above that of the source, owing to forward bias of the substrate-to-source junction by the reverse current of the drain-to-substrate junction. Therefore, the substrate is often connected to the source. However, the substrate-channel junction acts as a gate on the channel, in the manner of the junction of the JFET, and is sometimes used as a second gate. The effect on drain current I_D of reverse-biasing the substrate-to-source junction of a MOSFET is shown in Fig. 10-22, and the effect on g_m is similar to that on I_D. Reverse-biasing the substrate-source junction also reduces the drain-to-source capacitance and causes it to be less dependent on drain voltage.

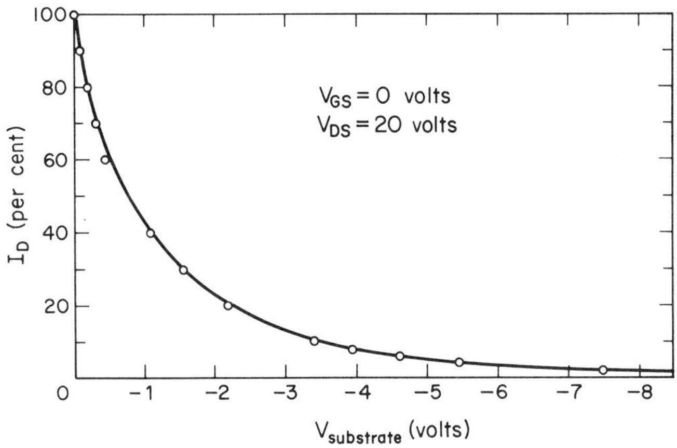

Fig. 10-22. Effect of substrate bias on the drain current of an *n*-channel depletion-type MOSFET.

The dual-gate MOSFET. Although the substrate of a depletion-type MOSFET can be used as a gate, it is not an insulated gate. By dividing the metal gate of Fig. 10-17 into two sections, and providing connection to each, the result is a MOSFET having a series arrangement of two separate channels, each channel having an independent control gate. Like the tetrode JFET, the dual-gate MOSFET is useful in frequency-mixing, frequency-conversion, and automatic-gain-control applications. Transfer characteristics of a typical type 3N141 dual-gate *n*-channel depletion-type MOSFET are shown by Fig. 10-23.

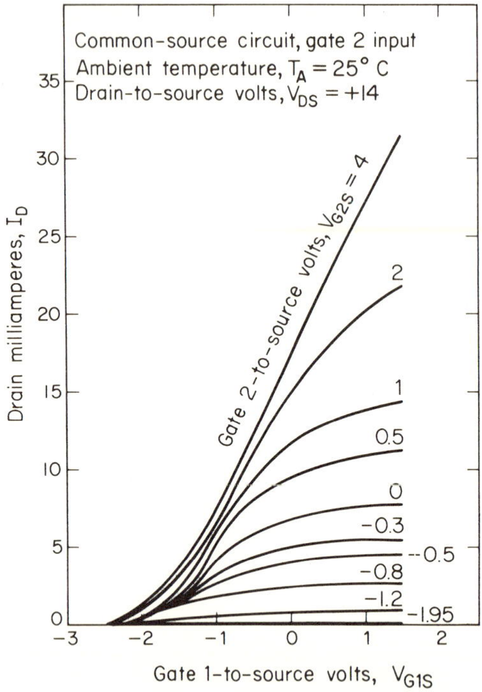

Fig. 10-23. Transfer characteristics of a typical type 3N141 dual-gate *n*-channel depletion-type MOSFET (courtesy Radio Corporation of America).

MOSFET equivalent-circuit model. For frequencies low enough that the capacitances can be considered open circuits, the JFET circuit models of Fig. 10-6 are applicable to the MOSFET, provided that the source-contact resistance is negligible. An "all-frequency" circuit model is shown in Fig. 10-24. The equivalent-circuit elements have the following significance:

r_{gs} *and* r_{gd}. These resistances represent the leakage paths through and across the insulating oxide from gate to source and from gate to drain, respectively. They are usually very high, their parallel combination typically being of the order of 10^{15} ohms or more.

C_c *and* r_c. These elements represent a lumped approximation of the active-channel resistance and the metal gate. Voltage v_c across C_c performs the charge control of the MOSFET. At low frequencies, v_c is equal to input

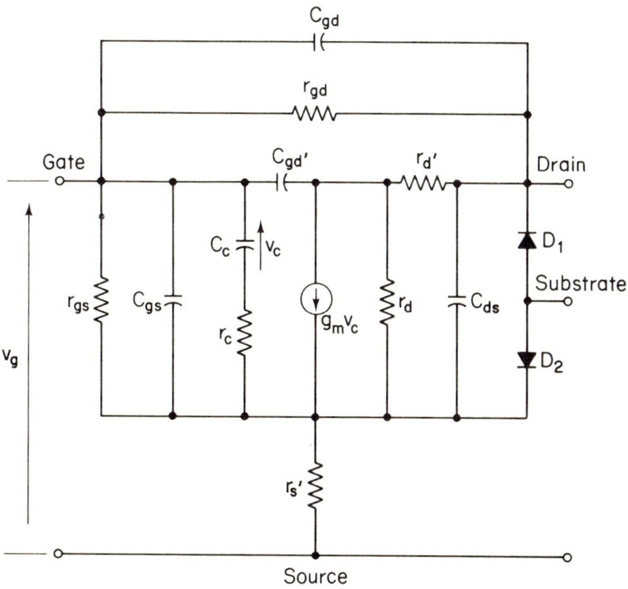

Fig. 10-24. Common-source equivalent circuit of the *n*-channel MOSFET. For the *p*-channel device, D_1 and D_2 are inverted.

voltage v_g. Channel resistance r_c is the reciprocal of the slope of the i_D-v_{DS} characteristic in the triode region below pinch-off.

g_m *and* r_d. These elements are the gate-to-drain transconductance (forward transfer admittance) and dynamic drain resistance of the JFET model of Fig. 10-6.

$r_{d'}$ *and* $r_{s'}$. These resistances represent the portion of the channel resistance that is not modulated by the gate electrode. They approach zero in a "full-gate" (enhancement-type) MOSFET. Ideally $r_{s'}$ is negligible, as indicated earlier.

C_{gd}, C_{gs}, and C_{ds}. These elements are the case and interlead capacitances. In addition, C_{ds} includes the capacitances related to diodes D_1 and D_2.

$C_{gd'}$. This element is the intrinsic gate-to-drain capacitance. As indicated earlier, $C_{gd'}$ is higher for the "full-gate" (enhancement-type) MOSFET than for the "offset-gate" (depletion-type) device.

D_1 *and* D_2. These diodes are formed by the junction of the drain and source regions to the substrate.

Substrate. As indicated earlier, the substrate forms a *pn* junction with the channel and acts as a second gate. For this reason, it should be at a-c ground potential and at a d-c potential that minimizes the potential difference between it and the gate.

Problems with MOSFETs. Because of their extremely high input resistance, MOSFETs are subject to damage by the accumulation of excess static charge. Therefore, their leads should be shorted, say by the shipping foil or washer, except when they are being tested or are in actual use. Unnecessary handling should be avoided, and the devices should be picked up by the can instead of the leads. Because of possible damage by voltage transients, MOSFETs should not be inserted in or removed from circuits with the power on. (The latter precaution is good practice for all semiconductor devices.) This problem has been alleviated in some device types by means of a "built-in" breakdown diode (or diode-resistor network) connected between the gate and the substrate.

When an enhancement-type MOSFET has been biased ON for an extended time, it may be found that the threshold voltage has shifted; in an extreme situation, the device may not turn OFF when the gate bias is removed.* This effect was quite common in early devices and delayed the practical application of enhancement-type MOSFETs. It is believed to be due to the presence of a built-in electric field in the silicon dioxide, as a result of the migration of ions in the very strong electric field, and is less likely in depletion-type devices that have weaker electric fields in the oxide layer, owing to their smaller gate voltages. Although SiO_2 is impervious to most contaminants, which accounts for its widespread use in the passivation of semiconductor devices, sodium is a significant exception. Since sodium ions are positive, positive gate voltage causes them to migrate toward the SiO_2-silicon interface, where they act like an applied voltage that lowers the threshold voltage. Any negative ions present in the SiO_2 layer would have a similar affect on a *p*-channel MOSFET. When gate voltage is removed, the ions migrate away from the interface, and the device tends to regain its original threshold voltage. When this effect was first discovered, threshold-voltage drifts as high as 5 V, or more, were observed. It is now possible to limit the drift to less than 0.6 V by the use of more stable SiO_2 layers and by *silicon-nitride passivation.*

Figure 10-25 shows the cross section of a silicon-nitride-passivated MOSFET. Since SiO_2 layers free from contaminants such as sodium can be grown, the silicon-nitride layer prevents contamination during subsequent processing steps, silicon nitride being impervious to sodium. Because of the higher dielectric constant of silicon nitride, nitride-passivated MOSFETs tend to have lower threshold voltages.

When compared with JFETs, MOSFETs are found to be somewhat noisier. Depletion-type MOSFETs have been found to be much more subject to damage by nuclear radiation than JFETs, but enhancement-type MOSFETs have even greater radiation tolerance than JFETs.†

* Reference 5, pp. 210–214; Reference 9, pp. 82–85.

† For "Radiation Tolerance of Field-Effect Transistors," see Reference 5, pp. 176–185.

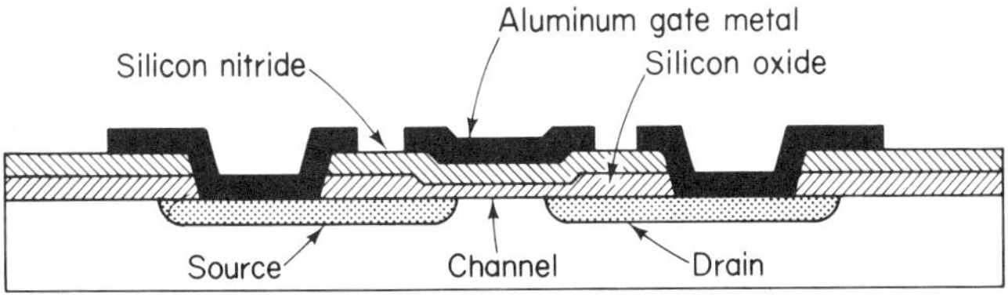

Fig. 10-25. Cross section of a silicon-nitride-passivated enhancement-type MOSFET.

10-3 SMALL-SIGNAL (LINEAR) AMPLIFIER STAGES*

Since the field-effect transistor and the bipolar transistor are duals, they can be used in the same general applications. Therefore, JFETs and MOSFETs can be used as amplifiers operated with any of the three device terminals common to the input and output circuits. Figure 10-26 shows the three basic FET amplifier configurations. These are the common-source arrangement, the common-drain or source-follower arrangement, and the common-gate configuration. The FET symbol shown in Fig. 10-26 indicates the use of an n-channel depletion-type MOSFET, but JFETs of either type and all types of MOSFETs can be used in each configuration by making proper bias provisions. Bias arrangements for the several types of field-effect transistors are considered in Sec. 10-4, and we shall see that those of Fig. 10-26 are not necessarily the most attractive ones for depletion-type devices.

10-3.1 The Common-Source Amplifier

The common-source configuration is the amplifier circuit most frequently used with field-effect transistors. It is analogous to the BJT common-emitter amplifier and has been considered qualitatively in Secs. 10-1 and 10-2 (see Figs. 10-1, 10-6, 10-16, and 10-20).

At frequencies such that the gate-to-channel capacitive reactance of a MOSFET is large compared to the channel resistance, the equivalent-circuit models of Fig. 10-6 may be used to represent both JFETs and MOSFETs, provided that leakage and contact resistances have negligible effects. These models will be used in our consideration, and Fig. 10-27 shows the current-source model in an equivalent circuit for the common-source stage of Fig. 10-26. In this circuit, coupling capacitor C_c and bias-supply internal impedances are assumed to have negligible effects, and load impedance Z_L

* Reference 4, pp. 51–57; Reference 5, pp. 253–276; Reference 6, pp. 53–71; Reference 7, pp. 21–34; Reference 10, pp. 99–111; Reference 11, pp. 119–124.

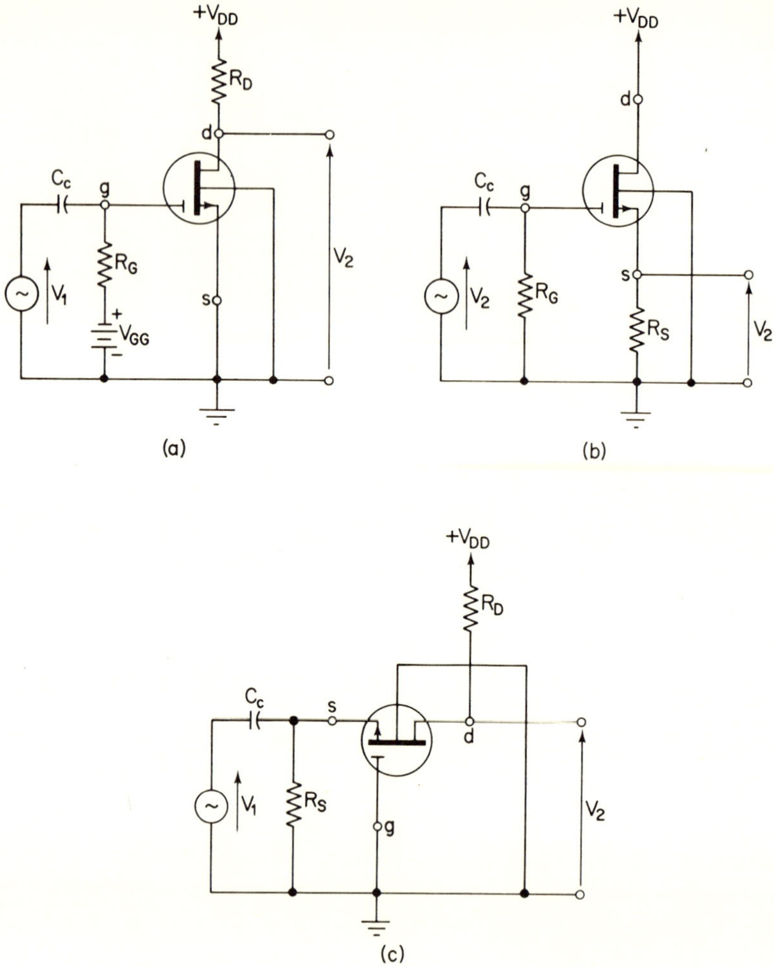

Fig. 10-26. Basic field-effect transistor circuit configurations. (a) Common source. (b) Common drain or source follower. (c) Common gate.

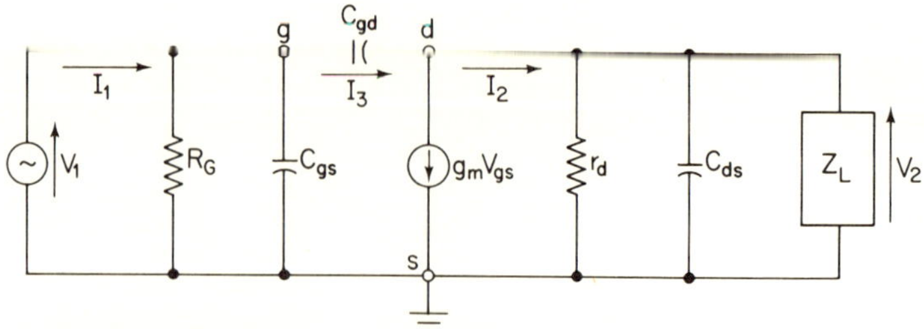

Fig. 10-27. Small-signal equivalent circuit of the common-source amplifier stage of Fig. 10-26(a).

is the parallel combination of drain resistor R_D and any load connected to the output terminals.

Voltage gain. From Fig. 10-27, we see that output voltage V_2 is given by

$$V_2 = \frac{I_2}{Y_L + g_d + j\omega C_{ds}} \tag{10-29}$$

where $Y_L = 1/Z_L$ and $g_d = 1/r_d$, and "feedback current" I_3 is

$$I_3 = j\omega C_{gd}(V_1 - V_2) \tag{10-30}$$

where V_1 is the input voltage. Considering the currents at node d, we obtain

$$I_2 = I_3 - g_m V_{gs} = I_3 - g_m V_1 \tag{10-31}$$

since $V_{gs} = V_1$. Solving Eqs. (10-29), (10-30), and (10-31) for voltage gain A_V yields

$$A_V \equiv \frac{V_2}{V_1} = \frac{-g_m + j\omega C_{gd}}{Y_L + g_d + j\omega(C_{ds} + C_{gd})} \tag{10-32}$$

At low frequencies, where interelectrode capacitances have negligible effect, Eq. (10-32) becomes

$$A_V = \frac{-g_m}{Y_L + g_d} = \frac{-g_m Z_L}{1 + g_d Z_L} = -g_m Z'_L \tag{10-33}$$

where $Z'_L \equiv r_d \,||\, Z_L$. From Eq. (10-33), the following conclusions may be reached regarding voltage gain A_V of a common-source field-effect transistor amplifier stage:

1. For $Z_L \ll r_d$, A_V is proportional to Z_L.
2. For $Z_L \gg r_d$, $A_V \cong g_m r_d = \mu$, indicating A_V upper limits of the order of several hundred for typical field-effect transistors.
3. For resistive Z_L, the input and output voltages are of opposite phase when considered with respect to the common terminal.

Input admittance. From Fig. 10-27, we see that input admittance Y_i is given by

$$Y_i \equiv \frac{I_1}{V_1} = \frac{V_1(g_G + j\omega C_{gs}) + I_3}{V_1} \tag{10-34}$$

where $g_G = 1/R_G$. From Eq. (10-30), we obtain

$$I_3 = j\omega C_{gd}(V_1 - A_V V_1) \tag{10-35}$$

since $V_2 = A_V V_1$, and substituting in Eq. (10-34) for I_3 yields

$$Y_i = g_G + j\omega[C_{gs} + C_{gd}(1 - A_V)] \tag{10-36}$$

Since A_V may be a large negative number [see Eq. (10-33)], we see from Eq. (10-36) that gate-to-drain capacitance C_{gd} can have a major effect on common-source FET input admittance.

Input capacitance. Consider a field-effect transistor stage having a resistive load R_L and operating at frequencies low enough that interelectrode capacitances have negligible effect on voltage gain A_V. From Eq. (10-33),

$$A_V = -g_m R'_L \tag{10-37}$$

where $R'_L = r_d || R_L$; combining Eqs. (10-36) and (10-37) yields

$$Y_i = g_G + j\omega[C_{gs} + C_{gd}(1 + g_m R'_L)] \tag{10-38}$$

Equation (10-38) indicates that the input admittance of the stage is the conductance of R_g in shunt with input capacitance C_i, where

$$C_i = C_{gs} + C_{gd}(1 + g_m R'_L) \tag{10-39}$$

This increase in input capacitance C_i above the gate-to-source capacitance C_{gs} is known as the *Miller effect.*

Output impedance. The output impedance Z_o of the common-source stage of Fig. 10-26(a) is the impedance seen "looking into" the output terminals with input voltage V_1 reduced to zero. From the equivalent circuit of Fig. 10-27, Z_o is seen to be $r_d || R_D$ at low frequencies. Thus R_o, the low-frequency value of Z_o, is given by

$$R_o = \frac{r_d R_D}{r_d + R_D} \tag{10-40}$$

10-3.2 The Common-Drain Amplifier or Source Follower

The source follower is analogous to the BJT emitter follower and, like the emitter follower, is useful for impedance transformation, being characterized by high input impedance and low output impedance. An a-c equivalent circuit for the source follower of Fig. 10-26(b) is shown by Fig. 10-28,

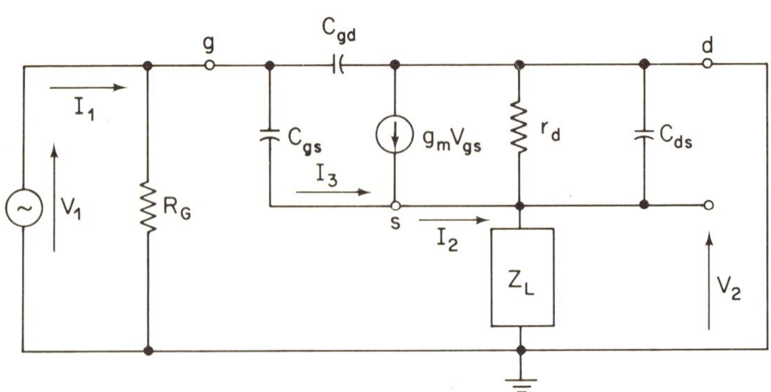

Fig. 10-28. Small-signal equivalent circuit of the source follower of Fig. 10-26b.

in which Z_L is the parallel combination of source resistor R_S and any load connected to the output terminals.

Voltage gain. From Fig. 10-28, we see that output voltage V_2 is given by

$$V_2 = \frac{g_m V_{gs} + I_3}{Y_L + g_d + j\omega C_{ds}} \tag{10-41}$$

where

$$I_3 = j\omega C_{gs}(V_1 - V_2) \tag{10-42}$$

and

$$V_{gs} = V_1 - V_2 \tag{10-43}$$

From these expressions, we obtain

$$V_2 = \frac{(g_m + j\omega C_{gs})(V_1 - V_2)}{Y_L + g_d + j\omega C_{ds}} \tag{10-44}$$

from which voltage gain A_V is

$$A_V \equiv \frac{V_2}{V_1} = \frac{1}{1 + \dfrac{Y_L + g_d + j\omega C_{ds}}{g_m + j\omega C_{gs}}} \tag{10-45}$$

From Eq. (10-45), we see that the voltage gain of the source follower is less than unity. For low frequencies, Eq. (10-45) becomes

$$A_V = \frac{g_m}{g_m + Y_L + g_d} \tag{10-46}$$

indicating that the maximum possible source-follower gain is

$$A_{V,\max} = \frac{g_m}{g_m + g_d} \tag{10-47}$$

for a very large Z_L. The voltage gain for these conditions is nearly unity, and there is no phase difference between the input and output voltages. That is, the potential of the source "follows" that of the gate.

Input admittance. From Fig. 10-28, we see that input admittance Y_i is given by

$$Y_i \equiv \frac{I_1}{V_1} = \frac{V_1(g_G + j\omega C_{gd}) + I_3}{V_1} \tag{10-48}$$

From Eq. (10-42), we obtain

$$I_3 = j\omega C_{gs}(V_1 - A_V V_1) \tag{10-49}$$

which, substituted in Eq. (10-48), results in

$$Y_i = g_G + j\omega[C_{gd} + C_{gs}(1 - A_V)] \tag{10-50}$$

But for capacitances C_{gs} and C_{gd} being interchanged, this expression is identical to Eq. (10-36) for the common-source stage. However, the two configurations have quite different input admittances, owing to their different voltage gains A_V. The source follower is seen to have much the lower input admittance, since its typical A_V values are slightly less than unity. For low frequencies, source follower input capacitance is seen to be

$$C_i = C_{gd} + C_{gs}(1 - A_V) \qquad (10\text{-}51)$$

Output impedance. "Looking into" the output terminals of the source-follower stage of Fig. 10-26(b) with input voltage V_1 reduced to zero, we "see" an active network, since source $g_m V_{gs}$ is not controlled exclusively by V_1. We may "measure" the output impedance by applying an a-c voltage V_2 and determining V_2/I_2, where I_2 is the current supplied by V_2. The resulting low-frequency equivalent circuit is shown by Fig. 10-29.

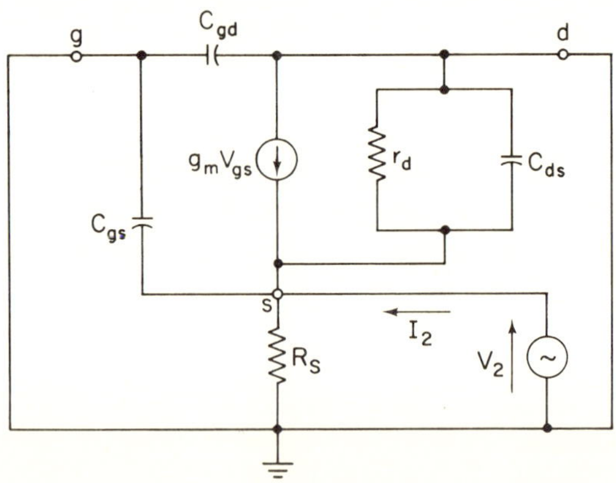

Fig. 10-29. Small-signal equivalent circuit of the source follower of Fig. 10-26(b) arranged for determination of output impedance.

From Fig. 10-29, we see that

$$I_2 = [g_S + g_d + j\omega(C_{ds} + C_{gs})]V_2 - g_m V_{gs} \qquad (10\text{-}52)$$

Since $V_{gs} = -V_2$, Eq. (10-52) becomes

$$I_2 = [g_m + g_S + g_d + j\omega(C_{ds} + C_{gs})]V_2 \qquad (10\text{-}53)$$

from which

$$Z_o \equiv \frac{V_2}{I_2} = \frac{1}{g_m + g_S + g_d + j\omega(C_{ds} + C_{gs})} \qquad (10\text{-}54)$$

For large values of R_S, $(g_S + g_d)$ is small compared to g_m and the low-frequency value of Z_o is given by

$$R_o \cong \frac{1}{g_m} \qquad (10\text{-}55)$$

Since typical g_m values may be of the order of millimhos, R_o may be of the order of hundreds of ohms, which is an order of magnitude or more below values typical of the common-source stage. Thus we see that the source follower is characterized by low output impedance.

10-3.3 The Common-Gate Amplifier

An equivalent circuit for the common-gate stage of Fig. 10-26(c) is shown by Fig. 10-30. By summing currents at node d, we obtain

$$g_m V_{gs} + (V_2 - V_1)(g_d + j\omega C_{ds}) + V_2(Y_L + j\omega C_{gd}) = 0 \qquad (10\text{-}56)$$

Recognizing that $V_{gs} = -V_1$, we obtain from Eq. (10-56)

$$A_V \equiv \frac{V_2}{V_1} = \frac{g_m + g_d + j\omega C_{ds}}{Y_L + g_d + j\omega(C_{ds} + C_{gd})} \qquad (10\text{-}57)$$

At low frequencies, Eq. (10-57) becomes

$$A_V = \frac{g_m + g_d}{Y_L + g_d} \qquad (10\text{-}58)$$

which is similar to the common-source voltage-gain expression [Eq. (10-33)]. Thus, like the common-source configuration, the common-gate amplifier is capable of high voltage gain, the two configurations having essentially the same gain, provided that $g_d \ll g_m$. Unlike the common-source configuration, no phase reversal occurs in the common-gate stage. Comparison of the equivalent circuits of Figs. 10-27 and 10-30 discloses that the two configurations have the same output impedance, provided that source V_1 has negligible impedance.

Input admittance. The input admittance of the common-gate amplifier can be determined by summing currents of node s of Fig. 10-30. Thus

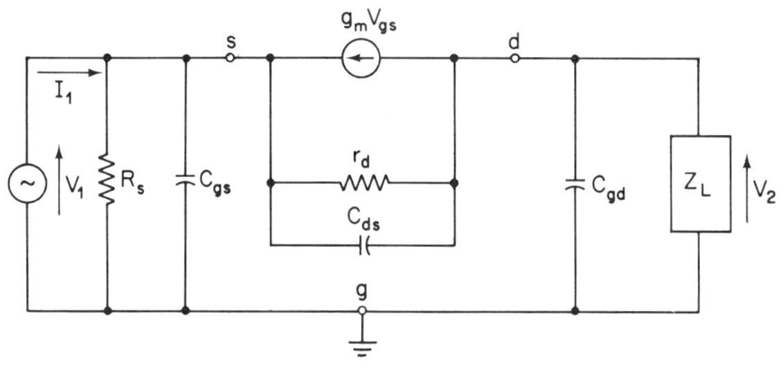

Fig. 10-30. Small-signal equivalent circuit of the common-gate amplifier stage of Fig. 10-26(c).

$$I_1 = V_1(g_S + j\omega C_{gs}) - g_m V_{gs} + (V_1 - V_2)(g_d + j\omega C_{ds}) \qquad (10\text{-}59)$$

Recognizing that $V_{gs} = -V_1$ and $V_2 = A_V V_1$, we obtain

$$Y_i \equiv \frac{I_1}{V_1} = g_S + g_m + g_d(1 - A_V) + j\omega[C_{gs} + C_{ds}(1 - A_V)] \qquad (10\text{-}60)$$

At low frequencies, the principal term of Eq. (10-60) is g_m, resulting in an input resistance of approximately $1/g_m$. Because of its low input impedance, the common-gate configuration has little application. If an amplifier having low input impedance is desired, a BJT stage is usually preferred.

10-4 BIAS ARRANGEMENTS*

The selection of the quiescent operating point for a FET amplifier stage involves considerations similar to those of Sec. 8-5 for the BJT. The primary requirement is that average values of drain voltage and current be able to accommodate desired superimposed signal components without excessive distortion. The voltage gain desired and the variation in drain characteristics among FETs of the same type may also be determining factors. We shall examine several biasing circuits.

10-4.1 Biasing Depletion-Type FETs

Figures 10-6(a) and 10-26(a) show *n*-channel depletion-type FETs in common-source amplifier stages in which gate-bias voltages are provided by fixed d-c supplies. This is known as fixed gate bias, and is an arrangement that is quite impractical for several reasons. First, it requires a second d-c supply, since the gate and drain voltages are of opposite polarity, and this is undesirable. Perhaps the most serious objection to fixed gate bias is that the gate voltage required for a particular drain current or the drain current for a particular gate bias varies greatly from unit to unit of the same type. For example, specifications for the type 2N4222 *n*-channel silicon JFET (a device having rather close tolerances) indicate that the gate voltage V_{GS} for a drain current of 0.5 mA, at a drain voltage of 15 V, ranges between -2.0 V and -6.0 V. For a drain voltage of 15 V, the zero-gate-voltage drain current I_{DSS} may be as high as 15 mA or as low as 5 mA. Thus it is unlikely that all 2N4222 devices could be accommodated in a common-source stage having a single fixed gate bias. Furthermore, many FET types have much greater I_{DSS} and V_{GS} spreads than those of the 2N4222.

* Reference 4, pp, 57–59, Reference 5, pp. 277–281; Reference 6, pp. 53–60; Reference 7, pp. 34–36; Reference 11, p. 124.

Source bias. If a resistor R_S is inserted in series with the source of a depletion-type FET, and the gate of the FET is at d-c ground potential by way of gate resistor R_G as shown in Fig. 10-31, d-c current through R_S develops a voltage drop that is applied between the gate and the source. The polarity of this

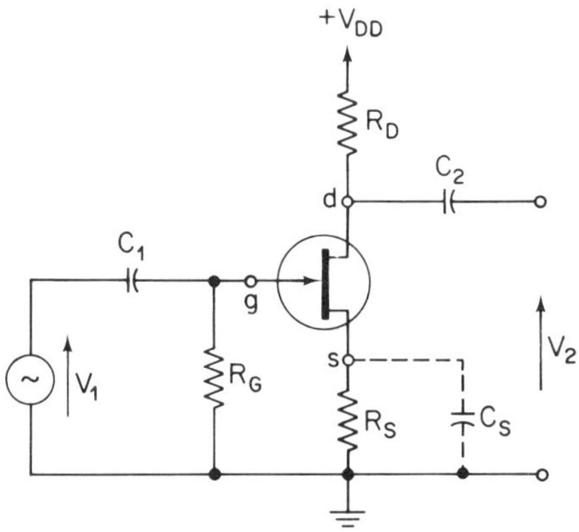

Fig. 10-31. *n*-channel depletion-type FET common-source amplifier arranged for self-bias.

voltage is seen to be such that the gate is reverse biased and the FET is said to be *self-biased.* In this circuit, as well as in the fixed-bias arrangement of Fig. 10-26(a), R_G is in shunt with the input impedance of the FET. In the interest of high input impedance, R_G should be as large as possible but not so large that the d-c voltage drop developed across it by gate-leakage current is appreciable compared to the voltage across R_S. As gate-leakage current increases with temperature, the gate-bias voltage magnitude will decrease because of voltage drop across R_G. It should be noted, however, that any change in source current will result in a change in d-c voltage across R_S in a direction that attempts to correct for the original change. In addition to reducing the effect of leakage current change, this d-c degeneration or bias stabilization reduces the effects of unit-to-unit variation in FET characteristics.

In addition to the d-c component of current flowing through R_S, an a-c component occurs when an input signal is applied. In order to investigate the effect of R_S on a-c conditions, let us consider the mid-frequency equivalent circuit of the amplifier of Fig. 10-31, as shown by Fig. 10-32. Since R_S and R_D are in series with the FET drain-source circuit, we choose the voltage-source model of the FET for our consideration. Capacitances are not shown, since at mid-frequencies coupling capacitors may be considered to be short circuits and interelectrode capacitances act as open circuits.

From Fig. 10-32, we see that the a-c component I_d of drain current is given by

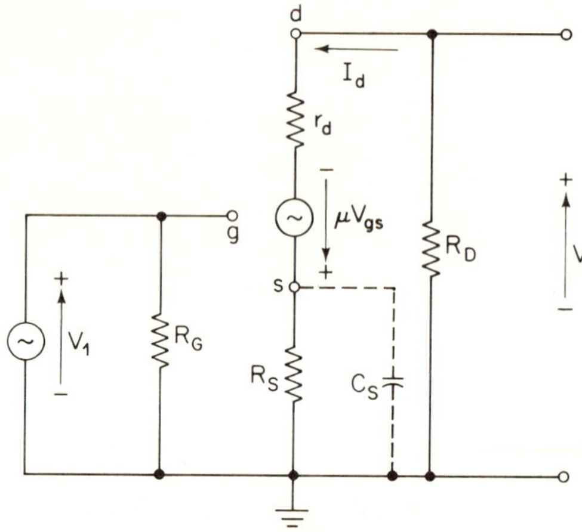

Fig. 10-32. Small-signal mid-frequency equivalent circuit of the self-bias circuit of Fig. 10-31.

$$I_d = \frac{\mu V_{gs}}{r_d + R_S + R_D} \tag{10-61}$$

where

$$V_{gs} = V_1 - I_d R_S \tag{10-62}$$

Solving these expressions for I_d, we obtain

$$I_d = \frac{\mu V_1}{r_d + R_D + (\mu + 1)R_S} \tag{10-63}$$

which indicates that the effect of R_S on a-c drain current is as though a resistance of $(\mu + 1)R_S$ were added to the circuit. This occurs because signal voltage developed across R_S reduces the gate-source signal, as indicated by Eq. (10-62). Since output voltage V_2 is given by $V_2 = -I_d R_D$, the voltage gain is seen to be

$$A_V \equiv \frac{V_2}{V_1} = \frac{-\mu R_D}{r_d + R_D + (\mu + 1)R_S} \tag{10-64}$$

The reduction in voltage gain caused by R_S is a-c degeneration similar to that caused by the emitter resistor of the bias-stabilized common-emitter BJT amplifier of Fig. 8-1. As in the common-emitter stage, this a-c degeneration may be reduced without affecting d-c degeneration (or bias stabilization) by connecting a capacitor across the troublesome resistor. With R_S shunted by a capacitor C_S of such size that almost all of I_d bypasses R_S, Eq. (10-64) becomes

$$A_V \cong \frac{-\mu R_D}{r_d + R_D + (\mu + 1)/j\omega C_S} \tag{10-65}$$

From Eq. (10-65), we see that A_V is the same as for fixed gate bias if

$$\frac{\mu + 1}{\omega C_S} \ll r_d + R_D \tag{10-66}$$

Equation (10-66) proves to be of little help in the selection of a suitable value of C_S, because μ and r_d of a high-quality FET are nebulously high in the pinch-off region. However, the ratio μ/r_d is quite finite, being equal to g_m, and can be used to determine the required value of C_S. When Eq. (10-66) is satisfied, V_{gs} of Fig. 10-32 is essentially equal to input V_1 because the voltage across C_S is very small compared to V_{gs}. Therefore, C_S should be selected so that

$$\frac{g_m V_{gs}}{\omega C_S} \ll V_{gs} \tag{10-67}$$

or

$$\frac{g_m}{\omega C_S} \ll 1 \tag{10-68}$$

at the lowest frequency of operation. These expressions are valid because I_d of Fig. 10-32 is essentially $g_m V_{gs}$ if r_d is very large relative to $(R_D + R_S)$.

The required value of R_s for the self-bias arrangement of Fig. 10-31 is simply

$$R_S = -\frac{V_{GS}}{I_D} \tag{10-69}$$

where V_{GS} and I_D are quiescent values of gate voltage and drain current at the desired operating point. This results in an appropriate R_S value for a particular FET, but there is no assurance that it will be satisfactory for all FETs of a particular type. A preferred procedure might be to choose an I_D value and determine R_S from Eq. (10-69) and the *typical* common-source static transfer characteristic (i_D versus v_{GS} for fixed v_{DS}) provided by the manufacturer. Strictly speaking, the dynamic transfer characteristic (i_D *versus* v_{GS} for fixed drain-circuit supply voltage) should be used, but there is little difference in the two transfer characteristics for operation in the pinch-off region, where i_D is largely independent of v_{DS}. When such an R_S value is used, the quiescent values of I_D and V_{DS} will deviate from the design values, depending on the characteristics of the particular FET, and the limits of this deviation can be obtained by plotting the gate-bias load line [I_D *versus* V_{GS} from Eq. (10-69)] on the coordinates of static transfer characteristics deduced from I_{DSS} and V_{GS} limits specified by the manufacturer. Drain resistor R_D and drain-circuit supply voltage V_{DD} can then be selected so that the quiescent operating point is in the pinch-off region for all FETs meeting the manufacturer's specifications. If the limits of I_D are excessive, R_D may be too small to provide adequate voltage gain, or the quiescent operating point may not accommodate the desired superimposed signal components. In either of these events, a smaller I_D value might be chosen for the typical FET, with a correspondingly larger R_S and reduced I_D limits, or some other bias arrangement may be used.

As an example, suppose a typical type 2N4222 FET is to have a quiescent I_D of 4 mA in the circuit of Fig. 10-31. From the typical transfer characteristic of Fig. 10-33, we see that a V_{GS} of -2.0 V is needed, requiring an R_S of 500 ohms. Also shown by Fig. 10-33 are the bias load line for $R_S = 500$

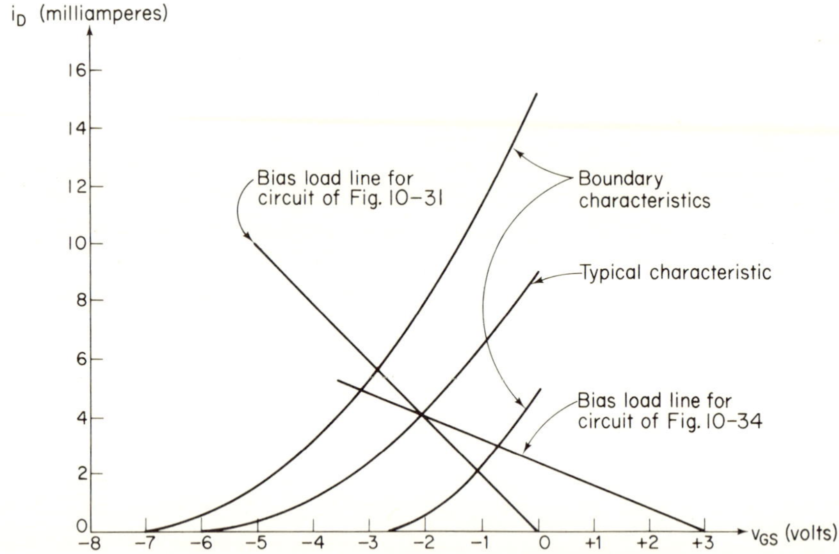

Fig. 10-33. Transfer characteristics for type 2N4222 FETs. The typical characteristic is courtesy of Motorola Semiconductor Products, Inc. and the boundary characteristics are approximated from I_{DSS} and V_{GS} specifications. The bias load lines are plots of Eqs. (10-69) and (10-70).

ohms and approximate transfer characteristics that are obtained from I_{DSS} limits of 5 mA and 15 mA and V_{GS} limits of -2.0 V and -6.0 V for an I_D of 0.5 mA, the latter being obtained by assuming transfer characteristics of all 2N4222 units to have the same general shape. From intersections of the bias load line with the boundary characteristics, we see that I_D may range between limits of approximately 2 mA and 6 mA for 2N4222 units meeting manufacturer's specifications.

Although limits of I_{DSS} and V_{GS} are usually specified for depletion-type FETs, typical static transfer characteristics may not be provided by the manufacturer. In fact, FET types having specification sheets completely devoid of characteristic curves are not unusual. For these devices, approximate transfer-characteristic boundaries may be obtained through use of the square-law relationship

$$I_{DS} = I_{DSS}\left(1 - \frac{V_{GS}}{V_P}\right)^2 \tag{10-23}$$

where V_P is the value of V_{GS} for drain-current pinch-off. Source-bias resistor R_S may then be determined from a bias load line drawn so that the limits of I_D are adequately restricted.

Combining fixed bias and self bias. As already suggested, the changes in I_D with changes in FET characteristics can be reduced in the self-bias arrangement by using a larger source resistance R_S, resulting in a more nearly horizontal bias load line. Increasing R_S decreases I_D of a particular FET; if the decrease is undesired, it can be prevented by adding positive fixed bias to the gate to compensate for the increased source bias. This combination of fixed bias and source bias is shown by Fig. 10-34(a). For this arrangement, the bias load line is a plot of the equation

$$V_{GS} = V_{GG} - I_D R_S \tag{10-70}$$

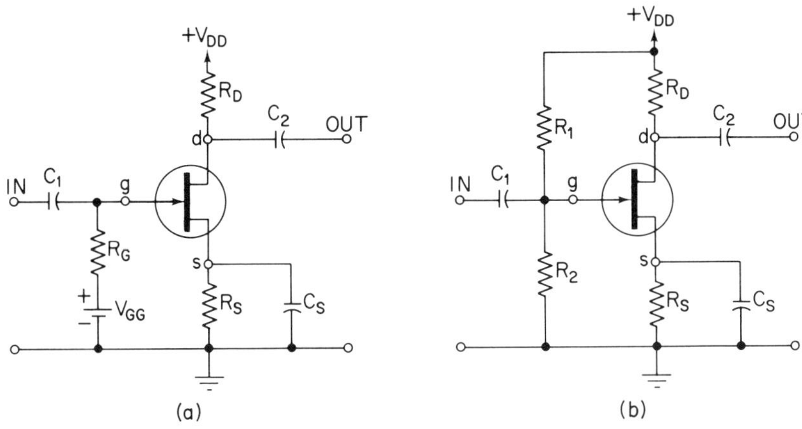

Fig. 10-34. (a) Biasing an *n*-channel depletion-type FET with fixed bias V_{GG} in addition to the source bias resulting from R_S. (b) A single power-supply arrangement equivalent to that of (a).

A load line for $V_{GG} = 3$ V and $R_S = 1250\ \Omega$ plotted on the transfer characteristics of Fig. 10-33 shows that I_D of the typical 2N4222 is the same as for $V_{GG} = 0$ V and $R_S = 500\ \Omega$, but the maximum deviation of I_D is approximately half that for the self-bias arrangement. Clearly, the larger the values of V_{GG} and R_S, the better the bias stabilization, but stabilization is improved at the expense of efficiency.

A circuit that can satisfy Eq. (10-70) without requiring a second power supply is shown in Fig. 10-34(b). For this circuit,

$$V_{GG} = \frac{R_2 V_{DD}}{R_1 + R_2} \tag{10-71}$$

if gate-leakage current is negligible. The bias stabilization arrangement of Fig. 10-34(b) is seen to be identical to that of Fig. 8-3 for a common-emitter BJT stage.

Bias stabilization can be further improved by arranging for V_{GG} to vary with FET characteristics as in the circuit of Fig. 10-35(a). In this arrangement, a change in drain-to-ground voltage resulting from an I_D change causes V_{GG} to vary in a direction that attempts to correct for the original change. This circuit is seen to be identical to the BJT bias-stabilization arrangement of Fig. 8-9, and, like the BJT circuit, it has low input impedance because of a-c degeneration caused by R_1. That is, the input impedance of the FET is shunted by a resistance of $R_1/(1 - A_V)$ in parallel with R_2, owing to the Miller effect (Sec. 10-3.1). The input impedance of the circuit can be increased by connecting a resistor in series with the bias network, as shown by Fig. 10-35(b). Typically, resistor R_3 can be of the order of megohms, depending on the gate-leakage current of the FET.

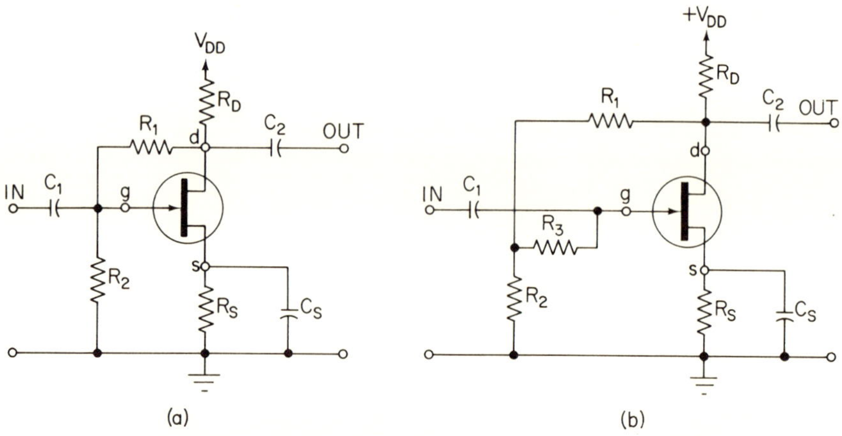

Fig. 10-35. (a) Modification of the circuits of Fig. 10-34 to improve bias stabilization by providing a variable gate-bias voltage V_{GG}. (b) Use of a resistor in series with the bias network to increase the resistance shunting the input impedance of the FET.

Source-follower biasing. If the source-follower circuit of Fig. 10-26(b) is to have a voltage gain approaching unity, source resistor R_S proves to be much too large for proper biasing of the FET. When the source follower is preceded by a common-source stage, as in Fig. 10-36(a), the stage may be designed so that its drain-to-ground voltage counteracts most of the d-c

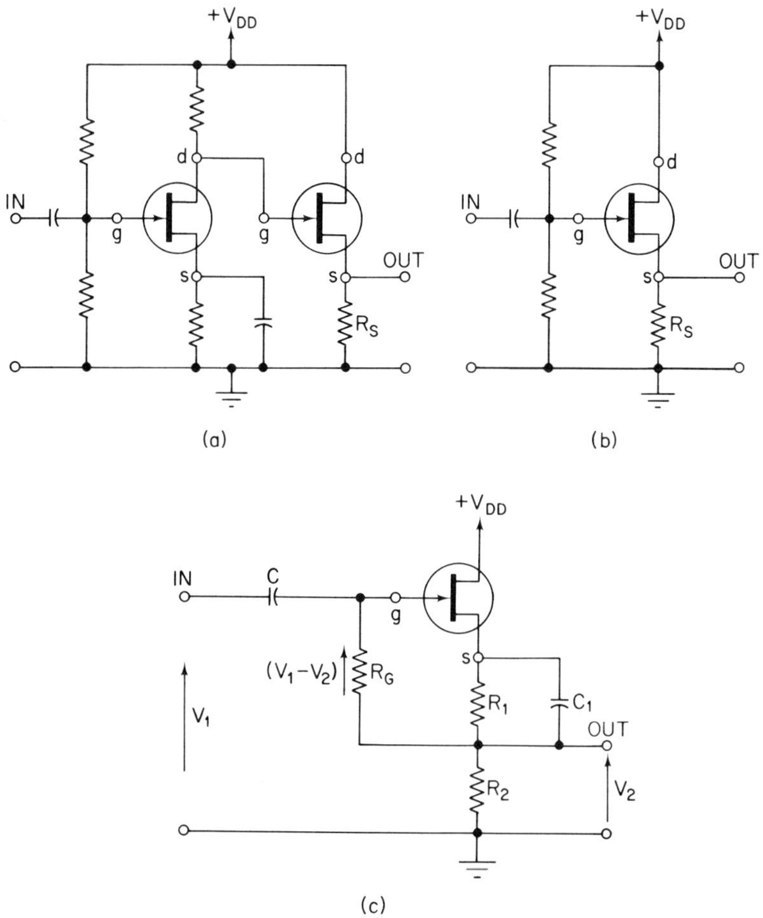

Fig. 10-36. Bias arrangements for *n*-channel depletion-type FET source followers.

voltage across R_S, resulting in proper source-follower bias. If no stage precedes the source follower, the counteracting voltage may be obtained from d-c supply V_{DD}, as in Fig. 10-36(b). Figure 10-36(c) shows a biasing arrangement that can be used for a particular FET or for FET types having V_{DSS} and V_{GS} limits suitable for self-bias. In this arrangement, gate bias is provided by the d-c voltage drop across R_1, which is bypassed by C_1 to eliminate a-c degeneration.

Let us now see how characteristics of the circuit of Fig. 10-36(c) compare with those of the basic source follower of Fig. 10-26(b). Using the voltage-source model of the FET, and considering FET capacitances to be open circuits and coupling and bypass capacitors to be short circuits, we may show that voltage gain A_V is

$$A_V = \frac{V_2}{V_1} = \frac{(\mu R_G + r_d)R_2}{(\mu + 1)R_G R_2 + r_d(R_G + R_2)} \tag{10-72}$$

If μR_G is large compared to r_d and R_G is large compared to R_2, as is usually true in the practical circuit, Eq. (10-72) becomes

$$\frac{V_2}{V_1} = \frac{\mu R_2}{(\mu + 1)R_2 + r_d} \tag{10-73}$$

By substituting $g_m r_d$ for μ in the expression, it is found to be in agreement with Eq. (10-46) for the voltage gain of the basic source follower.

With ideal source bypassing and *negligible source impedance*, the output impedance of the circuit of Fig. 10-36(c) is seen to be R_G in shunt with the output impedance of the basic source follower. However, unlike the basic arrangement, source impedance has a degrading effect on output impedance in that output impedance increases with increased source impedance.

The circuit of Fig. 10-36(c) proves superior to the basic source follower, when input impedances are compared. Since the voltage across R_G is $(V_1 - V_2)$ or $V_1(1 - A_V)$, the effective resistance shunting the input impedance of the FET is $R_G/(1 - A_V)$. For A_V values approaching unity, this resistance is seen to be quite large compared to R_G, the resistance shunting the FET in the basic circuit.

10-4.2 Biasing Enhancement-Type FETs

Unlike depletion-type FETs, the enhancement-type MOSFET requires forward gate bias for amplifier operation, as pointed out in Sec. 10-2. Since gate and drain bias potentials are of the same polarity, both may be provided by a single d-c power supply, as in the fixed-bias arrangement of Fig. 10-37(a). Like depletion-type-device characteristics, the spread in characteristics of enhancement-type FETs is too large for the use of fixed gate bias unless it is tailored to a particular FET. For example, the gate-source threshold voltage of a type 2N4351 *n*-channel enhancement-type silicon MOSFET may be as high as 5 V or as low as 1 V for a drain current of 10 μA at a drain-source voltage of 10 V.

Self-bias by use of a source resistor R_S cannot be used to establish an operating point for the enhancement-type FET, because voltage drop across R_S tends to reverse-bias the gate, but it can be combined with fixed bias to provide bias stabilization, as shown by Fig. 10-37(b). Bias stabilization is also present in the circuit of Fig. 10-37(c), in which $V_{GS} = V_{DS}$, since any change in drain current results in a gate-voltage change that attempts to correct for the original change. If it is desired that $V_{GS} < V_{DS}$, a voltage divider may be used as in Fig. 10-37(d). The circuits of Figs. 10-37(c) and 10-37(d), like that of Fig. 10-35(a), have reduced input impedance because

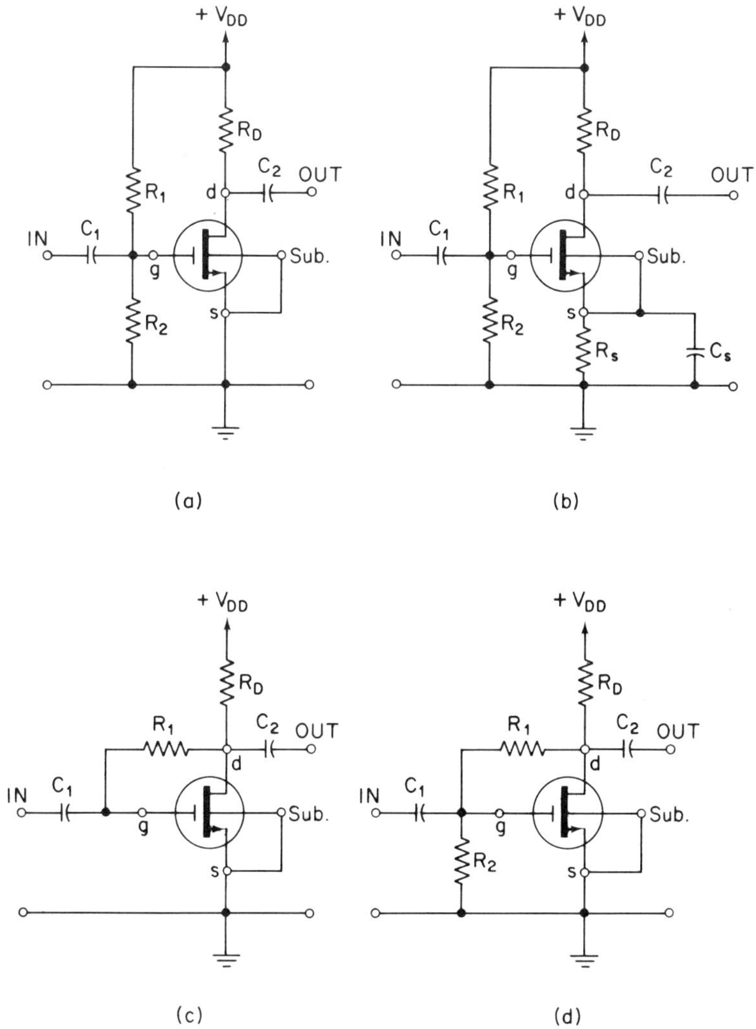

Fig. 10-37. Biasing arrangement for enhancement-type *n*-channel MOSFET amplifier stages. (a) Fixed gate bias. (b) Fixed bias-source bias combination. (c) Gate bias equal to drain bias. (d) Gate bias less than drain bias.

of a-c degeneration, the effect of R_1 being as though the input impedance of the stage were shunted by $R_1/(1 - A_V)$, owing to the Miller effect. The input impedance of each circuit of Fig. 10-37 can be increased by connecting a resistor in series with the bias network, as shown by Fig. 10-35(b). For MOSFETs, this series resistance can be very large (many megohms) because of the extremely small gate-leakage current.

10-5 THE FET AS A SWITCH*

In competing with the BJT in the area of switching applications, the FET has inherent limitations such as higher ON resistance and relatively low speed. For these reasons, it was believed that FETs had little likelihood of displacing BJTs in digital systems. However, this has been changed by the advent of enhancement-type MOSFETs. MOSFET switching circuits are less complex, relative to BJT circuits, consume less power, and occupy much less space in integrated form (see Sec. 11-11). For these reasons, MOSFETs are displacing BJTs in most lower-speed digital applications. In some applications, such as chopper circuits for converting d-c signals into a-c signals, the absence of an offset voltage is an important FET advantage; that is, unlike BJT i_C–v_{CE} characteristics, FET i_D–v_{DS} characteristic curves pass through the origin.

10-5.1 Nonregenerative Switching Circuits

Figure 10-38(a) shows an *n*-channel depletion-type FET (either a JFET or a MOSFET) in an inverter circuit. This arrangement is identical to the BJT circuit of Fig. 9-11(a), suggesting that depletion-type FETs may be substituted for BJTs in each logic circuit of Fig. 9-11. When v_A is zero,

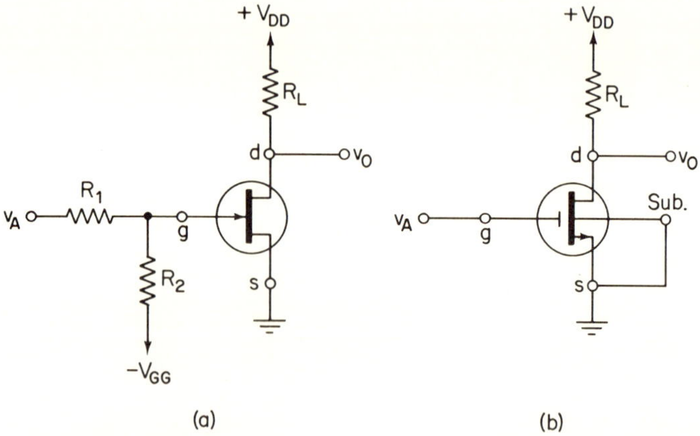

Fig. 10-38. FET inverter circuits. (a) Use of a depletion-type JFET (or MOSFET). (b) Use of an enhancement-type MOSFET.

* Reference 4, pp. 84–92; Reference 5, pp. 312–340; Reference 6, pp. 156–158, 237–241; Reference 7, pp. 93–110; Reference 10, pp. 121–140.

gate-bias voltage $-V_{GG}$ reduces the drain current to essentially zero so that $v_o = V_{DD}$. With v_A positive, drain current through R_L results in $v_o \ll V_{DD}$. The advantages of this circuit, relative to the BJT inverter, relate to the absence of a FET gate-current requirement for turn-on. The loading of source v_A is much less, since R_2 can be quite large, and a greater number of similar circuits can be driven in parallel by output v_o. That is, FET logic circuits have larger *fan-outs*. For the same number of circuits driven by v_o, R_L of the FET inverter can be much larger, owing to the smaller load-current requirement. Therefore, FET logic circuits require less power.

The advantages of an enhancement-type MOSFET inverter may be clearly seen by comparing the circuit of Fig. 10-38(b) with those of Figs. 9-11(a) and 10-38(a). The absence of OFF-mode gate bias permits direct coupling of circuit stages without the necessity for voltage-level shifting networks. Because of the superiority of the enhancement-type device in switching applications, depletion-type FETs find little use in digital systems.

Figure 10-39 shows enhancement-type-MOSFET inverter circuits that

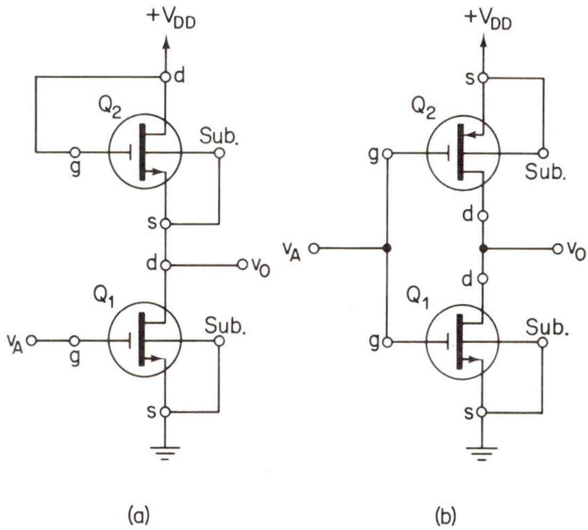

Fig. 10-39. Enhancement-type MOSFET inverter circuits using MOS drain resistors. (a) Use of MOSFETs of the same general type. (b) Use of complementary MOSFETs.

use active elements as "drain resistors." MOSFETs of the same general type are used in Fig. 10-39(a), although they should be selected so that Q_1 has much the larger saturation drain current I_{DS} for a particular gate voltage (say $V_{GS} = V_{DD}$), and complementary devices are used in Fig. 10-39(b).

Load device Q_2 of Fig. 10-39(a) is seen to have equal gate and drain potentials and to be in a conducting stage for output v_o such that

$$V_{DD} - v_o > V_{th} \tag{10-74}$$

where V_{th} is the gate threshold-voltage of Q_2. Clearly, Q_2 would be non-conducting if

$$V_{DD} - v_o < V_{th} \tag{10-75}$$

If input v_A rises to approximately $+V_{DD}$, Q_1 turns ON, decreasing v_o and increasing the conduction of Q_2. For this condition, the current is limited to I_{DS} of Q_2, which is less that I_{DS} of Q_1, so Q_1 operates in the triode region, and v_o is very small relative to V_{DD}. When v_A falls to zero, Q_1 turns OFF and v_o rises, approaching $(V_{DD} - V_{th})$ as a limit, since Q_2 must be in a conducting state if v_o exists. Load device Q_2 is called a *MOS resistor*. Since MOSFETs are considerably more costly than conventional resistors, there is little reason for using the circuit of Fig. 10-39(a) with discrete components. However, the exact opposite is true for integrated components, and the circuit is widely used in integrated form. NOR- and NAND-gate circuits using MOS resistors (in integrated form) are shown in Figs. 11-41 and 11-42.

In the circuit of Fig. 10-39(b), the MOS resistor of n channel enhancement-type MOSFET Q_1 is p-channel enhancement-type device Q_2. When v_A has a large positive value, say $+V_{DD}$, Q_1 is biased to a high conducting state, but Q_2 is cut off because its gate-source voltage is zero. Thus output voltage v_o is essentially zero when $v_A = +V_{DD}$. On the other hand, if v_A is brought to zero, Q_1 is cut off but Q_2 has high drain-to-source conductance because of its large negative gate-to-source voltage. Thus, but for the effect of leakage current of the cut-off unit, v_o is equal to $+V_{DD}$ when $v_A = 0$. A remarkable property of this circuit is that, in either its ON or OFF state, the MOSFETs conduct only leakage currents and very little power is dissipated. This small static power dissipation does not change if the circuit is used to control other MOSFET stages, since they draw no average gate current. However, larger currents do flow during switching transients, owing to circuit capacitance, producing dissipation that can become appreciable at high switching rates. It should be noted that grounding the source of Q_2, rather than that of Q_1, results in a circuit that operates with negative values of V_{DD}, v_A, and v_o. This dual-polarity property of complementary-symmetry MOS (CMOS) logic circuits proves to be quite useful.

The complementary MOSFET NAND gate. The advantages of the complementary MOSFET inverter also pertain to other CMOS logic circuits such as the three-input NAND gate of Fig. 10-40. In this circuit, parallel-connected p-channel MOSFETs serve as load devices for series-connected n-channel units, and each input is either at 0 V or $+V_{DD}$. Unless all inputs are at $+V_{DD}$, it is seen that at least one n-channel MOSFET will be OFF and at least one p-channel unit will be ON, and the output will be $+V_{DD}$. When

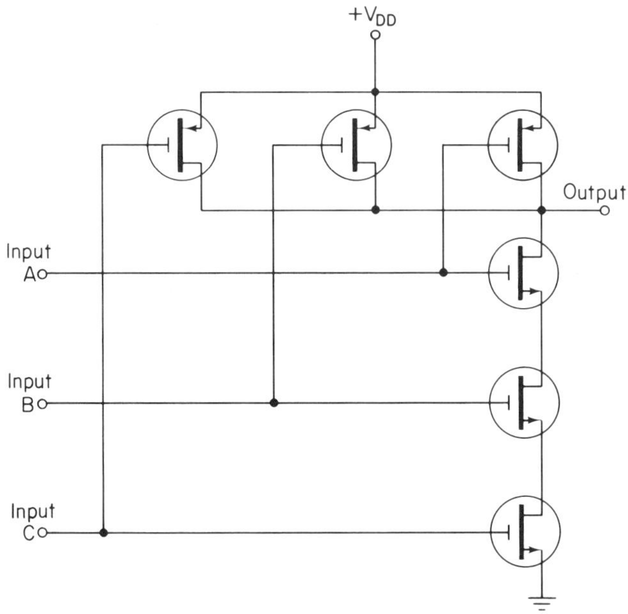

Fig. 10-40. The complementary MOSFET NAND gate. Not shown are FET substrates that are normally connected to their respective sources (in discrete circuits).

all inputs are at $+V_{DD}$, all the *n*-channel devices will be ON and all the *p*-channel units will be OFF. Only then will the output fall to 0 V. Thus the circuit performs as a NAND gate. If it is followed by a complementary inverter stage, the combination is an AND gate. Comparison of the complementary and noncomplementary NAND gate circuits of Figs. 10-40 and 11-42 shows that the complementary circuit requires two extra MOSFETs. This is a small price (in integrated circuitry) for the saving in required power.

The complementary MOSFET NOR gate. A three-input complementary NOR gate is shown in Fig. 10-41. In this circuit, the *p*-channel devices are connected in series and the *n*-channel units are in parallel. If any input is at $+V_{DD}$, one *n*-channel device will be ON and one *p*-channel unit will be OFF, and the output will be 0 V. For all inputs at 0 V, all *n*-channel devices are OFF and all *p*-channel units are ON, and the output is $+V_{DD}$. Again, a complementary inverter stage may be employed if an OR gate is desired. Like the NAND gate, the complementary NOR gate requires two more MOSFETs than in the noncomplementary circuit, as may be seen by comparing Figs. 10-41 and 11-41.

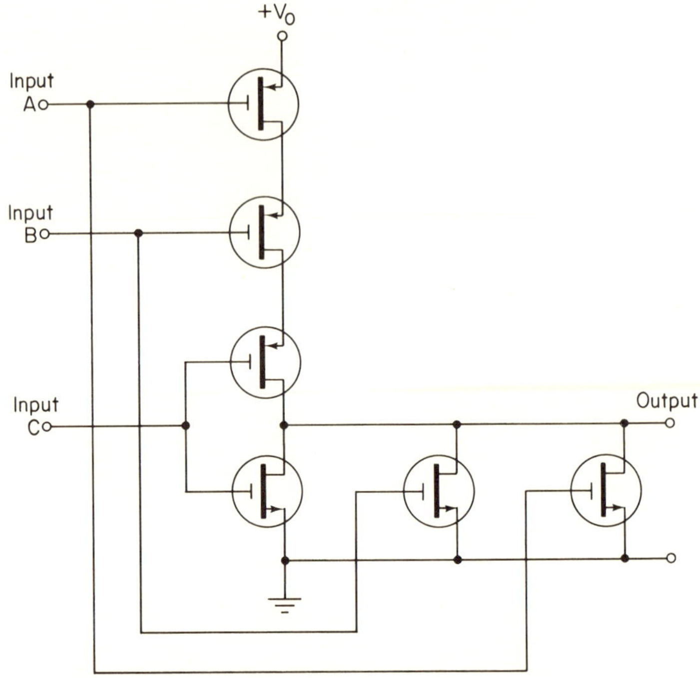

Fig. 10-41. The complementary MOSFET NOR gate. Not shown are FET substrates that are normally connected to their respective sources (in discrete circuits).

10-5.2 Regenerative Switching Circuits

Like comparable BJT circuits, FET multivibrators are formed by cascading a pair of inverter circuits and supplying the input of the first by the output of the second. Since the depletion-type MOSFET or JFET may be thought of as a BJT having an extremely high h_{FE}, depletion-type FET multivibrator circuits are much like those using BJTs, as may be seen by comparing the JFET circuits of Fig. 10-42 with the BJT circuits of Figs. 9-12, 9-19, and 9-20. The only difference in the circuits is that the JFET gates of the monostable and astable multivibrators are returned (through the timing resistors) to ground, since no gate bias is required for JFET ON-mode operation.

The bistable circuit of Fig. 10-42(a) is so designed that if either device is ON, its low drain voltage enables V_{GG} to bias the gate of the other unit to pinch-off (or beyond), resulting in a high drain voltage that overcomes V_{GG} and keeps the first device ON. The circuit may be triggered and operated

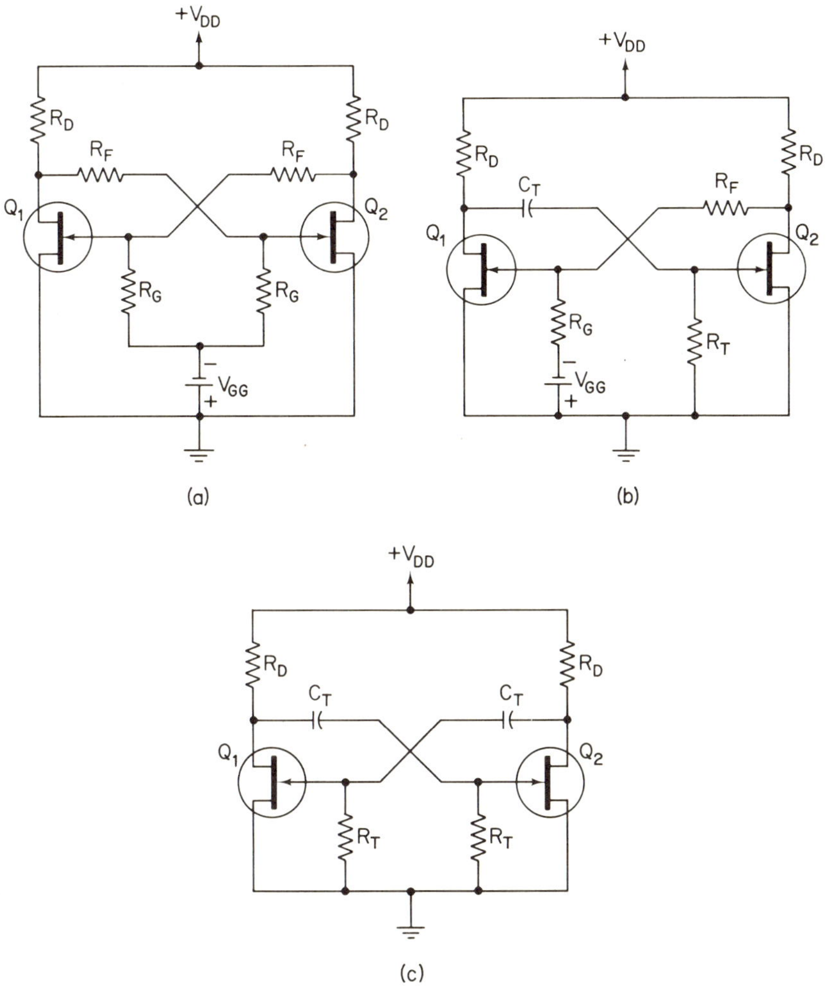

Fig. 10-42. Basic JFET multivibrator circuits. (a) Bistable circuit. (b) Monostable circuit. (c) Astable circuit.

as a binary by triggering methods like those for BJT flip-flops, as illustrated in Fig. 9-15.

In the JFET monostable circuit of Fig. 10-42(b), Q_2 is ON because it lacks gate bias, and its low drain voltage causes Q_1 to be OFF. If the gate of Q_2 is momentarily made negative enough for turn-off, Q_1 turns ON and Q_2 is held OFF by the charge of C_T. This operating state continues until C_T discharges to a voltage below the pinch-off voltage of Q_2; at this time, Q_2 begins to conduct and the circuit reverts to its stable state.

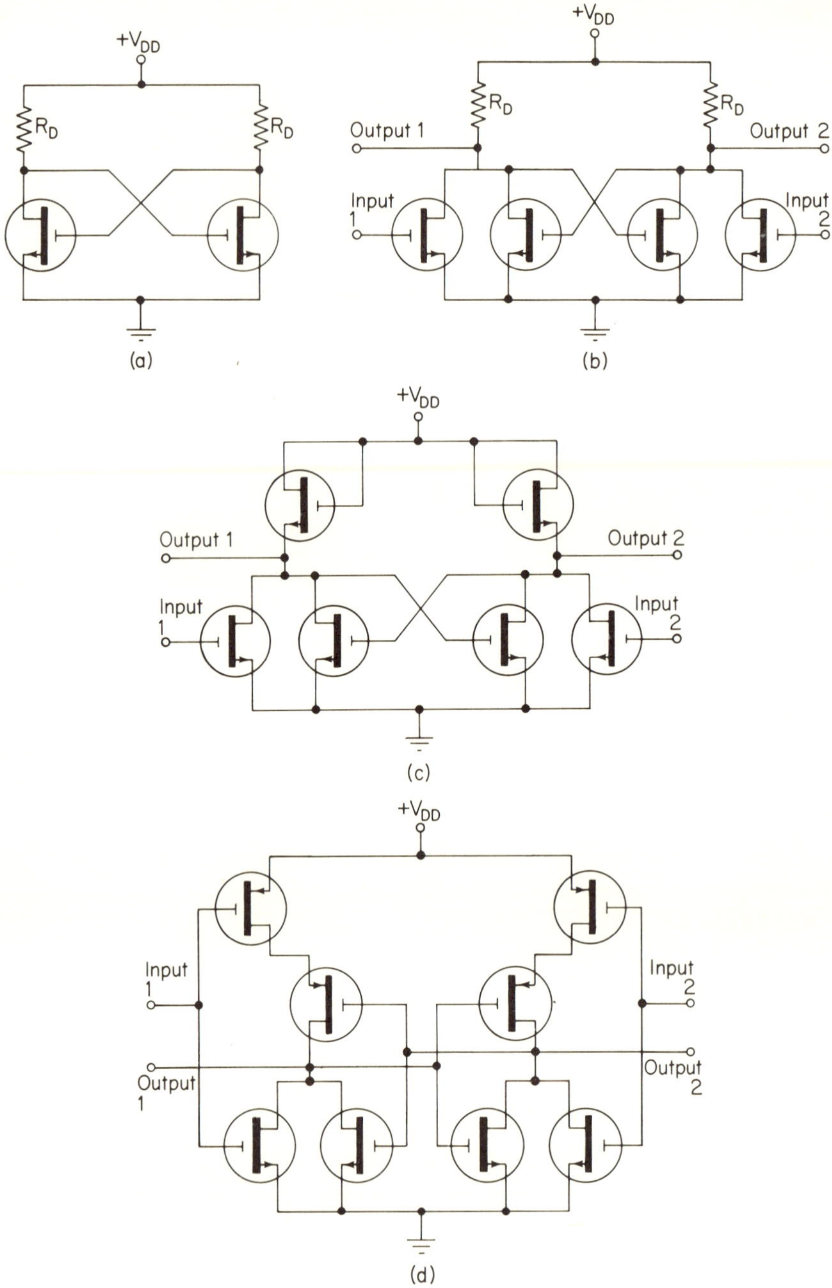

Fig. 10-43. Enhancement-type MOSFET flip-flops. (a) Basic circuit. (b) Circuit arranged for triggering. (c) Circuit using MOS resistors. (d) Complementary circuit. (MOSFET substrates are connected to their respective sources in discrete circuits and grounded in integrated circuits.)

In the JFET astable circuit of Fig. 10-42(c), both devices would be ON but for the effect of timing capacitors C_T. Instead, each device turns ON and OFF periodically, each being turned OFF by the charge of C_T when the other unit turns ON. The essential difference between JFET and BJT monostable and astable multivibrators is that the timing capacitors in the JFET circuits discharge toward zero volts, rather than charging toward $+V_{DD}$, during the timing intervals. For this reason, BJT circuit Eqs. (9-21) and (9-26) do not apply to JFET circuits.

Enhancement-type MOSFET flip-flops. Like the inverter circuit, the enhancement-type MOSFET flip-flop is less complex than its JFET counterpart, as may be seen by comparing the basic circuits of Figs. 10-42(a) and 10-43(a). When either MOSFET of Fig. 10-43(a) is OFF, its drain potential is at $+V_{DD}$. Clearly, the other MOSFET must be ON, if its gate threshold voltage is less than $+V_{DD}$, and its low drain potential insures that the first unit is OFF. The flip-flop may be triggered by momentarily short-circuiting the gate-source terminals of the ON unit. Like the DCTL BJT flip-flop of Fig. 9-16, binary operation is difficult and circuit triggering is by use of amplifier MOSFETs, as illustrated in Fig. 10-43(b). Figure 10-43(c) shows a flip-flop using MOS resistors, and a complementary MOSFET flip-flop is shown in Fig. 10-43(d). The MOS units in the complementary circuit are so placed that, in either operating state, an OFF unit blocks each current path. Like CMOS gates, the complementary flip-flop requires additional MOSFETs.

REFERENCES

1 O. M. Stuetzer, "A Crystal Amplifier with High Input Impedance," *Proc. IRE*, Vol. **38**, No. 8 (August 1950).

2 W. Shockley, "A Unipolar Field-Effect Transistor," *Proc. IRE*, Vol. **40**, No. 11 (November 1952).

3 G. Dacey and I. M. Ross, "Unipolar Field-Effect Transistor," *Proc. IRE*, Vol. **41**, No. 8 (August 1953).

4 L. J. Sevin, Jr., *Field-Effect Transistors*. New York: McGraw-Hill Book Co., Inc., 1965.

5 J. T. Wallmark and H. Johnson, *Field-Effect Transistors*. Englewood Cliffs, N. J.: Prentice-Hall, Inc., 1966.

6 C. D. Todd, *Junction Field-Effect Transistors*. New York: John Wiley and Sons, Inc., 1968.

7 W. Gosling, *Field Effect Transistor Applications*. New York: John Wiley and Sons, Inc., 1965.

[8] B. L. Gregory and F. M. Smits, "A Comparison of Radiation Tolerance of Field-Effect and Bipolar Transistors," *IEEE Trans.*, Vol. **ED 12**, No. 5 (May 1965).

[9] S. R. Hofstein and F. P. Heiman, "The Silicon Insulated-Gate Field-Effect Transistor," *Proc. IRE*, Vol. **51**, No. 9 (September 1963).

[10] P. Richman, *Characteristics and Operation of MOS Field-Effect Devices.* New York: McGraw-Hill Book Co., 1967.

[11] R. H. Crawford, *MOSFET in Circuit Design.* New York: McGraw-Hill Book Co., Inc., 1967.

[12] O. Heil, "Improvements in or Relating to Electrical Amplifiers and Other Control Arrangements and Devices," British Patent 439, 457 (September 26, 1939).

PROBLEMS

10-1. An n-channel silicon JFET has a zero-gate-voltage channel height of 5×10^{-4} cm and a channel resistivity of 10 ohm-cm. Find the approximate magnitude of the pinch-off voltage V_P.

10-2. Repeat Prob. 10-1 for a p-channel silicon JFET having the same specifications.

10-3. Repeat Prob. 10-1 for a p-channel germanium JFET having the same physical structure but a channel resistivity of 5 ohm-cm.

10-4. Using values of V_P and I_{DSS} of the JFET of Fig. 10-5, predict several points on the transfer characteristic of Fig. 10-5.

10-5. A JFET having a I_{DSS} value of 10 mA has a drain current I_{DS} of 4 mA for $V_{GS} = -2$ V, $V_{DS} = 10$ V. What value of V_{GS} is required to reduce I_{DS} to essentially zero?

10-6. A JFET has a pinch-off voltage of -5 V. What drain-source voltage might be required for drain-current saturation ($i_D = I_{DS}$) if the gate-source voltage is -4 V?

10-7. The 2N2497 JFET of Fig. 10-2(b) is used in the common-source amplifier stage of Fig. P10-7. Determine V_{DD} and R_S for a quiescent operating point of $V_{GS} = 2$ V, $V_{DS} = -15$ V. If all capacitors have negligible reactances, graphically determine the expected voltage gain V_1/V_2.

10-8. Determine the equivalent-circuit parameters of the JFET of Prob. 10-7 and use the parameters to predict the voltage gain for the conditions of Prob. 10-7.

10-9. Determine the voltage gain that might be expected in the circuit of Prob. 10-7 if source bypass capacitor C_s is removed.

10-10. For all values of R_L, including zero, the signal voltage across C_s in Fig. P10-7 should be equal to or less than $0.10 V_{gs}$ at all frequencies above 100 Hz. Determine the value of C_s.

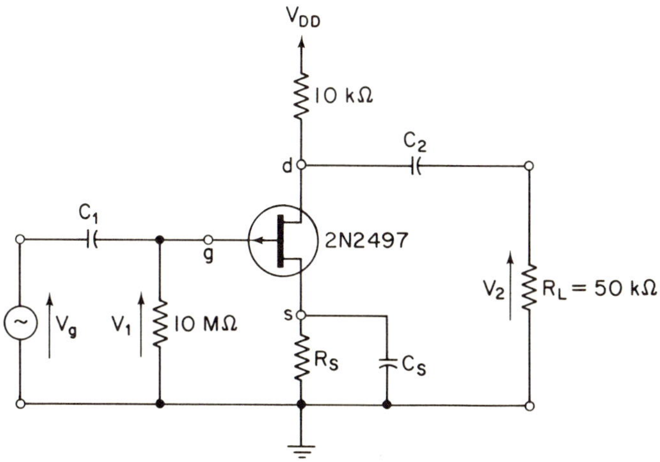

Figure P10-7

10-11. What value of coupling capacitance C_2 causes the 100-Hz voltage gain of the circuit of Fig. P10-7 to be 3 dB below the value for negligible C_2 reactance?

10-12. A JFET has a g_m of 4000 μmhos at the operating point $V_{GS} = 0$ V, $V_{DS} = 10$ V. If the pinch-off voltage magnitude is 5 V, what g_m might be expected at the operating point $V_{GS} = -3$ V, $V_{DS} = 10$ V?

10-13. The JFET in the amplifier stage of Fig. P10-13 has a g_m of 1000 μmhos, an r_d of 100 kΩ, a C_{gs} of 2 pF, a C_{gd} of 3 pF, and a C_{ds} of 2 pF. If the operating frequency is such that the magnitude and phase of voltage gain V_2/V_1 is unaffected by circuit capacitances, what is the input capacitance of the stage?

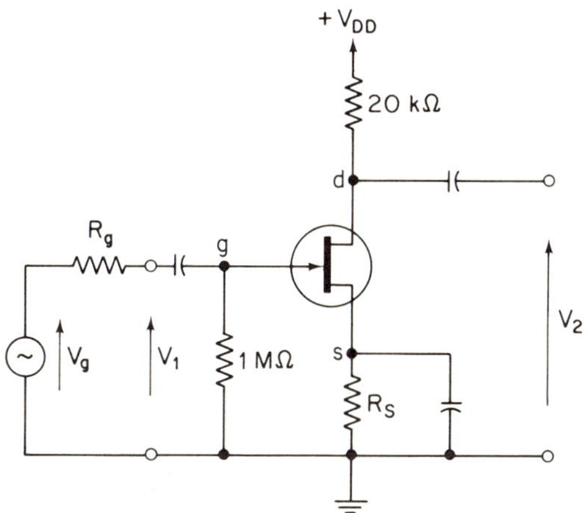

Figure P10-13

10-14. When the circuit of Fig. P10-13 is operated at a frequency of 100 kHz, V_2/V_1 is the same as in Prob. 10-13, but V_2/V_g is found to be 3 dB below the mid-frequency value. What is the value of source resistance R_g?

10-15. Determine the complex voltage ratio V_2/V_1 for the circuit of Prob. 10-13 operating at a frequency of 2.0 MHz.

10-16. The input admittance of the circuit of Prob. 10-13 may be represented as a resistance R_p in shunt with a reactance X_p. Determine R_p and X_p for the operating frequency of Prob. 10-15.

10-17. Determine the complex voltage ratio V_2/V_g of the circuit of Prob. 10-13 at 2.0 MHz if R_g is 10,000 Ω.

10-18. The JFET of Prob. 10-13 is used in the source-follower circuit of Fig. P10-18. Determine the voltage gain V_2/V_1, the input capacitance, and the output impedance for operation in the mid-frequency region.

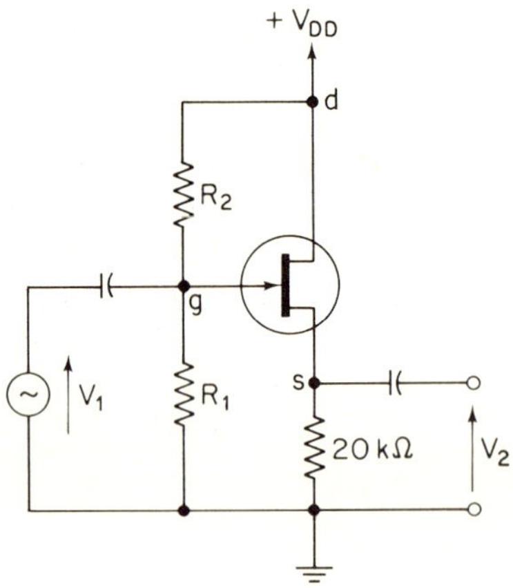

Figure P10-18

10-19. The JFET of Fig. 10-5 is used in the self-bias arrangement of Fig. P10-7. If R_S is 1 kΩ and V_{DD} is 20 V, what quiescent operating point might be expected?

10-20. The JFET of Fig. 10-5 is used in the source follower of Fig. P10-18. If $R_1 = R_2 = 2$ MΩ and $V_{DD} = 20$ V, what quiescent operating point might be expected?

10-21. Specifications of the 2N4222 *n*-channel JFET are 5 mA $< I_{DSS} <$ 15 mA and $-6\text{ V} < V_{GS} < -2\text{ V}$ for $I_{DS} = 0.5$ mA. Determine the quiescent operating-point range that might be expected in the bias arrangement of Fig. P10-21.

10-22. Repeat Prob. 10-21 for the bias arrangement of Fig. P10-22.

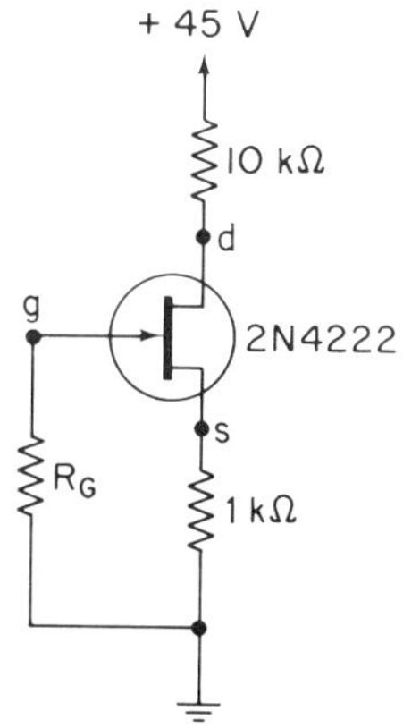

Figure P10-21

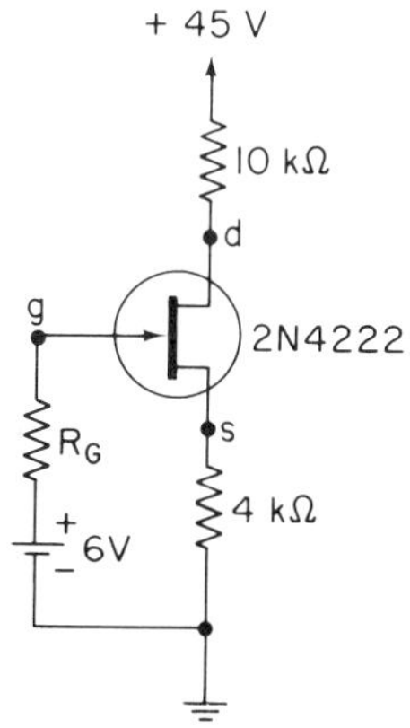

Figure P10-22

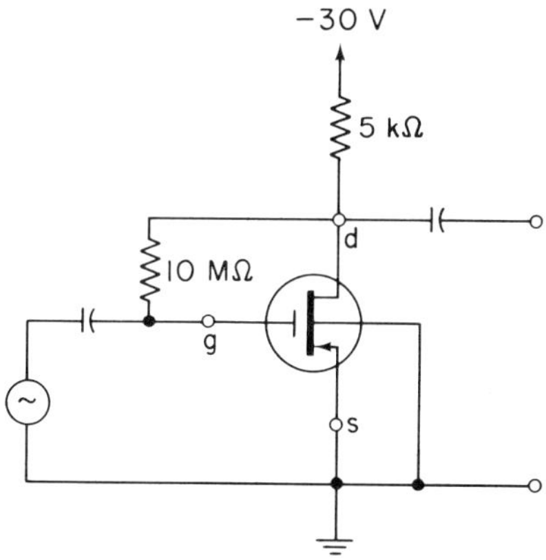

Figure P10-23

10-23. The enhancement-type MOSFET of Fig. 10-15 is used in the circuit of Fig. P10-23. What quiescent operating point is expected at a temperature of 25°C?

10-24. The MOSFET of Prob. 10-23 has a C_{gs} of 2 pF and a C_{gd} of 3 pF. If the input impedance is represented by a resistance R_p in shunt with a capacitance C_p, determine R_p and C_p at mid-frequencies.

10-25. Repeat Prob. 10-24 for R_D and $-V_{DD}$ values such that a-c voltage ratio V_{ds}/V_{gs} has a magnitude of 10.

10-26. Determine R_p and X_p of the input admittance of the circuit of Prob. 10-25 at a frequency of 1.0 MHz, if the load is such that V_{ds}/V_{gs} is $10\,\underline{/-135^\circ}$.

10-27. Derive Eq. (10-72) for the voltage gain of the source-follower circuit of Fig. 10-36(c).

11

Integrated Circuits

An *integrated circuit* (IC) is "a combination of interconnected circuit elements inseparably associated on or within a continuous *substrate*," the substrate (of an integrated circuit) being "the supporting material upon which or within which an integrated circuit is fabricated or to which an integrated circuit is attached."* Integrated circuits consist of both active and passive elements and represent the culmination of the circuit miniaturization effort that began with the development of the transistor and other semiconductor devices. For example, 100 or so interconnected logic circuits, each consisting of several transistors and resistors, can now be produced in a piece of silicon 0.1 in. on a side.[1] Prior to this development, great strides in miniaturization had resulted in a single such circuit mounted on a ceramic substrate about 0.5 in. on a side. In addition to this tremendous reduction in physical size, integrated-circuit techniques have resulted in a great reduction in circuit cost and a significant improvement in operating reliability, the latter being largely due to the need for fewer connecting wires.

The material considered in earlier chapters is basic to the understanding,

* These definitions and those that follow are IEEE Standard Definitions (Copyright 1966 by the Institute of Electronics and Electrical Engineers).

[1] S. Tribwasser, "Large-Scale Integration and the Revolution in Electronics," *Science*, Vol. **163**, No. 3866 (January 31, 1969).

design, and judicious use of integrated circuits, since the processes used to produce such circuits are adaptations of those for fabricating discrete semiconductor devices. In this chapter, we consider the fabrication of integrated circuits as well as their nature, characteristics, and uses.

11-1 TYPES OF INTEGRATED CIRCUITS

Integrated circuits are classified according to their applications and by the methods employed in their fabrication. General application classes are linear and nonlinear ICs, linear ICs being used for analog functions and nonlinear ICs for digital-type (switching) functions. General fabrication types include *monolithic* ICs, *film* ICs, *hybrid* ICs, and *multichip* ICs. In addition, various subtypes are indicated by prefixing modifiers to the general type designations. For example, there are junction-isolated and dielectric-isolated monolithic ICs, single-crystal monolithic ICs, thin-film and thick-film ICs, silicon-chip tantalum thin-film hybrid ICs, etc. Monolithic integrated circuits receive major consideration in this chapter, since they are the most important type at present and are basic to other types of ICs, as may be seen from their definitions.

The word "monolithic" is derived from the Greek words *monos*, meaning "single," and *lithos*, meaning "stone." Thus a monolithic circuit is built into a single stone, or semiconductor chip, and is defined as "an integrated circuit whose elements are formed in situ upon or within a semiconductor substrate with at least one of the elements formed within the substrate." Monolithic circuit elements include resistors, capacitors, diodes, transistors (BJTs), and FETs.

A film IC is "an integrated circuit whose elements are films formed in situ upon an insulating substrate." Film circuit elements include resistors, capacitors, and thin-film transistors, the latter being a type of insulated-gate FET considered in Sec. 11-10.4.

A hybrid IC is "an integrated circuit consisting of a combination of two or more integrated circuit types or of one integrated circuit type and discrete elements." Hybrid ICs are of particular importance because of circuit elements, such as inductors and large capacitors, that do not lend themselves to fabrication by monolithic or film techniques.

A multichip IC is, as the name implies, "an integrated circuit whose elements are formed on or within two or more semiconductor chips which are separately attached to a substrate." Multichip integrated circuits have limited use, typical applications being those that require a small number of identical circuits. They also find use in the development of prototype equipment that may later be converted to monolithic ICs.

11-2 MONOLITHIC-CIRCUIT COMPONENTS AND THEIR ISOLATION[2,3,4,5]

As indicated earlier, a monolithic IC is fabricated within a chip of semiconductor material, usually silicon[*,6], with the resistors, capacitors, diodes, transistors, etc. (as required) built into the chip. The semiconductor is either single-crystal material or single-crystal islands on a polycrystalline substrate, depending on the method for electrical isolation of the circuit components.

The essential features of the *npn* transistor used in single-crystal monolithic integrated circuits are shown in Fig. 11-1. Two significant differences between this transistor and its discrete counterpart (see Fig. 5-2) are the location of the collector contacts and the existence of a second *p* region. All contacts to monolithic components are on the top surface, in order to facili-

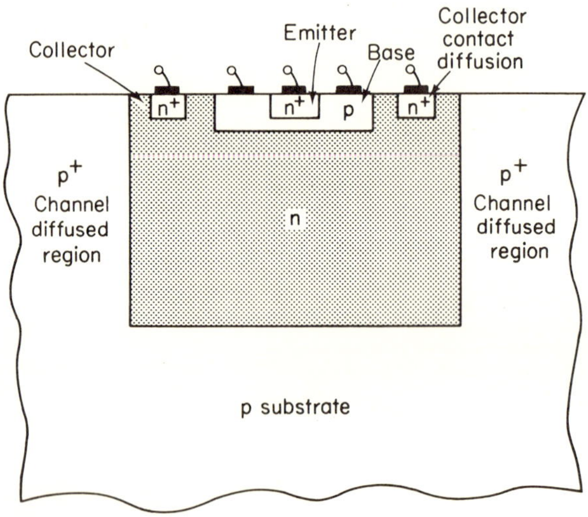

Fig. 11-1. Cross section of a transistor for single-crystal monolithic integrated circuits. Note that it contains three junctions and four semiconductor layers.

* Silicon is the only material used commercially at present, but this may soon change.

2 L. Stern, *Fundamentals of Integrated Circuits.* New York: Hayden Book Co., Inc., 1968.

3 R. M. Warner and J. N. Fordemwalt, *Integrated Circuits.* New York: McGraw Hill Book Co., Inc., 1965.

4 C. S. Meyer, D. L. Lynn, and D. J. Hamilton, *Analysis and Design of Integrated Circuits*, New York: McGraw Hill Book Co., Inc., 1968.

5 H. C. Lin, *Integrated Electronics.* San Francisco: Holden-Day Inc. 1967.

6 A. Reisman, "Germanium ICs Point the Way Towards Picosecond Computers," *Electronics*, Vol. **42** No. 5 (March 3, 1969).

tate interconnections, and the additional *p* region is the substrate which the transistor shares with all other integrated components. The *pn* junction formed by the collector region and the substrate constitutes a diode that, when reverse biased, isolates the transistor from all other components. This form of isolation is illustrated in Fig. 11-2, where the cross-sectional view

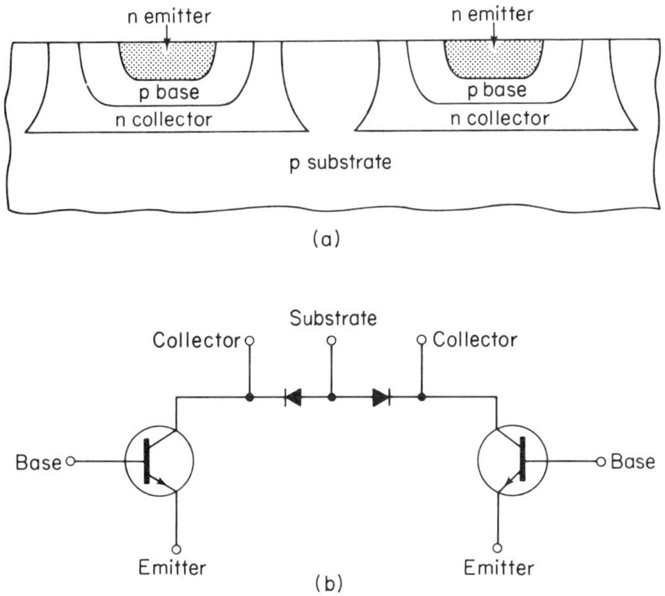

Fig. 11-2. Isolation of monolithic transistors by back-to-back diodes. (a) Cross-sectional view of a semiconductor chip containing two transistors. (b) Equivalent circuit of the arrangement of (a).

of a semiconductor chip containing two transistors is shown together with the equivalent circuit of the structure. From the equivalent circuit, we see that series-connected back-to-back diodes exist between transistor collectors and that maintaining the *p*-type substrate more negative than either collector region causes the transistors to be separated by the high d-c resistance of a reverse-biased diode. However, there is capacitance coupling between the components, owing to capacitance associated with the isolating *pn* junctions.

It might appear from the circuit of Fig. 11-2(b) that a potential difference between the collector regions always results in reverse-biasing of one of the diodes and that isolation occurs regardless of the potential of the substrate. However, this reasoning fails to recognize that isolation will be lost if both diodes become forward biased, or that a virtual short circuit may develop through *pnpn*-diode action if the substrate becomes positive with respect to a transistor emitter.* Therefore, the substrate is normally connected to the most negative point in the circuit, back-biasing both back-to-back diodes and reducing the capacitance coupling between circuit components.

The location of the collector contacts of the monolithic transistor results in higher collector bulk resistance than that of the discrete device. This resistance is somewhat reduced by the n^+ diffused contacts shown in Fig. 11-1 and may be further reduced by an n^+ layer buried in the collector region (see Fig. 11-4). The n^+ contacts are formed during the impurity diffusion that forms transistor emitter regions. After all diffusions have been completed, a layer of silicon dioxide (not shown in Figs. 11-1 and 11-2) exists over the surface of the semiconductor wafer. This layer of glass protects or passivates the integrated circuit so that it is not contaminated by its environment.

Monolithic diodes are obtained by using the semiconductor layers produced during the impurity diffusions that form transistor base and emitter regions; also, transistors may be connected to form diodes. Resistors are usually obtained by using single semiconductor layers in which current flows in a direction parallel to the *pn* junctions; also, thin films of resistive material may be deposited on top of the silicon-dioxide layer to form resistors. Capacitors are usually made by depositing a conducting material over the oxide layer, which serves as the dielectric; also, *pn*-junction capacitances may serve as capacitors.

The cross-sectional view of a single-crystal monolithic logic circuit containing typical components (a resistor, two diodes, and a transistor) is shown by Fig. 11-3(a); Fig. 11-3(b) shows the circuit schematic.[7] The resistor is a layer of *p*-type material with an *n*-region for isolation. The diodes also have an isolating *n*-region, and consist of a *p*-layer as a common anode and n^+ regions as cathodes. The *npn* transistor is seen to be similar to that of Fig. 11-1. As suggested earlier, the transistor collector region and the *n*-type isolation regions of the other components are formed at the same time. The transistor base region, the anode region of the diodes, and the resistor region are also formed simultaneously, as are the emitter region of the transistor, the collector contacts, and the cathode regions of the diodes. The metallic contacts to the components and the interconnections are also formed at the same time, as pointed out in Sec. 11-3.

Dielectric isolation. The capacative coupling between components of integrated circuits using back-to-back diode isolation may have a detrimental effect on circuit performance, particularly at high frequencies. A much smaller coupling capacitance exists in ICs using an alternate isolation scheme known as dielectric isolation, which surrounds each component by a solid

* This phenomenon is considered in Sec. 12-3.

[7] A. B. Phillips, "Monolithic Integrated Circuits," *IEEE Spectrum*, Vol. **1**, No. 6 (June 1964.)

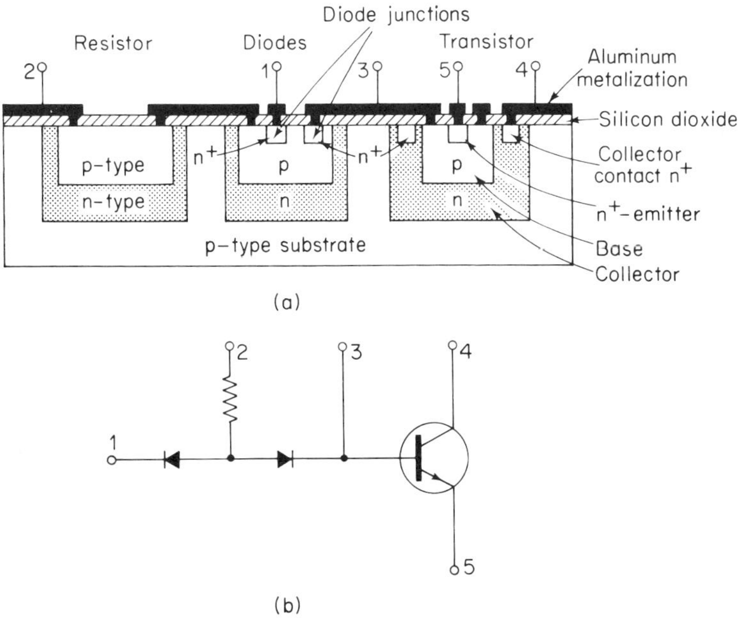

Fig. 11-3. (a) Cross-sectional view of a single-crystal monolithic circuit containing a resistor, two diodes, and a transistor. (b) Schematic diagram of the monolithic circuit of (a). (From Reference 7, p. 85.)

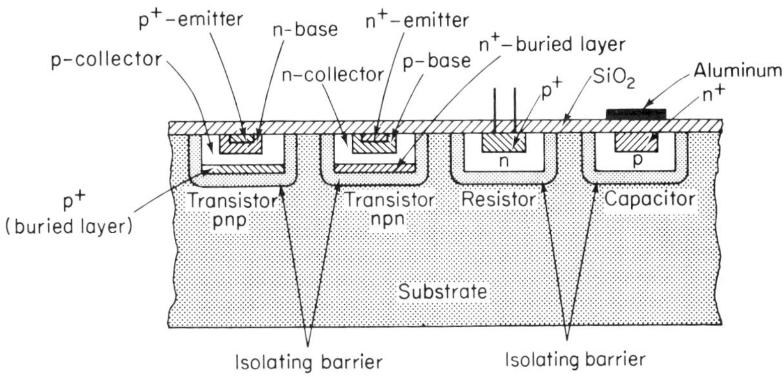

Fig. 11-4. A dielectrically isolated monolithic integrated-circuit structure. (From Reference 4, p. 27.)

dielectric layer that isolates it both physically and electrically from other components and from the common substrate. Such a circuit is also known as a *multiphase* monolithic integrated circuit and is illustrated in Fig. 11-4.*

* See Reference 3, p. 168, or Reference 4, p. 27.

The multiphase circuit has additional advantages, such as much higher breakdown voltages between circuit components and easy fabrication of *npn* and *pnp* transistors on the same substrate. The principal disadvantage is the higher cost resulting from extra processing steps that are required.

The transistors of Fig. 11-4 are seen to utilize buried high-conductivity layers in order to reduce collector bulk resistance. Also shown in Fig. 11-4 is a monolithic capacitor that uses the silicon-dioxide layer as its dielectric, the n^+ layer as the bottom plate, and an aluminum layer as the top plate.

11-3 FABRICATION OF SINGLE-CRYSTAL MONOLITHIC CIRCUITS

The basic technology for fabricating monolithic integrated circuits was developed for and is used in the production of diffused silicon diodes and transistors of the passivated planar type. In discrete device production, many similar devices are formed on a silicon wafer, which is then scribed and separated into discrete components. Relatively minor modifications of the technology permits the formation of dissimilar devices such as resistors, diodes, and transistors which, with interconnections, constitute a monolithic circuit. Many such circuits may be formed on a silicon wafer which can then be separated into individual circuits. The technology for producing passivated planar silicon devices, whether they be discrete components or monolithic circuits, consists of the proper sequence of four basic processes:

1. Photo resist and oxide masking.
2. Impurity diffusion into silicon.
3. Epitaxial growth of silicon containing impurities.
4. Metallization.

Photo Resist and Oxide Masking. Because doping impurities diffuse much less readily in silicon dioxide than in silicon, a thin layer of SiO_2 can be used on a silicon wafer to form a mask through which impurity diffusion can occur.* Such a mask is formed by coating the wafer with an SiO_2 layer and then using an etching process to remove the SiO_2 from areas where diffusion is desired. This etching process is implemented through use of a photographic technique that is begun by coating the SiO_2 layer with a light-sensitive material that becomes highly resistant to the action of a corrosive etchant, following exposure to ultraviolet radiation. Kodak Photo Resist (KPR) is the

* Although the discussion throughout this chapter refers to the use of silicon, SiO_2 masking is also employed in germanium ICs, as pointed out in Reference 6.

trade name of material so widely used that the photo-resist process is commonly referred to as the KPR process.

After being coated with photo resist, the silicon wafer is exposed to ultraviolet light through a mask that prevents the exposure of areas where impurity diffusion is to occur. A two-step etching process removes first the unexposed areas of photo resist and then the SiO_2, cutting holes in the SiO_2 layer. The exposed photo resist is then removed and the wafer is ready for the diffusion process. During diffusion, a new layer of SiO_2 can be grown on the wafer by performing the process in an oxidizing atmosphere. Thus, at the termination of each such diffusion cycle, the entire wafer is covered by SiO_2, which can then be used to form a mask for the succeeding diffusion. In an alternate approach, diffusion is performed in a reducing atmosphere that prevents oxide growth during the diffusion cycle. A layer of SiO_2 is then deposited for masking purposes. This requires an extra step, but it reduces the impurity content of the SiO_2 layer and provides better surface-condition control.

Diffusion. The formation of *pn* junctions by the vapor-diffusion process is introduced in Sec. 2-1. This is known as *constant-source* diffusion, in which the wafer is placed in a furnace at a temperature of the order of 1000°C in the presence of an atmosphere bearing the desired doping impurity. If impurity atoms in the vapor are of sufficient number, the impurity-atom density at the surface of the wafer will be determined by the solid solubility of the impurity in silicon. As impurity atoms diffuse into the silicon, more atoms form at the surface, and the surface density is constant and equal to the solid-solubility density. If diffusion were to proceed for a very long time, the silicon would become uniformly doped at this level, but for normal diffusion times the impurity concentration decreases exponentially with distance from the surface and the impurity-penetration depth depends on the diffusion time, as shown by Fig. 11-5(a).

In a type of diffusion known as *limited-source* diffusion, a fixed number of impurity atoms per unit area are deposited on the wafer surface and the wafer is heated so that impurity atoms diffuse into the silicon. In this arrangement, impurities are not replaced at the surface, as they diffuse into the wafer, and the surface concentration decreases as diffusion proceeds. Impurity profiles for two diffusion times are shown by Fig. 11-5(b), the area under the two curves being the same, since it represents the total amount of impurity being diffused. Both types of diffusion may be employed at different stages in the fabrication of a monolithic circuit.

We see from Fig. 11-5 that a characteristic of both diffusion techniques is the variation of impurity concentration with depth into the silicon. Since the crystal is normally predoped with a type of impurity opposite that being diffused, an exception being the formation of n^+ or p^+ contacts on *n*-type

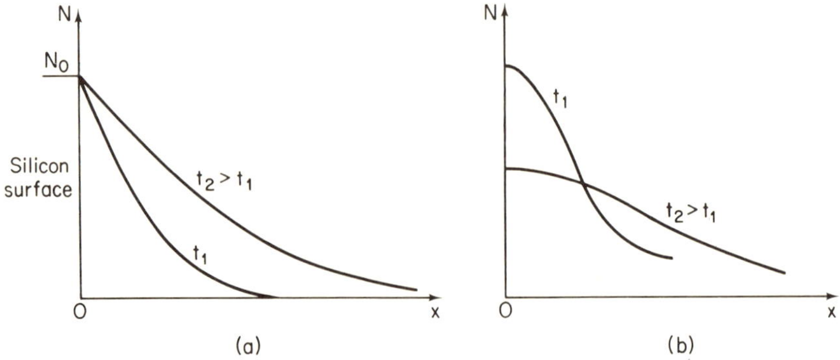

Fig. 11-5. Impurity atom concentration N as a function of distance x into a silicon wafer for two values t_1 and t_2 of diffusion time using: (a) Constant-source diffusion. (b) Limited-source diffusion.

or p-type material, the diffused wafer contains both donor and acceptor impurity atoms. Whether a region of the crystal is n-type or p-type is determined by the excess of donor or acceptor atoms. Thus, in producing a change in material type, say from n-type to p-type material, the diffusion cycle must be arranged so that the density of acceptors in the region of interest exceeds that of the donor atoms already present. This is known as *compensation*. It is clear that the number of times that a region can be compensated is limited by the solid solubility of the impurity atoms. A further limitation is the degradation of such material properties as carrier mobility and lifetime each time compensation occurs. In practice, the maximum number of times that a region can be compensated by diffusion, without serious difficulty, has been found to be about three.

Epitaxy. A process known as epitaxy permits greater flexibility with regard to impurity concentration than is provided by diffusion techniques. For example, a uniform impurity concentration can be much more readily obtained, and the problems associated with an excessive number of impurity cycles can be avoided. In this process, silicon wafers are placed in a furnace at a temperature of about 1200°C, and a vapor containing H_2 and $SiCl_4$ is passed over them. During the ensuing chemical reaction, HCl is formed and silicon atoms are deposited on the wafers. These atoms arrange themselves along the axes of the crystalline structure of the wafers, and new layers of single-crystal silicon are grown. By introducing an impurity into the vapor, impurity atoms can be made to segregate and enter into the crystal growth, forming a layer with uniform impurity concentration. If the impurity content of the vapor is varied during crystal growth, other impurity distributions can be obtained, and the net concentration can be changed from one type to the other without the use of compensation. Junctions formed by the epitaxial process are usually nonlocalized, since the process does not lend itself as readily to the formation of junctions in selected regions as does the diffusion process.

The epitaxial process can be used to grow a silicon layer on a wafer coated with silicon dioxide. Such a layer is polycrystalline in nature, because of the random crystalline orientation of the SiO_2, but it may have a useful function. For example, the substrate of the dielectrically isolated monolithic circuit of Fig. 11-4 may be polycrystalline silicon produced by an epitaxial process.

Metallization. In order to provide contacts to, and interconnections among, components fabricated on the substrate, a metal evaporating and etching technique is employed. After using the photo-resist process to remove SiO_2 from the areas that are to be contacted, the wafer is placed in a vacuum system. Aluminum is then evaporated in the system and condenses over the entire surface of the wafer. When an aluminum layer about 1 μm thick has been deposited, the wafer is removed from the system and a photo-resist process used to remove aluminum from all but the contact areas and the desired interconnection paths. Metal wires may then be bonded to appropriate places on the aluminum to facilitate making external connections.

Circuit fabrication steps. Now that the basic processes employed in making monolithic circuits have been described, we are in a position to follow the fabrication of such a circuit. The sequence in which the processes may be used is shown by the flow chart of Fig. 11-6; accompanying cross sections show the formation of a transistor, since it has the most complex cross section of components usually found in a monolithic circuit. Figure 11-6 illustrates the *epitaxial-diffused* fabrication system—a system that has largely superseded those used prior to development of the epitaxial process.

The process begins with a wafer of single-crystal p-type silicon on which an SiO_2 layer has been grown. The photo-resist process (mask 1 of Fig. 11-6) is used to select regions where "buried layers" of heavily doped n-type material are to be formed in order to reduce the collector resistance of transistors that will be fabricated subsequently.

After the buried-layer diffusion, all oxide is removed, and an n-type layer is grown epitaxially to serve as the collector region of each transistor and as the region in which other components will be formed. An oxide layer is then grown on the surface, and the photo-resist process (mask 2) is used to select regions for *isolation diffusion*. During this cycle, acceptor impurities are diffused through the n-type layer to join the p-type substrate, converting regions of the layer to p-type material. This separates the layer into islands of n-type material "floating" on the p-type substrate. Each island will sub-

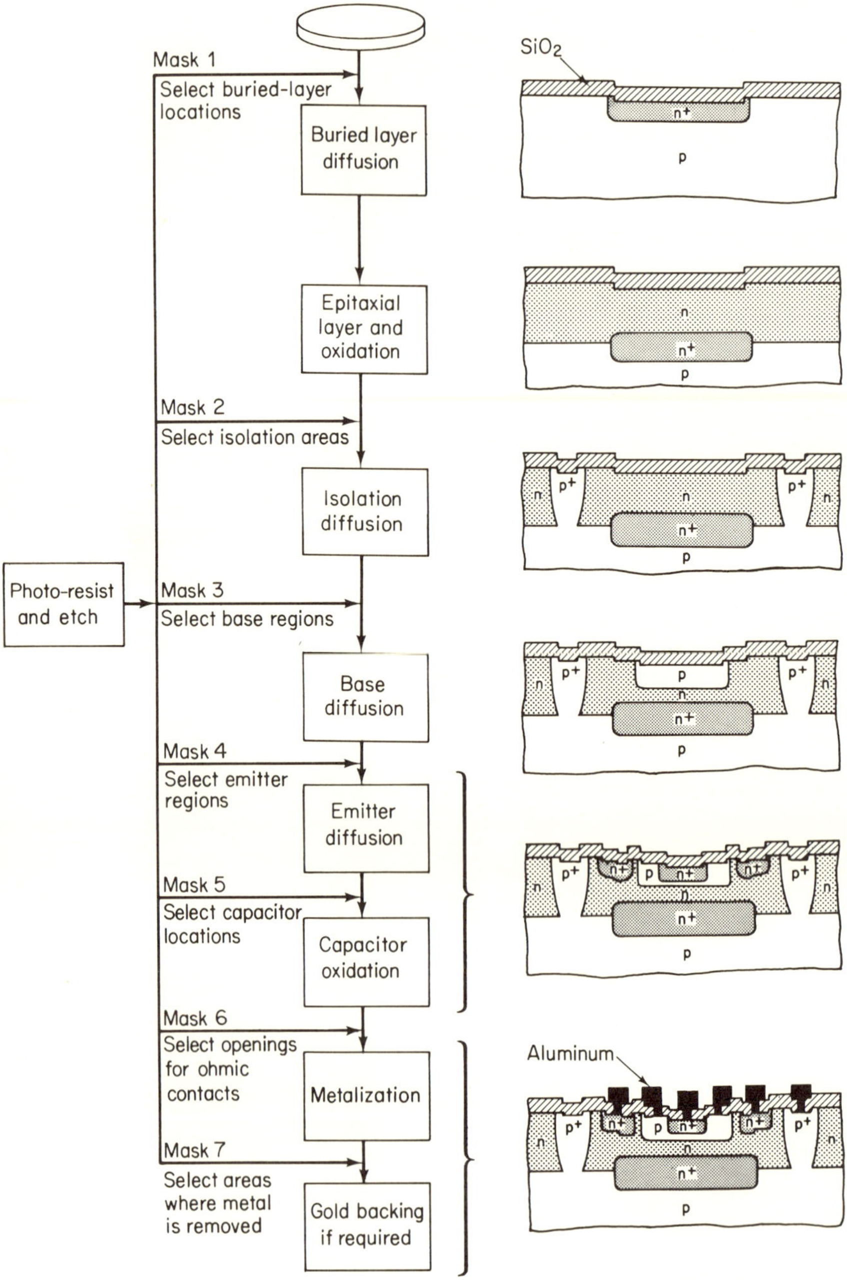

Fig. 11-6. Flow chart showing the sequence of processes used in the fabrication of a single-crystal monolithic integrated circuit. (From Reference 4, p. 21.)

sequently contain a circuit component electrically isolated by back-to-back *pn* junctions between adjacent *n* regions. During isolation diffusion, more SiO_2 is grown for future mask formation.

Next, the photo-resist process (mask 3) is used to select areas where transistor base regions and resistors are to be formed by the diffusion of acceptor impurities. Ater this diffusion, the photo-resist process (mask 4) is used to remove SiO_2 from areas where transistor emitters, transistor collector contacts, and the bottom plates of capacitors are to be formed. These heavily doped n^+ regions are then formed by donor-impurity diffusion.

Capacitor fabrication is continued by using the photo-resist process (mask 5) to remove SiO_2 from the n^+ regions that are to serve as capacitor plates, and then regrowing an SiO_2 layer of the proper thickness to serve as the dielectric of the capacitors.

The photo-resist process (mask 6) is next used to remove SiO_2 from areas that are to be contacted by aluminum metallization. Aluminum is then evaporated over the entire surface of the wafer, making contact with the silicon in the exposed areas. In addition to contacts and interconnections, the upper plate of each capacitor is formed by this process. A photo-metal-etch-resist process (mask 7) is then used to remove undesired metal from the wafer.

The final step shown by Fig. 11-6 is the application of gold backing, if it is required, in order to make contact between the header and the substrate. In this step, gold is evaporated onto the back of the wafer and then alloyed into the silicon. After application of gold backing, each integrated circuit is tested and defective ICs are marked. Since the cost of the packaged IC may be closer to the cost of the package than that of the chip, testing precedes packaging.

Following the testing operation, the wafer is scribed and broken into individual chips. After defective ICs are discarded, individual chips are bonded to headers and connected to package leads in a variety of ways. These connections have been made by wire-bonding processes, 1-mil-diameter wire being used to link package leads with metallized pads on the chip, but other techniques are coming into use.[8]

During oxide growth, as in the steps of Fig. 11-6, the silicon for the SiO_2 is provided by the silicon wafer. Therefore, each time a diffusion process is performed in an oxidizing atmosphere a small amount of silicon is removed from the diffused region. This is illustrated, to an exaggerated scale, in the cross sections of Fig. 11-6 by the steps in the silicon.

The flow chart of Fig. 11-6 fails to indicate the time required for the series of processes. A time chart, showing the typical time required for each

[8] L. Curran, "In Search of a Lasting Bond," *Electronics*, Vol. 41, No. 24 (November 25, 1968).

step in volume production of ICs, is shown in Fig. 11-7.[9] It is of interest to note that the entire process takes place over a period of months.

It should be recognized that the fabrication steps required for a complicated circuit are just those shown in Figs. 11-6 and 11-7 for an integrated transistor, since all components are made simultaneously. Thus the flow

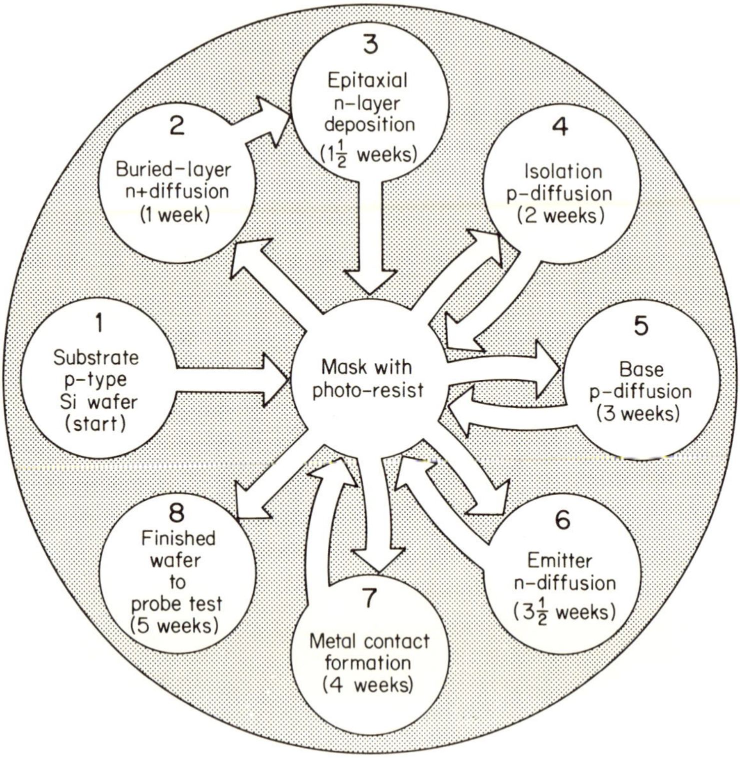

Fig. 11-7. Time chart for the flow chart of Fig. 11-6. (From Reference 9.)

chart of Fig. 11-6 also applies to circuits such as the flip-flop shown schematically in Fig. 11-8. A photomicrograph of the monolithic circuit and an element-location diagram are shown in Fig. 11-9. The 20 transistors, 16 resistors, 2 capacitors, and 2 diodes of the circuit are contained by a silicon chip 0.06 in. by 0.08 in. (60 by 80 mils) in cross section.* In comparison, the dimensions of a typical discrete transitor chip are 20 mils by 20 mils, this being the smallest chip that can be conveniently handled in device fabrication.

[9] T. A. Longo, "IC Processing: Healthy, Still Growing," *Electronics*, Vol. 41, No. 22 (Oct. 28, 1968).

* See Reference 4, p. 24.

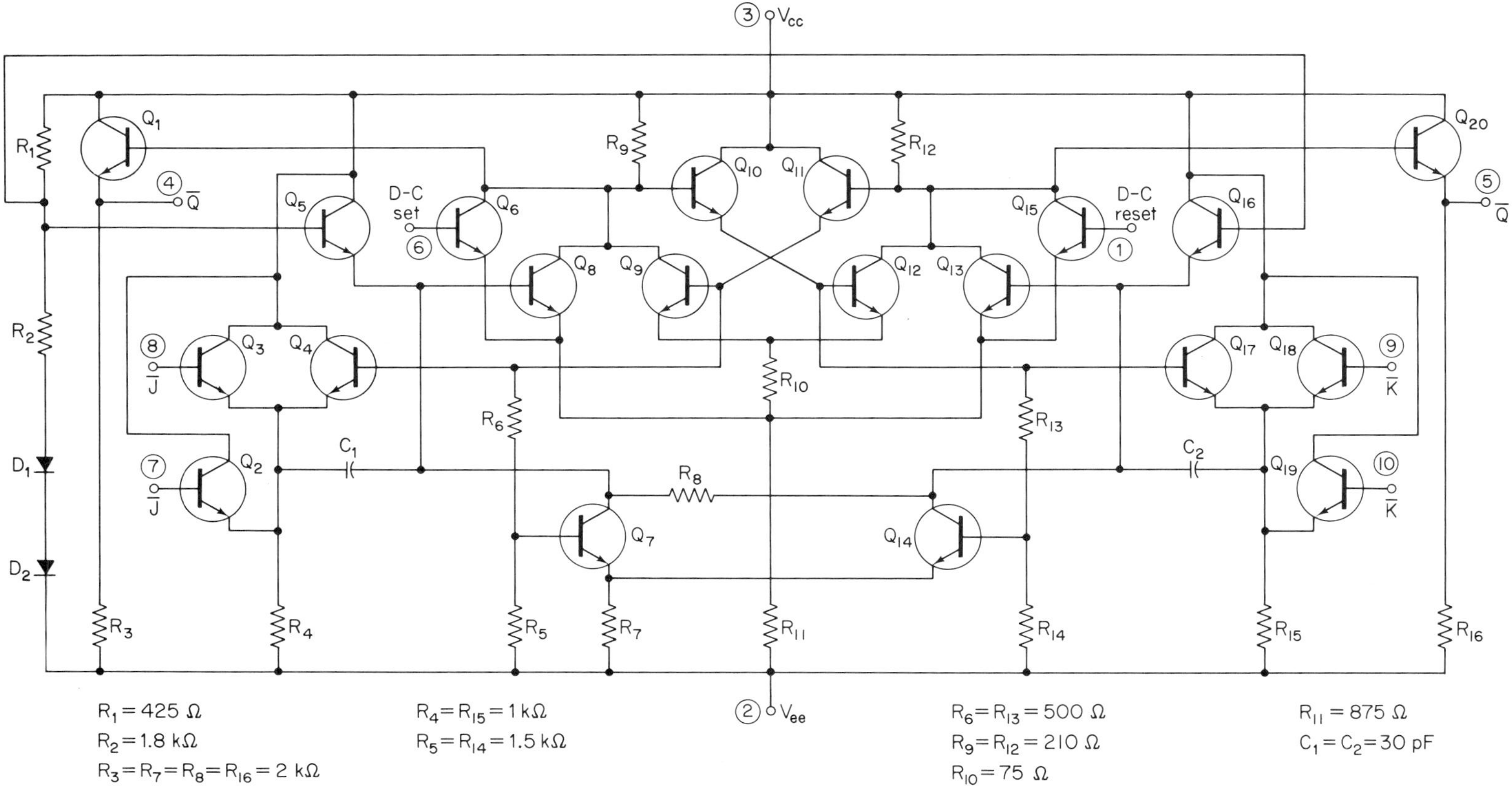

Fig. 11-8. Circuit diagram of the flip-flop of Fig. 11-9. (From Reference 4, p. 22.)

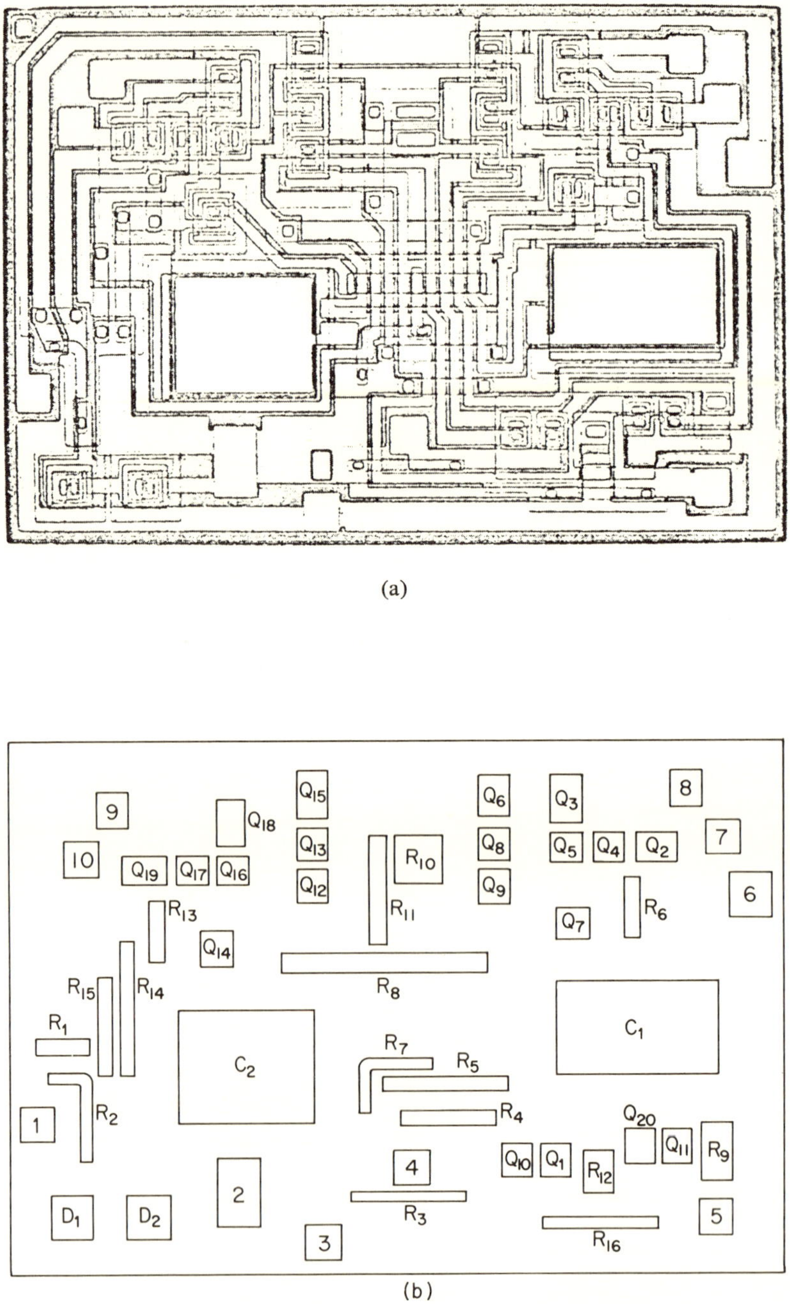

Fig. 11-9. Photomicrograph and element-location diagram of the monolithic flip-flop of Fig. 11-8. (From Reference 4., p. 23).

Large-scale integration (LSI). Clearly, the metallization step of Fig. 11-6 can be used to make interconnections among circuits on a wafer, and these circuits need not all be of the same type. Therefore, the fabrication steps of Fig. 11-6 can be used to form an entire system or subsystem on a single silicon chip. If the number of circuits in the system is large, say 50 to 100 or more as required in computers or in systems performing computerlike functions, the technology is referred to as *large-scale integration* (LSI). Somewhere between systems of this magnitude and those of one circuit per chip lies the border line between LSI and MSI (medium-scale integration). The usage of these terms has been such that this border line is rather nebulous.

Large-scale integration implies putting the maximum possible amount of a system on the largest usable wafer. The extent to which LSI is feasible in practice is limited to both technological and economic restraints[10]. Since none of the processes used in monolithic-circuit fabrication can be carried out with 100 percent yield, the larger the number of circuits per chip, the smaller the yield. Equally important is the fact that silicon, like any other material, is limited in the variety of functions that it can perform well. For example, the highly precise, stable resistors and capacitors needed for many analog functions are not provided by silicon. Thus, at different levels in different systems, all systems become hybrid arrangements of interconnected material subsystems that are alike or different for economic or performance reasons.

11-4 FABRICATION OF DIELECTRIC-ISOLATED MONOLITHIC CIRCUITS[11,12,*]

The use of dielectric isolation of monolithic-circuit components is introduced in Sec. 11-2 and illustrated by Fig. 11-4. Although dielectric isolation is superior to *pn*-junction isolation in almost all respects, its higher cost has restricted its use to applications that require this superiority. The cost differential has been due to the extra processing steps required by the dielectric system, to the smaller component-packing densities obtained, and to the low yields achieved. However, improved processes said to be cost com-

[10] J. A. Morton, "Strategy and Tactics for Integrated Circuits," *IEEE Spectrum*, Vol. **6**, No. 6 (June 1969).

[11] R. M. Burger and R. P. Donovan, *Fundamentals of Silicon Integrated Circuit Technology, Volume I*. Englewood Cliffs, N. J.: Prentice Hall, Inc., 1967, pp. 465–469.

[12] R. M. Burger and R P. Donovan, *Fundamentals of Silicon Integrated Circuit Technology, Volume II*. Englewood Cliffs, N.J.: Prentice-Hall, Inc., 1968, pp. 407–411.

* Reference 3, pp. 166–173; Reference 4, pp. 26–28; Reference 5, pp. 323–326.

petetive with junction isolation may soon result in the exclusive use of dielectric isolation in new system designs.[13]

The dielectric-isolation scheme that has been most used in commercial monolithic circuits employs SiO_2 as the isolating barrier between components on a polycrystalline silicon substrate. In this approach, one begins with a wafer of single-crystal silicon on which an oxide layer has been grown. The SiO_2 layer is to serve as a mask for the etching process to follow. A photo-resist process is then used to remove SiO_2 from selected areas of the wafer. After removal of all photo-resist materials, the wafer is subjected to an etching procedure during which a chemical medium attacks the exposed silicon but not the SiO_2. This is known as *isolation-channel etching* and results in the formation of channels surrounding oxide-covered areas. The SiO_2 is then removed, and the result is as shown in the cross-sectional view of Fig. 11-10(a). In the final circuit, the channels are filled with substrate material, and the surrounded silicon areas contain the circuit components.

Following isolation-channel formation and SiO_2 removal, a new oxide layer is grown over the top surface of the wafer of Fig. 11-10(a). This SiO_2 layer is to be the isolating barrier between components of the final circuit. The wafer is then placed in an epitaxial furnace, where a layer of silicon is grown on the SiO_2 surface. Although the layer is grown epitaxially, it is polycrystalline in nature because of the random crystalline orientation of the SiO_2 material. This polycrystalline silicon fills the isolation channels and forms the substrate of the final circuit. The structure after this step is shown by Fig. 11-10(b).

The next step in the fabrication process is the removal of excess single-crystal material. This is performed by using a lapping machine that removes layers of original single-crystal silicon until SiO_2 is exposed on those regions of the surface that were the bottoms of the isolation channels. At this stage, the wafer is as shown in Fig. 11-10(c), and may be thought of as a relatively large "sea" of polycrystalline material in which "float" individual islands of single-crystal silicon, each island being isolated from the sea by the SiO_2. Circuit components are now formed in the isolated islands, and contacts and interconnections are made, by using processing steps that have been described for the single-crystal monolithic circuit (see Sec. 11-3). A final result is shown by the cross-section of Fig. 11-4.

Now that the fabrication of both junction-isolated and dielectric-isolated monolithic circuits has been described, we are in a better position to appreciate their differences. The superiority of dielectric isolation has been pointed out in Sec. 11-2 and is verified by the data of Maxwell, et al.[14], which shows

[13] D. F. Allison, A. P. Youmans, and T. H. Wong, "New Etchant Puts Dielectric Isolation in the Groove," *Electronics*, Vol. 42, No. 10 (May 12, 1969).

[14] D. A. Maxwell, R. B. Beeson, and D. F. Allison, "The Minimization of Parisitics in Integrated Circuits by Dielectric Isolation," *IEEE Trans. on Electron Devices*, Vol. ED-12, (Jan. 1965).

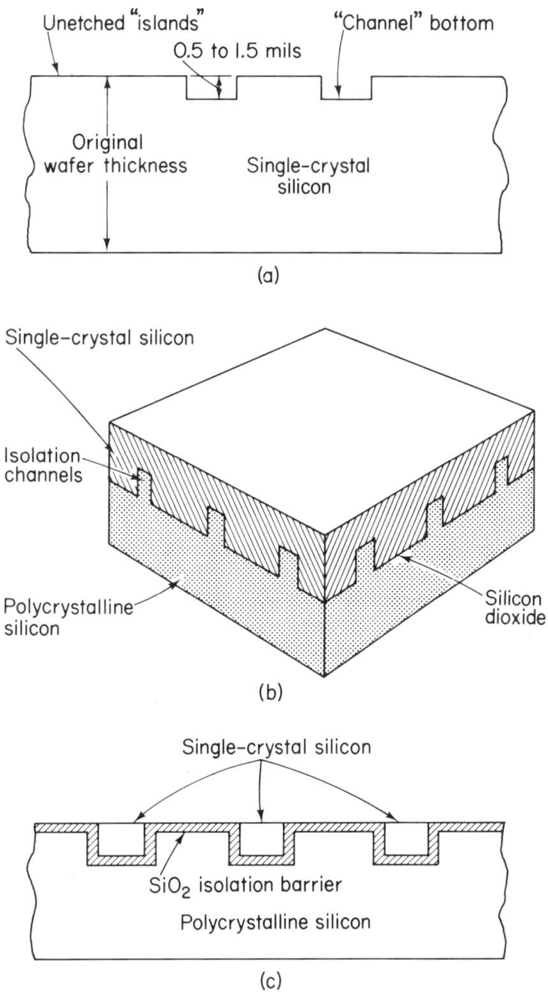

Fig. 11-10. Fabrication of a dielectric-isolated monolithic circuit. (a) Cross-sectional view of single-crystal wafer after isolation-channel etching. (b) Dielectric-isolated structure after growth of polycrystalline layer. (c) Cross-sectional view of dielectric-isolated structure after removal of excess single-crystal material. (From Reference 4, p. 27.)

the capacitance between junction-isolated components to be ten times that for polycrystalline silicon/SiO_2 isolation. Maxwell, et al. also found the leakage current to be reduced by a factor of 10^7 by use of SiO_2 isolation.

The greater flexibility of the dielectric-isolation scheme makes possible the formation of components having radically different characteristics. With the SiO_2 barrier, the resistivities of isolated islands can be selectively changed by long diffusions without concern over back diffusion from *p-isolation* regions, facilitating the fabrication of both *npn* and *pnp* transistors on the same substrate. Since SiO_2 also acts as a barrier to the diffusion of gold, which occurs readily in silicon unless impeded by such a barrier, the minority-carrier lifetimes of the islands can be selectively reduced by gold diffusion. This facilitates the formation, on the same chip, of both fast and charge-control diodes and of both high-frequency and low-frequency transistors. Higher-voltage diodes and transistors are also possible with SiO_2 isolation,[15] since the devices can be formed in high-resistivity grown silicon rather than in a relatively low-resistivity epitaxial layer.

The advantage of dielectric isolation that has justified its higher cost is its greater resistance to radiation. Radiation that causes junction isolation virtually to cease, owing to minority-carrier generation in the junction regions, generates relatively few carriers capable of crossing the dielectric barriers. Thus dielectric isolation has been used primarily in "radiation-hardened" monolithic circuits.

The low yield of SiO_2-isolated monolithic circuits has been largely due to inadequate control of the dimensions of the isolation channels. Excessive channel widths have also been responsible for relatively large spacing of circuit components and component-packing densities lower than those for junction isolation. These problems appear to have been alleviated through use of a directional etchant that forms narrow v-shaped channels, making possible circuits having significantly reduced component spacings and improved production yields.* Figure 11-11 shows the cross section of a transistor of such an SiO_2-isolated circuit and contrasts it with a junction-isolated device. The component-packing density of a circuit fabricated by using v-shaped isolation channels is seen to be significantly greater than that of a junction-isolated circuit.

Many dielectric-isolation schemes have been developed, but few have worked out well enough to be put in high-volume commercial production. In the silicon-on-sapphire (SOS) approach,[16] a single-crystal silicon film is grown on a single-crystal sapphire substrate, and component locations are isolated by etching away the intervening silicon to form an array of silicon islands. In a planar-air isolation technique[17] used commercially in the fabrication of diode arrays, components are isolated by air in addition to SiO_2. Finally, several methods using glass and/or ceramic isolating barriers

15 H. R. Camenzind, B. Polata, and J. Kocsis, "ICs Break Through the Voltage Barrier," *Electronics*, Vol. **42**, No. 7 (March 31, 1969).

16 A. Miller, "Silicon-On-Sapphire Approach Affords Freedom and Flexibility," *Electronics*, Vol. **40**, No. 4 (February 20, 1967).

17 H. Clausen and R. B. Rusert, "Isolation Problems Get an Airing", *Electronics*, Vol. **41**, No. 10 (April 29, 1968).

* Reference 13, p. 112.

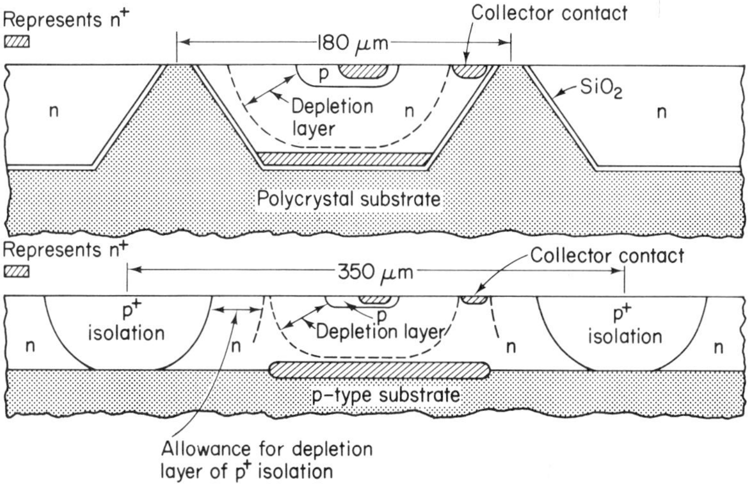

Fig. 11-11. Comparison of a dielectrically isolated transistor (above) of a circuit fabricated using v-shaped isolation channels with a junction-isolated device (below). Dimensions are in micrometers. (From Reference 13.)

have been developed.[18] Compared to junction isolation, dielectric-isolation technology is quite young, and many advances are yet to be expected.

11-5 INTEGRATED RESISTORS

Integrated-circuit components are introduced in Sec. 11-2 and illustrated in Figs. 11-3 and 11-4. We now consider them in somewhat greater detail, beginning with resistors.

Two general types of integrated resistors are *diffused resistors*, obtained by utilizing the bulk resistivity of one of the diffused silicon regions, and *thin-film resistors* obtained by techniques compatible with monolithic processes. These are linear resistors. In addition, the nonlinear *MOS resistor*, obtained by using the MOSFET structure, finds use in switching circuits, as pointed out in Sec. 11-11.1.

11-5.1 Sheet Resistance R_s

Since all integrated resistors employ thin layers or sheets of material, it is advantageous to define a quantity known as *sheet resistance*. Consider

[18] J. A. Amick, A. W. Fisher, A. I. Stoller, and W. H. Schilp, Jr., "Isolation Techniques for Fabricating Integrated Circuits", *RCA Review*, Vol. **29**, No. 4 (December 1968).

flow of current through a sheet of material, as in Fig. 11-12. The resistance R to current I is given by

$$R = \frac{\rho l}{A} = \frac{\rho l}{yw} \tag{11-1}$$

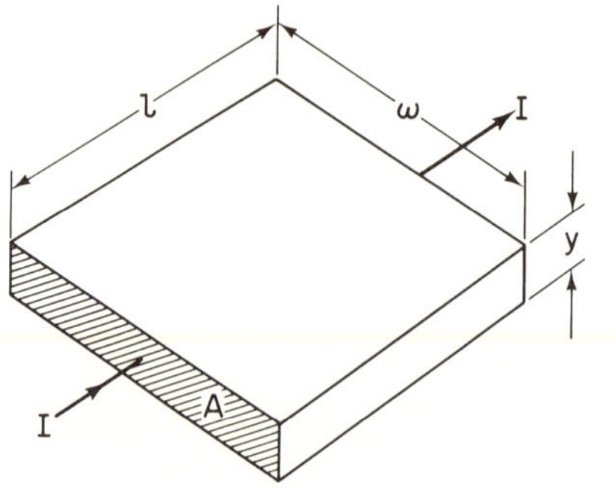

Fig. 11-12. Pertaining to the concept of sheet resistance.

where ρ is the average resistivity of the material and l, w, and y are sheet length, width, and thickness, respectively. Since ρ and y are properties of the particular sheet of material, Eq. (11-1) can be written

$$R = R_s \frac{l}{w} \tag{11-2}$$

where R_s is a constant given by

$$R_s = \frac{\rho}{y} \tag{11-3}$$

R_s is seen to be a characteristic of the sheet and is known as the *sheet resistance*. From Eq. (11-2), it is seen that $R = R_s$ when $l = w$, indicating that the resistance for a square sheet is the sheet resistance regardless of the size of the square. Therefore, R_s is expressed in ohms per square. However, it should be noted that "square" is not part of the unit of R_s, but serves as an identifying label. For a particular R_s, we see from Eq. (11-2) that the design of a resistor reduces to a problem in geometry, the value of the resistance being determined by the ratio of length to width.

11-5.2 Diffused Resistors*

The sheet resistance of typical base and emitter diffusion layers are 200 Ω per square and 2.5 Ω per square, respectively. Either may be used

* Reference 2, pp. 73–78; Reference 3, pp. 255–264; Reference 7, pp. 87–88.

for a diffused resistor, depending on the magnitude of the desired resistance. For example, suppose a 4000-ohm resistor is required. From Eq. (11-2), l/w is given by

$$\frac{l}{w} = \frac{R}{R_s} \tag{11-4}$$

and is seen to be either $\frac{20}{1}$ or $\frac{1600}{1}$, depending on the diffusion layer used. Clearly, the resistor should be fabricated from the 200 Ω per square, p-type, base-diffusion region, and could be obtained by employing a pattern 20 mils long by 1 mil wide or by any pattern of the same proportions. However, pattern dimensions less than 1 mil are not ordinarily used, owing to process-control limitations.

Figure 11-13(a) shows a possible mask layout for the diffused resistor of the preceding example. The 20-mil by 1-mil area pattern is used, with enlargement of the ends to provide for the ohmic contacts. Since junction isolation is employed, edges of the n-type region containing the p-type resistor are shown 2 mils away from the resistor area to allow for spreading

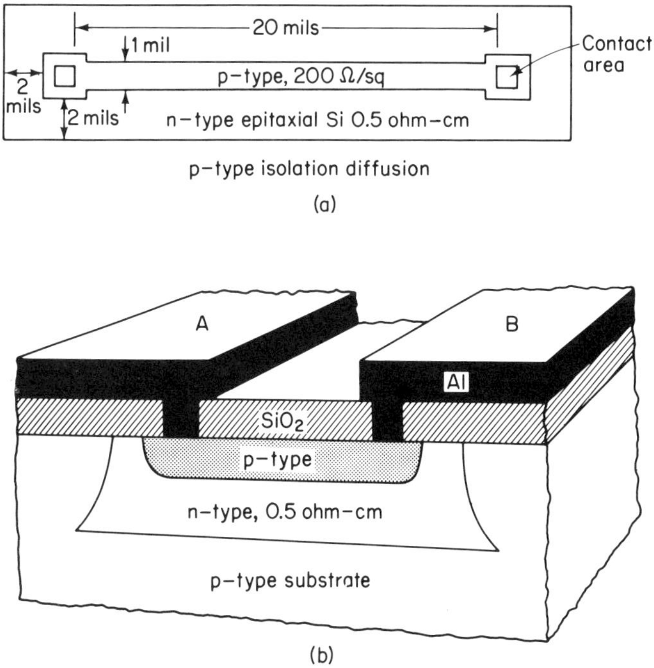

Fig. 11-13. Diffused silicon resistor. (a) Top view showing dimensions of a 4000-Ω resistor. (b) Cross-sectional view of a typical junction-isolated monolithic resistor. (From Reference 3, p. 257.)

of the p-type isolation diffusion. The cross-sectional view of a typical junction-isolated diffused resistor is shown by Fig. 11-13(b).

Resistance values and tolerance. The range of resistance values obtainable is limited by the area required and the sheet resistances available. For p-type diffusion, R_s can vary from 10 to 500 Ω per square. In addition, n-type diffusion can provide values as low as 2.5 Ω per square. However, the desired characteristics of monolithic transistors usually formed simultaneously with resistances are such that sheet resistances are limited to values of 2.5 and 100 to 300 Ω per square. These R_s values permit a possible resistance range of 2.5 Ω to 100 kΩ, but a more practical range from an economic point of view is from 100 Ω to 30 kΩ.

The accuracy of the final resistance value depends on the control of the diffusion process and the accuracy of the oxide mask through which diffusion occurs. Manufacturing tolerances have been about 20 percent, with respect to absolute values, but ratios of resistance have been held to within 5 percent. This suggests that, wherever possible, the design of integrated circuits should emphasize resistance ratios rather than absolute values.

Equivalent circuit. A circuit model of a typical junction-isolated diffused resistor is shown in Fig. 11-14 and is seen to include distributed parasitic

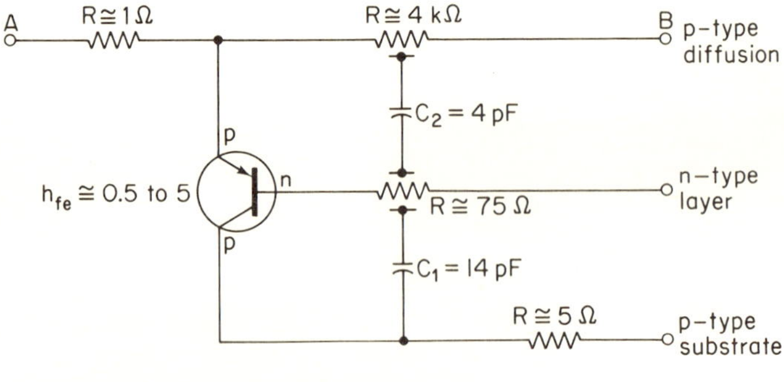

Fig. 11-14. Equivalent circuit of a junction-isolated diffused resistor such as that of Fig. 11-13. (From Reference 3, p. 257.)

capacitances and a parasitic *pnp* transistor. The capacitances are those of junctions with the n-type isolation layer, and the transistor has the n-type layer as the base, the substrate as the collector, and the resistor material as the emitter. The transistor has a rather low h_{fe} (typically from 0.5 to 5), owing to the relatively thick base, but may be responsible for excessive resistor-to-substrate leakage current unless it is cut off by reverse-biasing the emitter-base junction. This may be accomplished by connecting the n-type isolation layer to the *most positive* point in the circuit. The parasitic transistor is further evidence of the inferiority of junction isolation relative to dielectric isolation.

11-5.3 Thin-Film Resistors*

Diffused resistors prove unsatisfactory in some applications because they are affected by nuclear radiation, because the required values or tolerances cannot be obtained, because the variation of resistance with temperature is excessive, or because the parasitic elements (with junction isolation) are unacceptable. In such cases, resistors may be formed from a thin film of metal (usually nichrome) or tin oxide deposited on the surface of the final SiO_2 layer of the monolithic circuit. After masked etching to produce the desired geometry, the resistor is covered by an insulating layer through which apertures are opened for ohmic contacts. Typical sheet resistances are from 40 to 400 Ω per square for nichrome and from 80 to 4000 Ω per square for tin oxide, resulting in resistance values of from about 20 Ω to 50 kΩ and 40 Ω to 500 kΩ, respectively. The temperature coefficient of nichrome resistors is positive, like that of diffused-silicon resistors, but is an order of magnitude smaller. Tin-oxide resistors have negative temperature coefficients of resistance that may be either smaller or larger than those of diffused resistors, depending on the sheet resistance. Both thin-film types have superior resistance tolerances, relative to diffused resistors.

11-6 INTEGRATED CAPACITORS

Capacitors of integrated circuits may be either *junction capacitors*, obtained by utilizing the transition capacitance of a reverse-biased *pn* junction, or *thin-film capacitors*, consisting of two conductors separated by an insulator (dielectric).

11-6.1 Junction Capacitors†

The cross-section view of a typical junction-isolated junction capacitor is shown in Fig. 11-15(a). The capacitor is formed by junction J_2 between the upper *p*-region and the *n*-region. An additional junction J_1 exists between the *n*-region and the substrate, forming a parasitic capacitance. The equivalent circuit of the junction capacitor is shown in Fig. 11-15(b), where C_2

* Reference 2, pp. 84–85; Reference 3, pp. 264–267; Reference 5, pp. 460–470.

† Reference 2, pp. 78–81; Reference 3, pp. 247–252.

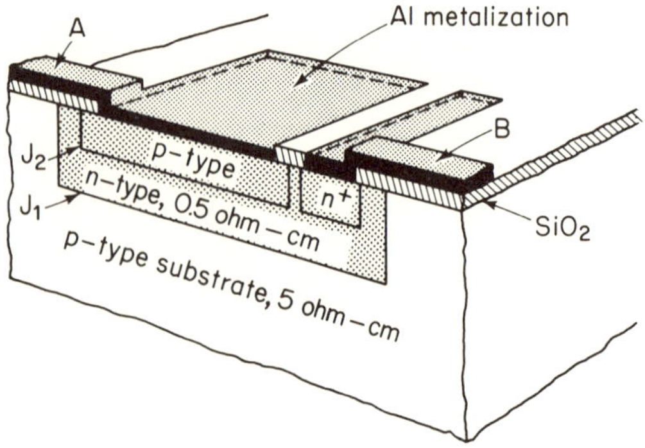

(a)

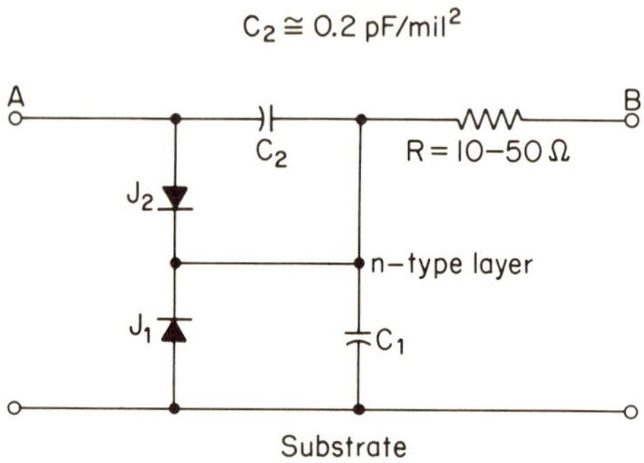

(b)

Fig. 11-15. (a) Cross-sectional view of a junction-isolated junction-type monolithic capacitor. (b) Equivalent circuit of the capacitor of (a). (Courtesy of Motorola, Inc.)

is the desired capacitance, C_1 is the parasitic capacitance, R is a parasitic resistance resulting from the bulk resistance of the region, and the two diodes are the "idealized" diodes of the two junctions. Clearly, C_2 should be as large as possible relative to C_1 in order to obtain maximum transfer of signal from point A to point B. It is also clear that the substrate should be at the most negative potential in the circuit in order to minimize C_1 and isolate the capacitor from the other components by keeping diode J_1 reverse biased. Since J_2 must also be reverse biased, *junction capacitor* C_2 *is polarized.*

11-6.2 Thin-Film Capacitors*

The usual dielectric materials of thin-film capacitors are silicon dioxide, aluminum oxide (Al_2O_3, also known as alumina), and tantalum oxide

* Reference 2, pp. 81–82; Reference 3, pp. 252–255; Reference 7, p. 89.

(Ta_2O_5), the choice depending on the capacitance required. Silicon-dioxide films can be used readily with monolithic circuits, since SiO_2 is formed as a part of the monolithic process, but other dielectric materials (used for larger capacitances) require special processing steps for fabricating the capacitor on top of the monolithic circuit.

Silicon-dioxide capacitors. A junction-isolated SiO_2 capacitor is shown by Fig. 11-16(a), where the top plate is aluminum metallization and the bottom plate is the heavily doped n^+ region formed during the emitter diffusion. The equivalent circuit of the SiO_2 capacitor is shown in Fig. 11-16(b), where

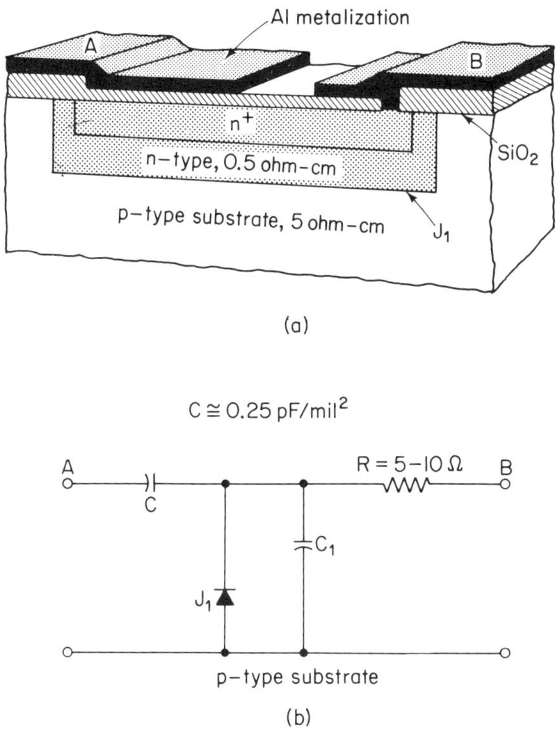

Fig. 11-16. (a) Cross-sectional view of a junction-isolated silicon-dioxide monolithic capacitor. (b) Equivalent circuit of the capacitor of (a).

C is the desired capacitance, R is a parasitic resistance resulting from the bulk resistance of the n^+ region, and C_1 and J_1 denote the parasitic capacitance and idealized diode of the isolation junction. A dielectric-isolated SiO_2 capacitor (see Fig. 11-4) has no substrate diode and a much smaller parasitic capacitance.

The silicon-dioxide capacitor has several advantages over the junction capacitor: it is nonpolar, in that the applied voltage may be of either polarity, thus giving greater circuit flexibility; the capacitance is independent of applied-voltage magnitude; it has lower dissipation (higher Q) because of its lower parasitic resistance.

Metallic-oxide capacitors. Compatible thin-film metallic-oxide capacitors are fabricated by depositing a film of either aluminum or tantalum over the SiO_2 layer covering the monolithic wafer. The oxide dielectric is then either deposited on the aluminum film or formed on the tantalum film, as the case may be, and covered by a film of the same metal to form the other plate. An aluminum-oxide (alumina) thin-film capacitor is shown in Fig. 11-17. It too has a parasitic capacitance to the substrate, but this capacitance can be kept quite small by using a relatively thick SiO_2 layer under the capacitor.

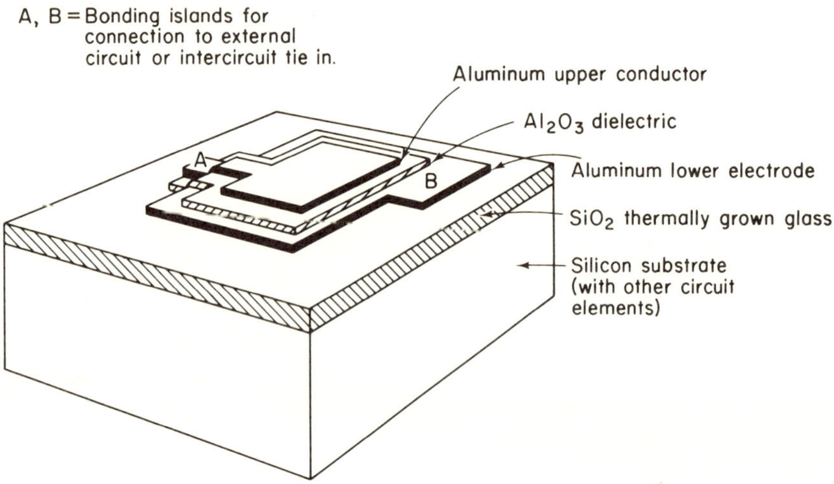

Fig. 11-17. Thin-film aluminum-oxide capacitor compatible with monolithic circuits. (From Reference 3, p. 254.)

Capacitances employed in integrated-circuit practice have upper limits determined by their area requirements. These limits are of the order of hundreds of picofarads for SiO_2 and junction capacitors and thousands of picafarads for metallic-oxide capacitors. When larger capacitances are required, a hybrid circuit is used with the capacitor mounted either inside or outside the integrated-circuit package.

In general, capacitors are avoided in integrated circuits because even very small capacitances require relatively large areas. For example, a 10-pF monolithic capacitor requires an area about three times that occupied by a transistor. Thus, amplifier stages are usually direct-coupled, emitter and drain circuits are unbypassed, etc.

11-7 INTEGRATED INDUCTORS*

Practical inductors are by far the most difficult components to fabricate by integrated techniques. Since inductors are desirable for frequency-selective circuits, much effort has been expended in this area with results that have been most discouraging. Typical of the state of the art is a 2-μH inductor, having a Q of about 50 at a frequency of 50 MHz, obtained by thin-film techniques. This relatively low-quality inductor consists of a 20-turn flat spiral, with an outside diameter of 0.33 in., deposited on glass. Because of the large area required for such a small inductance, greatly improved results must be achieved in order to avoid having to place almost all inductors external to an integrated circuit. Clearly, inductors should be avoided whenever possible in the design of systems using monolithic circuits.

11-8 MONOLITHIC DIODES†

Monolithic diodes are usually obtained by using transistor structures in one of five possible arrangements of the two available junctions. Each arrangement yields a diode having different characteristics, providing considerable freedom for monolithic-diode design.

The five different diode structures of a junction-isolated monolithic circuit, together with their equivalent circuits, are shown in Fig. 11-18. The first arrangement is the emitter-base diode, with the collector shorted to the base; the second is the emitter-base diode with the collector open; the third is the collector-base diode with no emitter diffusion; the fourth is the collector-base diode with the emitter shorted to the collector (in effect, two diodes connected in parallel); the fifth is the collector-base diode with the emitter shorted to the base.

Comparable electrical characteristics of the five possible diodes of a typical monolithic structure are listed in the table of Fig. 11-18. Characteristics such as breakdown voltage and storage time are seen to be highly dependent on the diode arrangement, but forward voltage and other characteristics are much less so.

Breakdown voltage. The difference in breakdown voltages is simply that between the emitter-base and collector-base junctions, low emitter-base breakdown voltage being characteristic of the diffused transistor because of its relatively low base-region resistivity. Therefore, arrangements 1, 2, and 4

* Reference 3, pp. 267–268.

† Reference 2, pp. 72–73; Reference 3, pp. 195–207; Reference 5, pp. 229–236; Reference 7, pp. 89–92.

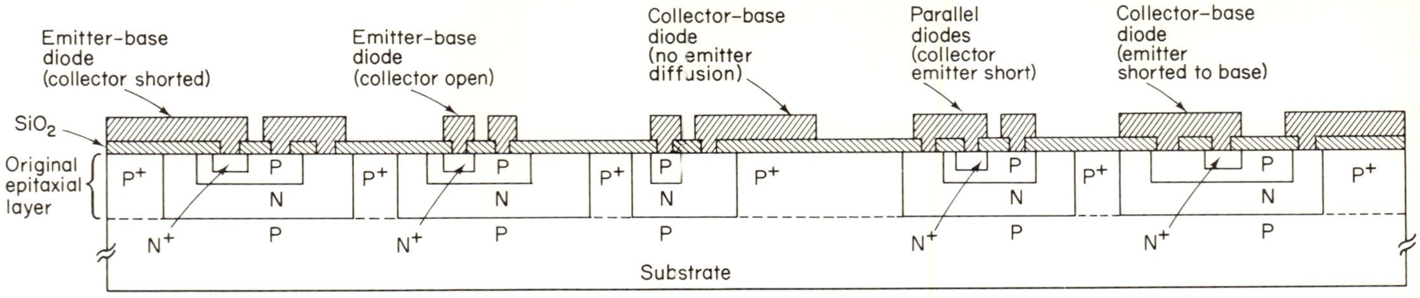

(a)

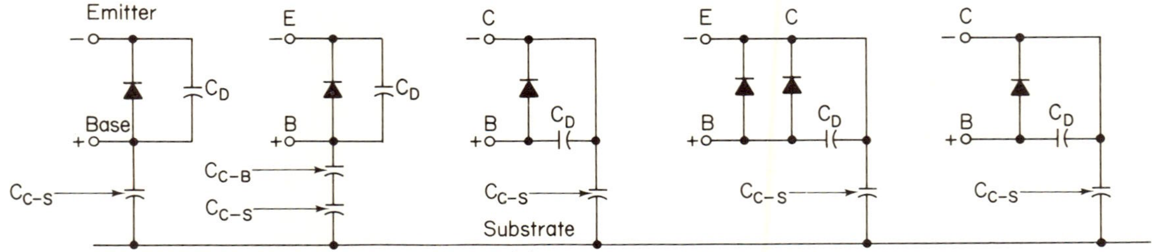

(b)

Breakdown voltage	7 V	7	55	7	55
Diode capacitance	0.5 pF	0.5	0.7	1.2	0.7
Parasitic capacitance	2.9 pF	1.2	2.9	2.9	2.9
Storage time	9 nsec	56	85	100	53
Forward voltage	0.85 V	0.96	0.95	0.92	0.94

(c)

Fig. 11-18. (a) The basic diode connections of the monolithic transistor. (b) Equivalent circuits of the diodes of (a). (c) Comparable electrical characteristics of the diodes of (a). (From Reference 2, p. 72.)

result in breakdown voltages that are relatively low but adequate for most switching applications.

Storage time. The storage times of the diode arrangements differ markedly, depending on the amount of charge stored in the collector region during forward bias. This charge exists, in varying degrees, in all arrangements except the one having the collector-base junction shorted (arrangement 1). Thus, arrangement 1 has minimum storage time. Maximum collector-charge storage (and maximum storage time) occurs when the collector is shorted to the emitter (arrangement 4), so that both junctions are heavily forward biased when the diode is ON. Clearly, arrangement 1 should be used when a high-speed diode is required, since the storage time is almost an order of magnitude less than that of any other arrangement. The other arrangements may be used for charge-storage diodes.

Forward voltage. Although forward voltage does not differ greatly among the diode arrangements, it is significantly less for the emitter-base diode with the collector-base junction shorted (arrangement 1). The reason for this is transistor action that causes the base-region voltage drop to be considerably smaller than that of any other arrangement.

Diode capacitance. The transition capacitance is that of the emitter-base junction, or of the collector-base junction, or of the sum of the two, depending on the diode arrangement. The emitter-base diode has the smaller capacitance, because of its smaller area, even though its capacitance per unit area is much larger, owing to its more heavily doped junction regions.

Parasitic capacitance. The parasitic capacitance is that of the collector-substrate junction, except for the emitter-base diode with the collector open (arrangement 2), which has two series junctions between the diode and the substrate. For dielectric isolation, the parasitic capacitance would be much smaller and essentially the same for all arrangements.

Parasitic-transistor action. Like the junction-isolated resistor (see Fig. 11-13), the diodes of Fig. 11-17 may be adversely affected by parasitic *pnp* transistors having the substrate, collector, and base regions acting as transistor collectors, bases, and emitters, respectively. Parasitic transistors are not shown in the equivalent circuits of Fig. 11-18, owing to omission of the collector-substrate diodes, but they can be seen to be present in every arrangement except that of the emitter-base diode with the collector-base junction shorted (arrangement 1). In this arrangement, the parasitic transistor is rendered inactive by its emitter-base junction short. This is another important advantage of arrangement 1, since the *pnp* transistor action can be quite harmful if it is not taken into account.

Summary. The diode of arrangement 1, in which the collector-base junction is shorted, is probably the most widely used configuration. Here we find the lowest forward voltage, the shortest storage time (by almost an order of magnitude), and no *pnp* transistor action to the substrate. Only the need for a higher breakdown voltage, or the requirement of a charge-control device, is likely to indicate a different arrangement. In these arrangements, gold doping can be used to reduce the parasitic transistor h_{fe} to virtually zero. Also, this problem, and that of parasitic capacitance, can be alleviated by the use of dielectric isolation.

11-9 MONOLITHIC TRANSISTORS*

The comparison of a monolithic transistor with its discrete counterpart is facilitated by the cross sections of Fig. 11-19. Both transistors have epitaxial collectors, but they differ in that the discrete transistor is self-isolated, whereas the monolithic device is junction-isolated. Furthermore, collector current of the discrete device flows down through the collector contact on the bottom, but monolithic-transistor collector current flows along the collector in a thin sheet to the top collector contacts. Thus the collector series resistance of the monolithic transistor is considerably higher than that of a discrete device having the same collector-region resistivity. For this reason, lower epitaxial resistivities are employed in monolithic transistors. Typical resistivities in use are 0.5 ohm-cm and 0.1 ohm-cm, and we shall compare the properties of transistors fabricated by utilizing these materials.

Monolithic-transistor geometry. The geometry of a typical monolithic transistor, looking at the top surface of the device, is illustrated in Fig. 11-20. The emitter consists of a rectangle measuring 1 by 1.5 mils, and is diffused into a 2.5- by 4.0-mil base region. Ohmic contact to the base is made by two metallized stripes on either side of the emitter. The metallized stripe that completely surrounds the base region forms the contact to the collector region. The dashed-line rectangle represents the transistor substrate area and is 6.5 by 8 mils. Finally, the outermost rectangle surrounding the structure is the edge of the original silicon dioxide that was used to mask the *p*-type isolation diffusion. From this, we see that the *lateral-diffusion* distance under the SiO_2 is 1 mil. Since diffusion proceeds in three dimensions, this is in agreement with the indicated depth of the substrate junction, one mil being about 25 microns (micrometers).

Table 11-1 summarizes the electrical properties of this transistor for both the 0.5- and the 0.1-ohm-cm collector resistivities. The 0.1-ohm-cm

* Reference 2, pp. 70–72; Reference 3, pp. 174–195; Reference 4, pp. 15–18; Reference 7, pp. 92–95.

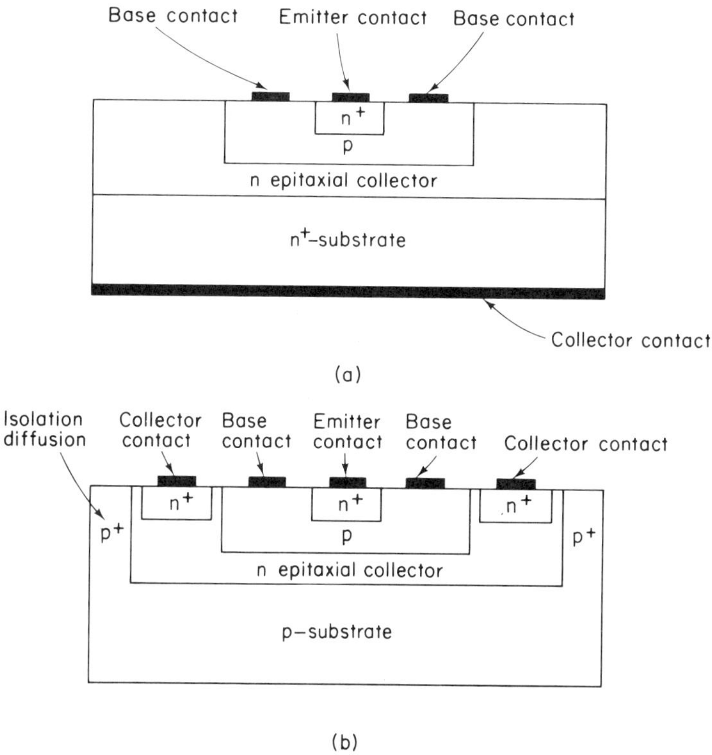

Fig. 11-19. Comparison of the cross-section of a discrete planar-epitaxial transistor (a) with that of a monolithic transistor (b). (From Reference 7, p. 92.)

device has lower collector series resistance R_{CS} and a correspondingly lower collector-emitter voltage $V_{CE(\text{sat})}$ for the transistor in saturation. However, this improvement is achieved by sacrificing collector breakdown voltage and accepting larger transition capacitances.

The 0.1-ohm-cm collector transistor is gold-doped in order to reduce minority-carrier lifetime and decrease the amount of stored charge when the transistor is in saturation. This accounts for its higher transition frequency f_T. It is of interest to compare the f_T values listed in Table 11-1 with those of very similar discrete transistors. The 2N918 transistor closely resembles the 0.5-ohm-cm transistor described in Table 11-1, and the 2N709 is very similar to the gold-doped 0.1-ohm-cm monolithic device. Both transistor types have f_T values about double those of their monolithic counterparts, since they have less parasitic capacitance. Presumably dielectric-isolated transistors compare more favorably with discrete devices.

Buried layer. The technique of using a heavily doped layer in the collector region to reduce collector series resistance is rather recent and is not employed

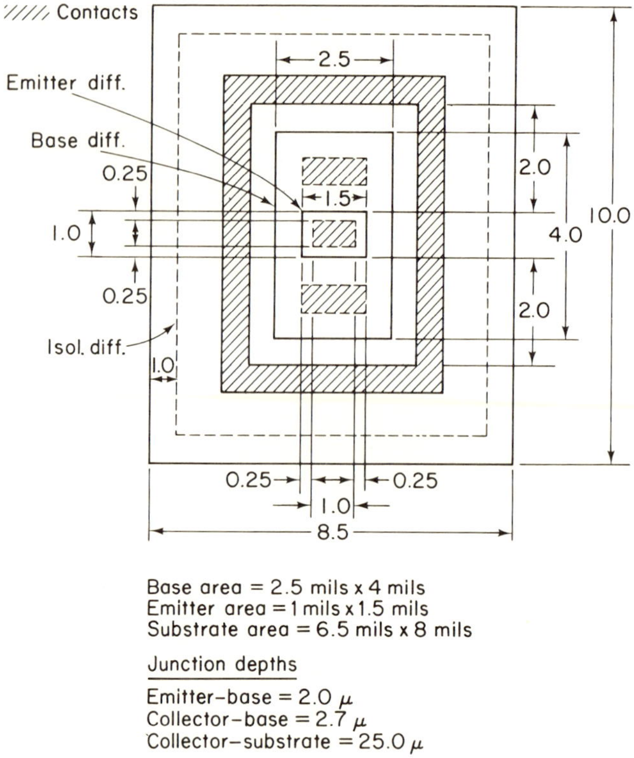

Fig. 11-20. Typical double-base-stripe geometry of a monolithic transistor. Surface dimensions are in mils. (From Reference 3, p. 177.)

Table 11-1 Characteristics for 1- by 1.5-mil double-base stripe monolithic transistors*

Transistor parameter	0.5 ohm-cm	0.1 ohm-cm†
BV_{CBO}, V	55	25
BV_{EBO}, V	7	5.5
BV_{CEO}, V	23	14
C_T, (forward bias), pF	6	10
C_T, at 0.5 V, pF	1.5	2.5
C_T, at 5 V, pF	0.7	1.5
h_{FE} at 10 mA	50	50
R_{CS}, Ω	75	15
$V_{CE(\text{sat})}$ at 5 mA, V	0.5	0.26
V_{BE} at 10 mA, V	0.85	0.85
f_T at 5V, 5 mA, MHz	440	520

* From Reference 3, p. 193.

† Gold-doped.

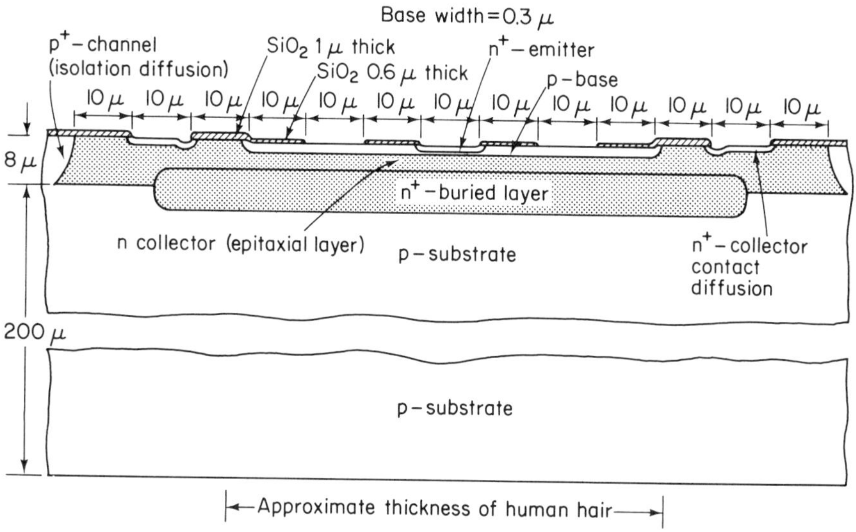

Fig. 11-21. Cross-section, *to correct scale*, of a buried-layer monolithic transistor. Dimensions are in micrometers. (From Reference 4, p. 16.)

in the transistors described in Table 11-1. The cross-section of a buried-layer monolithic transistor is shown, *to correct scale*, in Fig. 11-21. The buried layer is seen to be quite thick, relative to the collector layer, and in shunt with the high-resistivity collector region, thereby reducing collector series resistance R_{CS} by about an order of magnitude. The major advantage of the buried-layer structure is that, for a particular R_{CS}, higher collector-region resistivities can be used to increase collector breakdown voltage and decrease capacitance.

Unlike cross sections drawn to an exaggerated scale, as in Fig. 11-19, Fig. 11-21 correctly shows the relative thickness of each semiconductor layer. The transistor is seen to be formed in the "skin" of the wafer, occupying about 5 percent of the wafer thickness. Ohmic contacts are not shown, but their locations are indicated by openings in the SiO_2 layer.

11-10 INTEGRATED FIELD-EFFECT TRANSISTORS

An advantage of field-effect transistors is that they lend themselves more readily to fabrication by integrated-circuit techniques than do bipolar junction transistors. Monolithic JFETs require fewer diffusion steps than BJTs, but may be formed simultaneously with BJTs when both are desired. In general, JFETs are used in analog (linear) ICs that may include BJTs. Integrated MOSFETs are easier to make than JFETs, and may be either monolithic or silicon-on-sapphire (SOS) devices of either the enhancement or depletion type. The thin-film transistor (TFT) is also an insulated-gate FET that lends itself to use in integrated circuits. Enhancement-type integrated MOSFETs are used primarily in digital (switching) applications.

11-10.1 Monolithic Junction-Type FETs*

Monolithic junction-type FETs have many possible versions, depending on the fabrication technique employed and the method of isolation. One of the simplest designs is the double-diffused JFET shown by the cross-section of Fig. 11-22(a). The source, gate, and drain regions are identified by *s*, *g*, and *d*, respectively; *L* indicates the channel length; the direction of

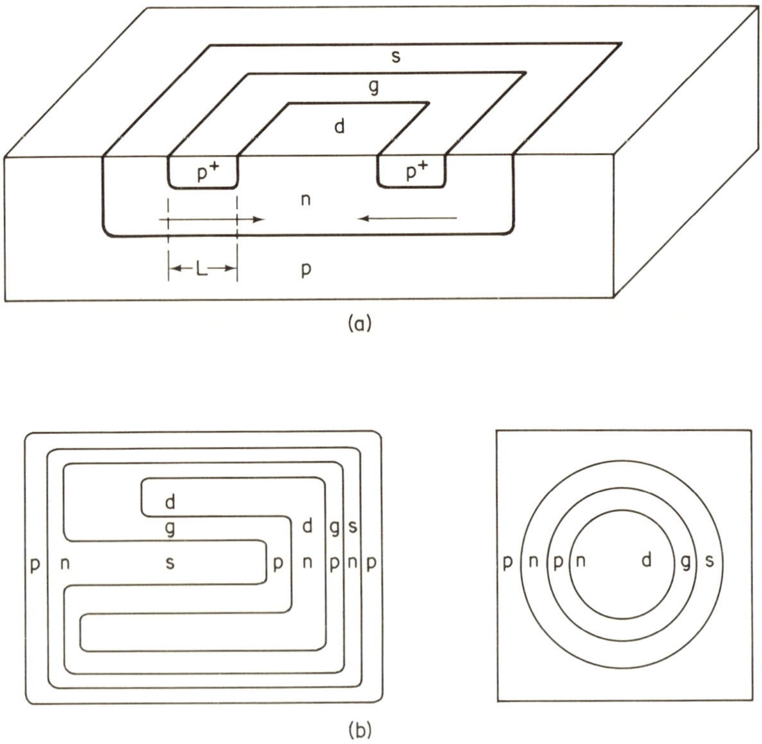

Fig. 11-22. Monolithic *n*-channel junction-type FET. (a) Cross-section. (b) Additional topological designs.

* Reference 3, pp. 223–228; Reference 5, pp. 250–269; Reference 12, pp. 402–421.

carrier flow from source to drain is indicated by arrows. The device may be constructed in a variety of topological forms, as shown in Fig. 11-22(b). An essential feature of every form is the prevention of current flow from drain to source *around* the gate. This is achieved in the structures of Fig. 11-22 by the gate diffusion closing on itself. Another scheme is used in the device shown in Fig. 11-26.

The devices of Fig. 11-22 are constructed by an *n*-type diffusion into a *p*-type wafer to form the channel region. This is followed by a *p*-type diffusion to form the gate. The region surrounded by the gate diffusion (*d* in Fig. 11-22) may be used as either the drain or source region since, for a gate periphery large compared with the channel length, interchanging the drain and source will have negligible effect on d-c characteristics. A-C characteristics may differ, however, owing to parasitic effects.

For this construction, the upper gate is isolated from the lower gate, which is the *p*-type substrate, resulting in a dual-gate structure. The lower-gate region is lightly doped, relative to the diffused channel, so channel width is relatively unaffected by the potential of the lower gate. This permits construction of additional devices on the same substrate, without an added isolation junction, as shown in Fig. 11-23. The two JFETs of Fig. 11-23

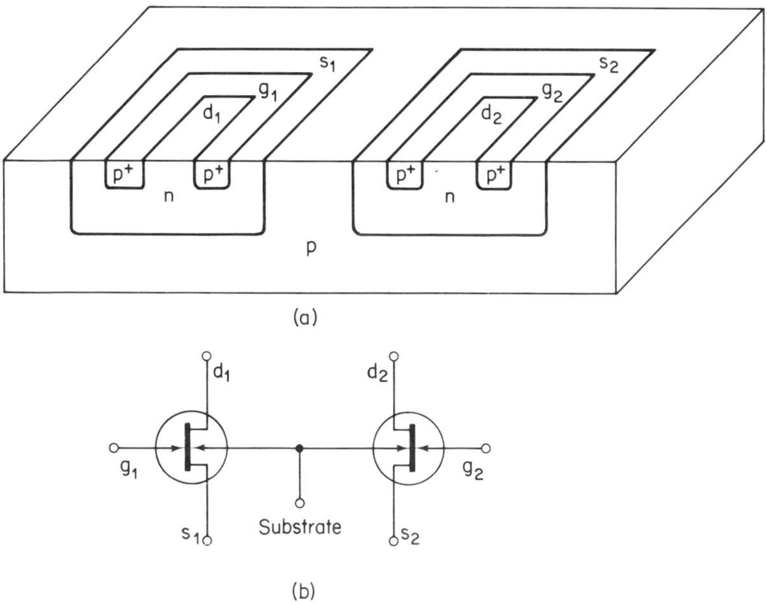

Fig. 11-23. Two double-diffused, *n*-channel, monolithic, junction-type FETs on the same substrate. (a) Structure cross-section. (b) Equivalent circuit.

are not completely isolated, owing to their common lower gate, but there will be no interaction between them if the substrate potential is fixed.

The same device structure can be formed by using an n-type epitaxial layer on a p-type substrate, as shown in Fig. 11-24. This structure is fabri-

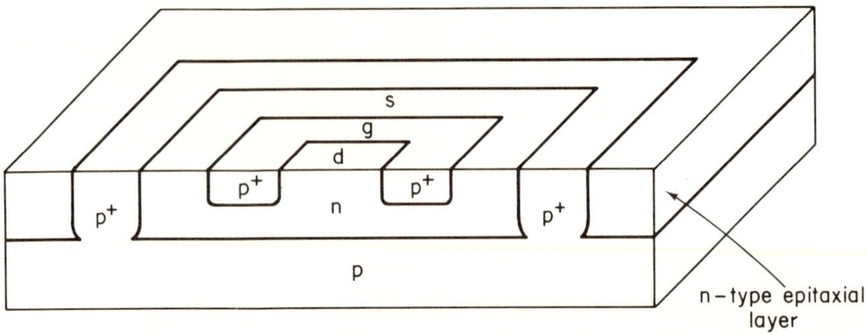

Fig. 11-24. Epitaxial-diffused n-channel junction-type FET.

cated by making a p^+ isolation diffusion through the n-region to the substrate and then making a second p^+ diffusion to form the upper gate. The isolation scheme of Fig. 11-23 can be employed, provided that the substrate layer is lightly doped relative to the epitaxial n layer.

The device structures considered so far achieve isolation through use of a common lower gate. Devices having isolated lower gates require the use of an additional junction and a third diffusion. Figure 11-25 shows a triple-diffused monolithic n-channel JFET formed on an n-type substrate. Gates g_1 and g_2 may be connected together for increased g_m, and isolation is

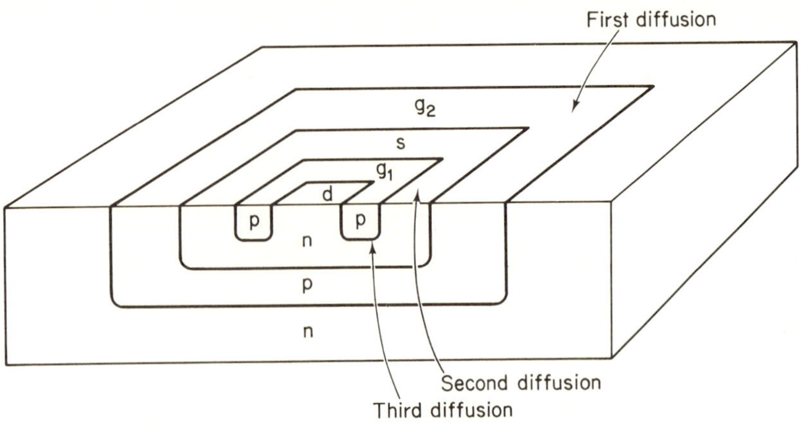

Fig. 11-25. Junction-isolated, triple-diffused, n-channel, junction-type FET.

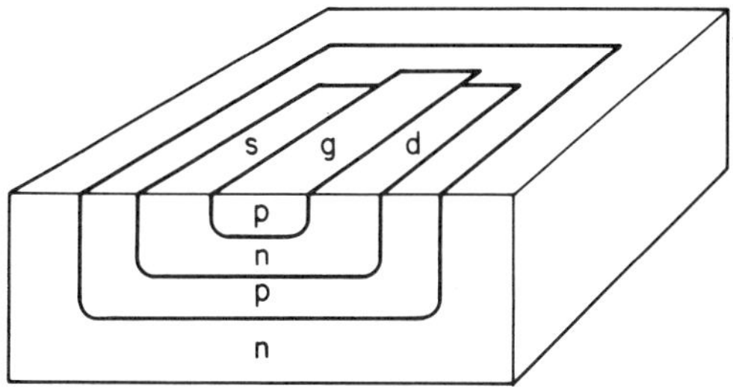

Fig. 11-26. Junction-isolated, triple-diffused, *n*-channel, junction-type FET having gates connected together internally.

achieved by connecting the substrate to the *most positive* point in the circuit.

A junction-isolated structure having gates connected together internally is shown in Fig. 11-26. Extending the upper gate all the way across the area of the *n*-type diffusion not only connects the gates together but also prevents current flow from drain to source around the gate. This construction results in a device that is symmetrical with respect to drain and source—a characteristic that may be advantageous in certain applications.

Simultaneous Fabrication of JFETs and BJTs. When the epitaxial-diffused process (see Sec. 11-3) is used to form a JFET simultaneously with other monolithic elements, the channel is formed during the base diffusion, the emitter diffusion forms the upper gate, and the lower gate is epitaxial material grown on the substrate at the beginning of the process. Figure 11-27 shows the cross section of a junction-isolated circuit having a *p*-channel JFET and an *npn* BJT fabricated by the same diffusion steps. Similarly, an *n*-channel JFET and *pnp* BJT can be fabricated simultaneously, beginning with a *p*-type epitaxial layer on an *n*-type substrate.

The compatibility of JFETs and simultaneously formed BJTs is an important consideration. When a JFET is formed using diffusions considered

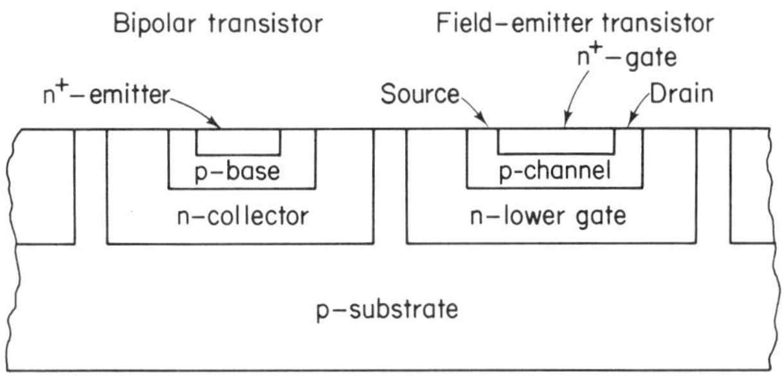

Fig. 11-27. Junction-isolated monolithic structure having a JFET and a BJT fabricated by the same diffusion steps.

optimum for BJT fabrication, the pinch-off voltage is found to be much too large for use in integrated circuits. This is because the resistivity of the base region is an order of magnitude or more below the channel resistivity for optimum JFET performance. Therefore, the only alternative is to accept the performance of BJTs made by diffusion steps for JFETs. The BJT characteristic degraded most by this process is the base-region reach-through voltage.

The JEFT structures shown in Figs. 11-22 through 11-27 are by no means the only basic configurations possible with diffusion and epitaxial techniques. Among those not considered are devices having vertical channels* and JFETs formed in wafers having two epitaxially grown layers[19]. In addition, dielectric-isolated structures† are not shown, since they are basically the same as those of Figs. 11-22 and 11-24.

11-10.2 Monolithic MOSFETs‡

The MOSFET is one of the simplest monolithic devices to fabricate, requiring only one noncritical diffusion to make contact to the source and drain ends of the channel. In addition, the device occupies less than 10 percent of the surface required by a BJT in an integrated circuit. These and other attractive features cause the monolithic MOSFET to be widely used in commercial integrated circuits.

Enhancement MOSFET. Figure 11-28 shows the topology and cross-section of a typical enhancement-type MOSFET structure. The device is formed on a lightly doped substrate by a heavily doped diffusion, of opposite conductivity, to form the drain and source. An insulating layer of silicon dioxide is then grown, and holes are etched for the source and drain contacts. The metal for these contacts, as well as for the gate electrode, is evaporated at the same time to complete the device. This procedure may be used to fabricate, on different substrates, both n-channel and p-channel devices.

Isolation of devices on the same substrate is achieved by the depletion region separating the drain, source, and channel from the substrate. For n-channel devices on a p-type substrate, the substrate should be connected to the most negative point in the circuit; for p-channel devices, the substrate is connected to the most positive point. For high-resistivity substrates (say over 100 ohm-cm), the depletion region between the channel and the substrate has little effect on device characteristics, and the source-to substrate

[19] G. C. Onodera, W. J. Corrigan, and R. M. Warner, Jr., "Silicon Field-Effect Transistor with Internal Epitaxial Channel," *Proc. IEEE*, Vol. **50**, (August 1962), p. 1824.

* Reference 12, p. 408 and Reference 5, pp. 265–269.

† Reference 12, pp. 407–408.

‡ Reference 3, pp. 229–231; Reference 12, pp. 421–434.

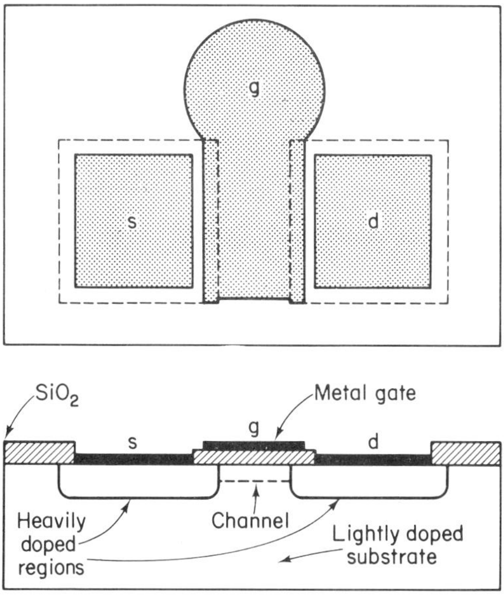

Fig. 11-28. Topology and geometry for an enhancement-type MOSFET. The structure may be used for either *n*-channel or *p*-channel devices. The channel is an inversion layer formed by a gate-to-substrate voltage of the proper polarity.

voltage may vary over a wide range. For low substrate resistivities (below 10 ohm-cm), device characteristics depend to some extent on source potential, and some other isolation technique may be desired. A different technique is mandatory when FETs are used with other monolithic devices, such as resistors and BJTs.

Figure 11-29 illustrates the use of an additional reverse-biased junction to provide isolation for an *n*-channel enhancement-type MOSFET. Since the substrate and source potentials are essentially the same when the device is ON, the two regions are connected together internally. A similar isolation arrangement is shown in Fig. 11-30 for a *p*-channel MOSFET fabricated simultaneously with an *npn* BJT. Unlike JFET, the MOSFET is formed by

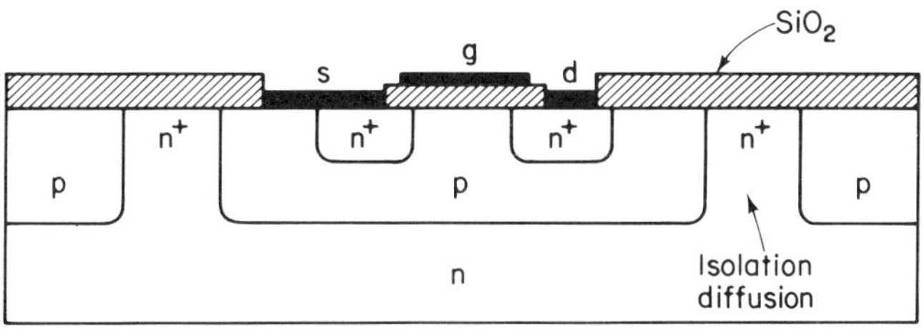

Fig. 11-29. Junction-isolated, *n*-channel, enhancement-type MOSFET.

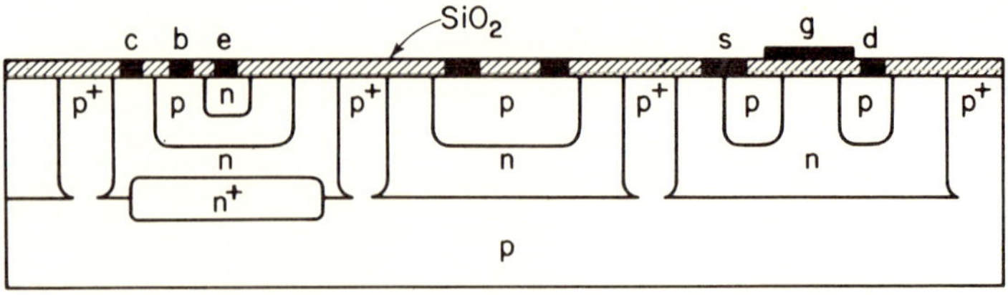

Fig. 11-30. Junction-isolated MOSFET formed simultaneously with other monolithic components.

using the *normal* base diffusion of the BJT, and no compromises are necessary.

Complementary MOSFETs.[20,*] The formation of *n*-channel and *p*-channel, enhancement-type MOSFETs on the same wafer is highly desirable, because of the attractiveness of complementary-symmetry switching circuits (see Sec. 10-5.1). Two techniques for producing complementary monolithic devices are shown in Fig. 11-31. In both methods, the *p*-channel device is

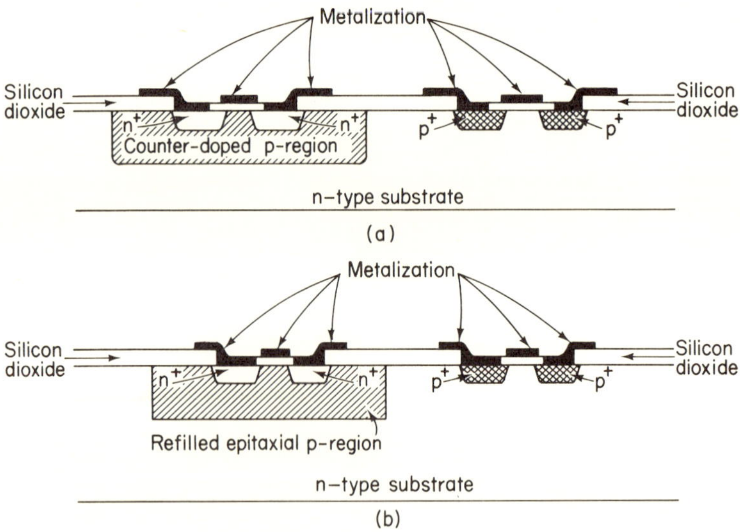

Fig. 11-31. Complementary MOSFET structures using: (a) the substrate counter-doping technique, (b) the epitaxial refill technique.

[20] P. Richman, *Characteristics and Operation of MOS Field-Effect Devices*. New York: McGraw Hill Book Co., Inc., 1967, pp. 132–134.

* Reference 12, pp. 424–428.

formed directly on a high-resistivity substrate, as in the structure of Fig. 11-28, and the *n*-channel FET is fabricated in a *p*-region formed on the substrate. The techniques differ in the formation of the *p*-region for the *n*-channel device, the region being formed either by a counter-doping diffusion [Fig. 11-31(a)] or by epitaxial growth to refill slots etched in the wafer [Fig. 11-31(b)]. Isolation is achieved by the depletion region in the *n*-type substrate, which is connected to the *most positive* point in the circuit. Complementary MOSFETs may also be produced by using dielectric-isolation and silicon-on-sapphire techniques (see Sec. 11-10.3).

Depletion MOSFETs. For depletion-type MOSFETs, a built-in channel exists under the silicon-dioxide layer for zero applied gate voltage, as pointed out in Sec. 10-2. The channel is formed by an inversion layer (layer of conductivity type opposite that of the substrate) that exists under the entire oxide surface. Therefore, like the monolithic JFET, either the source or drain must be enclosed by the gate to prevent drain-to-source leakage current around the gate. A possible topology for an *n*-channel device is shown in Fig. 11-32.* In addition to the n^+ diffusions that make contact to the channel,

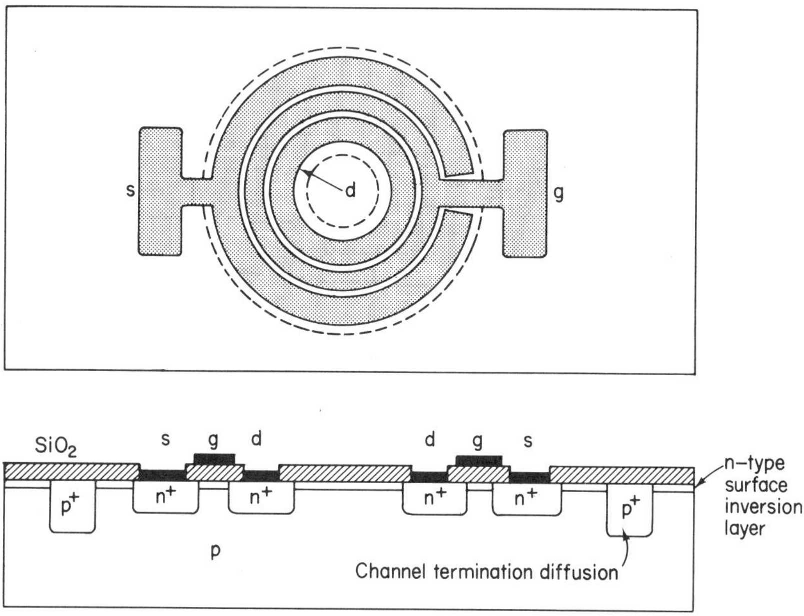

Fig. 11-32. Topology and geometry for *n*-channel depletion-type MOSFETs.

* At present, *p*-channel depletion-type MOSFETs have not been successfully made, owing to instabilities related to charges in the oxide layer.

a p^+ diffusion completely surrounds the FET to terminate the inverted layer and prevent leakage current between the FET and other devices on the substrate. Isolation is achieved in this structure by connecting the substrate to the *most negative* point in the circuit.

11-10.3 Silicon-On-Sapphire (SOS) MOSFETs[21]

In the silicon-on-sapphire dielectric-isolation scheme introduced in Sec. 11-4, integrated-circuit components are formed by the usual masking, oxidation, and diffusion techniques in islands of single-crystal silicon on an insulating substrate of sapphire. Figure 11-33 shows the structures and drain characteristics of three types of MOSFETs that have been made by the SOS process.* The nonjunction *n*-channel enhancement-type device of Fig. 11-33(c) is of particular interest, since it differs from devices (both discrete and integrated) considered so far. It depends for its operation on a thin film of high-resistivity silicon that forms a channel having high resistance unless its conductivity is enhanced by electrons attracted into the channel when the gate is positive. Such devices cannot be fabricated in bulk silicon.

The SOS process also lends itself to the fabrication of complementary enhancement-type MOSFETs on a common substrate, as illustrated in Fig. 11-34.† The *p*-channel device of this structure is seen to be a nonjunction MOSFET.

11-10.4 Thin-Film Transistors (TFTs)[22,23,‡]

The thin-film transistor is an insulated-gate FET that employs a structure and method of operation quite similar to the nonjunction SOS MOSFET, but differs in its fabrication technique and material makeup. The TFT is not a silicon device, nor are diffusions made during the fabrication process. Instead, thin films of different materials are deposited on top of each other to form the structure.

Two of several possible TFT structures are shown in Fig. 11-35. The

[21] C. W. Mueller and P. H. Robinson, "Grown-Film Silicon Transistors on Sapphire," *Proc. IEEE*, Vol. 52, pp. 1487–1490, (Dec., 1964).

[22] P. K. Weimer, "The TFT—A New Thin-Film Transistor," *Proc. IEEE*, Vol. 50, pp. 1462–1469 (June, 1962).

[23] J. T. Wallmark and H. Johnson, *Field-Effect Transistors, Physics Technology and Applications*. Englewood Cliffs, N.J.: Prenctice-Hall, Inc., 1966, pp. 216–251.

* Reference 16, p. 173.

† Reference 20, p. 134.

‡ Reference 5, pp. 472–479.

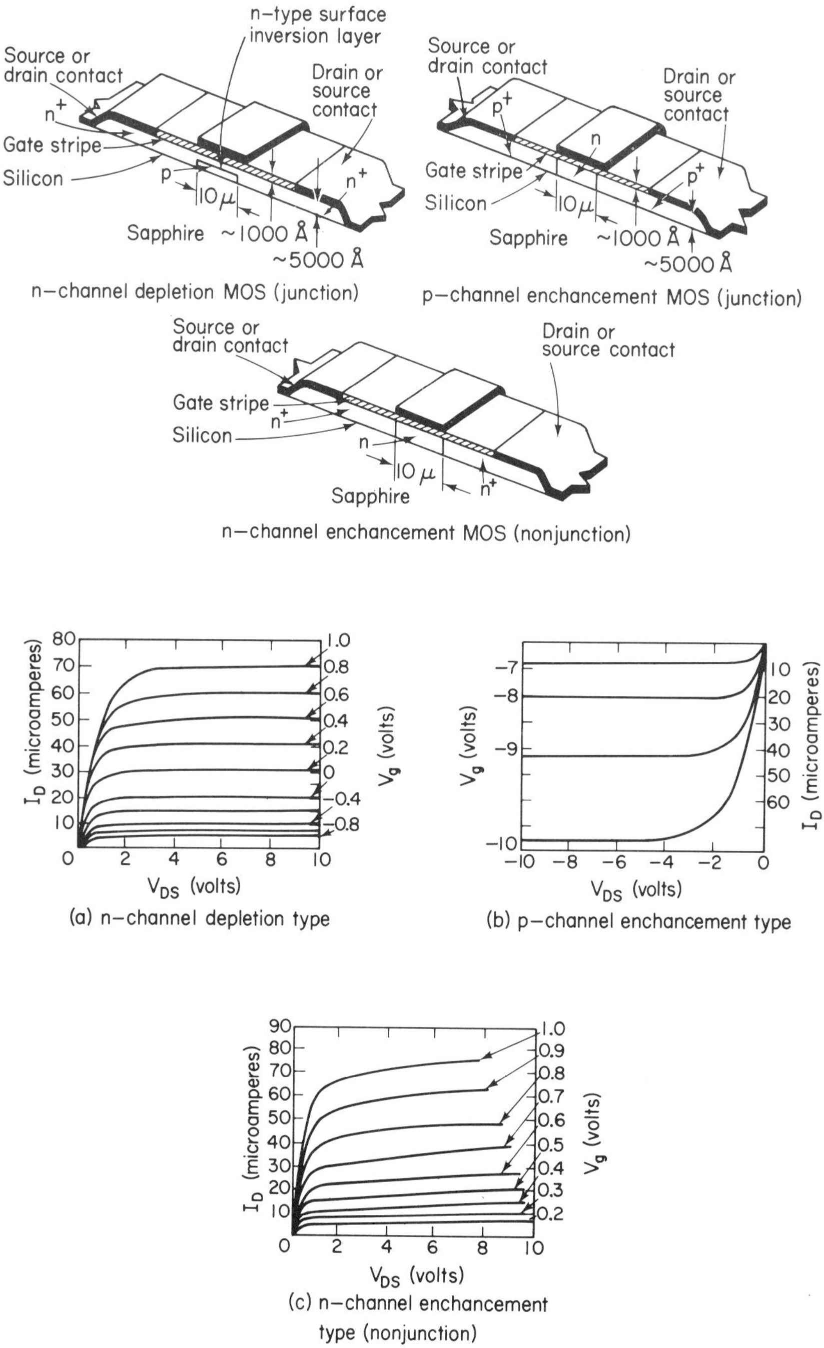

Fig. 11-33. Structures and characteristics of three types of MOSFETs made using the silicon-on-sapphire process. (From Reference 16, p. 173.)

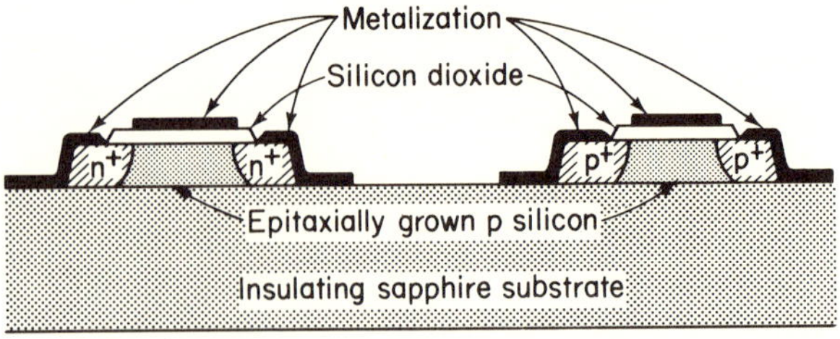

Fig. 11-34. A silicon-on-sapphire complementary MOSFET structure. (From Reference 20, p. 134.)

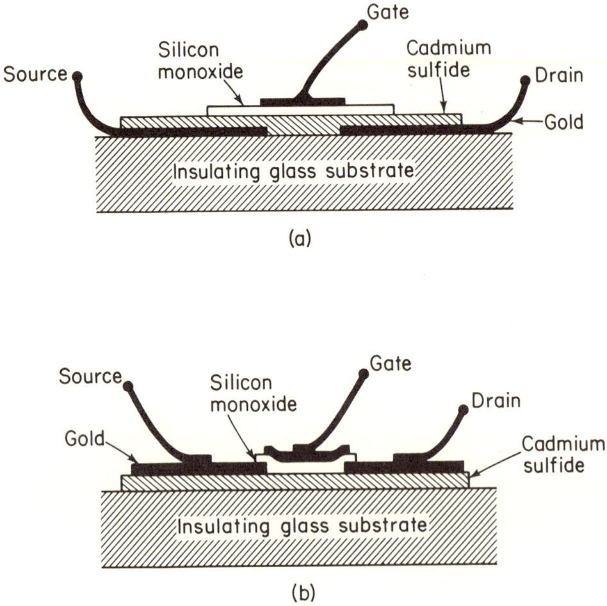

Fig. 11-35. Thin-film transistor structures.

semiconductor is an evaporated film of cadmium sulfide* (CdS), and the insulator is a deposited layer of silicon monoxide (SiO). Metal electrodes formed by gold evaporation are used as gate, drain, and source regions. The evaporations either take place through masks that prevent coating the substrate in undesired areas, or the films are deposited uniformly over the entire substrate and then selectively removed from the undesired areas. The isolation of components is excellent, as a result of the insulating glass substrate. Not shown in Fig. 11-35 is an encapsulating overcoat, consisting of a

* Other semiconductor materials used in TFTs are cadmium selenide (CdSe)—also an *n*-type material—and tellurium (Te), a *p*-type material used in *p*-channel TFTs.

nonporous layer of material such as silicon dioxide, that may be deposited over the top of the structure for protection against contamination.

The operation of the TFT is much like that of the nonjunction SOS MOSFET. If a high-resistivity (say above 20 ohm-cm) CdS or CdSe layer is used, little drain current flows when there is no gate voltage and the TFT operates in the enhancement mode. As in the *n*-channel MOSFET, positive

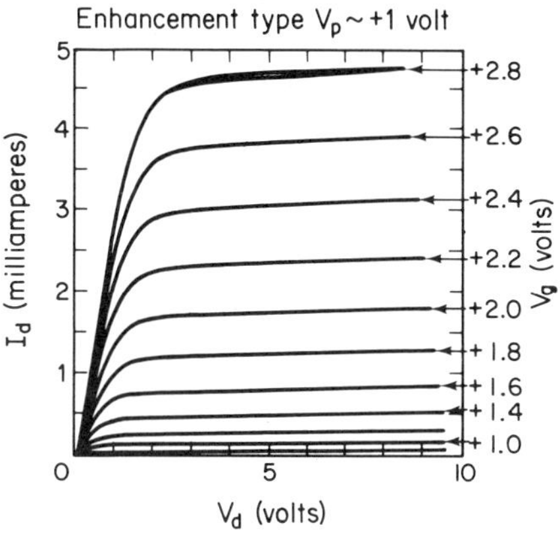

(a) Operating characteristics for an enhancement-type cadmium sulfide thin-film transistor. Drain current is plotted against voltage for different values of gate voltage. The source is grounded.

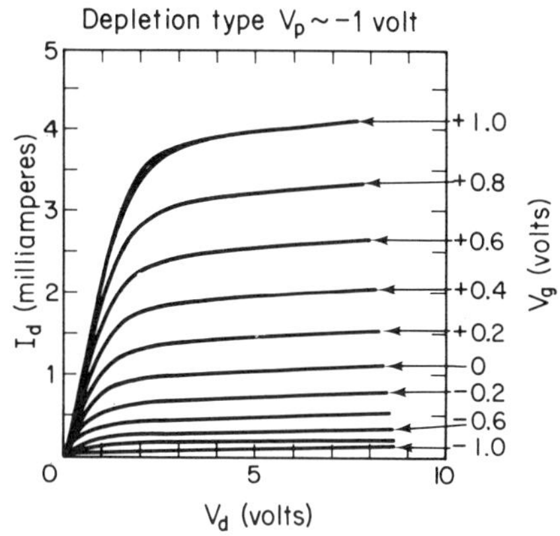

(b) Operating characteristics for a depletion-type cadmium sulfide thin-film transistor. Such a unit can be operated in either the enhancement or depletion mode.

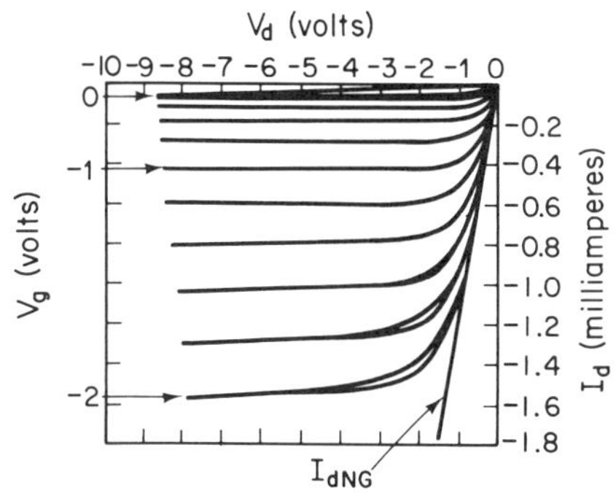

(c) Drain characteristics of a p-type tellurium TFT. The straight line labeled I_{dNG} is the current-voltage curve for an identical tellurium TFT from which the gate was omitted.

Fig. 11-36. Drain characteristics of thin-film transistors. (a) *n*-channel enhancement type. (b) *n*-channel depletion type. (c) *p*-channel enhancement type. (From Reference 23, p. 225.)

gate voltage attracts additional electrons into the semiconductor film and increases its conductivity. If a lower-resistivity semiconductor film is used, the drain current for zero gate voltage is significant and the result is a depletion-type TFT. Like the depletion-type MOSFET, the device can be operated with either positive or negative gate voltage that either enhances or depletes channel carrier density. Drain characteristics of experimental TFTs are shown in Fig. 11-36.*

Thin-film technology has long promised big advantages over silicon monolithics for low-cost, uniform, large-area integrated circuits, since the fabrication process is simpler and devices can be formed on glass substrates several inches on a side in arrays much too large for formation on silicon wafers. However, these promises have not been fulfilled, owing to TFT instabilities that have precluded their use in commercial thin-film ICs. Like early enhancement-type MOSFETs, TFTs have been subject to gate threshold-voltage drifts of the order of volts during operating periods of a few hours to a few days. After tracing the trouble to the SiO gate insulator and failing to obtain significant improvement in SiO devices, researchers were led to the use of aluminum-oxide (Al_2O_3) as insulation. New TFTs using Al_2O_3 as the gate insulator, in the inverted structure illustrated in Fig. 11-37, have been found to be extremely stable devices[24].

These new TFTs have excellent characteristics for digital applications. Their high transconductance (about 1000 μmhos) and low threshold voltage (less than 1 V), together with the availability of complementary devices, suggests the early realization of commercial active-element thin-film ICs. For example, an experimental thin-film circuit containing 360 CdSe TFTs, 180 field-effect diodes (obtained by connecting TFT gates and drains together), 360 nichrome resistors, and 180 capacitors has been life-tested successfully.[25] Another experimental thin-film IC has a total of 1320 CdSe TFTs

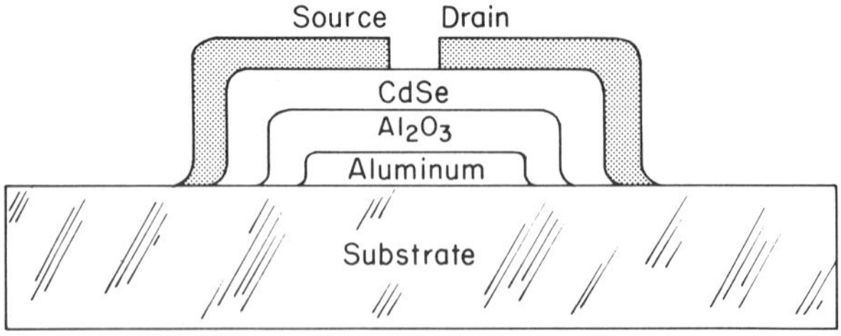

Fig. 11-37. Inverted thin-film transistor structure using aluminum oxide as the gate insulator. (From Reference 24, p. 91.)

* Reference 23, p. 225.

[24] A. Waxman, "Thin-Film Transistors Don't Have To Be Drifters," *Electronics*, Vol. **41**, No. 6 (March 18, 1968).

[25] P. K. Weimer, G. Sadasiv, J. E. Meyer, L. Meray-Horvath, and W. S. Pike, "A Self-Scanned Solid State Image Sensor," *Proc. IEEE*, Vol. **55**, No. 9, (September 1967), p. 1591.

and 792 Te TFTs on a single glass substrate, pairs of CdSe and Te TFTs comprising complementary inverters.*

11-11 MOS INTEGRATED CIRCUITS[26]

Since the development of field-effect transistors lagged that of bipolar transistors, the development of BJT and FET ICs has proceeded in like manner. With BJT ICs firmly entrenched before the advent of discrete FETs, it seemed unlikely that the lower-speed, higher ON-mode resistance FET had much to offer digital integrated circuitry. However, this was changed by the enhancement-type MOSFET with its simpler circuitry, fewer fabrication processes, lower power requirement, and smaller surface area. As a result, MOS ICs are replacing BJT ICs in lower-speed computers and in parts of high-speed computers, such as memories, that operate at relatively low speeds. In fact, it has been predicted that MOS ICs will eventually capture at least 50 percent of the digital integrated-circuit market.†

Examination of the physical layout, structure, and schematic diagram of a MOS integrated circuit discloses the total lack of conventional components. That is, MOS ICs contain only MOS devices, there being no individual resistors, capacitors, or diodes.‡ Because of this, we single out MOS IC structures and layouts for consideration, while noting that other ICs (bipolar ICs, SOS ICs, thin-film ICs, etc.) are merely combinations of individual integrated components considered in preceding sections. In this section, we discuss MOS devices as resistors, inverters, and gates—the basic building blocks of MOS integrated circuits.

11-11.1 The MOS Resistor[27, 28, §]

The use of an enhancement-type MOSFET in place of the drain resistor in a MOS inverter is considered in Sec. 10-5.1 and illustrated in Fig. 11-38.

* Reference 24, p. 93.

† *Electronic News*, June 23, 1969, Sec. 2, p. 8.

‡ As used at present, the term "MOS integrated circuit" refers to digital (switching) circuits containing only enhancement-type MOS devices (except for protective diodes across input gates).

§ Reference 20, pp. 121–126.

[26] D. E. Farina and D. Trotter, "MOS Integrated Circuits Save Space and Money," *Electronics*, Vol. **38**, No. 20 (October 5, 1965).

[27] L. Vadasz, "The Use of MOS Structure for the Design of High-Value Resistors in Monolithic Integrated Circuits," *IEEE Trans. on Electron Devices*, Vol. **ED-13**, No. 5 (May 1966) pp. 459–465.

[28] R. H. Crawford, *MOSFET in Circuit Design*. New York: McGraw Hill Book Co., Inc., 1967, pp. 87–92.

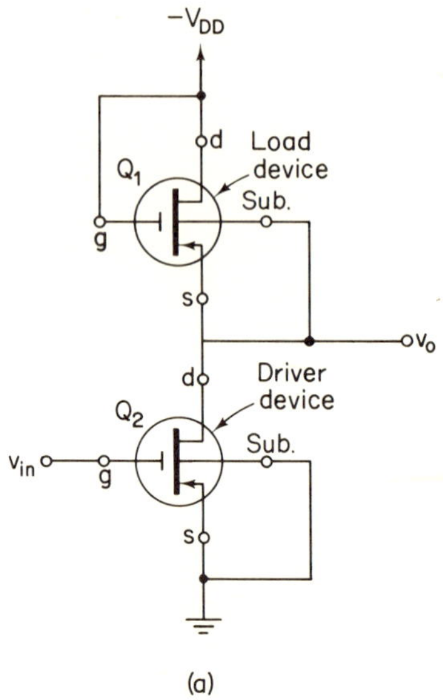

Fig. 11-38. Inverters using *p*-channel enhancement-type MOSFETs as drain resistors. (a) Discrete circuit schematic. (b) Schematic for an integrated circuit.

Such a MOSFET is known as *MOS resistor*. There is little reason for using MOS resistors in discrete circuitry (except for complementary-symmetry circuits), but there is no reason for any other type resistor in MOS integrated circuits. The MOS resistor occupies far less surface area than the conventional diffused resistor, requires no additional diffusion steps for fabrication, and provides the high resistance that is highly desired for MOS circuits but impractical if diffused resistors are used. The latter may be seen from the sheet resistances, which are typically 200 Ω per square for diffused resistors and 20,000 Ω per square for MOS resistors. Thus, a 100-kΩ MOS resistor requires approximately 5 square mils of active area, in a pattern 5 mils long by 1 mil wide, compared with 500 square mils in a 500-by-1 mil pattern for a comparable diffused resistor. The impracticality of the latter structure is quite clear.

Two possibilities for connecting MOSFET substrates are shown in Fig. 11-38. In Fig. 11-38(a), the substrate of each FET is connected to the source (the usual arrangement when discrete devices are used), but both substrates are grounded in Fig. 11-38(b). The latter arrangement is mandatory in MOS ICs, since all devices share a common substrate that is grounded in order to achieve isolation. (That is, the *n*-type substrate is connected to the most positive point in the circuit.)

The MOS resistor of Fig. 11-38, referred to as the load device, is seen to be an enhancement-type MOSFET operated with equal gate and drain potentials. The locus of operation of the load device may be plotted on its drain characteristic, as illustrated in Fig. 11-39(a), showing that it operates in the pinch-off region. Using this locus (recognizing that the drain voltage of the driver device is the difference between the supply voltage and the load-device voltage and that the device currents are the same), we may obtain the path of operation (load line) on the driver-device drain characteristic, as shown in Fig. 11-39(b). The nonlinear operating path indicates that the load device is nonlinear, but this is of no concern in digital circuitry.

When the driver device is switched ON by gate voltage $-v_{G1}$, operation is shown at point P_1. A driver device having a channel geometry providing low resistance, relative to that of the load device, is assumed. The voltage across the driver device is then small compared to the supply voltage so that $v_{ON} \cong 0$ V. When the driver is switched OFF, operation is at point P_2, where v_{OFF} is the difference between the magnitudes of the supply voltage and the threshold voltage V_{th} of the load device. Thus, we see that the output voltage swing is less than the full supply voltage even though $v_{ON} = 0$ V. This reduction in output occurs because a voltage equal to V_{th} must be across the load device to initiate conduction.

The resistance of the load device may be obtained from its equivalent circuit. From Eq. (10-55), we find that the impedance seen looking into the source of a FET is approximately $1/g_m$. If, for example, $g_m = 10$ μmhos (for the load device, of course) we see that $R = 1/g_m = 100$ kΩ. This might be the load resistance when the driver device is ON. With the driver device OFF, the load resistance is much higher, owing to the lower load transconductance.

11-11.2 Integrated MOS Inverter*

Figure 11-40 shows the schematic diagram, a scale drawing of the physical layout, and a cross-section sketch of a practical, *p*-type, MOS integrated inverter. The schematic is identical to that of Fig. 11-38(b), and is reproduced for reference purposes.

In the physical layout [Fig. 11-40(b)], the load device Q_1 is at the top and the driver device Q_2 is at the bottom. The bottommost p^+ diffusion serves as the source of the driver device, the top p^+ diffusion is the drain of the load device, and the middle diffusion acts as both the load-device source and the driver-device drain. This utilization of a common diffused area for more than one function greatly reduces the chip area required by MOS circuits. Another space saving in MOS ICs occurs because no isolation

* Reference 28, pp. 92–103.
* Reference 28, p. 94.

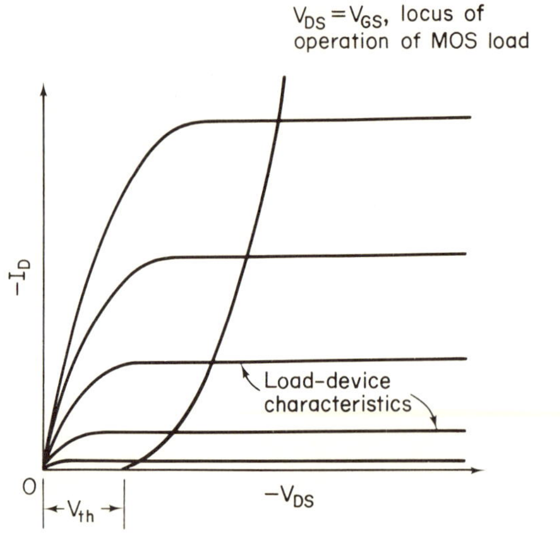

(a)

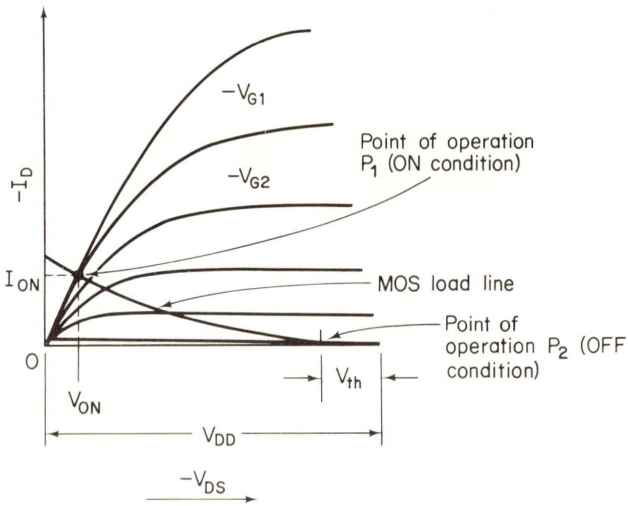

(b)

Fig. 11-39. Drain characteristics of the devices of Fig. 11-38. (a) Load-device characteristic, showing locus of operation for $v_{GS} = v_{DS}$. (b) Driver-device characteristic, showing path of operation (load line) for the MOS load of (a).

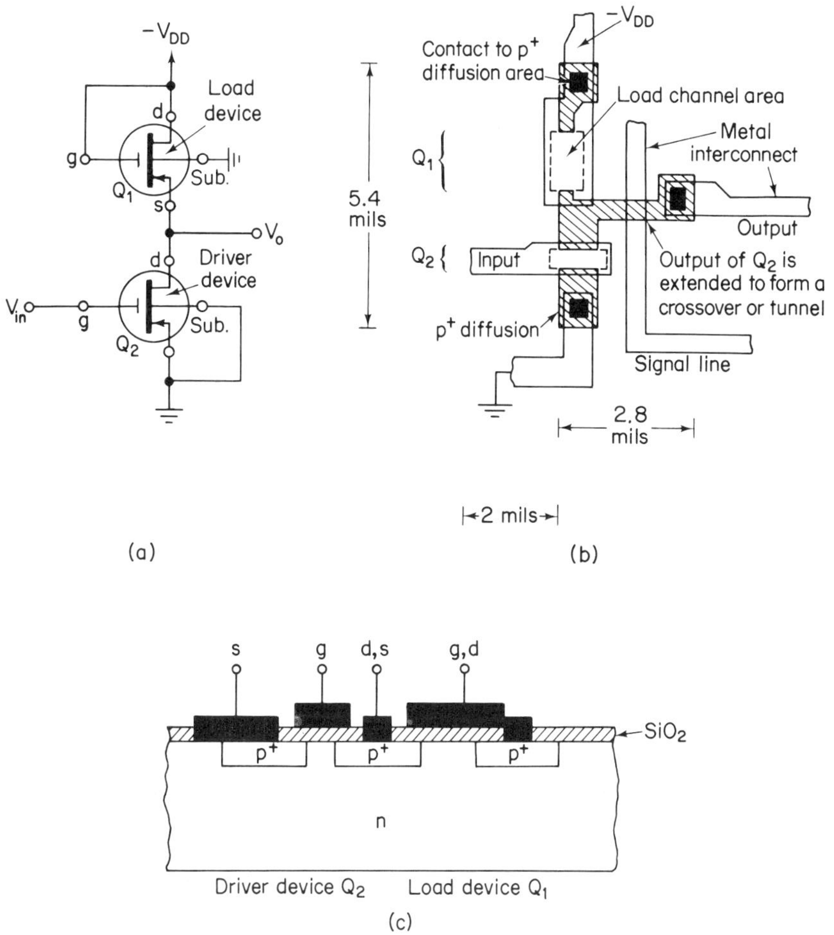

Fig. 11-40. Practical MOS inverter. (a) Schematic diagram. (b) Physical layout (to scale).* (c) Cross-section sketch.

diffusions are required between devices. These savings, plus the smaller surface area required by MOSFETs* and MOS resistors relative to BJTs and diffused resistors, result in MOS circuits that require about one-fifth the chip area of comparable BJT circuits.[29]

In the layout of Fig. 11-4(b), the gate and drain of the load device are connected together by the metallization, and the gate metal of each device

* The area requirement of a MOSFET, relative to the BJT, may be seen by comparing the scale drawings of Figs. 11-20 and 11-40(b).

[29] R. M. Warner, Jr., "Comparing MOS and Bipolar Integrated Circuits," *IEEE Spectrum*, Vol. **6**, No. 6 (June 1966).

extends past the p^+ diffusions to prevent fringing effects. The lateral extension of the middle diffusion is particularly significant. This is a crossover or tunnel (under the SiO_2 and metal interconnect) interconnect to the output metallization. Diffused interconnections (crossovers) constitute another important advantage of MOS ICs. Unlike those in BJT ICs, the resistive effects of diffused crossover regions (having sheet resistances of the order of 100 Ω per square) are negligible, since they are in series with large-value load resistors and high-impedance inputs. The different geometries of the channel areas of the MOS devices of Fig. 11-40 should be noted. Clearly, for equal gate voltages, the load device has much the larger channel resistance (lower g_m), as required for the operation illustrated in Fig. 11-39(b).

11-11.3 Integrated MOS Gates*

Figure 11-41 shows the schematic diagram and a possible physical layout of a three-input MOS NOR gate. The parallel arrangement of MOS devices, as shown in Fig. 11-41(a), lends itself to MOS integrated circuitry, in that the load-device source and drains of the driver devices can be a

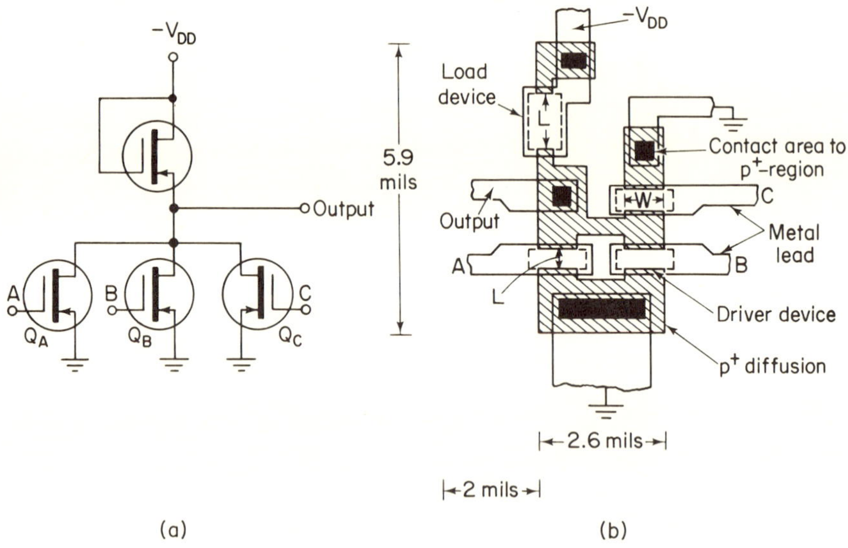

Fig. 11-41. Three-input MOS NOR gate. (a) Schematic diagram. (b) Physical layout. (Scale drawing from Reference 28, p. 104.)

* Reference 28, pp. 103–107.

single diffused region. This diffused-region sharing, as well as that of driver-device sources, is seen in the physical layout [Fig. 11-41(b)]. The separate drain region of driver device C is not a necessity, but it facilitates making connections to the circuit. Again, the geometry of the device channels should be noted. It is seen that the driver devices have the same L/W ratio, making them equal-g_m devices, and the larger L/W ratio of the load device results in a lower g_m and a larger channel resistance. The MOS devices are isolated by grounding the substrate—a connection not shown in the schematic [Fig. 11-41(a)], since its existence is understood.

NAND gate. The schematic diagram and a possible physical layout of a three-input MOS NAND gate are shown in Fig. 11-42. The four devices are seen to require only five diffused regions, each of three regions being shared by two devices. It is of interest to compare the driver-device channel geometry with that of the NOR gate. Since the devices are in series, their resistances add and, for the same v_{ON}, each resistance must be one-third that of a single device. Comparable L/W ratios in Figs. 11-41(b) and 11-42(b) are seen to have this relationship.

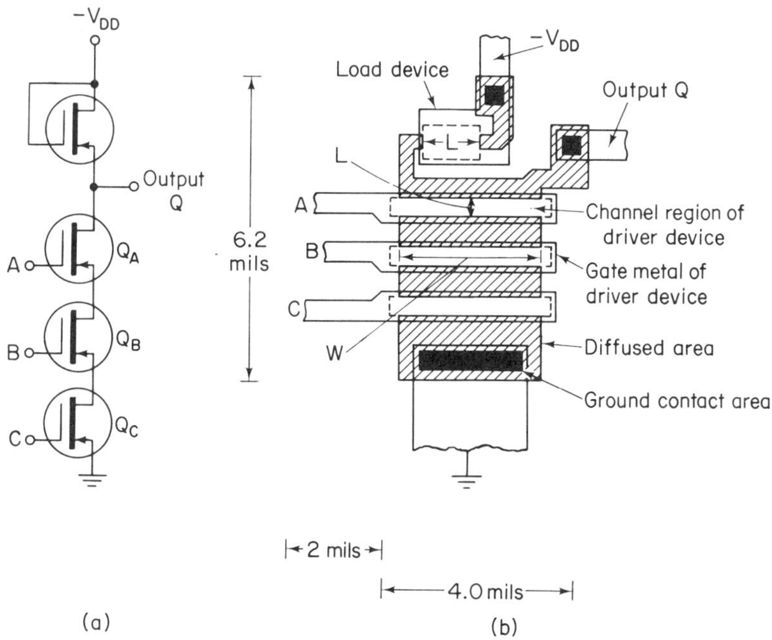

Fig. 11-42. Three-input MOS NAND gate. (a) Schematic diagram. (b) Physical layout. (Scale drawing from Reference 28, p. 105.)

11-11.4 Complementary-Symmetry MOS (CMOS) ICs[30]

The MOS integrated circuits of Figs. 11-40 through 11-42 use *p*-channel devices, as do most MOS ICs at present. This is because *p*-channel enhancement-type devices have been easier to fabricate than those of the *n*-channel type, as pointed out in Sec. 10-2. Problems in fabricating *n*-channel enhancement-type devices* and the added difficulty of forming *p*- and *n*-channel devices on the same substrate have delayed the use of complementary-symmetry MOS (CMOS) ICs. However, CMOS ICs are on the market and are said to be cost competitive with BJT ICs for military use now and for commercial use within two years.† The big advantage of complementary-symmetry switching circuits is their extremely low power requirement, as pointed out in Sec. 10-5.1. An added advantage is that, unlike *p*-type circuits, they operate from positive-voltage supplies (either polarity may be used) and interface more easily with positive-voltage bipolar circuits.

11-12 LINEAR INTEGRATED CIRCUITS[31,32,‡]

The development of linear (analog) integrated circuits has lagged that of digital circuitry for many reasons. Among these are the larger variety of components required, the supposed need for lower-tolerance components, and the existence of a ready-made market for digital ICs while linear ICs were not equally sought after. However, these problems have been solved, by and large, and linear IC sales are rapidly catching up with those of digital ICs. The answer to the component-variety requirement has been to combine integrated and discrete components either in hybrid ICs or in circuits having discrete components external to the IC. The component-tolerance problem is circumvented through use of an amplifier circuit having a different type of problem. The differential amplifier considered in Sec. 8-6 requires closely matched components (including active devices) that need not have close tolerances—a condition that is difficult to achieve in discrete circuitry, but readily accomplished in monolithic circuits. Thus, the differen-

* Now solved through use of silicon-nitride passivation.

† *Electronic News*, June 23, 1969, Sec. 2, p. 8.

‡ Reference 3, pp. 129–146; Reference 4, Part 3; Reference 12, pp. 258–261.

30 G. B. Herzog, "COS/MOS: The Best of Both Worlds," *Electronics*, Vol. **42**, No. 4 (February 17, 1969) pp. 109–113.

31 J. Eimbinder, *Linear Integrated Circuits, Theory and Practice*. New York: John Wiley & Sons, Inc., 1968.

32 RCA Linear Integrated Circuits Manual, Technical Series IC-41. Harrison, N.J.: Radio Corporation of America, 1967.

tial amplifier is the "heart" of most linear monolithic ICs.[33,34,*] Finally, the market developed as it became generally recognized that linear ICs are technically equivalent to, and more economical than, comparable discrete-component circuits. There are, after all, many more applications for linear ICs than for those of the digital type.

Linear monolithic ICs are basically direct-coupled amplifiers fabricated by using components, physical layouts, and structures such as those discussed in preceding sections. Therefore, our consideration is restricted to schematic diagrams and applications.

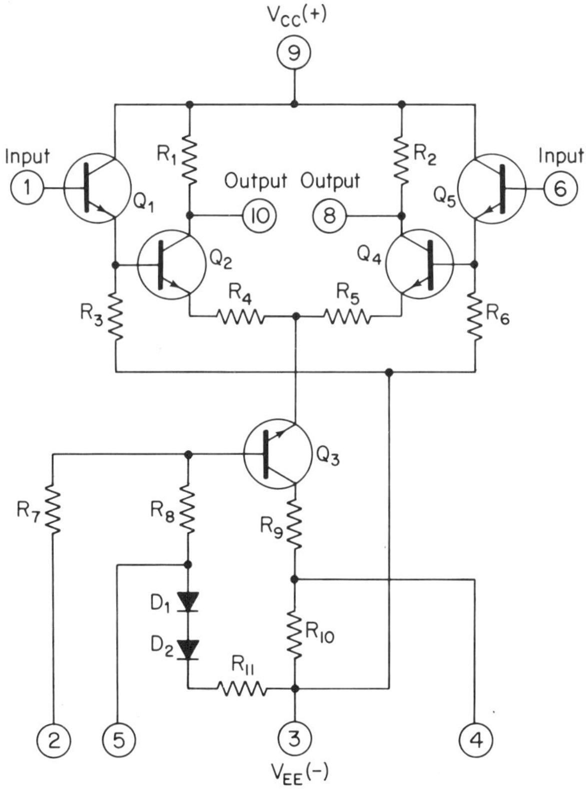

Fig. 11-43. Schematic diagram for the RCA CA3000 monolithic d-c amplifier. (From Reference 32, p. 114.)

* For information on single-ended (nondifferential-amplifier) ICs, see Reference 31, pp. 121–145.

[33] J. J. Robertson, "Linear ICs: Part 2, Heart of the Matter"; *Electronics* Vol. **40**, No. 17 (August 21, 1967).

[34] J. P. Keller, "Linear ICs: Part 3, Differential Amplifiers at Work," Electronics, Vol. **40**, No. 19 (September 18, 1967).

Most linear monolithic ICs in use at present consist of BJTs,* diodes, and diffused resistors in differential-amplifier arrangements like that of the CA3000 shown in Fig. 11-43.† The CA3000 is basically a single-stage differential amplifier (Q_2 and Q_4) with input emitter-followers (Q_1 and Q_2) and a constant-current source (or sink) Q_3 in the emitter-coupled leg. Transistor Q_3 replaces the large-value resistor usually found in discrete-circuit differ-

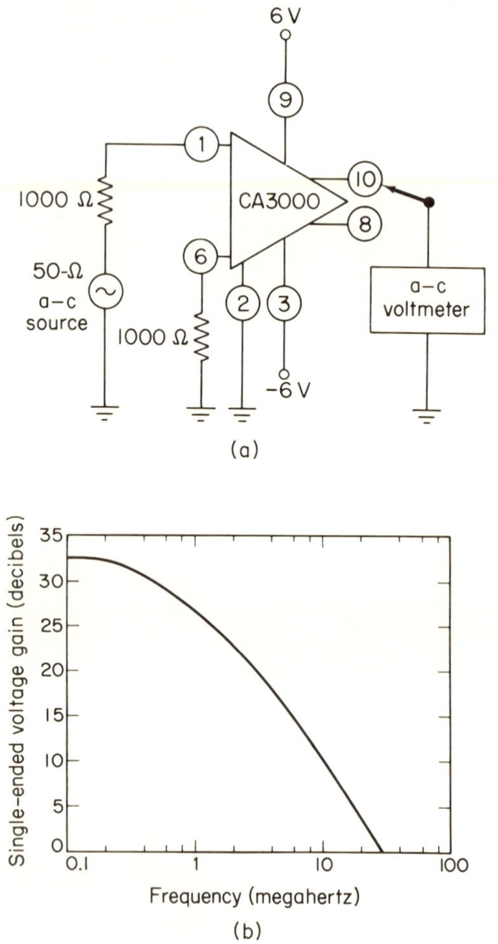

Fig. 11-44. Operation of the CA3000 as a single-ended amplifier. (a) Circuit arrangement. (b) Frequency response for the arrangement of (a). (From Reference 32, p. 120.)

* Since the bipolar IC was firmly entrenched before the advent of FETs, the use of FETs in linear ICs is just beginning.

† Reference 31, p. 71.

ential amplifiers; degenerative resistors R_4 and R_5 increase the linearity of the circuit, and diodes D_1 and D_2 are for purposes of temperature compensation. The CA3000 is supplied in a ten-lead TO-5 style package—the package style commonly used for a single low-power discrete BJT.

Figure 11-44(a) shows operation of the CA3000 as a single-ended a-c amplifier, and the frequency response of the arrangement is shown in Fig. 11-44(b). The gain of the amplifier may be reduced by applying to terminal 2 a negative voltage that reduces the base current of Q_3, lowering the d-c collector currents of Q_2 and Q_4. Automatic gain control (AGC) results when this voltage is proportional to amplifier output (through a rectifying-

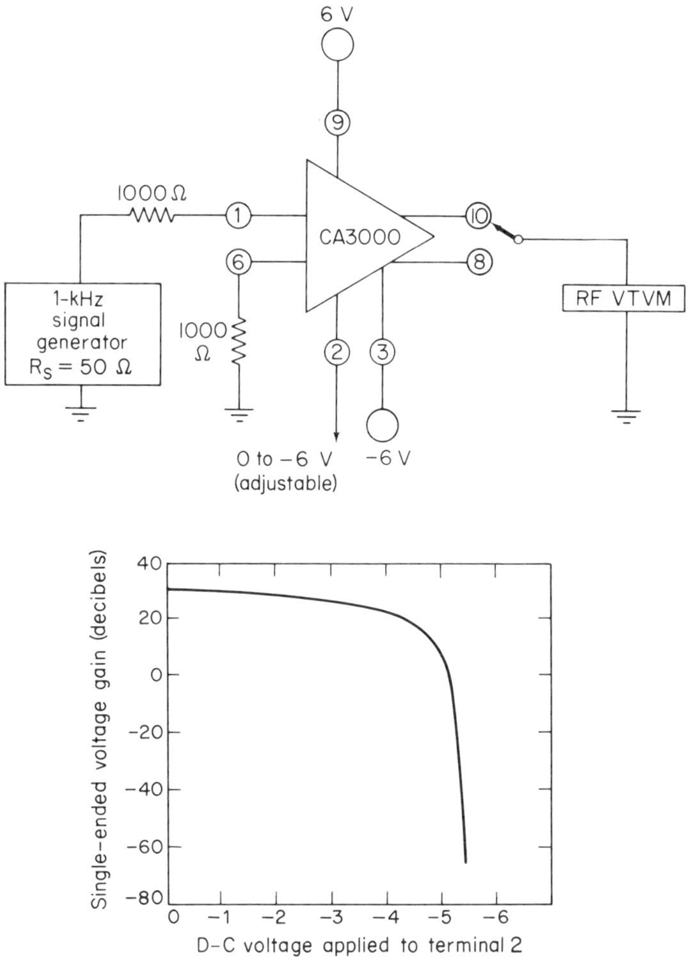

Fig. 11-45. Automatic gain control characteristic, in the test circuit shown, at a frequency of 1 kHz. (From Reference 32, p. 121.)

filtering process). The AGC characteristic of the CA3000 and the test circuit for its determination are shown in Fig. 11-45. Amplifier gain control may also be achieved by applying a voltage to terminal 5.

Applications of the CA3000. The gain-control capability of the CA3000 makes possible its use as a signal switch, squelchable amplifier, modulator, signal mixer, etc. With the addition of appropriate external components, it can also be used as an oscillator, a monostable multivibrator, a trigger circuit, or a narrow-band amplifer. We shall consider only amplifier applications.

The use of the CA3000 as a single-ended wide-band amplifier is shown in Figs. 11-44 and 11-45. Either terminal 8 or terminal 10 may be used for the output, depending on the desired phase relationship, or an output balanced with respect to ground may be obtained from terminals 8 and 10. The source impedance is not limited to 1050 Ω, but may be as high as 10,000 Ω without significantly affecting the frequency response, owing to the high-input-impedance emitter follower. In normal wide-band amplifier use, the inputs are returned to ground through approximately equal resistances.

Two CA3000 units may be operated in cascade to form an R-C coupled feedback amplifier, as in the arrangement of Fig. 11-46.* Alternate input terminal 6 of the first IC provides a convenient point for inserting the nega-

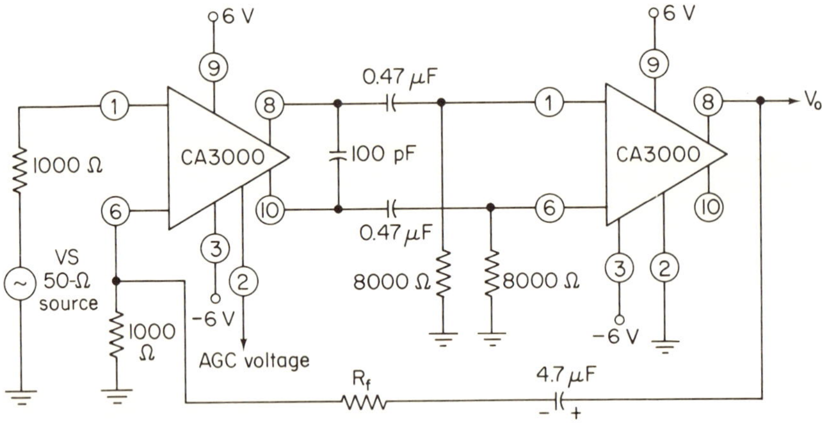

For $V_{AGC} = 0$

R_f	G_V
∞	63 dB (open loop)
100,000 Ω	40 dB
9000 Ω	20 dB

Fig. 11-46. Cascaded R-C coupled feedback amplifier using two CA3000 linear ICs. (From Reference 32, p. 126.)

* Reference 32, p. 126.

tive-feedback signal obtained through feedback resistor R_f from the output of the second IC. The effect of R_f on mid-frequency gain is also shown by Fig. 11-46. Figure 11-47 shows the frequency-response and AGC characteristics for no feedback (R_f of infinity) as well as AGC characteristics for two feedback conditions. The low-frequency roll-off of the amplifier is determined by the interstage coupling.

The high input and output impedances of the CA3000 make possible

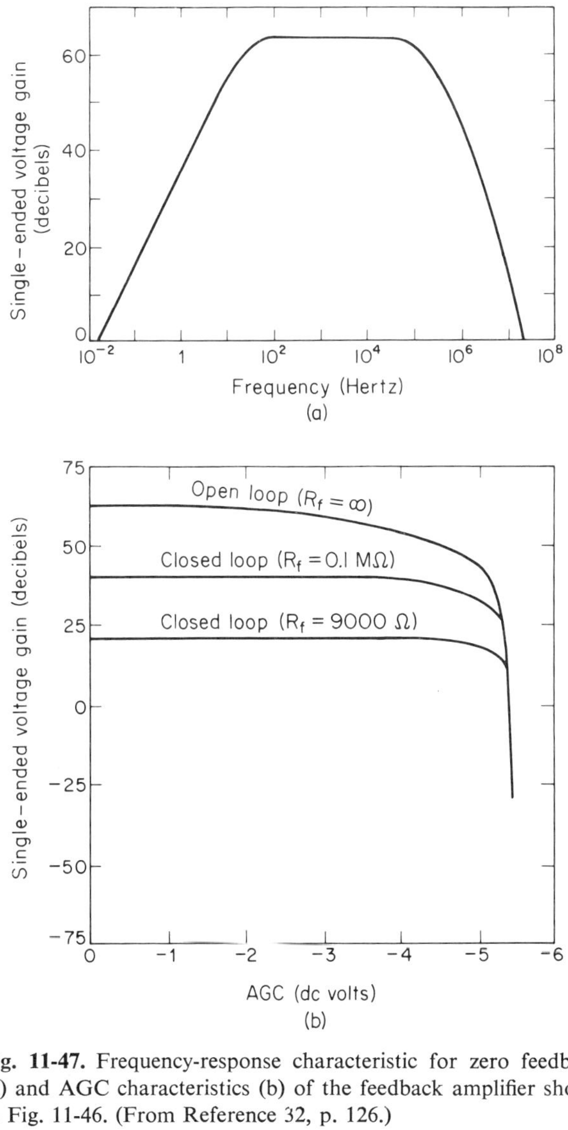

Fig. 11-47. Frequency-response characteristic for zero feedback (a) and AGC characteristics (b) of the feedback amplifier shown in Fig. 11-46. (From Reference 32, p. 126.)

its use in tuned-input tuned-output applications such as the narrow-band amplifier arrangement of Fig. 11-48. Figure 11-48 also shows the response curve of the amplifier when tuned to a frequency of 10 MHz.

In using the CA3000 as a narrow-band amplifier, the collector of the

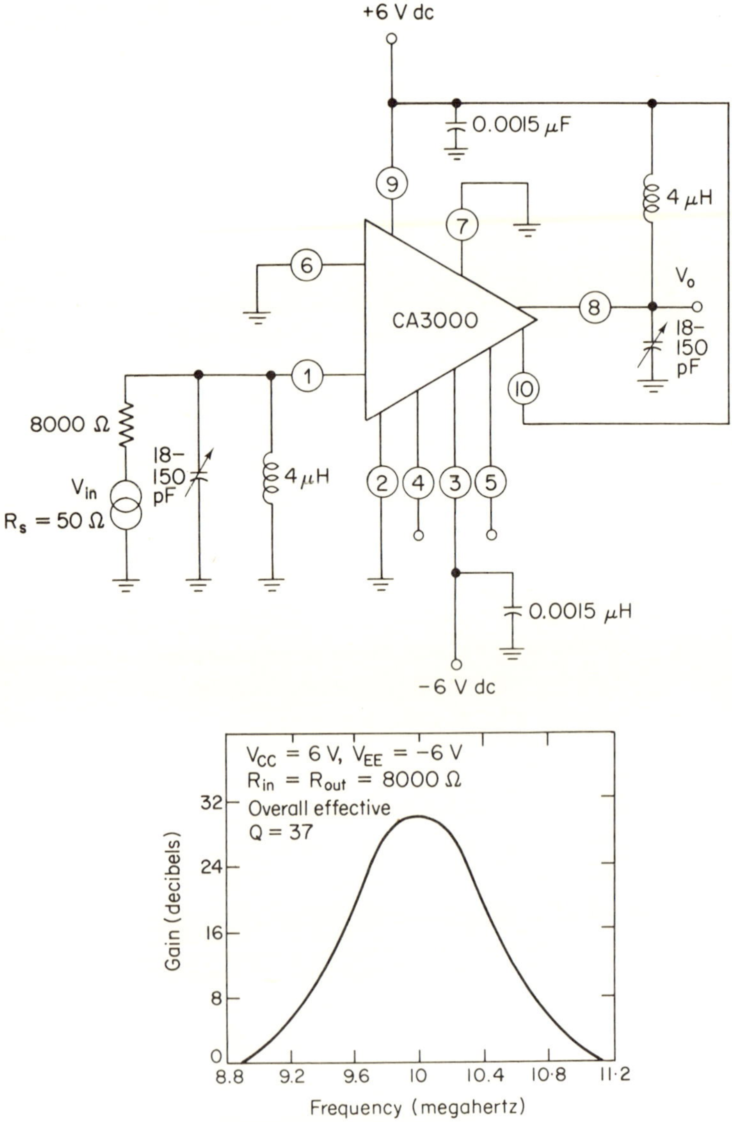

Fig. 11-48. Schematic diagram and frequency-response curve for a 10-MHz tuned-input, tuned-output narrow-band amplifier using a CA3000 linear IC. (From Reference 32, p. 128.)

driven transistor (terminal 10) is returned to a-c ground, shorting collector resistor R_1 (see Fig. 11-43), and collector resistor R_2 is shunted by the output-tuning inductor. Since neither resistor has a useful function, they are eliminated in linear monolithic ICs designed specifically for this type of application. The schematic of such an IC, the CA3005/CA3006, is shown in Fig. 11-49(a), and its arrangement as a narrow-band amplifier is illustrated in Fig. 11-49(b).

Operational amplifiers.[35,*] Instead of cascading ICs in order to obtain higher gain, as in the arrangement of Fig. 11-46, differential-amplifier stages may be cascaded on the same semiconductor chip. A high-gain (60 db, or more), high-input impedance (10 kΩ, or more), wide-band (dc to 1 MHz, or more), direct-coupled amplifier obtained in this way is known as a monolithic *operational amplifier*, because it can be used to perform various mathematical operations such as integration, differentiation, summation, and subtraction. Since it is also suitable for communications applications, the operational amplifier has been by far the most popular linear integrated circuit.

Figure 11-50 shows the schematic diagram of the CA3015 I-C operational amplifier.† The CA3015 consists basically of two differential amplifiers and a single-ended output circuit in cascade. Although it contains two diodes, ten BJTs, and 15 resistors, the CA3015 is supplied in a twelve-lead TO-5 style package and occupies the same space as a low-power discrete BJT.

The inputs to the CA3015 are applied to the bases of the pair of emitter-coupled BJTs Q_1 and Q_2 in the first differential amplifier, the inverting input (at terminal 3) being applied to the base of Q_2 and the noninverting input (at terminal 4) to that of Q_1.‡ The collector-to-collector output of these BJTs is directly connected to the bases of BJTs Q_3 and Q_4 in the second differential amplifier. Transistors Q_8, Q_9, and Q_{10} are used in an emitter-follower type of single-ended output circuit. The output of the second differential amplifier is directly connected to the base of Q_8, and the emitter circuit of Q_8 supplies the base drive of output BJT Q_{10}. Transistor Q_9 makes possible a small amount of signal gain in the output circuit through the connection between its emitter circuit and that of Q_{10}. The output of the operational amplifier is taken from the emitter of Q_{10} so that, *for no d-c input signal*, the d-c level at the output is essentially the same as that at the input. Transistors Q_6 and Q_7 serve as d-c constant-current sources for the two differential amplifier stages, and BJT Q_5 reduces common-mode error signals.

* Reference 32, pp. 73–113, 232–266.

† Reference 32, p. 197.

‡ An inverting-input signal results in an output voltage having a phase opposite that of the input (at low frequencies).

35 M. B. Leeds, "Linear ICs: Part 4, Inside the Operational Amplifier," *Electronics*, Vol. **40**, No. 21 (October 16, 1967).

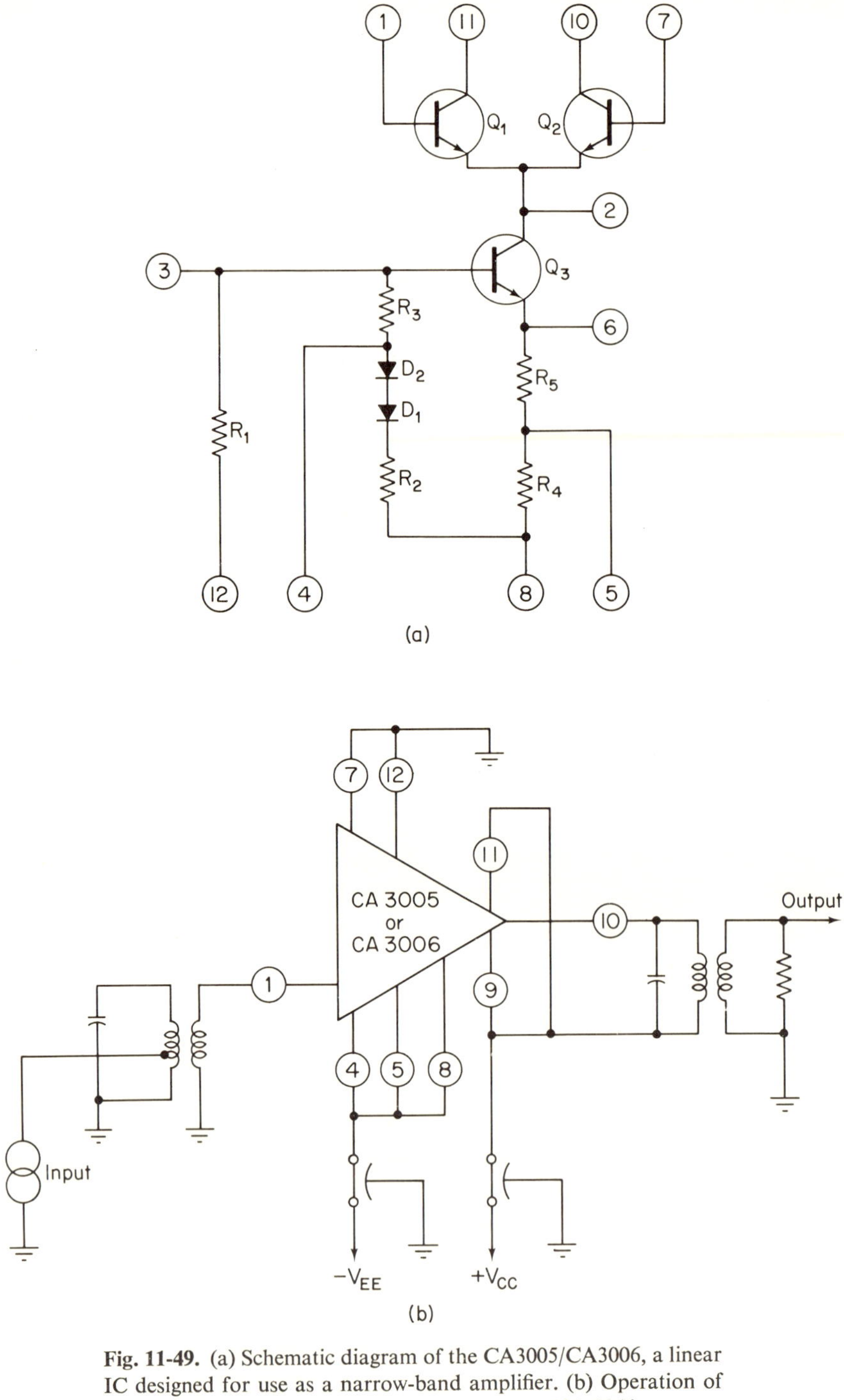

Fig. 11-49. (a) Schematic diagram of the CA3005/CA3006, a linear IC designed for use as a narrow-band amplifier. (b) Operation of the CA3005/CA3006. (From Reference 32, pp. 181–184.)

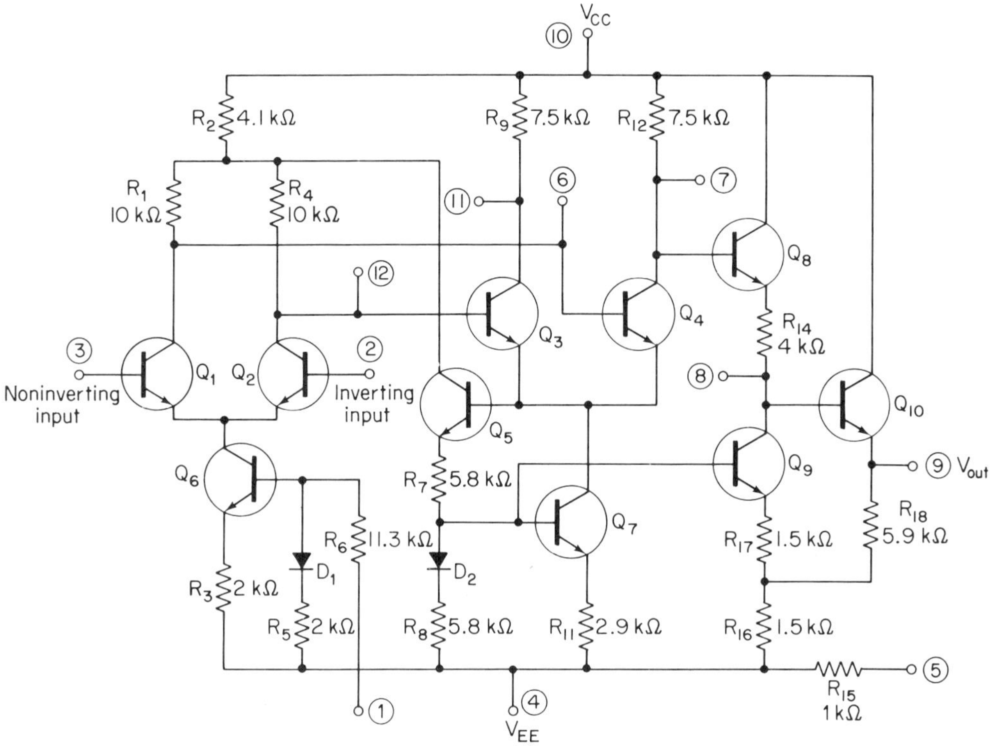

Fig. 11-50. Schematic diagram for the RCA CA3015 monolithic operational amplifier. (Courtesy Radio Corporation of America.)

Applications of the CA3015. We now briefly consider some represenative amplifier applications of the CA3015. In the normal use of an operational amplifier, negative feedback is employed to control amplifier gain and bandwidth. Negative feedback is achieved by applying to the inverting input (terminal 2) a signal obtained from the amplifier output; the amplifier input signal may be applied to either input, depending on whether or not phase inversion is desired.

Figure 11-51 shows the circuit diagram and frequency response of the CA3015 operating as a noninverting wide-band amplifier having a gain of approximately 50 dB.* For this arrangement, the amplifier has a 3-dB bandwidth of 3.5 MHz and a unity-gain frequency of 150 MHz. The 470 pF capacitor (for phase-lead compensation) is used to prevent instability (oscillation).

The bandwidth of the negative-feedback amplifier may be increased,

* Reference 32, p. 260.

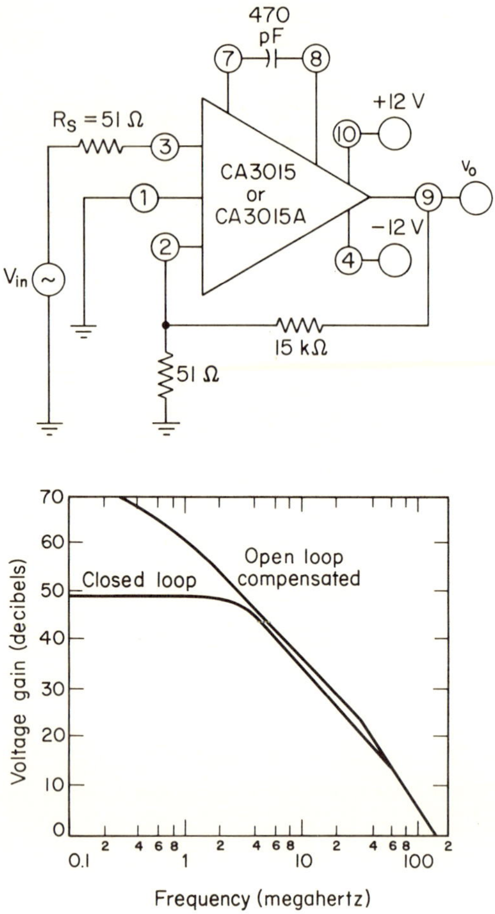

Fig. 11-51. Circuit arrangement and frequency response for a CA3015 operational amplifier used as a noninverting, wideband amplifier having a gain of approximately 50 dB. (From Reference 32, p. 260.)

at the expense of gain, by increasing the feedback, as shown in Fig. 11-52.* In comparing this arrangement with that of Fig. 11-51, it is seen that higher-value, unequal resistors are used between the input terminals and ground and that a pair of series RC combinations (for phase-lag compensation) are added. The input (source) resistances are larger in order to avoid undue amplifier loading by the feedback circuit and are unequal, so that the inputs are returned to ground through equal resistances. The phase-lag RC combinations are required for stability when the feedback is large. The peaking

* Reference 32, p. 261.

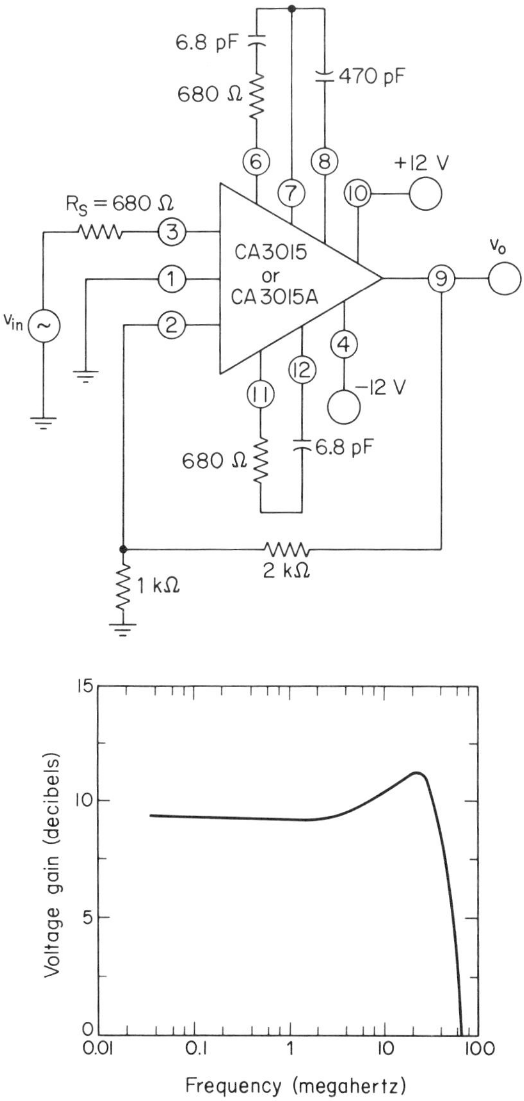

Fig. 11-52. Circuit arrangement and frequency response for a CA3015 operational amplifier used as a noninverting, wideband amplifier having a gain of approximately 10 dB. (From Reference 32, p. 261.)

of the response curve can be eliminated, at the expense of bandwidth, by using a larger-value phase-lag capacitor.

Figure 11-53 shows the CA3015 used as a narrow-band amplifier in an arrangement somewhat different from those of Figs. 11-48 and 11-49. Here

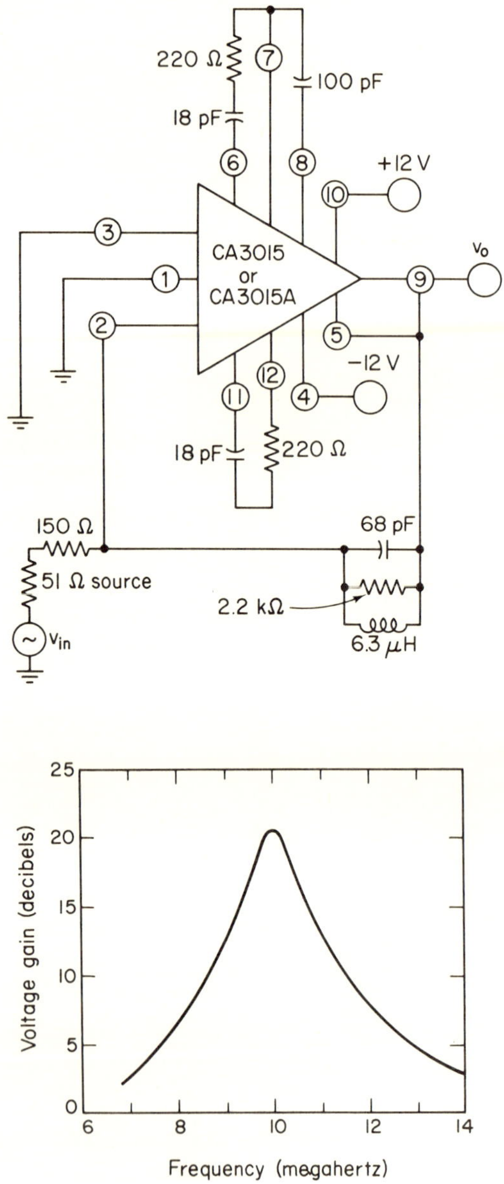

Fig. 11-53. Circuit arrangement and frequency response for a CA3015 operational amplifier used as a narrow-band amplifier. (From Reference 32, p. 262.)

the parallel-tuned circuit is in the feedback path, resulting in large feedback and low gain at all frequencies except those near the resonant frequency of the tuned circuit.

REFERENCES

[1] S. Tribwasser, "Large-Scale Integration and the Revolution in Electronics," *Science*, Vol. **163,** No. 3866 (January 31, 1969).

[2] L. Stern, *Fundamentals of Integrated Circuits.* New York: Hayden Book Co., Inc., 1968.

[3] R. M. Warner and J. N. Fordemwalt, *Integrated Circuits.* New York: McGraw-Hill Book Co., Inc., 1965.

[4] C. S. Meyer, D. L. Lynn, and D. J. Hamilton, *Anlysis and Design of Integrated Circuits.* New York: McGraw-Hill Book Co., Inc., 1968.

[5] H. C. Lin, *Integrated Electronics.* San Francisco: Holden-Day Inc., 1967.

[6] A. Reisman, "Germanium ICs Point the Way Towards Picosecond Computers," *Electronics*, Vol. **42**, No. 5 (March 3, 1969).

[7] A. B. Phillips, "Monolithic Integrated Circuits," *IEEE Spectrum*, Vol. **1,** No. 6 (June 1964).

[8] L. Curran, "In Search of a Lasting Bond," *Electronics*, Vol. **41,** No. 24 (November 25, 1968).

[9] T. A. Longo, "IC Processing: Healthy, Still Growing," *Electronics*, Vol. **41,** No. 22 (October 28, 1968).

[10] J. A. Morton, "Strategy and Tactics for Integrated Circuits," *IEEE Spectrum*, Vol. **6,** No. 6 (June 1969).

[11] R. M. Burger and R. P. Donovan, *Fundamentals of Silicon Integrated Device Technology, Volume I.* Englewood Cliffs, N. J.: Prentice-Hall, Inc., 1967.

[12] R. M. Burger and R. P. Donovan, *Fundamentals of Silicon Integrated Device Technology, Volume II.* Englewood Cliffs, N. J.: Prentice-Hall, Inc., 1968.

[13] D. F. Allison, A. P. Youmans, and T. H. Wong, "New Etchant Puts Dielectric Isolation in the Groove," *Electronics*, Vol. **42,** No. 10 (May 12, 1969).

[14] D. A. Maxwell, R. B. Beeson, and D. F. Allison, "The Minimization of Parisitics in Integrated Circuits by Dielectric Isolation," *IEEE Trans. on Electron Devices*, Vol. **ED-12** (January 1965).

[15] H. R. Camenzind, B. Polata, and J. Kocsis, "ICs Break Through the Voltage Barrier," *Electronics*, Vol. **42,** No. 7 (March 31, 1969).

[16] A. Miller, "Silicon-on-Sapphire Approach Affords Freedom and Flexibility," *Electronics*. Vol. **40,** No. 4 (February 20, 1967).

[17] H. Clausen and R. B. Rusert, "Isolation Problems Get an Airing," *Electronics*, Vol. **41,** No. 10 (April 29, 1968).

[18] J. A. Amick, A. W. Fisher, A. I. Stoller, and W. H. Schlip, Jr., "Isolation Techniques for Fabricating Integrated Circuits", *RCA Review*, Vol. **29,** No. 4 (December 1968).

[19] G. C. Onodera, W. J. Corrigan, and R. M. Warner, Jr. "Silicon Field-Effect Transistor with Internal Epitaxial Channels", *Proc. IEEE*, Vol. **50** (August 1962).

[20] P. Richman, *Characteristics and Operation of MOS Field-Effect Devices*. New York: McGraw-Hill Book Co., Inc., 1967.

[21] C. W. Mueller and P. H. Robinson, "Grown-Film Silicon Transistors on Sapphire," *Proc. IEEE*, Vol. **52** (December 1964).

[22] P. K. Weimer, "The TFT—A New Thin-Film Transistor," *Proc. IEEE*, Vol. **50** (June 1962).

[23] J. T. Wallmark and H. Johnson, *Field-Effect Transistors, Physics, Technology and Applications*. Englewood Cliffs, N. J.: Prentice-Hall Inc., 1966.

[24] A. Waxman, "Thin-Film Transistors Don't Have to be Drifters," *Electronics*, Vol. **41,** No. 6 (March 18, 1968).

[25] P. K. Weimer, G. Sadasiv, J. E. Meyer, L. Meray-Horvath, and W. S. Pike, "A Self-Scanned Solid-State Image Sensor," *Proc. IEEE*, Vol. **55,** No. 9 (September 1967).

[26] D. E. Farina and D. Trotter, "MOS Integrated Circuits Save Space and Money," *Electronics*, Vol. **38,** No. 20 (October 4, 1965).

[27] L. Vadasz, "The Use of MOS Structure for the Design of High-Value Resistors in Monolithic Integrated Circuits," *IEEE Trans. on Electron Devices*, Vol. **ED-13,** No. 5 (May, 1966).

[28] R. H. Crawford, *MOSFET in Circuit Design*. New York: McGraw-Hill Book Co., Inc., 1967.

[29] R. M. Warner, Jr., "Comparing MOS and Bipolar Integrated Circuits," *IEEE Spectrum*, Vol. **6,** No. 6 (June 1966).

[30] G. B. Herzog, "COS/MOS: The Best of Both Worlds," *Electronics*, Vol. **42,** No. 4 (February 17, 1969).

[31] J. Eimbinder, *Linear Integrated Circuits, Theory and Practice*. New York: John Wiley & Sons, Inc., 1968.

[32] *RCA Linear Integrated Circuits Manual*, Technical Series IC-41. Harrison, N. J.: Radio Corporation of America, 1967.

[33] J. J. Robertson, "Linear ICs: Part 2, Heart of the Matter," *Electronics*, Vol. **40,** No. 17 (August 21, 1967).

[34] J. P. Keller, "Linear ICs: Part 3, Differential Amplifiers at Work," *Electronics*, Vol. **40,** No. 19 (September 18, 1967).

[35] M. B. Leeds, "Linear ICs: Part 4, Inside the Operational Amplifier," *Electronics*, Vol. **40,** No. 21 (October 16, 1967).

PROBLEMS

11-1. A diffused silicon resistor having a pattern 10 mils by 2 mils has a resistance of 1100 Ω. Determine the sheet resistance of the silicon layer.

11-2. The acceptor concentration of a junction capacitor is 10^{15} atoms per cm^3 and is much smaller than the donor concentration. Determine the approximate area required for a 10 pF capacitor if the reverse barrier potential is 1 V.

11-3. A thin-film capacitor has a capacitance of 0.5 pF per mil^2. If the dielectric constant of silicon dioxide is 3.5, what is the thickness of the SiO_2 layer?

12

Thyristors and Related Devices

In general, the semiconductor control devices considered so far require continuous control-signal application for a particular conducting state. These devices may be classified as linear control devices, since conduction varies more or less linearly with control-signal level. Control devices of a class known as *thyristors* are bistable, having a high-impedance state and low-impedance state, and may switch states as a result of a control signal. Since no signal is required to maintain either state, the control signal may be of short duration, in the form of pulses, and relatively little control energy is required. A thyristor is "any semiconductor switch whose bistable action depends on *pnpn* feedback."* In this chapter, we consider characteristics and applications of various types of thyristors and of devices used in conjuction with them.

12-1 THE UNIJUNCTION TRANSISTOR (UJT)[1]

The UJT is a device having characteristics that lend it particularly well to thyristor control circuitry, in addition to many other applications. The

* "Nomenclature and Definitions of Semiconductor Devices," International Electrotechnical Commission, Geneva, Switzerland, IEC Document 47 (Central Office) 43, August 1964.

1 *GE Transistor Manual*, 7th ed. Syracuse, N.Y.: General Electric Co., 1964, Chapter 13.

construction of the UJT is similar to that of the JFET, as shown by the cross-section in Fig. 12-1(a). This is the bar structure, which is used in most commercial UJT types and consists of a bar of high-resistivity *n*-type silicon, called the *base*, with a *p*-type insert near the middle. The *p*-region, called the *emitter*, is formed by alloying an aluminum wire to the base. Ohmic connections to opposite ends of the bar are referred to as *base-one* and *base-two contacts*. Because of its two base contacts, the device was originally known as a *double-base diode*.[2]

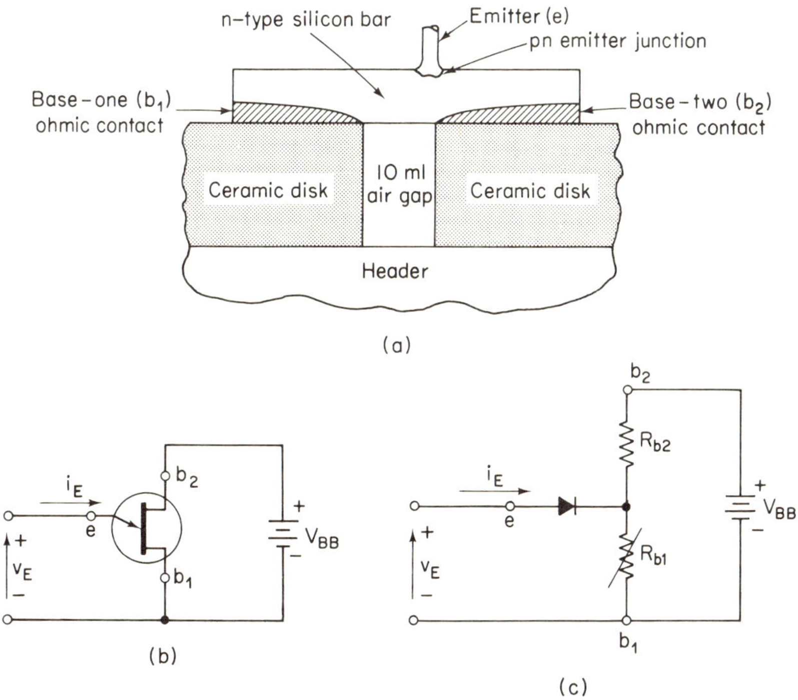

Fig. 12-1. Bar structure (a), circuit symbol (b), and simplified equivalent circuit (c) of a unijunction transistor. The normal base bias arrangement is shown in (b) and (c). (Fig. 12-1(a) is from Reference 1, p. 305.)

The principal constructional difference between the UJT and the JFET is that the emitter-junction area of the UJT is much smaller than the gate-junction area of the JFET. The chief operational difference between the devices is that the UJT emitter junction is forward biased during operation, whereas the JFET gate junction is normally reverse biased.

[2] I. A. Lesk and V. P. Mathis, "The Double-Base Diode: A New Semiconductor Device," *IRE Convention Record*, Part 6 (1953), pp. 2–8.

The symbol for the UJT and a simplified equivalent circuit are shown by Figs. 12-1(b) and 12-1(c). The resistance between base-one contact b_1 and base-two contact b_2 is the interbase resistance R_{bb}, which is represented by series resistances R_{b1} and R_{b2}. Resistance R_{b1} between b_1 and the emitter region is variable, being dependent upon emitter current, whereas R_{b2} between the emitter region and b_2 is fixed. For zero emitter current i_E, R_{bb} is between 5 kΩ and 10 kΩ for typical UJTs. The diode in the equivalent circuit represents the emitter diode of the UJT.

In normal operation, base-two is biased positively with respect to base-one by interbase bias voltage V_{BB}, as shown in Fig. 12-1. With zero emitter current, the silicon bar acts as a voltage divider and a fraction η of V_{BB} appears at the emitter. If emitter voltage v_E is less than ηV_{BB}, the emitter diode is reverse biased and a small negative emitter-leakage current flows. If v_E becomes greater than ηV_{BB}, the emitter diode is forward biased, permitting the flow of appreciable emitter current. This current consists primarily of holes injected into the silicon bar, owing to the high-conductivity p-region. As injected holes move through the bar toward the base-one contact, electrons flow into the emitter-base-one region from the base-one contact in order that electrical neutrality be maintained. The increased carrier density in the region decreases resistance R_{b1}, which reduces the emitter voltage required for forward-biasing the emitter diode. Thus an emitter-current increase causes an emitter-voltage decrease, resulting in a negative-resistance characteristic, as shown in Fig. 12-2.

The various regions of the emitter-current–emitter-voltage characteristic are indicated on the sketch of Fig. 12-2. Two characteristic points of particular interest are the *peak point* and the *valley point*. The region below the peak point is the *cutoff region*, in which the emitter diode is reverse biased and only a small leakage current flows. The region between the peak point and the valley point is the *negative-resistance region*, and that above the valley point is the *saturation region*, in which the dynamic resistance is positive. In the saturation region, the emitter characteristic approaches that of the emitter-to-base-one diode.

Peak point. The most important characteristic of the UJT is the peak point, since it determines the triggering voltage. The peak point voltage V_p depends upon the interbase voltage and the forward-bias voltage of the emitter diode, being given by

$$V_p = \eta V_{BB} + V_F \tag{12-1}$$

where η is the *intrinsic stand-off ratio* of the UJT and V_F is the emitter-diode voltage for a forward current equal to the peak-point current I_p. The peak-point voltage is somewhat temperature dependent, owing to the decrease in V_F with increase in temperature. Typical UJTs have η values ranging from

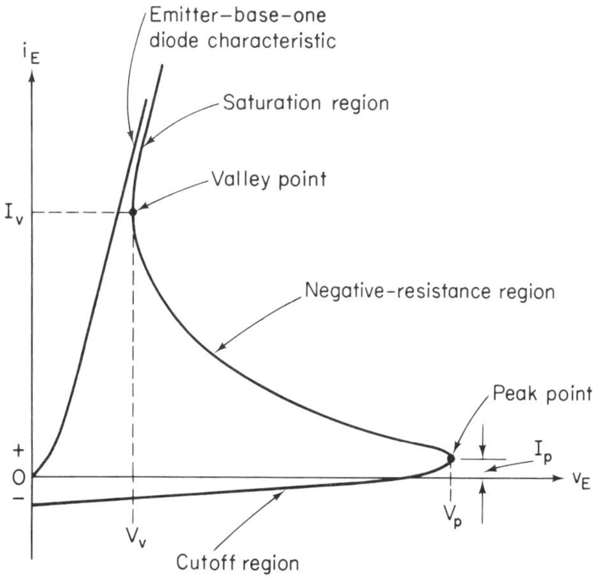

Fig. 12-2. Emitter-current – emitter-voltage characteristic of a unijunction transistor. Also shown is the characteristic of the emitter-to-base-one diode.

about 0.5 to 0.8, V_F values of the order of 0.5 V and I_p values from 2 μA to 25 μA (at 25°C), depending on the UJT type.

The emitter characteristic of a typical 2N490A UJT for each of several values of interbase voltage is shown in Fig. 12-3. The emitter-current scale is such that the cutoff region and the peak-point current are obscure, these currents being very small compared to the valley-point current and to currents in the saturation region.

The variation of resistance R_{b1} with emitter current for a typical type 2N492 UJT in a representative circuit is indicated in Table 12-1. This variation of R_{b1} with i_E causes a variation of base-two current i_{B2}, resulting in transistor action. A family of curves of i_{B2} versus v_{B2} for fixed values of

Table 12-1* Variation of R_{b1} with i_E
(Typical 2N492 UJT in a Representative Circuit)

i_E	R_{b1}	i_E	R_{b1}
0 mA	4600 Ω	10 mA	150 ohms
1	2000	20	90
2	900	50	40
5	240		

* Reference 1, p. 303.

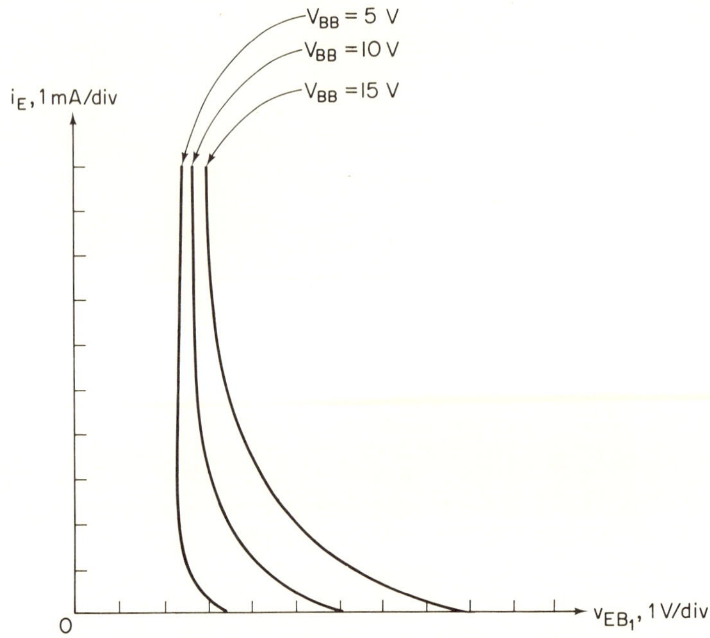

Fig. 12-3. Emitter current i_E versus emitter voltage v_{EB} for fixed values of interbase voltage V_{BB} for a type 2N490A UJT.

i_E is known as the interbase characteristic of the UJT. The interbase characteristic of a typical type 2N490A UJT is shown in Fig. 12-4. The V_{B2} intercept of the characteristic is the $i_E R_{b1}$ voltage drop for zero i_{B2}; the steep-slope (low-resistance) region of the characteristic is the result of injected carriers in the base-two-emitter region of the UJT; the high-resistance region is determined primarily by R_{B2}. The alpha of the UJT is defined as

$$\alpha = \left. \frac{\Delta i_{B2}}{\Delta i_E} \right|_{v_{B2}\text{ fixed}} \tag{12-2}$$

From Fig. 12-4, it can be seen that the 2N490 unit has alphas of the order of 0.25 at the operating point $I_{B2} = 2$ mA, $V_{B2} = 5$ V and 0.03 at $I_{B2} = 6$ mA, $V_{B2} = 5$ V. The alphas of UJTs are so much lower than those of BJTs that the devices are not competitive.

UJT structures. The relatively large bar ($8 \times 10 \times 60$ mils in the typical structure) used in the structure of Fig. 12-1 results in a device having a rather long turn-off time. Devices having higher speeds, as well as other improvements, are made by using the cube structure shown in Fig. 12-5. Figure 12-6 shows typical emitter characteristics for bar- and cube-structure UJTs. The cube-type UJT is seen to have a lower valley point and a higher negative resistance than the bar-type device.

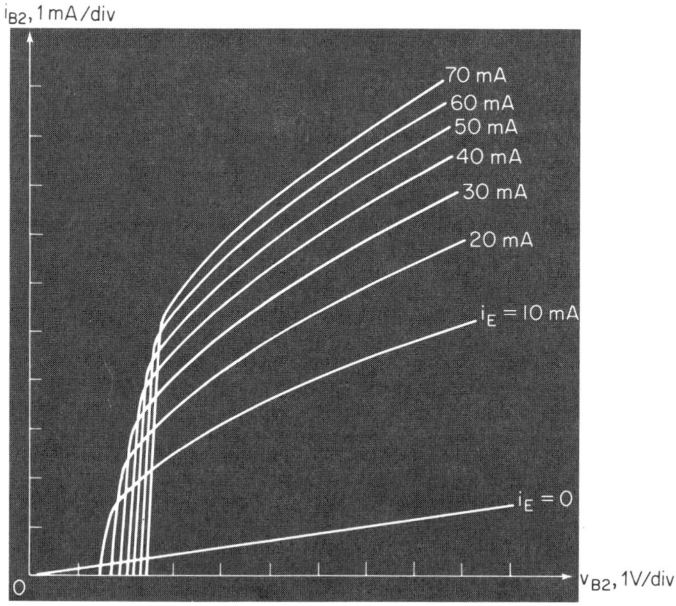

Fig. 12-4. Interbase characteristic of a typical type 2N490 UJT.

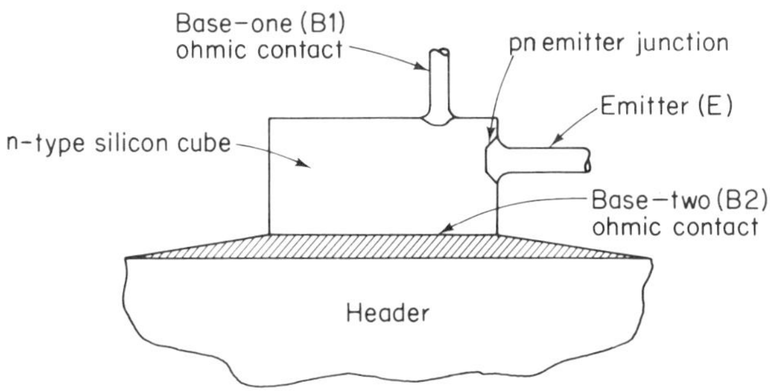

Fig. 12-5. Structure of a cube-type UJT. Typical cube dimensions are 13 × 17 × 17 mils. (From Reference 1, p. 305.)

The most recent UJT types are fabricated by using planar technology and have structures as shown in Fig. 12-7.[3] The *p*-type emitter region and n^+ base-contact regions are formed by diffusion. The structure of Fig. 12-7(b) is seen to lend itself to use in monolithic integrated circuits.[4] Planar-type UJTs have electrical characteristics similar to those of the cube type.

[3] W. R. Spoffard, Jr. and R. A. Stasior, "A Switch In Time," *Electronics*, Vol. **41**, No. 4 (February 19, 1968).

[4] L. S. Senhouse, Jr., "An Epitaxial Planar Structure for the Unijunction Transistor," *IEEE Trans. on Electron Devices*, Vol. **ED-16**, No. 2 (February, 1969).

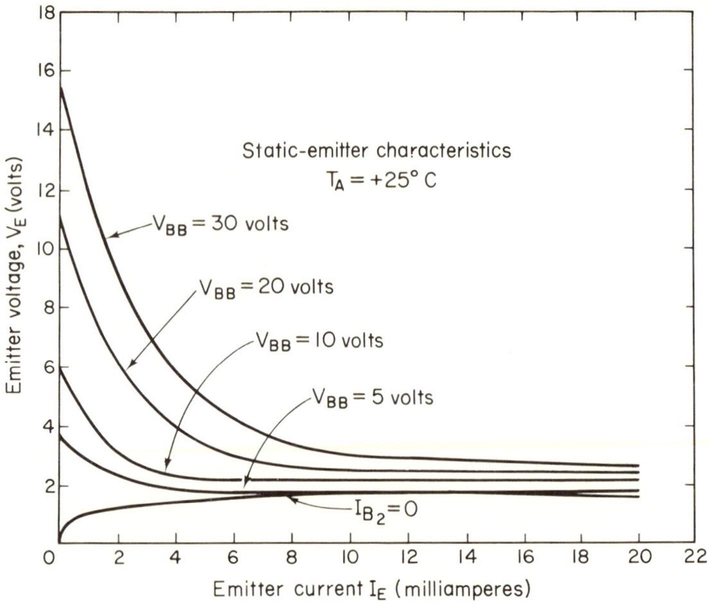

(a) Bar structure

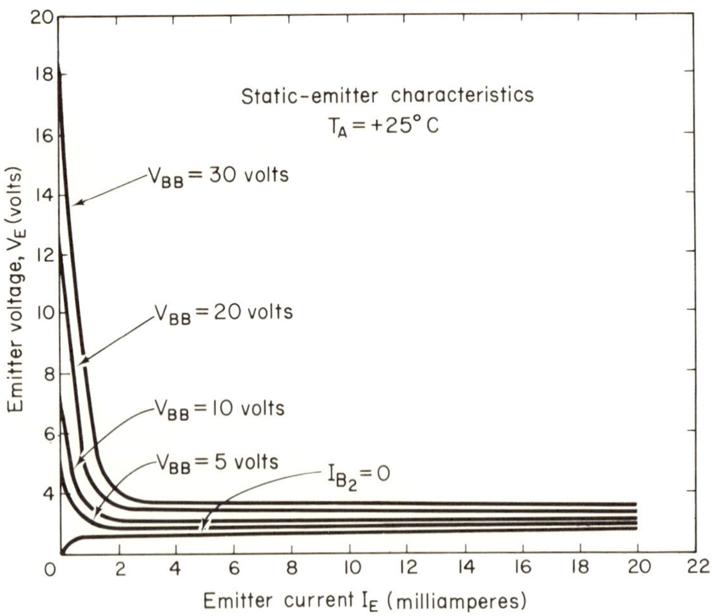

(b) Cube structure

Fig. 12-6. Typical static emitter characteristics for (a) bar-structure and (b) cube-structure UJTs. (From Reference 1, p. 306.)

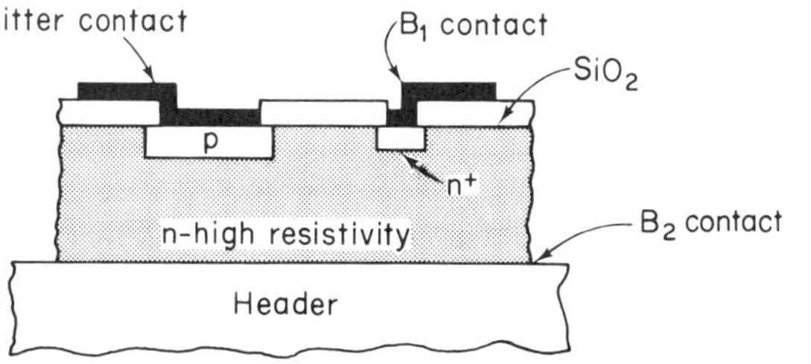

(a)

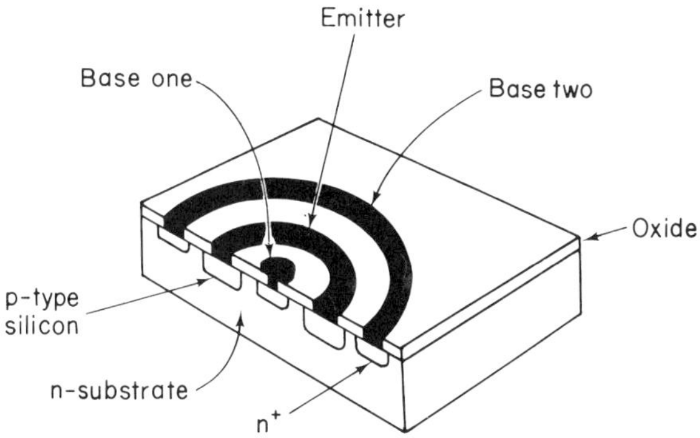

(b)

Fig. 12-7. Planar UJT structures. Typical device dimensions are $20 \times 20 \times 7$ mils. (From Reference 3, p. 116 and Reference 4, p. 162.)

The complementary UJT (CUJT). The planar technology lends itself to the fabrication of complementary UJTs having *p*-type bases and *n*-type emitters. The structure of a CUJT fabricated by using a *p*-type epitaxial layer on an *n*-type substrate is shown in Fig. 12-8(a). In this structure, the substrate serves as the emitter. The equivalent circuit of the device, superimposed on a sketch of the cross-section, is shown in Fig. 12-8(b). Unijunction-transistor action occurs through modulation of resistance R_{b1}, under the base-one contact, by negative emitter current. The characteristics of the CUJT are like those of the UJT, except that the currents and voltages are of opposite polarity. For most applications, the polarity is unimportant. Commercial CUJTs include the General Electric D5K1 and D5K2. These devices are fabricated by using standard *npn* planar passivated transistor-processing techniques and are said to have superior stability and good uniformity.*

* D5K1 Specification Sheet. General Electric Co.

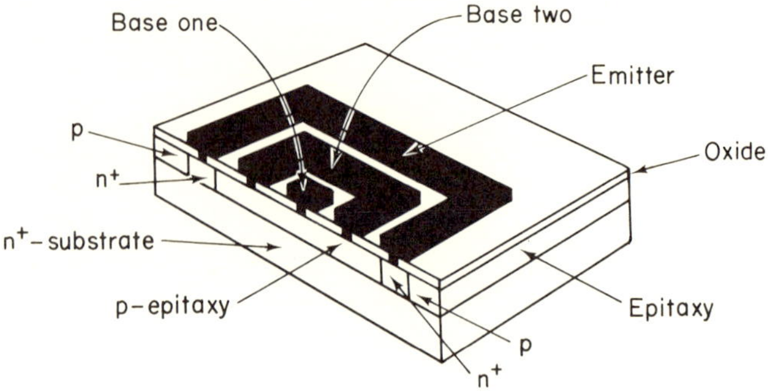

(a)

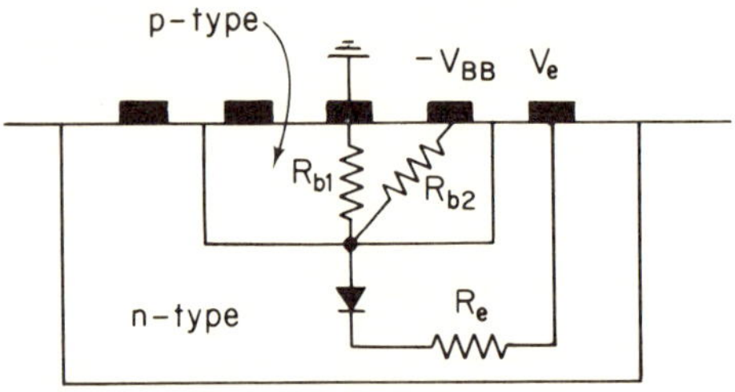

(b)

Fig. 12-8. The complementary (*p*-type) UJT (CUJT). (a) Epitaxial planar structure. (b) Equivalent circuit. (From Reference 4, p. 162.)

12-2 APPLICATIONS OF UJTs

Although the low alpha of the UJT precludes its use in most conventional transistor applications, the bistable emitter characteristic of the device makes it extremely useful. We now consider some representative applications.

Relaxation oscillator. The relaxation-oscillator circuits of Fig. 12-9 are basic circuits for applications such as sawtooth generators, pulse generators, and timing circuits. Their use as trigger circuits is considered in Sec. 12-6 in conjunction with silicon-controlled rectifiers. The complementary-UJT circuit of Fig. 12-9(b) is seen to be merely the inverted version of the standard circuit.

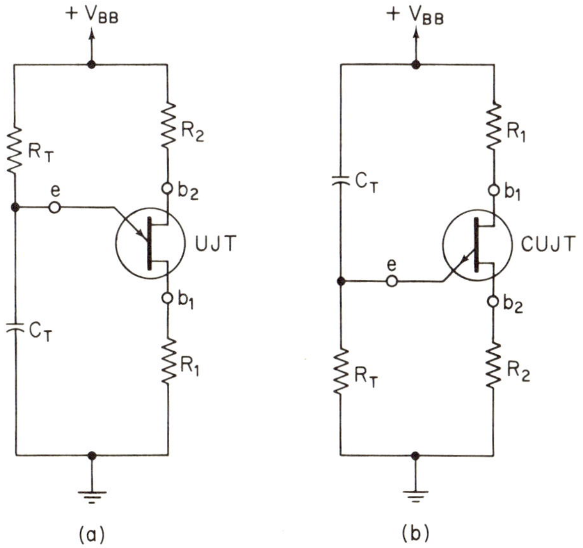

Fig. 12-9. Basic UJT relaxation oscillator circuits. (a) Standard UJT circuit. (b) Complementary UJT circuit.

At the beginning of an operating cycle of the relaxation oscillator, the charge on capacitor C_T is low and the emitter is reverse biased and nonconducting. As C_T charges through resistor R_T, the emitter voltage v_E increases exponentially. When v_E reaches the peak-point value V_p, the emitter becomes forward biased and emitter-to-base-one resistance R_{b1} begins to decrease. Capacitor C_T then discharges through the emitter, driving R_{b1} to a low value, and the emitter voltage falls rapidly. With C_T discharged, R_{b1} increases, the emitter ceases to conduct, and the cycle is repeated. For small values of R_1 and R_2, the frequency of oscillation is determined largely by the time constant R_TC_T and the intrinsic stand-off ratio η, being given approximately by

$$f \cong \frac{1}{R_TC_T \ln(1-\eta)} \tag{12-3}$$

In order for oscillation to occur in the relaxation-oscillator circuits of Fig. 12-9, R_T must neither be too large to pass sufficient current to trigger the UJT nor so small that the UJT fails to turn off. That is, load lines drawn on the emitter characteristic must intersect the characteristic above the peak point in the negative-resistance region. The limits of R_T are seen to be given by

$$\frac{V_{BB}-V_p}{I_p} > R_T > \frac{V_{BB}-V_v}{I_v} \tag{12-4}$$

For typical UJTs, the range of R_T encompasses about three orders of magnitude, say from 3 kΩ to 3 MΩ.

The upper limit of the frequency of oscillation is determined by the turn-off time of the UJT, which is the time required for the base-one-emitter region of the transistor to be depleted of stored charge. This limit is of the order of 100 kHz, depending on the UJT type.

If *synchronization of the oscillator* is required, this can be accomplished by applying either positive pulses at the emitter terminal or negative pulses at the base-two terminal.

Sawtooth-wave generator. Since the wave form of the emitter voltage in the basic relaxation oscillator is a fair approximation of a sawtooth wave, this signal may be coupled to a load by means of a direct-coupled emitter follower, as shown in Fig. 12-10. If the linearity obtained with this arrangement is inadequate, provision can be made for maintaining a constant charging current through timing capacitor C_T.*

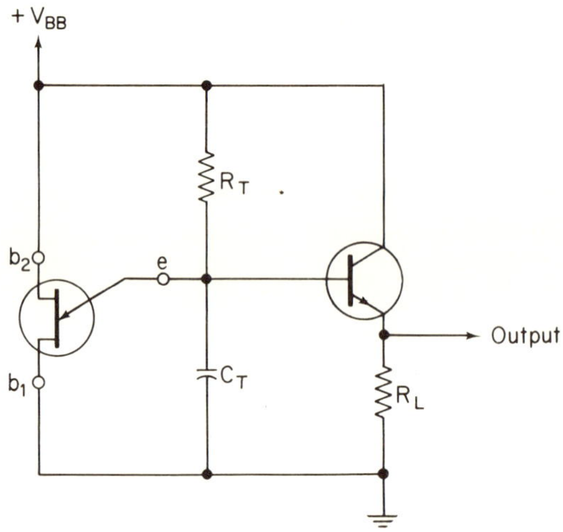

Fig. 12-10 UJT sawtooth generator.

Pulse-generator circuits. Since current pulses flow in the emitter, base-one, and base-two circuits of the relaxation oscillator each time the UJT conducts, each circuit current can be utilized to supply an output pulse, as shown in Fig. 12-11.

* Reference 1, pp. 318–319.

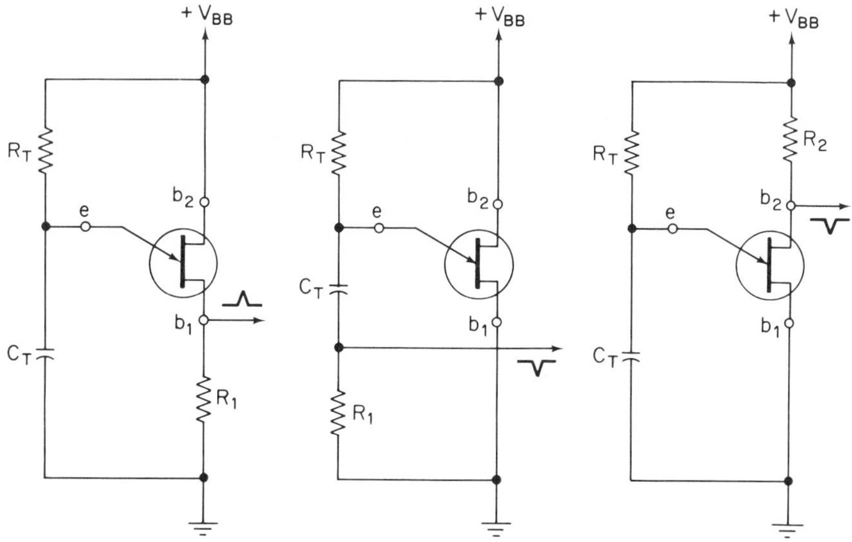

Fig. 12-11. Use of the UJT as a pulse generator.

Time-delay relay. The UJT can be used to obtain delay in the operation of a relay, as shown in Fig. 12-12. When switch S is closed, timing capacitor C_T charges to the peak-point voltage of the UJT, triggering the UJT. Capacitor-discharge current through relay R causes R to close, closing two sets of contacts. One set maintains the closed condition, and the other set is for control purposes.

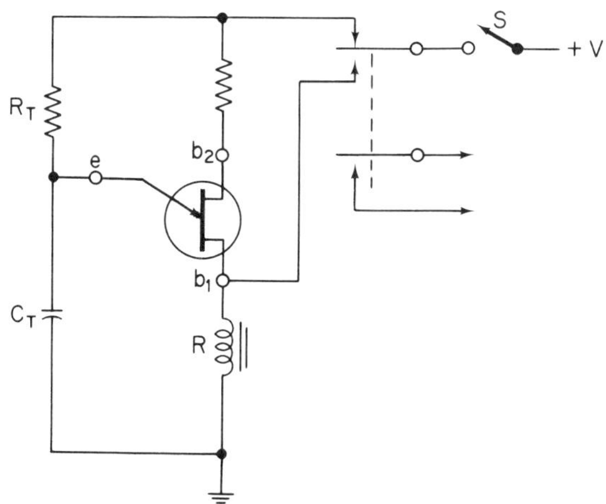

Fig. 12-12. Use of the UJT to delay the operation of a relay.

12-3 THE *pnpn* DIODE

The *pnpn diode* is basic to a series of semiconductor devices having four or more discrete semiconductor regions. Therefore, much of the material of this section is applicable to related devices such as silicon-controlled rectifiers and switches considered in later sections.

The structure of a *pnpn* diode, sometimes called a *four-layer* or *Shockley diode*, is somewhat as shown by Fig. 12-13(a). The polarity of the applied

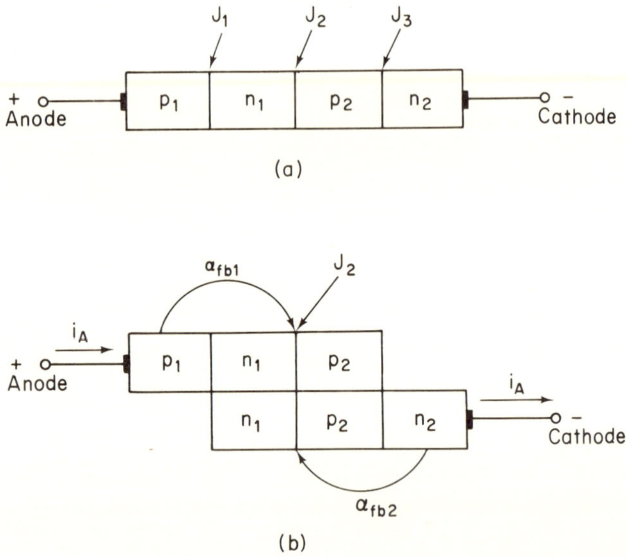

Fig. 12-13. (a) Structure of a *pnpn* diode. (b) Two-transistor representation of the *pnpn* diode in which α_{fb1} is the common-base alpha of the *pnp* section and α_{fb2} that of the *npn* section. The polarity of applied voltage for operation in the forward direction is indicated.

voltage for operation in the forward direction is also shown, the polarity being such that junctions J_1 and J_3 are forward biased and J_2 appears to be reverse biased. As we shall see, J_2 may be either reverse biased or forward biased, depending on the operating stage of the diode. If J_2 is reverse biased, the flow of appreciable current is blocked and the diode is said to be in a *forward-blocking* or OFF state. With J_1 forward biased and J_2 reverse biased, J_2 acts as the collector junction of a *pnp* transistor operating in the active mode with the p_1-region acting as the emitter, the n_1-region as the base, and the p_2-region as the collector. In addition, owing to the forward bias of J_3, reverse-biased J_2 acts as the collector junction of an *npn* transistor having the n_2 region as the emitter, the p_2 region as the base, and the n_1 region as the collector. Thus the *pnpn* diode may be considered to consist of two transistors, as indicated in Fig. 12-13(b).

With the p_1n_1 junction (J_1) forward biased, holes injected into the n_1-region diffuse through the region and enhance the minority-hole flow through back-biased J_2. The fraction of J_1 current that reaches J_2 is α_{fb1}, the common-base alpha of the $p_1n_1p_2$-transistor section of the diode. In addition, the current through J_2 is enhanced by electrons injected into the p_2-region, owing to the forward bias of J_3, and the fraction of J_1 current that reaches J_2 is designated as α_{fb2}. The alphas of the two transistor sections of the diode and the directions of associated carrier flow are indicated in Fig. 12-13(b).

From the preceding discussion, the current through junction J_2 is seen to have two components in addition to the reverse-leakage current I_{CO}. Since anode current i_A and the current through J_2 are identical, as are i_A and the currents through junctions J_1 and J_3, we may write

$$i_A = I_{CO} + \alpha_{fb1}i_A + \alpha_{fb2}i_A \tag{12-5}$$

where $\alpha_{fb1}i_A$ is the current of holes from the p_1 region and $\alpha_{fb2}i_A$ is the current of electrons from the n_2 region. From Eq. (12-5), we obtain

$$i_A = \frac{I_{CO}}{1 - (\alpha_{fb1} + \alpha_{fb2})} \tag{12-6}$$

which indicates that i_A increases without limit if $(\alpha_{fb1} + \alpha_{fb2})$ is equal to unity. This expression is similar to Eq. (5-17) for the collector current of a bipolar transistor operated with its base open-circuited.* Just as we concluded that collector-emitter breakdown of the bipolar transistor occurs when conditions are such that α_{fb} is equal to unity, we reach a similar conclusion for forward operation of the *pnpn* diode.

Shorted-emitter structure. *pnpn* devices are usually designed so that the alpha of one transistor section is reasonably close to unity, and not significantly dependent upon operating conditions, whereas that of the other section is quite small at very low currents and highly dependent upon current. This may be accomplished through use of a long n_1-region that results in the *pnp* transistor section having a low base-transport factor, or by use of the "*shorted-emitter*" *structure* shown in Fig. 12-14. In this structure, most of the hole current collected by J_2 flows laterally above emitter junction J_3 until the voltage drop across the p_2 region "shorting" J_3 becomes large enough to cause J_3 to emit electrons into base region p_2. The result is an emitter having an emitter efficiency that is essentially zero at very low currents but increases with current through the structure. In a structure having a long n_1-region, alpha of the *pnp* transistor section increases with current as a result of increase in base-transport factor.

* With an open base, the collector current i_C of a bipolar transistor is given by $i_C = I_{CBO}/(1 - \alpha_{fb})$.

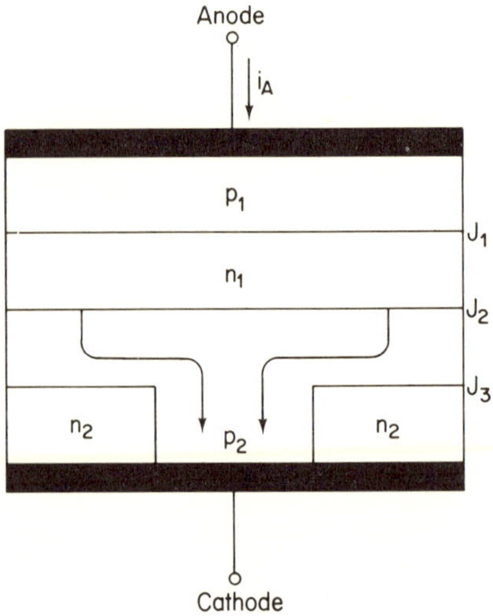

Fig. 12-14. *pnpn* device having a shorted-emitter structure. For low values of anode current i_A, most of the hole current collected by junction J_2 flows laterally above emitter junction J_3 as indicated by arrows in the figure.

For low forward voltages, the quantity $(\alpha_{fb1} + \alpha_{fb2})$ is significantly below unity, current i_A is low, and the diode is in a forward-blocking, high-impedance, OFF state. As the forward voltage is increased, i_A increases, causing $(\alpha_{fb1} + \alpha_{fb2})$ to increase, which further increases i_A. Additional increase in voltage causes $(\alpha_{fb1} + \alpha_{fb2})$ to reach unity, breakdown occurs, and i_A increases abruptly. The anode voltage and current that establish this condition are known as the *forward-breakover values.*

Since $(\alpha_{fb1} + \alpha_{fb2})$ depends upon both anode current and collector voltage (the voltage across back-biased junction J_2), the collector voltage required to maintain $(\alpha_{fb1} + \alpha_{fb2})$ equal to unity decreases with increase in current. Thus the anode voltage decreases as anode current increases following breakdown, and the $i_A - v_{AK}$ characteristic has a negative-resistance region, as shown by Fig. 12-15. The decrease in collector voltage with increase in anode current continues until the collector voltage reaches zero and reverses direction, the collector junction becoming forward biased. Forward biasing of collector junction J_2 (see Fig. 12-13) occurs as a result of holes injected into the n_1-region and electrons injected into the p_2-region, raising the minority-carrier densities at the J_2 edges above the equilibrium values. When J_2 becomes forward biased, the anode-to-cathode voltage drop v_{AK} is given approximately by

$$v_{AK} = v_{J1} - v_{J2} + v_{J3} \tag{12-7}$$

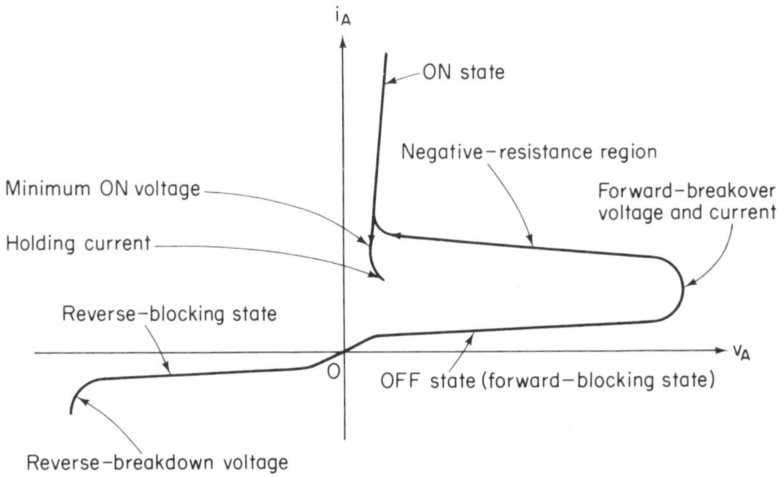

Fig. 12-15. The conduction regions of a *pnpn* diode.

the algebraic sign of v_{J2} being opposite to that of the other junction voltages owing to forward bias. Thus the voltage drop across the diode is less than the drop across two forward-biased *pn* junctions, but somewhat more than that across one junction, owing to the collector-junction bias being less than that of the other two junctions, and the diode is in its low-impedance, conducting, or ON state. Typical silicon *pnpn* diodes have ON voltages of the order of 1 V. Because of its low impedance in the ON state, *the diode must be operated with a series resistance large enough to limit the anode current to a safe value.*

When the diode is operating at low levels of current in the ON state, the resistive drops across the various regions of the device are small compared to the net junction voltage, and the anode voltage is given by Eq. (12-7). As the ON current is increased, these drops become significant and the diode has positive dynamic resistance, as shown by Fig. 12-15.

If the anode current is decreased, following establishment of the ON state, operation moves from the positive-resistance region into a negative-resistance region, and further reduction in current causes the diode to revert to the OFF state. The minimum current required to maintain the diode in the ON state is known as the *holding current*. Commercial silicon *pnpn* diodes have holding currents of the order of milliamperes.

When a negative voltage is applied to the anode for operation in the reverse direction, junctions J_1 and J_3 (see Fig. 12-13) are reverse biased and block current, whereas J_2 is forward biased. Thus J_2 acts as the emitter junction of both transistor sections of the diode, but the absence of a common collector junction prevents switching into an ON state. If the reverse voltage is large enough, reverse breakdown occurs at a voltage equal to the sum of the two collector-emitter breakdown voltages, and the breakdown characteristic is similar to that of a bipolar transistor operating with an open base.

The nature of the anode-current–anode-voltage characteristic of a *pnpn* diode, and the definitions of several device terms, are indicated by the sketch of Fig. 12-15. The conduction characteristic of a typical commercial diode is shown by Fig. 12-16.

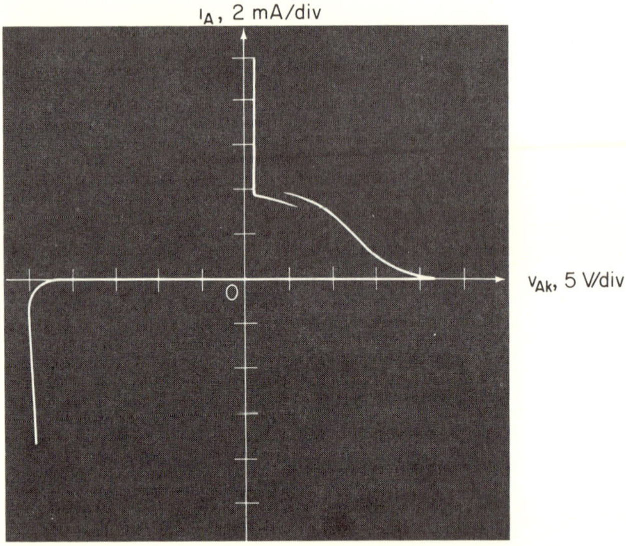

Fig. 12-16. Anode–current–anode voltage characteristic of a typical type 4E20-3 silicon *pnpn* diode.

Rate effect or dv/dt turn-on. Anode voltages less than the forward-breakover value cause a *pnpn* device to switch ON if the rate of voltage increase is great enough. Because of collector-junction capacitance *C*, a current $C\,dv/dt$ flows through the emitter junctions upon application of anode voltage v. If this current is large enough, $(\alpha_{fb1} + \alpha_{fb2})$ becomes unity and turn-on occurs. Turn-on by the dv/dt mechanism may occur for dv/dt values as low as *a few volts per microsecond*, the value depending upon the junction capacitance and the forward-breakover current of the device. The dv/dt rating of a thyristor is an important circuit-design consideration.

12-4 APPLICATIONS OF *pnpn* DIODES

In order to provide an indication of the circuit behavior of *pnpn* diodes, we consider some typical applications.

Sawtooth oscillator. In the circuit shown by Fig. 12-17, d-c supply voltage $+V$ is greater than the forward-breakover voltage of the diode, and resistance R is such that V/R is *less than the holding current*. When switch S is closed, capacitor C charges until the potential of point a reaches the forward-breakover voltage. The diode then switches ON, and point a is suddenly brought down to approximately ground potential as C discharges through the diode. When C is discharged, the diode switches to the OFF state, because of the inability of R to pass sufficient current to maintain operation in the ON state, and the process is repeated. The frequency of oscillation depends upon the time constant RC and the ratio of the supply voltage to the forward-breakover voltage of the diode. The upper limit of the frequency depends upon diode turn-off time, which is of the order of microseconds, owing to the time required for depletion of the charge stored in the base regions of the transistor sections during conduction.

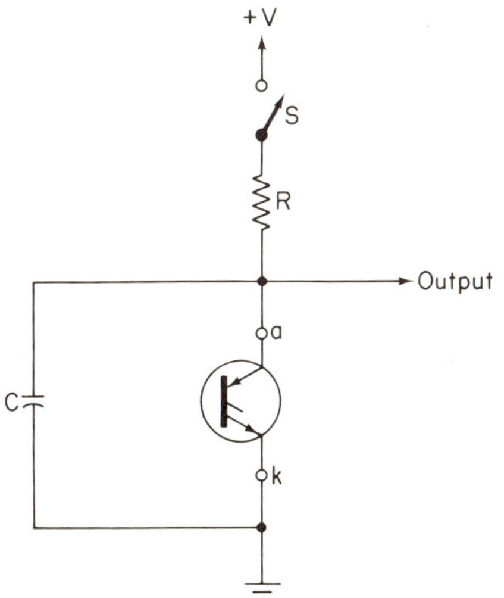

Fig. 12-17. *pnpn*-diode sawtooth oscillator.

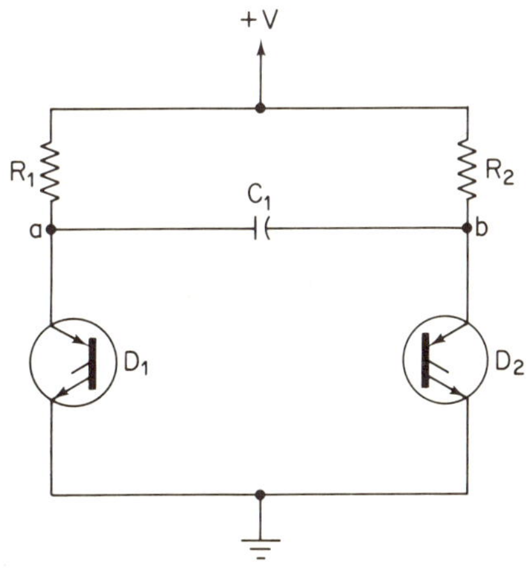

Fig. 12-18. *pnpn*-diode astable multivibrator circuit.

Astable multivibrator. In the astable or free-running multivibrator circuit of Fig. 12-18, d-c supply V is greater than the forward-breakover voltage of the diodes, and R_1 and R_2 are such that V/R_1 and V/R_2 are *greater than the holding currents* of diodes D_1 and D_2, respectively. When the circuit is activated, one diode switches ON, delaying turn-on of the other diode. If diode D_1 is ON, point a will be at approximately ground potential as long as the current passed by R_1 is greater than the holding current of D_1. Commutating capacitor C_1 charges through resistor R_2 and diode D_1, causing the potential of point b to rise until it reaches the forward-breakover voltage of D_2. When this occurs, D_2 switches ON, and point *b* is suddenly brought down to ground potential. Since point *a* is negative with respect to *b* at the instant of switching, bringing point *b* to ground potential causes the anode of D_1 to be negative until C_1 can discharge. This reverse-bias turns D_1 OFF and the D_1-OFF-D_2-ON condition continues while C_1 discharges and is charged in the opposite direction through R_1 and D_2. When C_1 is charged to the forward-breakover voltage of D_1, D_1 switches ON, the voltage across C_1 turns D_2 OFF, and the cycle is completed.

The frequency of operation is determined by the time constants R_1C_1 and R_2C_1 and by the ratio of the supply voltage to the forward-breakover voltages of the diodes. The circuit can be made symmetrical, or it can be caused to operate longer in one condition than in the other by making the time constants unequal or by using diodes having unequal forward-breakover voltages.

Monostable multivibrator. In the monostable multivibrator circuit of Fig. 12-19, diode D_1 has a forward-breakover voltage less than d-c supply voltage $+V$, and the forward-breakover voltage of diode D_2 is greater than $+V$.* Resistors R_1 and R_2 are selected so that V/R_1 and V/R_2 are greater than

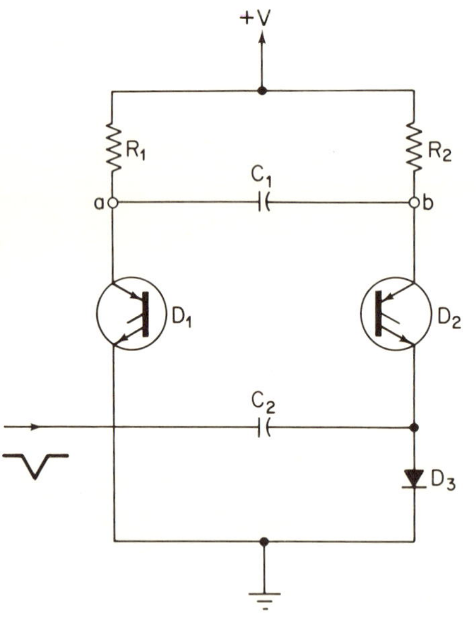

Fig. 12-19. *pnpn*-diode monostable multivibrator circuit.

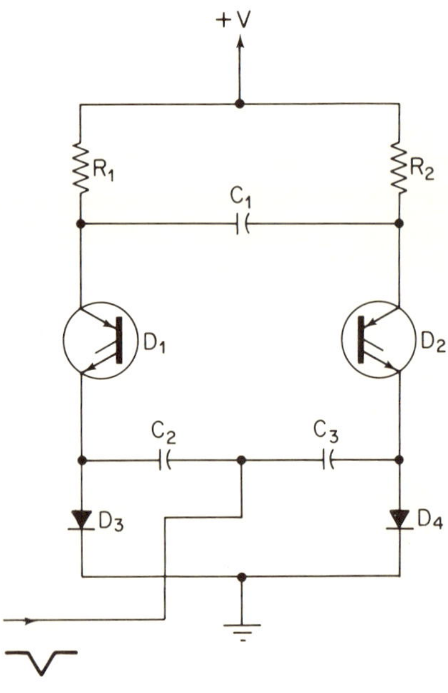

Fig. 12-20. *pnpn*-diode bistable multivibrator circuit.

* If diodes having significantly different breakover voltages are not available, a Zener diode may be connected in series with D_2.

the holding currents of D_1 and D_2, respectively. Thus D_1 is normally ON and D_2 is normally OFF.

A negative-going trigger pulse injected via capacitor C_2 causes the anode voltage of D_2 to rise to the forward-breakover value, switching D_2 ON. When D_2 turns ON, D_1 is switched OFF through the coupling action of C_1, and the D_1-OFF-D_2-ON state continues until C_1 charges through R_1 and D_2 to the forward-breakover voltage of D_1 and the circuit reverts to its normal stable state. The duration of the D_1-OFF-D_2-ON state is determined by the time constant R_1C_1 and the ratio of the supply voltage to the forward-breakover voltage of D_1.

The trigger pulse is developed across a diode (D_3) rather than a resistor, since the diode provides a high impedance for the trigger pulse but presents a low impedance to diode-D_2 current.

Bistable multivibrator. In the bistable circuit of Fig. 12-20, both *pnpn* diodes have forward-breakover voltages greater than d-c supply voltage $+V$, and R_1 and R_2 are small enough that V/R_1 and V/R_2 are greater than the holding currents of diodes D_1 and D_2, respectively. Diodes D_3 and D_4 are employed to provide high impedances across which the negative trigger pulses are applied. Each trigger pulse, inserted via capacitors C_2 and C_3, causes the OFF diode to switch ON, turning the other diode OFF. Capacitor C_1 must be large enough to insure positive switching so that D_1 and D_2 are never ON simultaneously. The magnitude of C_1 depends on the diode currents, ranging from a value of the order of 0.01 μF for currents of the order of milliamperes to values of the order of microfarads for currents of the order of amperes. The magnitude of coupling capacitors C_2 and C_3 is not critical, typical values being of the order of 0.01 μF.

12-5 THE SILICON-CONTROLLED RECTIFIER (SCR)[5,6,7]

If connection is made to the p_2-region of a $p_1n_1p_2n_2$ device, as shown in Fig. 12-21, current through the p_2n_2 junction via this contact, *known as the gate electrode*, increases α_{fb} of the *npn*-transistor section of the device. If the gate current is large enough, the sum of the alphas of the two transistor sections exceeds unity, and the device switches into the ON state.

[5] F. E. Gentry, F. W. Gutzwiller, N. Holonyak, Jr., and E. E. Von Zastrow, *Semiconductor Controlled Rectifiers: Principles and Applications of p-n-p-n Devices*. Englewood Cliffs, N.J.: Prentice-Hall, Inc. 1964.

[6] *GE SCR Manual*, 4th Ed. Syracuse, N.Y.: General Electric Co., 1967.

[7] *Westinghouse Silicon Controlled Rectifier Designer's Handbook*, Youngwood, Pa.: Westinghouse Electric Corp., 1963.

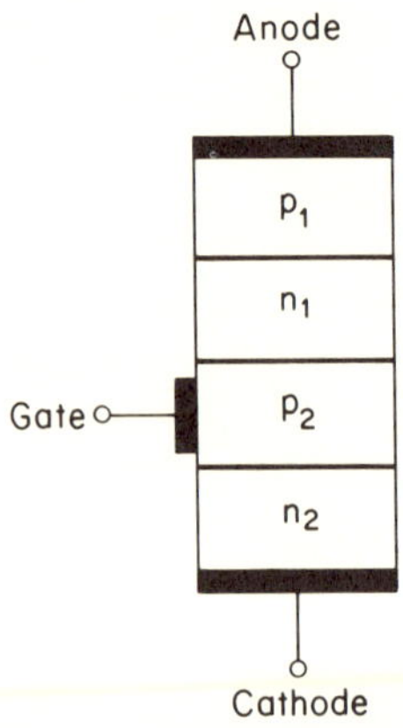

Fig. 12-21. Silicon-controlled rectifier structure.

Since devices of this type are usually fabricated of silicon* and are widely used as rectifiers, they are commonly known as *silicon controlled rectifiers* (SCRs). (The name *gate-controlled reverse-blocking thyristor* is also used.)

A two-transitor representation of the SCR, in which the n_1- and p_2-regions are shown in sections with connecting leads, is shown in Fig. 12-22(a). Figure 12-22(b) shows base and collector currents of the two transistors, which are represented symbolically. From Fig. 12-22(b), we see that for a gate current I_g flowing into the base of the *npn* transistor, the collector current of the *npn* unit is $\alpha_{fe2} I_g$. Since the collector current of the *npn* transistor is the base current of the *pnp* transistor, the collector current of the

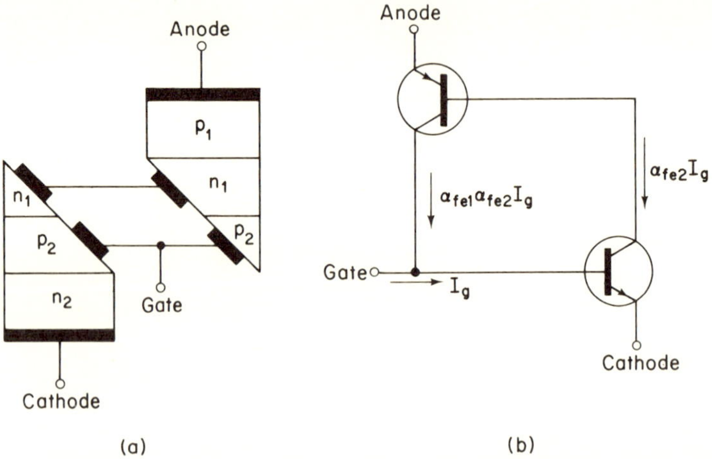

Fig. 12-22. Two-transistor representation of the silicon-controlled rectifier.

* Germanium *pnpn* devices exist, but they are quite uncommon.

pnp unit is $-\alpha_{fe1}\alpha_{fe2}I_g$. With the collector of the *pnp* transistor connected to the base of the *npn* unit, the transistor combination supplies its own gate current if $\alpha_{fe1}\alpha_{fe2}$ is equal to unity. That is, if

$$\alpha_{fe1}\alpha_{fe2}I_g = I_g \tag{12-8}$$

no external gate current is required. This is an unstable condition, of course, and either the application of gate current causes the device to switch ON, owing to $\alpha_{fe1}\alpha_{fe2}$ exceeding unity, or the removal of gate current causes the device to revert to the OFF state because $\alpha_{fe1}\alpha_{fe2}$ is less than unity.

The condition

$$\alpha_{fe1}\alpha_{fe2} > 1 \tag{12-9}$$

for instability or switching into the ON state in the SCR representation of Fig. 12-22(b) is identical to the condition

$$\alpha_{fb1} + \alpha_{fb2} > 1 \tag{12-10}$$

as may be shown by substituting in Eq. (12-9) for α_{fe} in terms of α_{fb}. Validity of the two-transistor representations of Fig. 12-22(b) may be verified by connecting discrete *pnp* and *npn* transistors in this configuration, resulting in an arrangement having characteristics of the *pnpn* device.

The effects of gate current on the forward anode-current–anode-voltage characteristic of a typical low-power SCR are shown by Fig. 12-23. Gate

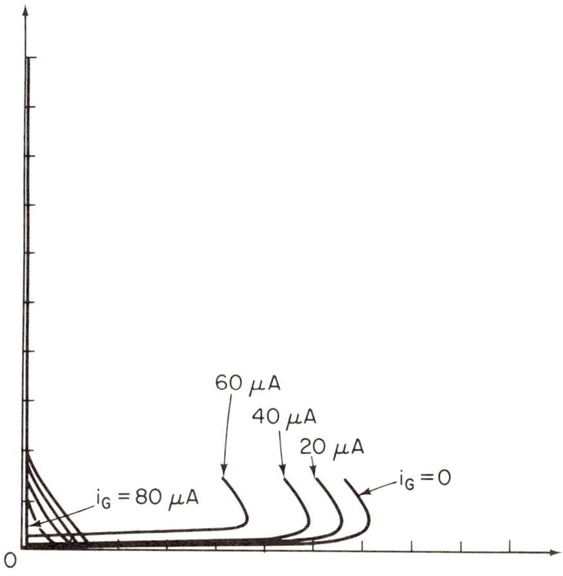

Fig. 12-23. Anode-current–anode-voltage characteristic of a typical low-power silicon-controlled rectifier for each of several gate-current values.

current is seen to increase the anode current for anode voltages in the OFF or forward-blocking state and to decrease the forward-breakover voltage. If the gate current is sufficiently high (80 μA or higher for the device of Fig. 12-23), the forward-breakover region ceases to exist, and the device characteristic is much like that of a *pn* junction.

Turn-on mechanism. In normal use, the SCR is in the forward-blocking state when gate current is applied. If we consider the device to consist of parallel *pnpn* filaments, as shown by Fig. 12-24, gate current initially turns ON only filament Δ_1 in contact with the gate electrode, the other filaments continuing to block current. At this moment, the complete load current flows into the small cross-section area of one filament, resulting in extremely high current density and large forward-voltage drop. Current through the J_2-region of this filament flows in a pattern depicted by the broken lines of Fig. 12-24, current flowing laterally into adjacent filaments. The lateral component of current in adjacent filament Δ_2 acts as a gate current, causing

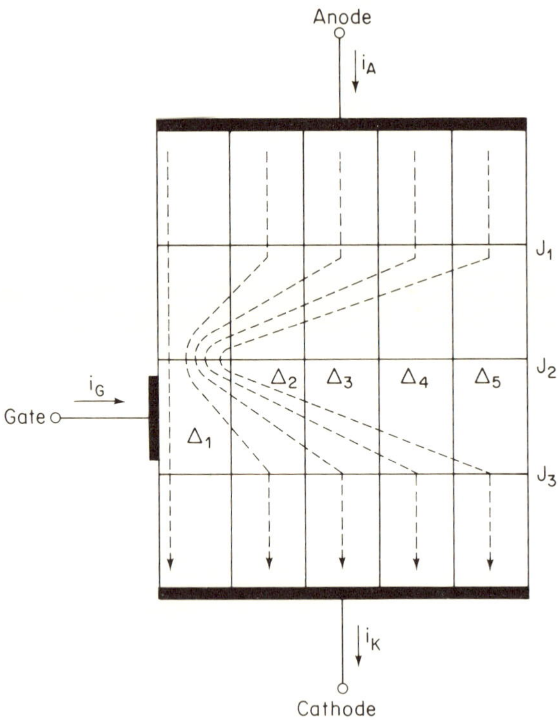

Fig. 12-24. Filamentary representation of a silicon-controlled rectifier. Broken lines depict current paths when filament Δ_1 is turned on by gate current i_G, but the other filaments continue to block current.

it to conduct and supply an effective gate current to filament Δ_3, and action spreads rapidly throughout the structure. As the conducting area increases, the current density decreases accordingly, decreasing the forward drop and the power dissipated in the structure. Because of the time required for the entire rectifier structure to turn ON, there is a limit to the rate of anode-current increase that can occur without overheating. This is the di/dt rating of the rectifier. Within limits, the J_2-area turned ON initially by gate current (the area of filament Δ_1 in Fig. 12-24) depends upon the magnitude of gate current i_G. Therefore, the larger i_G, the shorter the turn-on time and the higher the di/dt rating. For example, a type 2N3653 SCR operating at 60 pulses per second has di/dt ratings of 400 A per s and 10 A per s for gate currents of 1.0 A and 0.25 A, respectively. For these ratings, the gate current consists of 1.5 μs pulses having a rise time of not more than 0.10 μs. Both of these gate currents are much larger than that required to merely turn on the rectifier, the typical value being specified as 0.08 A for an anode voltage of 6 V. The di/dt rating is particularly important for devices operating in circuits without appreciable inductance, since the rate of anode-current rise during turn-on is limited by circuit inductance.

Turn-off mechanism. As in the *pnpn* diode, when the silicon-controlled rectifier is in the conducting state, electrons and holes are heavily injected into the two inner regions, and these carriers must be removed in order to return the device to the forward-blocking stage. Applying a reverse-bias voltage to the gate-cathode contacts is of little assistance in turning off the device, since doing so affects only the gate-cathode junction region near the surface.* Therefore, the gate loses control of the device after turning it ON. Turn-off requires reduction of the anode current below the holding current in order that injected carriers decrease through recombination. When the rectifier is used in an a-c circuit, turn-off is further assisted by the reverse half-cycle of applied voltage, which sweeps out stored charge, permitting blocking during the next half cycle. When the device is used as a switch in d-c circuits, turn-off is achieved by reversing the anode-cathode voltage by some means such as that employed in the *pnpn*-diode multivibrators of Figs. 12-18 through 12-20.

The turn-off time of an SCR is the shortest interval between the time when the forward current is zero and the time when the device is able to *block reapplied forward voltage* without turning ON. It should not be confused with the reverse-recovery time, which is the time required before the rectifier can block reverse voltage. Reverse recovery occurs when junctions J_1 and J_3 go into the blocking state, but this is not sufficient for complete recovery, since injected carriers still exist near center junction J_2. The reverse-recovery

* Small-area devices may be designed for gating off, as pointed out in Sec. 12-7.

time of typical SCRs is of the order of a few microseconds, whereas the turn-off time may be of the order of tens of microseconds, depending upon the forward current and junction temperature.

Comparison with power transistors. It is appropriate to compare SCRs with power transistors, since the two have comparable current capacities. In the power transistor, it is desirable that α_{FE} be high in order to achieve high current gain. Since this is accomplished by using very thin base regions, high current gain is incompatible with high voltage. In the controlled rectifier, thick base regions used to achieve low α_{FE} are compatible with high voltages, and the rectifier is inherently a higher voltage device. Another advantage of the SCR, relative to the power transistor, is the drive required for full conduction. In the interest of di/dt rating, the gate current of an SCR may be of the order of the base current of a transistor having a comparable current rating, but the average SCR drive current may be much smaller, since it need merely initiate conduction. The principal advantages of the power transistor are its higher frequency capability and its more nearly linear control characteristic.

12-6 APPLICATIONS OF SCRs

The SCR should not be construed as useful only for power rectification. Applications of these devices are numerous and highly diversified.* After considering triggering methods and circuits, we present a few representative applications.

12-6.1 Triggering Methods

For simple on-off switching of SCRs, a continuous d-c gate signal may be used. However, this method is inefficient, since gate current is not required after turn-on occurs.

Figure 12-25 shows a method of obtaining SCR gate current from the main supply whenever the anode is positive with respect to the cathode. When switch S_1 of Fig. 12-25(a) is closed, gate current increases with supply voltage v_{cd} until the rectifier turns on, whereupon the anode voltage drops to the conduction value and the gate current decreases to a low value. The SCR turns off during the negative half-cycle of v_{cd}, forward blocking ensues, and the process is repeated when v_{cd} again becomes positive. Resistor R limits the peak gate current and silicon diode D prevents reverse voltage from being applied to the gate-cathode junction during the negative half-cycle of v_{cd}. If desired, excessive reverse gate current can be prevented by con-

* See References 5, 6, and 7.

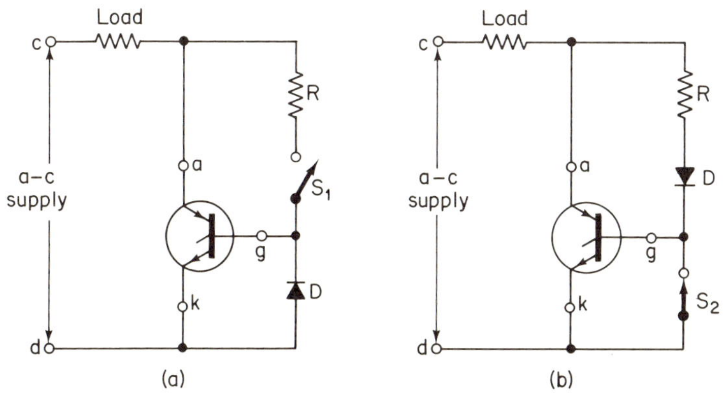

Fig. 12-25. Half-wave static-switch circuits. Closing S_1 in (a) and opening S_2 in (b) results in gate current during positive half-cycles of v_{cd}, turning on the SCR.

necting the diode in series with *R*, as in Fig. 12-25(b), in which conduction is initiated by opening switch S_2.

The circuits of Fig. 12-25 are half-wave, static-switch circuits. Figure 12-26 shows a back-to-back or inverse-parallel pair of SCRs used to switch an a-c load. In these circuits, the rectifiers are used to perform switching functions that could be performed by nonstatic devices such as mechanical and electromechanical switches. Static-switch circuits are devoid of contact sticking, bounce, and wear (problems that are characteristic of electromechanical relays), and are operable over a wider range of voltage than are comparable relays. In Fig. 12-26, resistors R_1 prevent SCR turn-on by spurious signals, and the other circuit elements have the functions of comparable elements in the circuits of Fig. 12-25. The control device can be either mechanical or electrical in nature. Rectifier ratings limit the amount of power that can be switched.

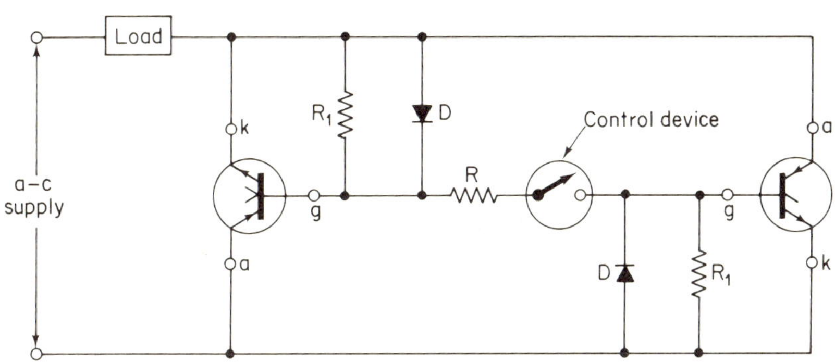

Fig. 12-26. SCRs used as an inverse-parallel or back-to-back a-c switch.

UJT trigger circuits. The UJT is an excellent device for use in SCR trigger circuits, owing to its low trigger current and stable trigger voltage. Figure 12-27 shows the basic UJT relaxation oscillator used as an SCR trigger cir-

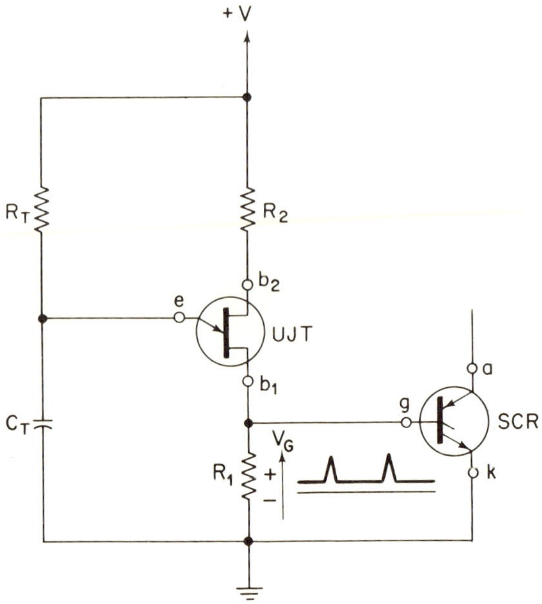

Fig. 12-27. Basic UJT relaxation-oscillator trigger circuit.

cuit. As pointed out in Sec. 12-2, timing capacitor C_T charges through timing resistor R_T until the UJT turns ON and discharges C_T through base-one resistor R_1 and the SCR gate-to-cathode circuit. The UJT then turns OFF, and the cycle is repeated. The voltage applied to the SCR gate consists of short-duration pulses produced by the C_T discharge superimposed on a constant value resulting from the interbase current, as indicated by the v_G wave form of Fig. 12-27. The value of R_1 should be low enough to prevent the constant component of gate voltage from exceeding the maximum value that *will not* trigger the SCR. Base-two resistor R_2 is used to provide compensation for the effect of temperature on the peak-point voltage of the UJT. If interbase voltage V_{bb} is fixed, the peak-point voltage decreases with temperature because of change in emitter-diode forward voltage V_F [see Eq. (12-1)]. With resistor R_2 in series with base two, the temperature coefficient of interbase resistance R_{bb} causes V_{bb} to increase with temperature, compensating for the decrease in V_F. In general, R_1 is less than 100 ohms and R_2 is of the order of a few hundred ohms.

As pointed out in Sec. 12-2, the UJT trigger circuit can be triggered

ON at any intermediate part of the cycle by reducing interbase voltage V_{bb}. This may be accomplished by use of a negative synchronizing pulse at base two. When an SCR operates from an a-c supply, it is required that the gate triggering pulses be synchronized with the supply. This is achieved in the circuit of Fig. 12-28, in which a full-wave rectified signal

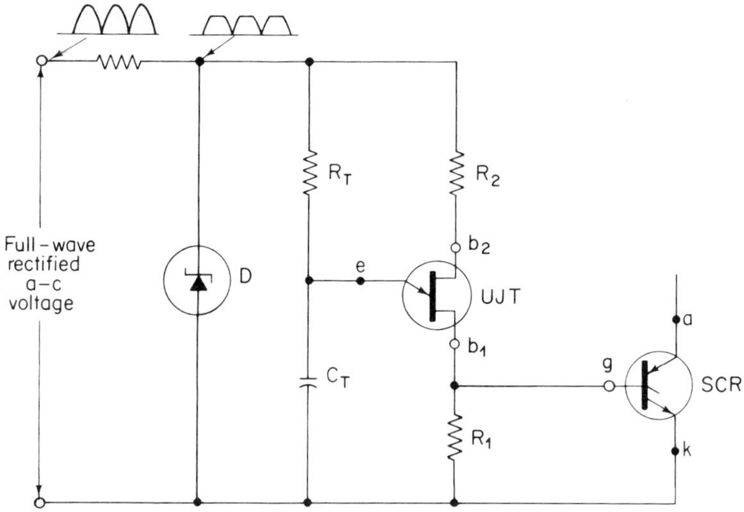

Fig. 12-28. UJT relaxation-oscillator trigger circuit for synchronization to an a-c supply.

obtained from the SCR a-c supply is used to provide power for, and a synchronizing signal to, the trigger circuit. A voltage-regulator (Zener) diode D is used to clip and regulate the peaks of the trigger-circuit supply voltage, as indicated in Fig. 12-28. At the beginning of each half-cycle of supply voltage, C_T charges through R_T until the UJT turns ON and an SCR gate pulse is produced. At the end of each half-cycle, the base-two voltage drops to zero, causing the UJT to turn ON again. Thus C_T is discharged at the beginning of each half-cycle and the circuit is synchronized with the a-c supply. The phase of the SCR gate pulse with respect to the supply voltage is controlled by adjusting R_T.

Avalanche-diode trigger circuits. Low-power diodes having negative-resistance switching characteristics can be used to trigger SCRs. Semiconductor devices of particular interest are three-, four-, and five-layer diode switches.

Three-layer diodes. The three-layer *unilateral trigger diode* is basically an *npn* silicon transitor without a base contact. As the "collector-emitter" voltage of the *npn* diode is increased, avalanche breakdown occurs when α_{fb} becomes unity. An increase in current following breakdown causes an increase in α_{fb}, decreasing the voltage required for unity alpha and avalanche breakdown. Thus the device has a negative-resistance region in which its voltage decreases several volts for a current increase of a few milliamperes. For example, the type TI-42 unilateral trigger diode has a forward-breakover voltage BV_F of 32 ± 4 V, a maximum forward-breakover current BI_F of 300 μA, and a minimum breakback voltage ΔV_F of 4 V for a forward current I_F of 2 mA.* The use of a unilateral trigger diode in SCR triggering is shown by the basic circuit of Fig. 12-29(a), in which the diode is interposed between

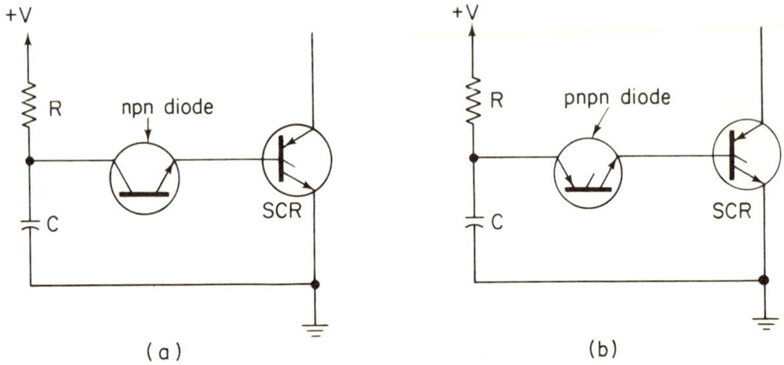

Fig. 12-29. The use of trigger diodes in SCR triggering. (a) Use of a unilateral three-layer diode. (b) Use of a four-layer diode.

a charged capacitor and an SCR gate. When the voltage of capacitor *C* reaches the forward-breakover voltage of the diode, the diode switches ON, causing the capacitor to discharge and trigger the SCR.

The circuit of Fig. 12-29(a) is designed for SCR triggering with a capacitor-voltage swing of a few volts. If a larger voltage swing is desired, the *npn* diode may be replaced by a *pnpn* diode, as shown in Fig. 12-29(b). The energy supplied to the SCR gate depends on the voltage swing during diode switching and the speed of the switching action.

If an *npn* trigger diode has essentially identical *n*-regions (that is, if it is bidirectional *npn* silicon transistor without a base contact), the result is a *bilateral* or a-c *trigger diode* having a conduction characteristic somewhat as shown in Fig. 12-30(b).† Typical bilateral trigger diodes have forward and reverse specifications comparable to the forward specifications of the TI-42 unilateral diode. In addition, bilateral diodes have a symmetry specification that sets an upper limit (typically 10 percent) to the difference between the

* At a temperature of 25°C.

† Bilateral diodes are also known as *diacs*.

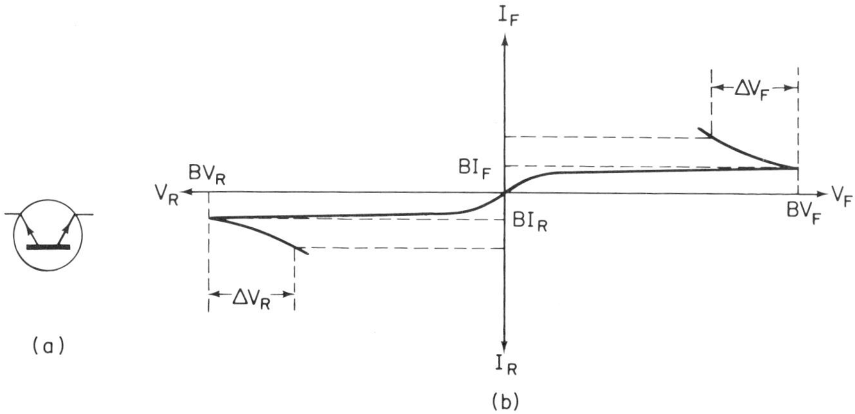

Fig. 12-30. Bilateral trigger diode (diac). (a) Circuit symbol. (b) Conduction characteristic.

forward and reverse breakover voltages. The use of a bilateral trigger diode in a full-wave SCR circuit is shown in Fig. 12-38. If a larger capacitor-voltage swing is desired, the three-layer bilateral diode may be replaced by a five-layer diode, which is the equivalent of an inverse-parallel pair of four-layer diodes, as pointed out in Sec. 12-8.

12-6.2 Phase-Control Circuits

Phase control is the process of connecting a load to an a-c supply for a controlled fraction of each cycle. This is an efficient means of controlling the average power to loads such as lamps, motors, d-c power supplies, etc. Control is achieved by governing the phase of the a-c supply wave at which the SCR is gated on. The SCR then conducts for the remainder of the half-cycle.

Basic phase-control circuits. Figures 12-31 and 12-32 show several basic forms of a-c phase control. In each circuit, the control signal is synchronized to the a-c supply and is adjustable in phase. Representative circuits, including provisions for control signals, will be presented later.

Figure 12-31 shows phase-controlled rectifiers of the half-wave, full-wave bridge, and full-wave biphase types. These circuits are used for loads that require or permit direct current. Several arrangements of SCRs with respect to conventional rectifiers may be used in bridge circuits, but the arrangement of Fig. 12-31(b) results in a common SCR cathode connection that simplifies control-signal circuitry.

Figure 12-32 shows phase-controlled circuits that provide full-wave

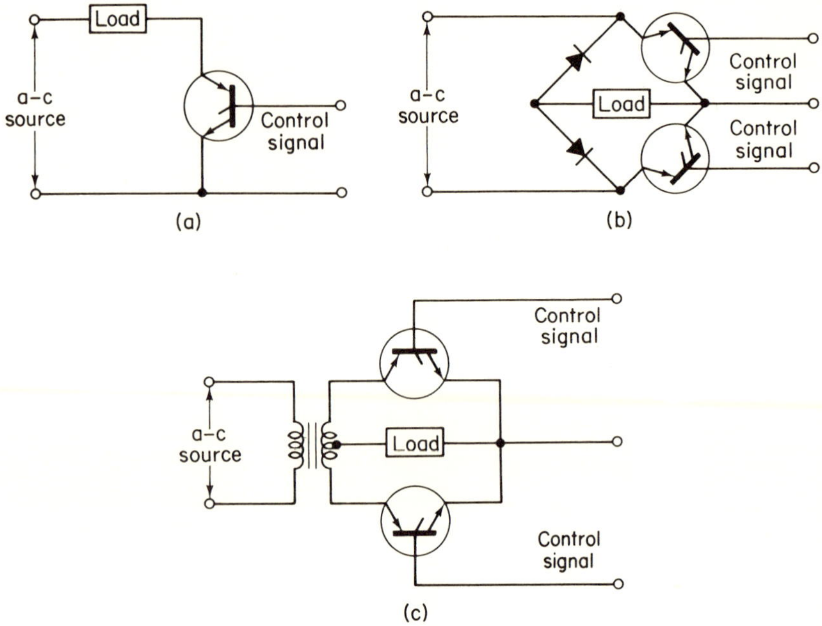

Fig. 12-31. Basic phase-controlled rectifier circuits. (a) Half- wave bridge. (b) Full-wave bridge. (c) Full-wave biphase.

control of a-c power. The circuit of Fig. 12-32(a) has the disadvantage of requiring isolated control signals. This can be alleviated by the addition of two conventional rectifiers, as in the arrangement of Fig. 12-32(b). The most flexible of the several circuits of Figs. 12-31 and 12-32 is the arrangement of Fig. 12-32(c), which uses a single SCR in conjunction with a bridge rectifier. This circuit can be used for the control of either a-c power or full-wave-rectified d-c power.

Commutation in a-c circuits. Commutation of SCRs in a-c circuits is usually no problem, owing to periodic reversal of the supply voltage. However, in some applications either insufficient time for turn-off or excessive dv/dt of reapplied forward voltage can lead to improper commutation. For example, in the inverse-parallel circuit of Fig. 12-33(a) operating with an inductive load, the current through one SCR reaches zero after supply-voltage reversal occurs, as shown by Fig. 12-33(b). The supply voltage then appears as forward voltage across the other SCR, being abruptly applied when the current reaches zero. The resulting large dv/dt of this forward voltage may cause premature turn-on of the SCR. This can be prevented by connecting a series RC "snubber" circuit in parallel with the SCR's. Capacitance C

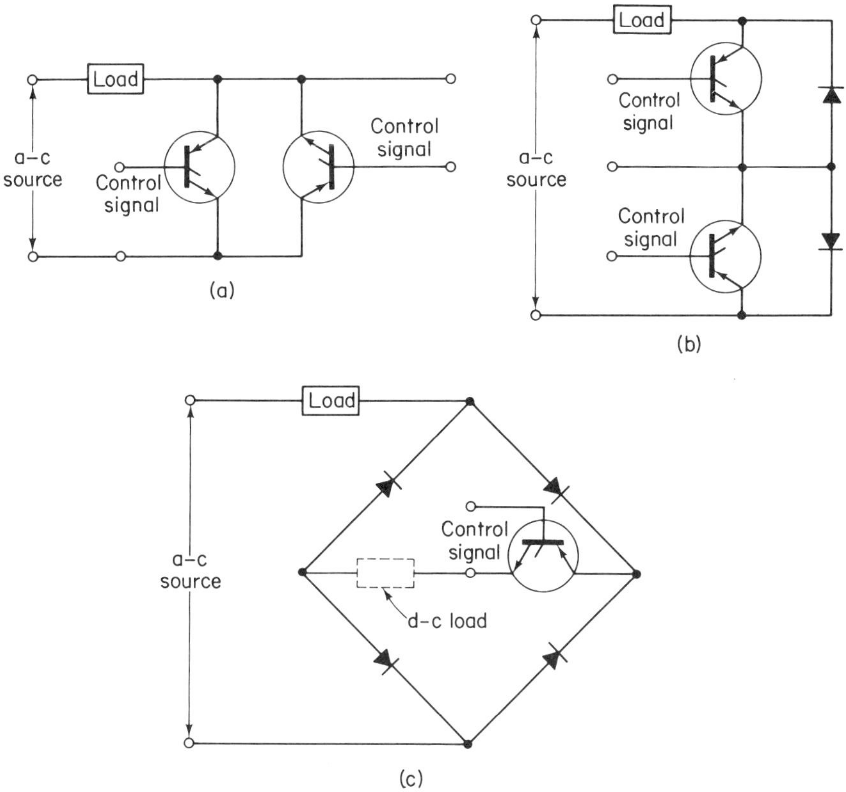

Fig. 12-32. Basic circuits for the control of a-c power. (a) Inverse-parallel arrangement. (b) The use of conventional rectifiers to eliminate the requirement of isolated control signals. (c) A circuit that permits the control of either a-c or d-c power.

reduces anode-voltage dv/dt and resistance R serves to damp out LC oscillation.

Inductive d-c loads often require the use of a "free-wheeling" diode to prevent feedback of power into the supply line during a portion of each half-cycle. If the circuit of Fig. 12-34(a) is operated without diode D and L is large enough that continuous current flows in the load, current flows through each SCR until the other SCR is triggered into conduction. With triggering of the SCRs delayed by the angle α, the load voltage is negative during the delay period, as shown by Fig. 12-34(b), and energy stored in L is returned through the SCRs to the a-c supply. With diode D across the load, the load voltage is clamped at essentially zero during the delay period, and the load current maintained by L flows through D during this period.

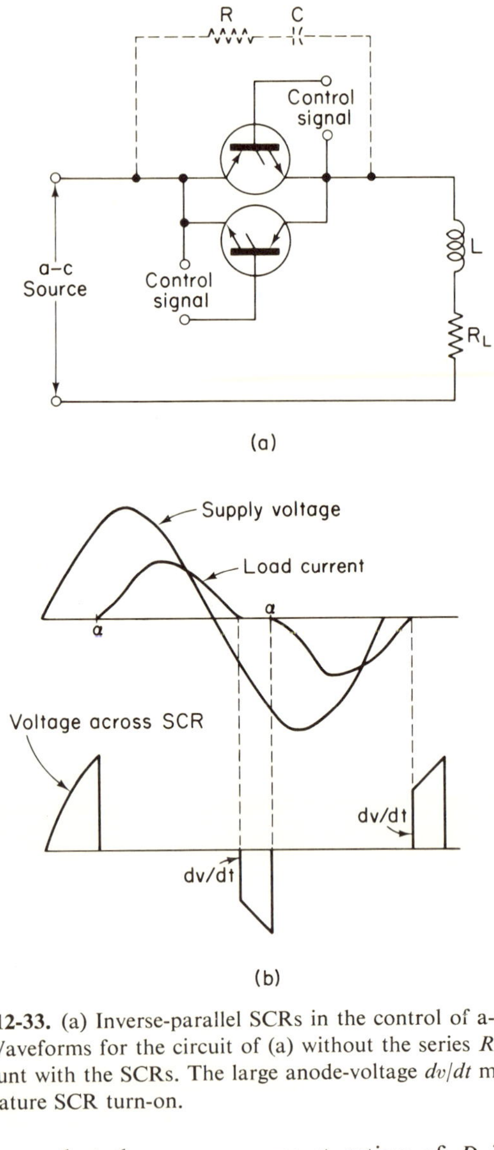

Fig. 12-33. (a) Inverse-parallel SCRs in the control of a-c power. (b) Waveforms for the circuit of (a) without the series *RC* circuit in shunt with the SCRs. The large anode-voltage *dv/dt* may cause premature SCR turn-on.

It may be shown that the average-current rating of *D* is one-fourth the maximum average load current. Wave forms of load voltage for operation without and with diode *D* are shown by (b) and (c) of Fig. 12-34.

Representative phase-control circuits. Two elementary half-wave phase-control circuits are shown by Fig. 12-35. In Fig. 12-35(a), the trigger-circuit supply voltage and the gate voltage are in phase, limiting the delay

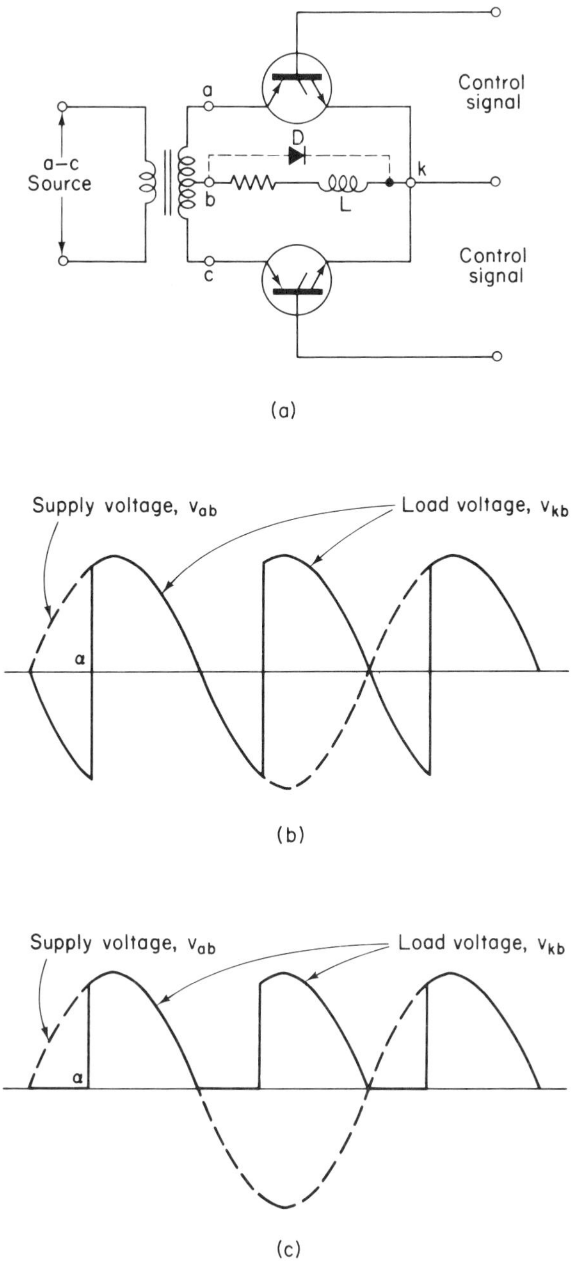

Fig. 12-34. (a) Full-wave rectifier supplying an inductive load. (b) Load-voltage wave form without free-wheeling diode *D*. (c) Load-voltage wave form with diode *D* in shunt with the load.

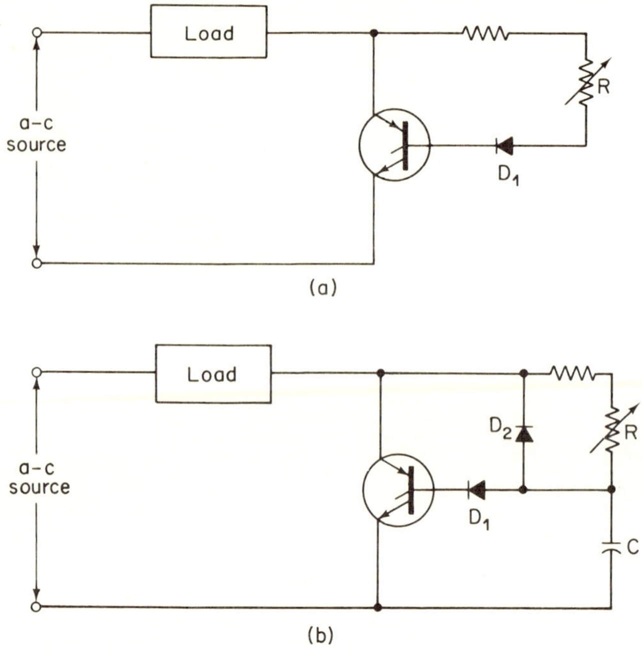

Fig. 12-35. Elementary half-wave phase-controlled rectifier circuits. (a) Circuit permitting a trigger-delay range of 0 to 90°. (b) Circuit permitting a range of 0 to 180°.

of triggering to a maximum value of 90° as control resistance R is increased to the maximum value that permits triggering. Diode D_1 blocks reverse gate voltage on the negative half-cycle of source voltage. In the circuit of Fig. 12-35(b), triggering may be delayed by a maximum of 180 deg, giving full half-cycle control of SCR conduction. During positive half-cycles of source voltage, capacitor C charges to the trigger voltage of the SCR in a time determined by the time constant RC and the anode voltage. During negative half-cycles, C discharges through diode D_2 and is reset for the next charging cycle.

Figure 12-36 shows a phase-controlled full-wave biphase rectifier circuit in which SCR gate signals are provided by a UJT trigger circuit synchronized to the a-c source (see Fig. 12-28). Diode D is a "free-wheeling" diode for operation with an inductive load. A similar triggering arrangement is used for the inverse-parallel SCRs of the a-c power control circuit of Fig. 12-37, except that the lack of a common SCR cathode terminal requires that the gate signals be applied through an isolation transformer. Another difference in Fig. 12-37 is the use of the rectified off-period SCR-anode voltage to provide the interbase supply voltage of the UJT. This arrangement avoids the use of

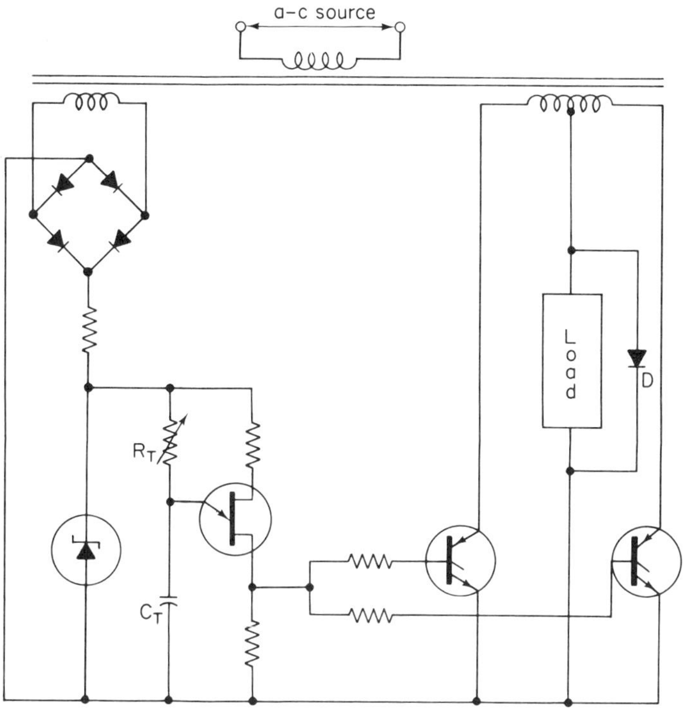

Fig. 12-36. Use of a unijunction-transistor trigger circuit to control a biphase full-wave rectifier.

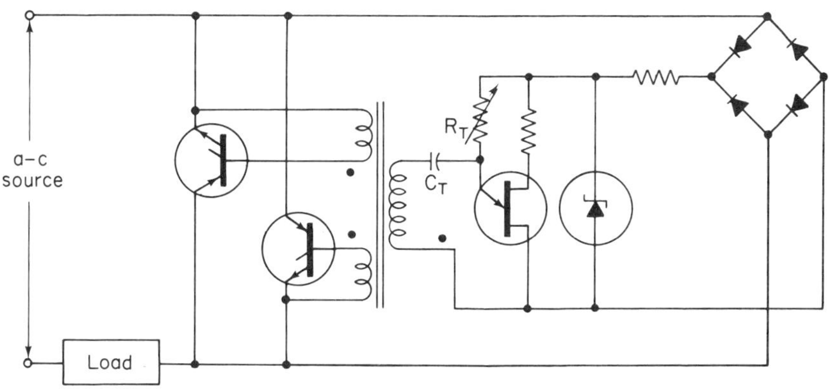

Fig. 12-37. Phase control of a-c power using inverse-parallel SCRs.

a transformer for the interbase supply by taking advantage of the lack of an interbase-voltage requirement after the UJT/turns on.

The use of a bilateral or a-c trigger diode to trigger an inverse-parallel pair of SCR's in an a-c power-control circuit is shown by Fig. 12-38. During

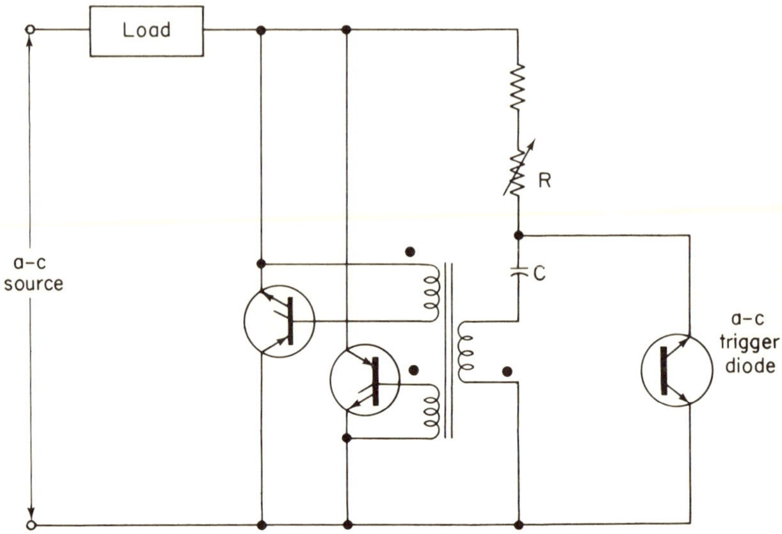

Fig. 12-38. Use of a bilateral or a-c trigger diode (trigger diac) in a full-wave phase-controlled SCR circuit.

each half-cycle, capacitor *C* charges to the breakover voltage of the trigger diode. When this voltage is reached, an abrupt drop in diode voltage causes *C* to discharge through the primary of the pulse transformer, resulting in the application of voltage pulses to the SCR gate-cathode terminals. The pulse transformer is so connected that SCR anode and gate voltages are of the same polarity. Trigger delay and a-c load power are controlled by adjusting resistance *R*, since *C* charges to the breakover voltage in a time determined by the time constant *RC* and the applied voltage.

12-7 THE GATE-CONTROLLED SWITCH (GCS)*

The gate-controlled switch (also known as the gate turn-off switch) is a three-terminal *pnpn* device that can be turned OFF as well as ON from its gate terminal. Compared to SCRs, GCSs are relatively low-power devices,

* Reference 5, pp. 124–129, 264–266, 292–294; Reference 6, p. 16; Reference 7, pp. 1–11, 4-22, 7-93.

having upper-current limits of the order of amperes. The GCS is a higher-speed device than the SCR, and can be used at frequencies up to about 300 kHz, an order of magnitude higher than the frequency range of the SCR. Compromises in the design of the GCS result in its requiring a larger gate current for triggering and having a larger holding current than SCRs of comparable current ratings.

Figure 12-39 shows the two-transistor representation of a three-terminal

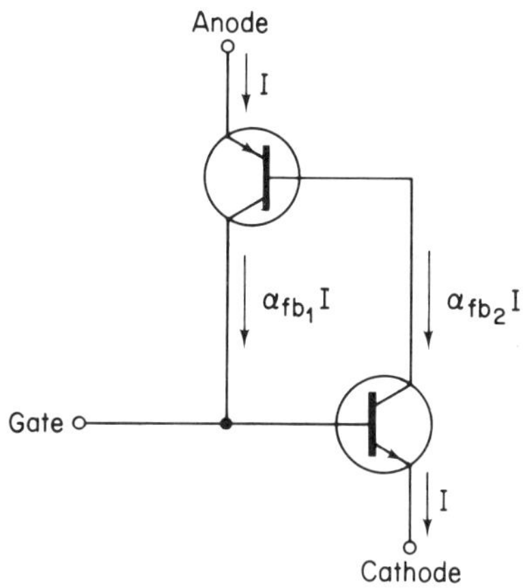

Fig. 12-39. Two-transistor representation of a three-terminal *pnpn* device operating in the ON state.

pnpn device, the electrode currents of each transistor being indicated for the ON condition. It appears from this representation that the device could be turned OFF by diverting *npn*-unit base current $\alpha_{fb_1}I$ out of the device through the gate terminal. In fact, devices having small cross-section areas can be gated OFF in this manner, but high-current devices cannot, since the gate affects only regions near the surface, as pointed out in the filamentary representation of Fig. 12-24. However, the turn-off time of relatively high-current SCRs can be significantly reduced by a negative gate current during the turn-off interval.

For an SCR, it is merely necessary that $(\alpha_{fb_1} + \alpha_{fb_2}) \geq 1$ for turn-on and during the ON period, and it matters little whether α_{fb_1} is larger or smaller than α_{fb_2}. However, if the device is to be gated OFF, α_{fb_1} should be much the smaller of the two alphas in order that $\alpha_{fb_1}I$ of the *npn* unit be relatively small. This enables the device to have an appreciable turn-off current gain, which is defined as the ratio of anode current I to the turn-off gate current. Typical GCSs have turn-off current gains of the order of 10.

Turn-off methods. Several methods of turning GCS units ON and OFF are shown in Fig. 12-40. In the series-capacitor circuit of Fig. 12-40(a),

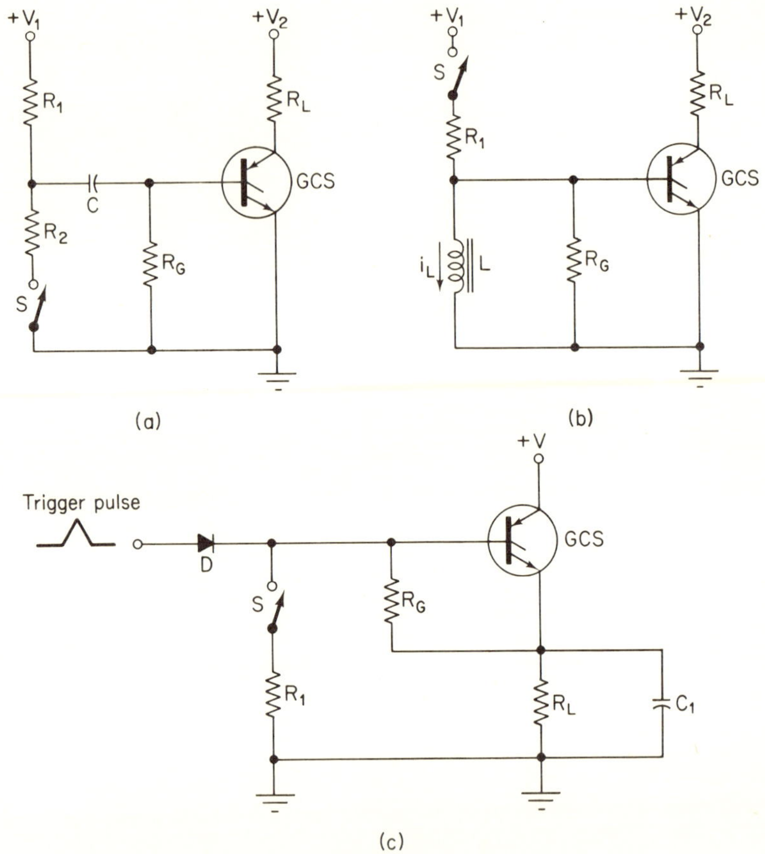

Fig. 12-40. Turn-off circuits for gate-controlled switches. (a) Series capacitor turn-off. (b) Parallel-inductor turn-off. (c) Load-in-cathode commutation.

opening switch S causes capacitor C to charge to V_1 through R_1 and the GCS gate-cathode circuit, turning the GCS ON. When S is closed, C discharges through R_2, resulting in a negative gate current that turns the GCS OFF. In the parallel-inductor turn-off circuit of Fig. 12-40(b), the GCS is turned OFF by interrupting d-c current i_L through inductor L. When switch S is opened, i_L is transferred from R_1 to the GCS gate-cathode circuit, resulting in a negative gate current that causes turn-off. The circuit of Fig. 12-40(c) employs load-in-cathode commutation. The GCS is turned ON by a short-duration, positive gate-current pulse applied through fast-recovery diode D, which quickly decouples the gate-trigger source. With the GCS ON, its gate and cathode terminals assume the potential of supply voltage V, and turn-off occurs when switch S is closed, diverting current *out* of the gate through current-limiting resistance R_1. Capacitor C_1 serves to provide a low-impedance bypass around the load for the triggering pulse and to aid the turn-off process by maintaining the load voltage as the cathode current falls. If the load impedance is sufficiently low to pass the trigger pulse, C_1 may be eliminated.

D-C chopper circuit. Figure 12-41 shows a representative GCS application that employs turn-on and turn-off methods similar to those of Fig. 12-40.

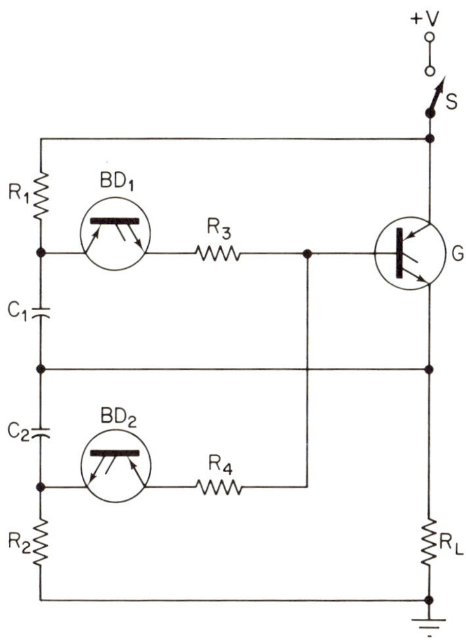

Fig. 12-41. Gate-controlled switch used as a d-c chopper.

When supply switch S is closed, the GCS blocks the applied voltage, and capacitor C_1 charges through resistances R_1 and R_L. The voltage across C_1 increases until it reaches the breakover voltage of *pnpn* diode BD_1, at which time BD_1 turns ON and C_1 discharges into the gate of the GCS, turning it ON. The applied voltage $+V$ then appears across load resistor R_L, causing capacitor C_2 to charge through resistance R_2. When the voltage of C_2 reaches the breakover voltage of *pnpn* diode BD_2, BD_2 turns ON and C_2 discharges out of the gate of the GCS, turning it OFF and completing the cycle of operation. By making R_1 and R_2 adjustable and ganging them so that R_2 decreases as R_1 is increased, the ratio of *on-period* to *off-period* can be controlled without change in frequency, resulting in the control of d-c load power. This circuit is representative of applications in which a single GCS can be used instead of two SCRs.*

12-8 THE BIDIRECTIONAL THYRISTOR†

One shortcoming of the SCR is the lack of a bidirectional characteristic; that is, SCR *on-state* current flow is limited to one direction. By extending the shorted-emitter concept shown in Fig. 12-14, symmetrical two-terminal and three-terminal switches can be fabricated. The two-terminal switch has the conduction characteristic of an inverse-parallel pair of *pnpn* diodes, and the three-terminal switch is the equivalent of two inverse-parallel SCRs. Trade name for the two-terminal a-c switch are the diac‡ (for a-c diode) and the biswitch, and the three-terminal switch is called a triac (for a-c triode) and the gated biswitch.

Bilateral diode switch. Figure 12-42 shows the structure of a bidirectional diode switch. The device is seen to consist of two *pnpn* sections in parallel but arranged in opposite order. Thus, when in the ON state, each section conducts current in a direction opposite that of the other section. Devices of this type are triggered by rapidly raising the applied voltage (*dv/dt* triggering), or by raising the applied voltage above the avalanche voltage of the blocking junctions.

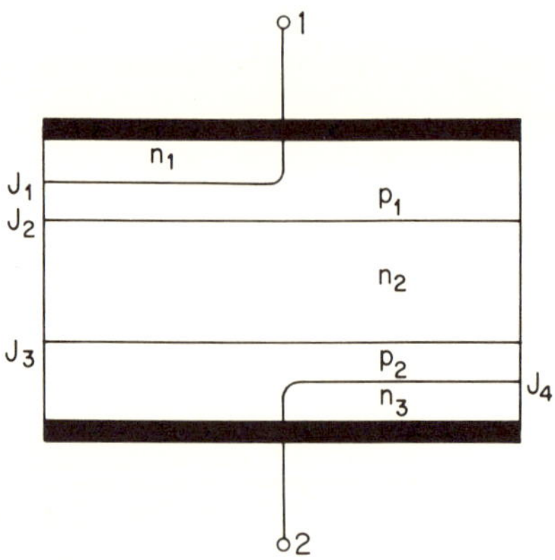

Fig. 12-42. Bilateral diode-switch (diac) structure.

* See Reference 6, pp. 237–243, for the use of SCRs in the control of d-c power.

† Reference 5, pp. 139–148, 244–246, 279; Reference 6, pp. 13–16, 133–148, 187–189, 216–223.

‡ The trade name diac is also used for the three-layer a-c trigger diode, which has a much higher ON voltage and is not a thyristor.

In order to visualize the operation of the device of Fig. 12-42, consider terminal 1 to be made positive with respect to terminal 2. As a result, junction J_2 is forward biased, junction J_3 is reverse biased, and holes injected into the n_2 region are collected by J_3. As the applied voltage is increased, the lateral voltage drop in the p_2-region becomes large enough to forward-bias junction J_4, causing electrons to be injected from n_3 into p_2. Further increase in voltage causes the $p_1n_2p_2n_3$ section to switch from the blocking state to the conducting state in the manner described in Sec. 12-3. For this polarity of applied voltage, current flow through junction J_1 is negligible, since voltage drop in the p_1 region causes J_1 to be reverse biased. This tends to restrict action to the right-hand section of the device, since any current flowing through the left-hand section must overcome the lateral resistances of the p_1- and p_2-regions.

If terminal 2 is made positive with respect to terminal 1, the $p_2n_2p_1n_1$ section of the device will switch in the same manner as that just described for the $p_1n_2p_2n_3$ section.

Figure 12-43 shows the basic circuit for the use of a bilateral diode in the

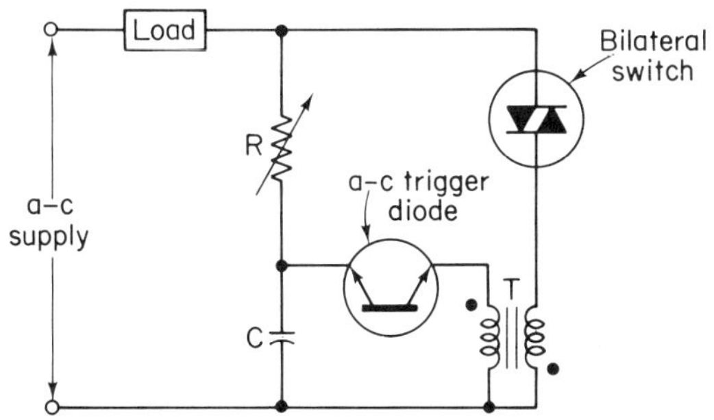

Fig. 12-43. Use of a bilateral diode switch (diac) in the control of a-c power.

control of a-c power. During each half-cycle, capacitor C charges through resistance R to the breakover voltage of the bilateral trigger diode. When the voltage of C reaches this value, the trigger diode turns on and C discharges through the primary of pulse transfomer T, increasing the voltage applied to the bilateral diode switch and triggering it ON. Load power is controlled by adjusting R, since C charges to the breakover voltage in a time determined by the time constant RC and the applied voltage.

Bilateral triode switch (triac). Judicious arrangement of the p- and n-regions of a multilayer device results in an a-c switch that can be triggered into conduction in a manner similar to that of the SCR. Figure 12-44 shows a possible electrode structure for such a device. When terminal 2 is made

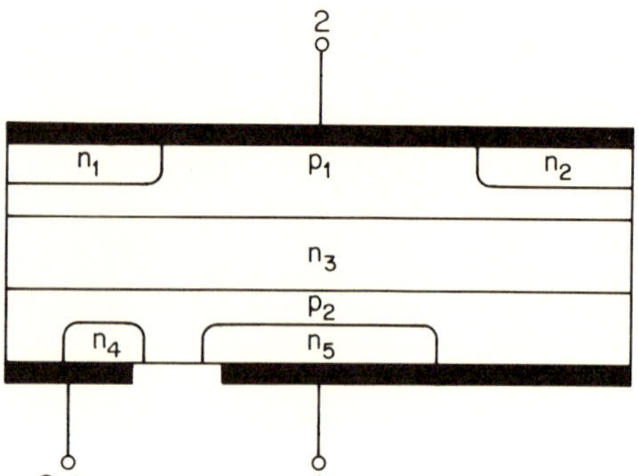

Fig. 12-44. Bilateral triode (triac) structure.

positive with respect to terminal 1, the p_1-, n_3-, p_2-, and n_5-regions act as an SCR that can be gated on in the usual manner (that is, by a positive gate current), owing to the gate contact with the p_2-region. For this polarity of V_{21} and the gate positive with respect to teminal 1, junctions p_1n_1, p_1n_2, and p_2n_4 are reverse biased and have negligile effect on device current.

When terminal 1 is biased positive with respect to terminal 2, the p_2-, n_3-, p_1-, n_1-, and n_2-regions act as a four-layer switch that can be turned on by remote-gate action as follows. If the gate electrode is made negative with respect to terminal 1, electrons are injected from the n_4-region into the p_2-region and diffuse toward the p_2n_3 junction, where they are collected. These collected electrons lower the potential of n_3 with respect to p_2 (that is, the forward bias of the p_2n_3 junction is increased) so that holes are injected from p_2 into n_3. These holes diffuse across n_3, are collected by the p_1n_3 junction, raise the potential of p_1 with respect to n_1 and n_2, and increase the forward bias of the p_1n_1 and p_1n_2 junctions. This causes electrons to be injected from n_1 and n_2 into p_1. These electrons diffuse toward the p_1n_3 junction, where they are collected and lower the potential of the n_3 region so that more holes are injected from p_2 into n_3. This process continues until the device turn ON. Thus the triac can be gated ON by negative gate current for negative V_{21} and by positive gate current when V_{21} is positive. These are *preferred* gate-current polarities. In reality, either polarity can turn the triac ON for each polarity of applied voltage, but the preferred polarities require smaller gate currents.

Figure 12-45 shows the use of the triac in the control of a-c power. During each half-cycle of applied voltage, capacitor C charges through resistance R to the breakover voltage of the trigger diode. When the voltage of C reaches this value, the trigger diode turns ON, resulting in a positive gate current for positive v_{21} and a negative gate current when v_{21} is negative. Load power is controlled by adjusting R, since R and C constitute a phase-

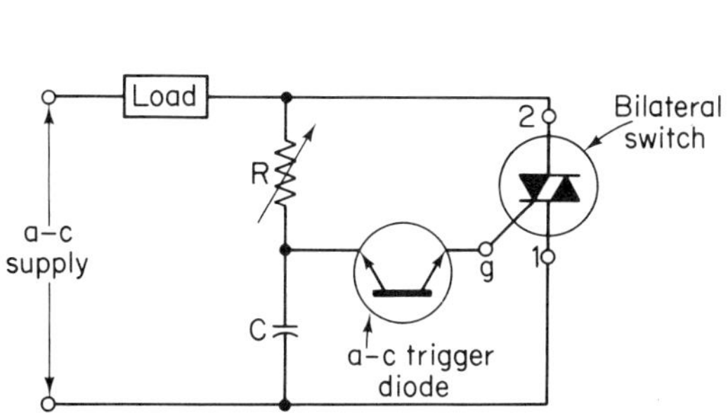

Fig. 12-45. Use of a bilateral triode switch (triac) in the control of a-c power.

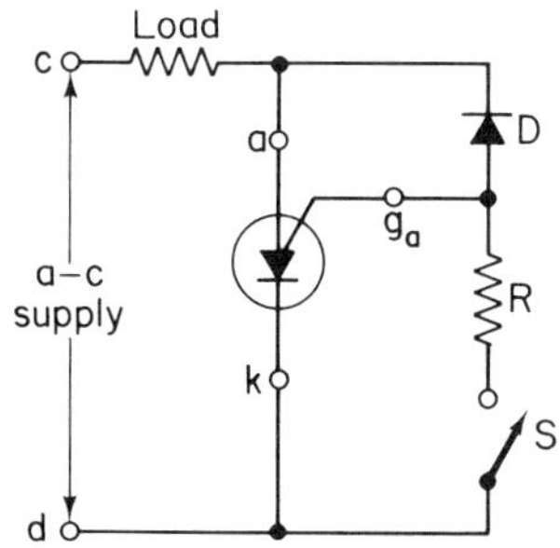

Fig. 12-46. Complementary SCR used as a half-wave static switch.

shift network. Because the circuitry is less complex, triacs have replaced SCRs in many a-c power-control applications.

12-9 MISCELLANEOUS THYRISTORS

The basic principles of thyristors and the most widely used devices are considered in the preceding sections. In this section, we briefly consider some additional thyristors.

12-9.1 The Complementary SCR

The complementary SCR is a $p_1n_1p_2n_2$ device having a connection to the n_1-region. This connection is known as the *anode gate*, and the device is turned ON by forward-biasing the p_1n_1 junction. That is, negative anode-gate current may cause turn-on, as in the half-wave static-switch circuit of Fig. 12-46. This circuit is seen to be the inverse of the conventional SCR static-switch circuit of Fig. 12-25(a). As in Fig. 12-25(a), closing switch S results in gate current during positive half-cycles of v_{cd}, turning on the SCR, but the gate current is negative. Most commercial SCRs are of the conventional (noncomplementary) type.

12-9.2 The Tetrode Thyristor*

The tetrode thyristor, also known as the silicon controlled switch (SCS), is a *pnpn* structure with all four semiconductor regions accessible. That is,

* Reference 1, pp. 391–435.

the device has an *anode gate* in addition to the *cathode gate* and may be used as a complementary SCR or as either a *pnp* or an *npn* transistor. When operated as a conventional SCR or GCS turned ON by positive cathode-gate current, the tetrode thyristor may be turned OFF by making the anode gate positive. Perhaps the most important use of the anode gate is the minimization of rate effect, or *dv/dt* turn-on, when anode voltage is applied. The anode gate makes it possible to charge the center-junction capacitance before anode-voltage application, so that anode *dv/dt* produces little charging current. This is illustrated in the relaxation-oscillator circuit of Fig. 12-47.

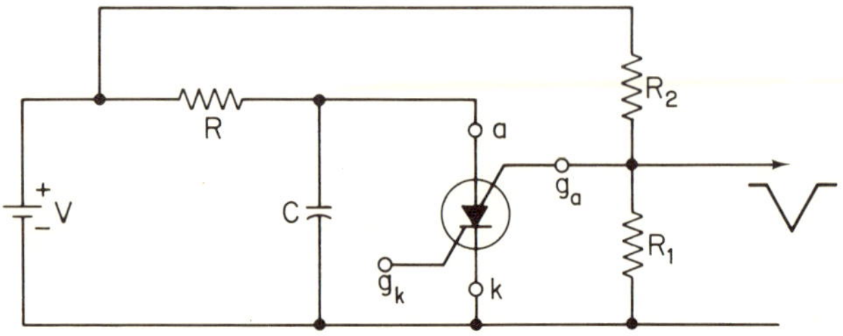

Fig. 12-47. Tetrode thyristor used as a relaxation oscillator in an arrangement that minimized *dv/dt* turn-on.

When the thyristor of Fig. 12-47 is OFF, the anode-gate voltage is approximately $VR_1/(R_1 + R_2)$, and anode current is negligible for anode voltages less than this value. As C charges through R, the anode voltage increases until it exceeds the anode-gate voltage, whereupon negative anode-gate current turns the thyristor ON, C discharges, and the device turns OFF (provided that V/R is less than the holding current). If synchronization is desired, the synchronizing signal may be applied to the cathode gate. Whether or not synchronization is used, the cathode gate may be returned to the cathode through a resistor in order to avoid turn-on by spurious signals. In general, tetrode thyristors are low-power devices having dissipation ratings less than one-half watt.

12-9.3 The Light-Activated Thyristor*

Light entering a forward-biased triode or tetrode thyristor in the vicinity of the reverse-biased center junction creates electron-hole pairs that are swept across the junction and increase the anode-to-cathode current. This current increases with increased light, increasing the alphas of the *pnp* and *npn* transistor equivalents of the structure, so turn-on may occur. The light threshold-level for turn-on depends on the forward voltage, the temperature, and the rate of change of forward voltage. The anode gate of the light-

* Reference 5, pp. 79–81, 191–192; Reference 6, pp. 1. 9, 289–300.

activated tetrode may be used to eliminate dv/dt turn-on, and the cathode gate of either the triode or the tetrode permits control of the light threshold-level by way of an external resistance connected between the gate-cathode terminals. This resistance bypasses current around the gate-cathode junction, increasing the threshold-level.

In order to obtain reasonable sensitivity to light, light-activated thyristors must be constructed so that they can be turned ON with very low current densities. This requires the use of a thin, small-area structure; therefore, light-triggered thyristors are relatively low-voltage, low-current devices. However, they may be used to trigger higher-power SCRs. Light-activated thyristors triggered by light-emitting diodes prove to be useful combinations.

12-9.4 The Reverse-Switching Rectifier[8]

The reverse-switching rectifier is a two-terminal *pnpn* device. When operated in the *reverse direction*, the device blocks current for voltages up to a level at which it switches to a low-impedance state (that is, the *pnpn* switch turns ON), where it remains until the current is reduced below the holding value. In the *forward direction*, the device conducts as a normal junction rectifier. Such characteristics might be expected of a *pnpn* device employing the shorted-emitter structure for both emitters, as shown in Fig. 12-48.

The reverse-switching rectifier is turned ON by raising the reverse voltage above the switching value. It is useful in pulse work because of its high di/dt and fast-switching capabilities and because of the clamping action of its forward characteristic.

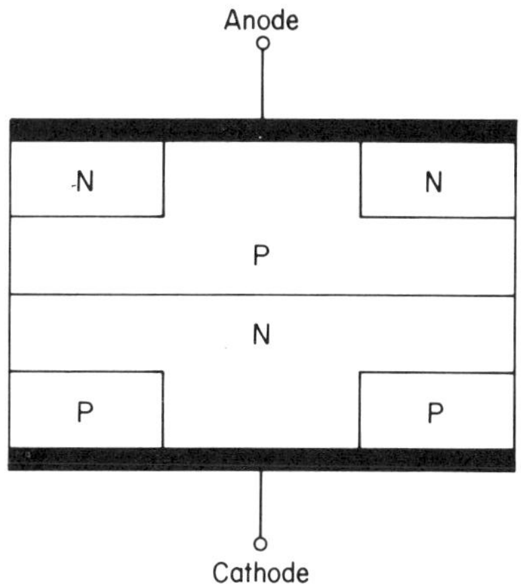

Fig. 12-48. Reverse-switching rectifier structure.

[8] Westinghouse Type 423 Reverse Switching Rectifier Specification Sheet.

12-9.5 The Programmable UJT (PUT)[9]

The programmable UJT is a UJT in name only, since it is a three-junction *pnpn* switch with an anode gate. However, the PUT may replace the UJT in typical applications, such as the SCR trigger circuit of Fig. 12-49(a). Figure 12-49(a) shows a UJT circuit, and Fig. 12-49(c) shows the same

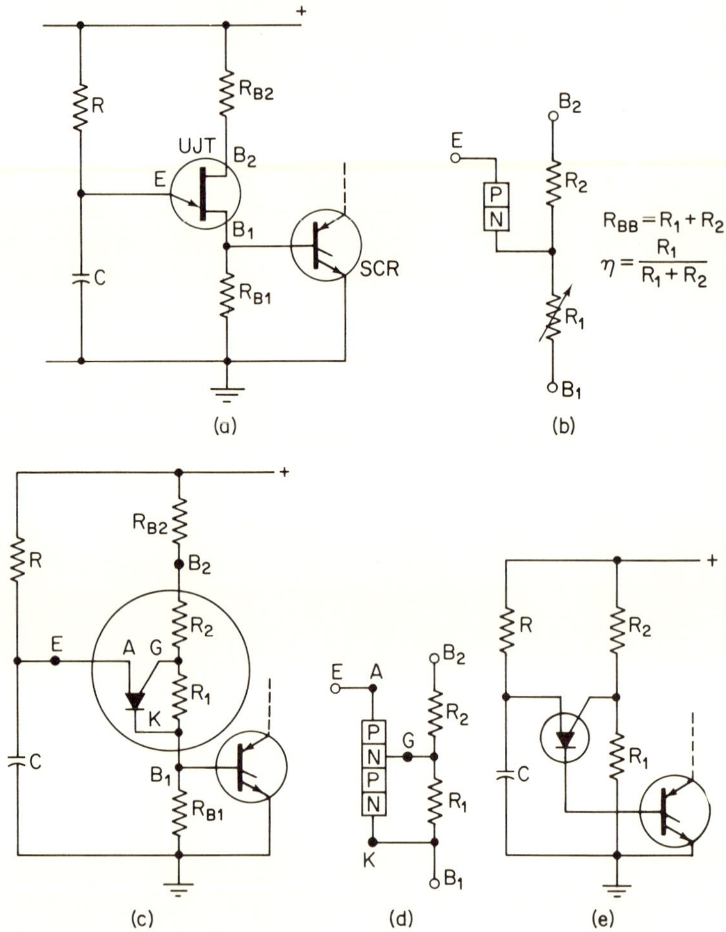

Fig. 12-49. Use of the programmable unijunction transistor (PUT) in place of the UJT. (a) Typical UJT circuit. (b) UJT equivalent circuit. (c) PUT replacing the UJT in the circuit of (a). (d) UJT equivalent circuit using a PUT. (e) Simplified typical PUT circuit. (From Reference 9.)

[9] General Electric Type D13T1/D13T2 Programmable Unijunction Transistor Specification Sheet.

circuit, except that the UJT is replaced by a PUT plus resistors R_1 and R_2. Comparing the equivalent circuits of Figs. 12-49(b) and 12-49(d), we see that each circuit has a *pn* diode connected to a voltage divider. When this diode of the UJT becomes forward biased, R_1 becomes strongly modulated to a lower resistance value, generating a negative resistance characteristic between the emitter (E) and base-one (B_1) terminals. For the PUT, resistors R_1 and R_2 control the voltage at which the diode (anode-to-gate) becomes forward biased. When the diode conducts, negative gate current causes the device to turn ON, generating a negative-resistance characteristic from anode to cathode that simulates the modulation of R_1 of the UJT. Resistors R_{B1} and R_{B2} are generally unnecessary when the PUT replaces a UJT, permitting the arrangement of Fig. 12-49(e). In this arrangement, R_1 is returned to ground so that the "interbase current" is kept out of the SCR gate. R_{B1} in the UJT circuit provides a gate bypass for UJT interbase current. R_{B2} of the UJT circuit is for temperature compensation and has no corresponding effect in the PUT circuit.

The advantages of the PUT, relative to conventional UJTs, stem from its programmability. By selecting R_1 and R_2, one can program UJT characteristics such as intrinsic stand-off ratio η, interbase resistance R_{BB}, peak-point current I_p, and valley-point current I_v. Clearly, η is determined by the ratio $R_1/(R_1 + R_2)$ and R_{BB} is $(R_1 + R_2)$. Large values of R_1 and R_2 result in low values of I_p and I_v, and high I_p and I_v values are obtained by using low-value resistors, as indicated in Fig. 12-50. A low-I_p high-I_v combination is often desired and may be obtained by using a silicon diode and an additional resistor, as in Fig. 12-50(c). The PUT is also usable as a complementary SCR.

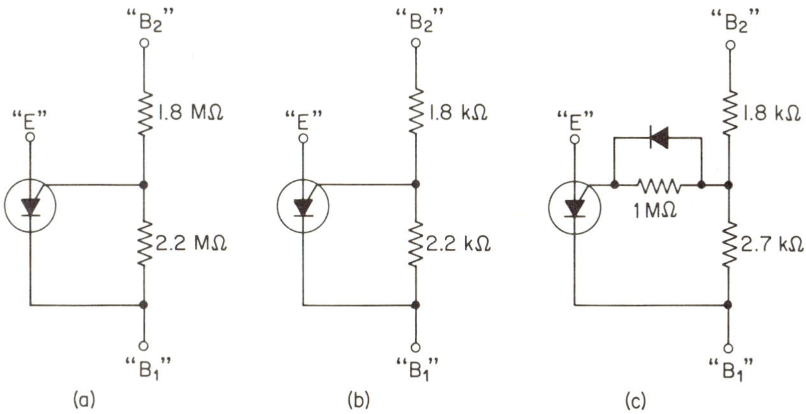

Fig. 12-50. UJT configurations using the programmable UJT. (a) Low I_p, low I_v. (b) High I_p, high I_v. (c) Low I_p, high I_v.

REFERENCES

[1] *GE Transistor Manual*, 7th ed. Syracuse, N. Y.: General Electric Co., 1964, Chapter 13.

[2] I. A. Lesk and V. P. Mathis, "The Double-Base Diode: A New Semiconductor Device," *IRE Convention Record*, Part 6 (1953), pp. 2–8.

[3] W. R. Spoffard, Jr. and R. A. Stasior, "A Switch In Time," *Electronics*, Vol. **41**, No. 4 (February 19, 1968).

[4] L. S. Senhouse, Jr., "An Epitaxial Planar Structure for the Unijunction Transistor," *IEEE Trans. on Electron Devices*, Vol. **ED-16**, No. 2 (February, 1969).

[5] F. E. Gentry, F. W. Gutzwiller, N. Holonyak, Jr., and E. E. Von Zastrow, *Semiconductor Controlled Rectifiers: Principles and Applications of p-n-p-n Devices*. Englewood Cliffs, N. J.: Prentice-Hall, Inc. 1964.

[6] *GE SCR Manual*, 4th ed. Syracuse, N. Y.: General Electric Co., 1967.

[7] *Westinghouse Silicon Controlled Rectifier Designer's Handbook*, Youngwood, Pa.: Westinghouse Electric Corp., 1963.

[8] Westinghouse Type 423 Reverse Switching Rectifier Specification Sheet.

[9] General Electric Type D13T1/D13T2 Programmable Unijunction Transistor Specification Sheet.

PROBLEMS

12-1. Specifications of the 2N490 UJT of Fig. P12-1 are $I_p \leqq 12\ \mu\text{A}$ and $I_V \geqq 8$ mA. With the aid of Fig. 12-3, determine the limits of R_T for oscillation.

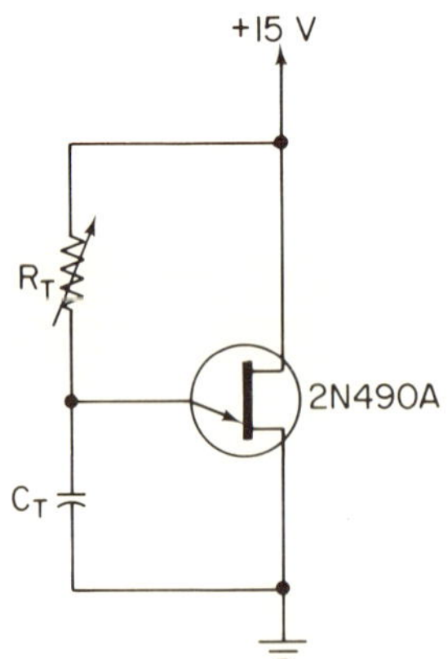

Figure P12-1

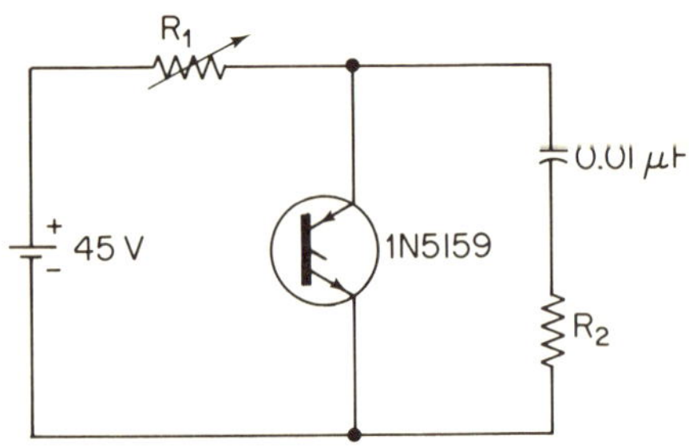

Figure P12-3

12-2. Derive Eq. (12-3) for the oscillating frequency of the circuit of Fig. P12-1.

12-3. The 1N5159 *pnpn* diode of Fig. P12-3 has forward current ratings of 180 mA average and 10 A peak. The holding current I_H is 1 mA $\leqq I_H \leqq$ 20 mA, the forward breakover voltage is 10 V $\leqq V_{BR,F} \leqq$ 12 V, and the forward switching current is less than 125 μA. Determine the range of R_1 for which oscillation may be expected. What value of R_2 should be used?

12-4. If the forward breakover voltage of the diode of Prob. 12-3 is 10 V, what is the approximate range of the oscillating frequency?

12-5. Two 1N5159 *pnpn* diodes having the specifications given in Prob. 12-3 are used in the astable multivibrator circuit of Fig. 12-18. If the d-c supply voltage is +45 V, what are the approximate limits of resistors R_1 and R_2 to assure the existence of oscillation and the avoidance of diode damage? What diode reverse-blocking-voltage rating is required?

12-6. Derive an expression for the approximate oscillating frequency of the circuit of Prob. 12-5 for the condition $R_1 = R_2 = R$.

12-7. The *pnpn* diodes of the bistable multivibrator of Fig. 12-20 have a forward average-current rating of 180 mA and holding currents ranging between 1 mA and 6 mA. If the d-c supply voltage is 18 V, determine the limits of resistors R_1 and R_2 and of the diode forward breakover voltage.

12-8. What average current rating is required for diode D of Fig. 12-34(a) if the average load current may be as large as 10 A?

Index

Abrupt junction 66
a-c trigger diode 518
Acceptor impurities 9
Active-mode operation 194, 333
Admittance, input 393, 395, 397
Admittance parameters 242
Alloy-junction BJT 189, 236
Alloy-junction process 66
Alpha, BJT 197, 220
  current dependence 228
  temperature dependence 227
  voltage dependence 232
Alpha, UJT 494
Alpha crowding 232
Alpha-cutoff frequency 257, 260
Alpha stability factor 313
Amplification factor, FET 317
Amplifier:
  common-base 209, 293
  common-collector 293, 295
  common-drain 394
  common-emitter 272, 296
  common-gate 397
  common-source 391
  feedback 478
  narrow-band 480
  operational 481
  RC-coupled 478
Analog ICs 474
AND circuit 348
Anode gate, SCR 533
Astable multivibrator:
  BJT 348, 356
  FET 415
  pnpn-diode 507
Atomic arrangements in solids 2
Automatic gain control 477
Avalanche breakdown 73
Avalanche-diode trigger circuits 517

Backward diode 156
Baird-Atomic Transistor Test Set 279
Band model 36
Bar structure, UJT 491
Bardeen, J. 188
Barrier potential 68, 79
Base, UJT 491
Base characteristic 218
Base current 197, 336
Base region 189
Base-region charge store 337
Base resistance 263
Base-transport factor 196, 221, 224
Base triggering 352
Base-width change 203, 227, 260, 264
Base-width modulation 203, 227
Bias arrangements:
  common-emitter 308
  depletion-type FETs 398
  emitter-follower 315
  enhancement-type MOSFETS 406
  fixed-bias 398
  self-bias 317, 399
  source-bias 399
  source-follower 404
  transformer-coupled stage 320
Bias dependence of alpha 227
Bias load line 402
Bias polarities, BJT 192
Bias stabilization:
  common-emitter 308
  common-source 399
  emitter-follower 315
  linear 320
  nonlinear 320
  source-follower 404
  transformer-coupled stage 320
Bias-stabilizer analysis 308
Bias-stabilizer design 310
Bidirectional thyristor 530
Bidirectional transistor 518
Bilateral BJT 237
Bilateral diode switch 530
Bilateral trigger diode 518
Binary operation 350
Bipolar transistor 188
Bistable multivibrator:
  BJT 348
  FET 412
  pnpn-diode 509
Biswitch 530
BJT 189
  notation 198
Blocking capacitor 315
Bond model 2
Boonton Radio RX Meter 281
Boonton Radio Transistor Test Set 280
Bootstrap circuit 316
Brattain, W. H. 188
Breakdown diode 75, 127, 390
Breakdown voltage:
  avalanche 74
  BJT-switch 335
  common-base 202
  common-emitter 214
  pn-junction 71
  Zener 73

Bridge measurement of BJT parameters 279
Bridge rectifier 164
Built-in electric fields 230, 233
Bulk breakdown 76
Buried layer:
  bipolar-transistor 451
  integrated-circuit 424
Bypass capacitor 313, 400

Capacitance:
  collector 256
  diffusion 103
  JFET 373
  junction-region 99
  Miller-effect 394
  MOSFET 385, 389
Capacitor:
  bypass 313
  coupling 315
  integrated 424, 443
  junction-type 424
  metallic-oxide 440
  silicon-dioxide 445
  speed-up 351
  thin-film 444
Carrier density 19
  intrinsic 21
  temperature dependence 21
Carrier-density product 9
Carrier-depletion region 68
Carrier injection 26
Carrier mobility 13, 15
Carrier recombination 29
Cascade BJT amplifier stages 305
Cat whisker 129
Channel resistance, MOSFET 389
Chopper, d-c 529
Circuit models:
  for BJTs 245, 256, 263
  for diodes 108
  for JFETs 372
  for integrated components 442, 444, 445, 448
  for MOSFETs 389
Clamping circuit, diode 181
Clipper, diode 179
CMOS:
  flip-flop 415
  integrated circuits 474
  inverter 410
Coherent light 148
Collector avalanche breakdown 214
Collector-base breakdown voltage 202, 214
Collector-base capacitance 283, 288
Collector-base resistance 265
Collector capacitance 256
Collector current, OFF mode 334
Collector diffusion capacitance 264
Collector efficiency 221, 226
Collector-emitter breakdown voltage 214
Collector-emitter voltage, ON mode 336
Collector injection 227
Collector-junction saturation current 196
  dependence on temperature 289
Collector region 189
Collector triggering 352
Commercial transistor test sets 279
Common base:
  alpha 197
  alpha-cutoff frequency 260
  amplifier 209, 293
  collector characteristic 199
  emitter characteristic 204
  h parameters 258
Common collector:
  amplifier 293, 295
  h parameters 250
Common-drain amplifier 394
Common emitter:
  alpha 211
  alpha-cutoff frequency 261
  amplifier 272, 296, 308, 311
  base characteristic 218
  collector characteristic 210
  h parameters 246
Common-gate amplifier 397
Common-mode gain 326
Common-source amplifier 391
Commutating capacitor 352
Commutation of SCRs 520
Compensation, semiconductor 428
Complementary:
  flip-flop 415
  MOS ICs 474
  MOS inverter 410
  MOSFETs 383, 460
  NAND gate 410
  NOR gate 411
  SCR 533, 537
  UJT 497
Conduction band 40
Conductivity 15
Conductivity-modulation 108, 136
  rectifier 136
Conductivity-temperature characteristics 21
Conductors, electrical 42
Constant-source diffusion 427
Contact potential 51, 79, 112
Continuity equation 32
Controlled-avalanche rectifiers 77
Coupling-capacitor requirement, BJT stage 315
Crossover interconnect, MOS IC 472
Crystal diode 121
Crystal growing 65
Cube structure, UJT 495
Current-feedback bias 317
Current gain:
  BJT stages 297, 300
  GCS turn-off 527
Current stability factor 309
Current-transfer ratio 246
Cutoff region 332, 492

d-c chopper, GCS 529
DCTL flip-flop 352

Degenerate semiconductor 151
Degeneration 308, 313, 399
Delay time:
  BJT 338
  diode 142
Density of states, effective 55
Depletion mode, MOSFET 379
Depletion-type MOSFET 380, 383, 461
Diac 518, 530
Diagraph, Rhode and Schwarz 281
di/dt rating, SCR 513
Dielectric isolation, ICs 424, 435
Differential (difference) amplifier 325, 474
Diffused BJT 236
Diffused-epitaxial BJT 192
Diffused interconnection 472
Diffused junction 66
Diffused-junction BJT 189
Diffused resistors 440
Diffusion 27
  impurity 427
Diffusion BJT 233
Diffusion capacitance 103, 264
Diffusion constant 28
Diffusion current 12, 27
Diffusion equation 30, 85, 92
Diffusion length 29, 85
Diffusion voltage 79
Diode:
  a-c trigger 518
  backward 156
  breakdown 75, 127, 390
  Esaki 150
  five-layer 519
  four-layer 502, 518
  free-wheeling 521
  hot-carrier 121, 131
  light-emitting 29, 148
  npn 517
  pnpn 502, 518
  point-contact 121, 129, 143
  three-layer 517
  tunnel 149
  Zener 76, 177
Diode circuit-model 108
Diode-clamping circuit 181
Diode equation 83, 114, 124
Diode limiter 179
Diode regulator 177
Diode surge current 176
Dipole charge layer 96, 112
Direct-coupled amplifier 382, 475
Direct-coupled flip-flop 352
Direct-coupled transistor logic 352
Divergence 30
Donor impurities 8
Doping 8
Double-base diode 491
Double-diffused BJT 235
Double-doped BJT 189
Drain characteristics 364
Drain resistance 371, 389
Drift current 12
Drift-field BJT 233
Drift transistor 233
Drift velocity 13
Dual-gate MOSFET 387
dv/dt triggering 530
dv/dt turn-on 506, 520, 534

Early effect 203
Einstein relationship 28
Electric field, built-in 230, 233
Electric field, *pn*-junction 96
Electrical conductivity 15
Electrical neutrality 7
Electron-pair bond 5
Electron volt 5
Electronic voltmeter 181, 183
Electrons:
  bound 3
  conduction 8
  free 3
  valence 3, 40
Emitter-bypass capacitor 313
Emitter capacitance 256
Emitter characteristic 206
Emitter diffusion capacitance 258
Emitter efficiency 196, 221
Emitter follower 293, 295, 315
Emitter region:
  BJT 189
  UJT 491
Emitter triggering 352
Energy-band model 36
Energy-gap determination 56
Energy gaps 40
  temperature characteristic 56
Enhancement mode 379
Enhancement-type MOSFET 380
Epitaxial-diffused ICs 429
Epitaxial layer 132, 192
Epitaxial rectifier 136
Epitaxy process 428
Equivalent-circuit models (*see* Circuit models)
Equivalent-circuit parameters, BJT 276
Equivalent input circuit 273
Esaki diode 150
Excess donor concentration 19
Excitation energy levels 38
Extinction angle 173
Extrinsic semiconductor 4, 8, 44, 57
Extrinsic-temperature range 21

f_T 261
Fall time 338
Fan-out 409
Feedback amplifier 478
Fermi brim 51
Fermi distribution 47
Fermi factor 48
Fermi level 48
Fermi level location:
  conductor 51

Fermi level location (*cont.*):
intrinsic material 54
n-type material 57
p-type material 59
FET 188, 362
FET amplifier stages:
common-drain 394
common-gate 397
common-source 391
source-follower 394
FET-bias arrangements 398
FET switch 408
Field-effect transistor (FET) 188, 362
depletion 383
enhancement 380
insulated-gate 379
integrated 453
junction 363
MOS 379
normally-on 380
normally-off 380
photoFET 374
Film integrated circuits 421
Five-layer diode 519
Fixed bias, FET 398
Flip-flop:
BJT 348
DCTL 352
FET 412
pnpn-diode 509
Form factor 161
Forward alpha 237
Forward bias 69, 92
Forward-blocking state 502
Forward breakover 504
Forward-current-transfer ratio 197, 211
Forward electron current 80
Forward hole current 80
Forward-recovery time 137
Forward transadmittance 389
Four-layer diode 502, 518
Free-running multivibrator 356
Free-wheeling diode 521
Full-wave rectifier 162, 176

Gain:
current 297, 300
power 297, 304, 306
voltage 298, 393, 395, 397
Gain control, automatic 477
Galena crystals 129
Gallium-arsenide tunnel diode 155
Gate:
AND 348
integrated MOS 472
NAND 346, 420, 473
NOR 346, 411, 472
OR 346
Gate-controlled switch (GCS) 526
Gate electrode 364, 509
Gate turn-off switch 526
Gated biswitch 530
General Radio Transfer Function and Immittance Bridge 281
General Radio Vacuum-Tube Bridge 279
Germanium:
intrinsic carrier density 56
properties of 57
Graded-base BJT 233
Graded junction 65, 101
Grown-diffused BJT 189
Grown-junction BJT 189
Grown-junction process 65

h-parameter conversion formulas 254
h-parameter frequency dependence 255
h-parameter measurement 275
h parameters 243
common-base 248
common-collector 250
common-emitter 246
typical values 250, 300
Half-power-supply-voltage principle 324
Half-wave rectifier 159
Hall effect 22
Hall mobility 24
Hall voltage 23
Heavy holes 24
High-frequency *h* parameters 255
High-frequency parameter measurements (BJTs) 281
Holding current:
gate-controlled switch 527
pnpn diode 505
Hole 5
Hole motion 7
Hot-carrier diode 121, 131
storage-delay time 143
Hot-probe method 11
Hybrid integrated circuits 421
Hybrid-π model 263
Hybrid-π parameters 244
determination 265
typical values 269

IGFET 362
Ignition angle 173
Impedance parameters 241
Impulse turn-off 346
Impulse turn-on 346
Impurities in semiconductors 8
Impurity diffusion 427
Impurity profile 19
Impurity scattering 12
Incoherent light 148
Incremental models (*see* Circuit models)
Inductors, integrated 447
Injection, minority carrier 26
Input admittance:
common-gate amplifier 397
common-source amplifier 393
source follower 395
Input capacitance:
common-source amplifier 394
source follower 396

Input resistance, BJT stages 297, 302
Insulated-gate FET 379
Insulators, electrical 42
Integrated capacitors 424, 443
Integrated-circuit substrate 420
Integrated-circuit types 421
Integrated circuits 420
  fabrication 426, 435
  linear 474
  monolithic 422
  MOS 467
  types 421
Integrated FETs 453
Integrated inductors 447
Integrated MOS gates 472
Integrated MOS inverter 469
Integrated resistors 439
Interbase characteristic 494
Interconnections, diffused 472
Intrinsic carrier density 9, 10, 21, 56
Intrinsic germanium and silicon, properties of 57
Intrinsic semiconductors 4, 43, 52, 57
Intrinsic stand-off ratio, UJT 492
Intrinsic temperature 10
Inverse alpha 237
Inversion layer, MOSFET 380
Inverter:
  BJT 346
  FET 408
  integrated MOS 469
Ionization energy 38
Ionization of impurity atoms 8
Isolation diffusion 429
Iterative connection 306
Iterative gain 305

JFET 362
  circuit model 372
  compared with BJT 378
  monolithic 454
  temperature effects 374
Junction breakdown 71
Junction capacitance 94, 99, 103
Junction capacitor 424, 443
Junction diode 121
  conduction characteristics 122
  switching transients 137
  temperature effects 127
Junction field-effect transistor (JFET) 362
Junction-isolated ICs 423
Junction laser 148
Junction region 68
  capacitance 94
  electric field 96
  width 94
Junction resistance 106
Junction temperature, BJT 320
Junction transistor 188
Junction voltage 80, 98

Large-scale integration 435
Laser 148
Lattice scattering 12
Lifetime, minority carriers 29, 87
Light-activated thyristor 534
Light-emitting diode 29, 148
Light holes 24
Light-intensity meter 147
Light-sensitive diode 145
Light-sensitive FET 374
Limited-source diffusion 427
Limiter, diode 179
Linear integrated circuits 474
Linear stabilization 320
Linearly graded junction 101
Logic circuits 346
Lower half-power frequency 315

Magnetoresistance 25
Majority carriers 9
Maxwell-Boltzmann function 48
Medium-scale integration 435
Metal-semiconductor contacts 110
  forward bias 113, 116
  ohmic 117
  rectifying 110, 115
  reverse bias 114, 117
Metallic-oxide capacitors 446
Metallic rectifiers 121, 133
Metallization process 429
Miller effect 394, 404, 407
Minority carriers 9
  distribution 89, 195
  injection 26, 92
  lifetime 29
  suppression 9, 58
Monolithic BJT geometry 450
Monolithic capacitor 426
Monolithic diodes 424
Monolithic integrated circuits 421
Monolithic JFET 454
Monolithic linear ICs 475
Monolithic MOSFET 458
Monolithic resistors 424
Monolithic transistor 422, 450
Monostable multivibrator 348
  BJT 354
  FET 413
  pnpn-diode 508
MOSFET 362, 379
  breakdown-diode 390
  circuit model for 388
  dual-gate 387
  flip-flop 415
  monolithic 458, 461
  problems with 390
  silicon-on-sapphire 462
MOS integrated circuits 467
MOS inverter, integrated 469
MOS NAND gate 473
MOS NOR gate 472
MOS resistor 410, 415, 467
MOS transistor 362
Multichip integrated circuits 421
Multiphase monolithic ICs 425

Multivibrator:
astable 348, 356, 415, 507
bistable 348, 412, 509
monostable 348, 354, 413, 508

n-channel JFET 363
n-type material 9
n-type semiconductor contacts 110
NAND circuit 346
NAND gate 410, 473
Narrow-band amplifier 480
Negative-resistance region, UJT 492
Neutrality, electrical 7
Nonjunction transistor 462
Nonlinear stabilization 320
Nonregenerate switching circuits 342, 408
NOR circuit 346
NOR gate 411, 472
Notation, voltage and current 198
np product 59
npn diode 517

Off-mode operation, BJT 334
base current 336
collector current 334
Offset geometry, MOSFET 385
Ohmic contacts 117
ON-mode operation, BJT 336
One-shot multivibrator 355
Open-circuit parameters 242
Operating point selection, BJT stage 324
Operational amplifier 481
OR circuit 346
Output impedance:
BJT amplifier 299, 303
common-gate amplifier 397
common-source amplifier 394
source follower 396
Oxide masking 426
Oxide passivation 191

p-channel JFET 363
p-type material 9
p-type semiconductor contacts 115
Parameter determination 265
Parameter measurement 275
Parameters:
admittance 242
h 243
hybrid-π 244, 265, 269
impedance 241
open-circuit 242
short-circuit 243
y 242
z 241
Parasitic capacitance 442
Parasitic elements in integrated circuits 442
Parasitic transistor 442
Passivated structure 191
Passivation 191, 391, 424
Pauli exclusion principle 41
Peak-inverse voltage 164
Peak point:
tunnel diode 154
UJT 492
Phase-control circuits 519
Phase-controlled a-c power 520, 531, 532
Phase-controlled rectifier 519
Photo current 145
Photo-resist process 426
Photo-voltaic voltage 146
Photodiode 145
PhotoFET 374
Piecewise linear characteristic 159
Pin rectifier 136
Pinch-off region 366
Pinch-off voltage 366, 367
Planar structure 191
Planar UJT 497
pn-junction:
barrier potential 79
carrier-density gradient 85
current-voltage characteristic 72, 83
diffusion voltage 79
drift and diffusion currents 90
electric field 96
electron and hole currents 89
energy-level representation 77
forming 68
forward bias 82, 92
forward current 80
majority-carrier diffusion current 90
photodiode 145
potential barrier 79
reverse bias 80
reverse current 80
reverse-saturation current 83, 84
pnpn diode 502, 518
action, IC 423
applications 506
multivibrators 507
oscillator 507
pn product 59
pnpn trigger diode 518
Point-contact diode 121, 129, 143
Point-contact transistor 188
Polycrystalline material 4, 436
Polyphase rectifier 165
Potential barrier 68, 79
Power gain 297, 304, 306
Programmable UJT (PUT) 538
Pulse-generator circuit, UJT 500
Pulse-steering diodes 352
Punch-through 77
voltage 203

Q meter 283
Quantum states 41

Radiation-hardened ICs 438
Rate effect 506, 534
Rate-grown BJT 189
Rate-grown junctions 65
R-C coupled amplifier 478

Reach-through 77, 217
  voltage 203, 237
Recombination 6, 29
Recombination centers 29
Rectification efficiency 161
Rectifier:
  biphase 162
  bridge 164
  definition 71, 158
  full-wave 162
  half-wave 159
  ideal 159
  phase-controlled 519
  polyphase 165
  reverse-switching 535
  silicon-controlled 509
  three-phase 165
Regenerative switching circuits 342, 348, 412
Regulated power supply 177
Regulator diode 178
Relaxation oscillator:
  trigger circuits 516
  UJT 498
Resistance, thermal 322
Resistor:
  MOS 467
Resistor (*cont.*):
  monolithic 424
  nonlinear 467
Reverse bias 70
Reverse-blocking thyristor 510
Reverse breakdown:
  pn junction 71
  pnpn diode 505
Reverse electron current 80
Reverse hole current 80
Reverse-recovery time 137, 513
Reverse-saturation current 70, 83, 84, 87
Reverse-saturation electron current 105
Reverse-saturation hole current 106
Reverse-switching rectifier 535
Rhode and Schwarz Diagraph 281
Ripple factor 161
Rise time, BJT switch 338
RX Meter, Boonton Radio Corp. 281

Saturation current, reverse 70, 83, 84, 87
Saturation region:
  BJT 213
  UJT 492
Saturation voltage 213, 324
Sawtooth-wave generator 500, 507
Scattering of electric charges:
  impurity scattering 12
  lattice scattering 12
  thermal scattering 12
Schottky-barrier diode 131
SCR 509
  applications 514
  complementary 537
Self bias, FET 399
Self-biased BJT stage 317
Semiconductor 1
  n-type 9
  p-type 9
Semiconductor contacts 110
  (*see also* Metal-semiconductor contacts)
Semiconductor models:
  energy-band 2, 36
  valence-bond 2
Semiconductor-type determination:
  Hall-effect method 24
  hot-probe method 11
Sheet resistance 439
Shockley diode 502
Short-circuit parameters 243
Shorted-emitter structure 503
Shunt-capacitor filter 171
Silicon:
  intrinsic carrier density 56
  properties of 57
Silicon-controlled rectifier (SCR) 509
Silicon-controlled switch (SCS) 533
Silicon-dioxide capacitors 445
Silicon-dioxide masking 426
Silicon-nitride MOSFET 391
Silicon-nitride passivation 391
Silicon-on-sapphire MOSFET 462
Single-crystal material 4
Single-diffused BJT 236
Six-phase rectifier 168
Snubber circuit, SCR 520
Solar cell 145
Source bias 399
Source follower 394
  biasing 404
Source region 364
Space-charge region 68
  capacitance 99
Speed-up capacitor 345, 351
Stability factor, current 309
Stabistor 123
Static switch, SCR 515
Stimulated emission 149
Storage delay time 142
Storage time, BJT switch 338
Substrate:
  integrated-circuit 421
  JFET 374
  MOSFET 380, 389
Substrate bias, MOSFET 387
Substrate potential:
  integrated-circuit 424
  MOSFET 383, 387
Surface breakdown 76
Surge current, diode 176
Sustaining voltage 217, 324
Switching circuits:
  nonregenerative 342, 408
  regenerative 342, 348, 412
Switching modes 332

Temperature-compensated
    voltage-reference diode 127
Temperature dependence of alpha 232

Temperature-sensitive resistors 320
Tetrode JFET 373
Tetrode thyristor 533
Thermal resistance 322
Thermal runaway 321
Thermal scattering 12
Thermal stability, BJT 320
Thin-film capacitor 444
Thin-film resistor 424, 443
Thin-film transistor 421, 462
Three-layer diode 517
Three-phase rectifiers:
  bridge 169
  full-wave 169
  half-wave 165
  zig-zag 167
Threshold voltage, MOSFET 380
Threshold-voltage shift, MOSFET 390
Thyristor 490
  bidirectional 530
  light-activated 534
  tetrode 533
Time delay 338
  fall 338
  relay, UJT 501
  rise 338
  storage-delay 142, 338
Transadmittance, FET 371
Transconductance, FET 371
Transfer characteristic, JFET 368
Transfer-function bridge 281
Transformer-coupled stages, stability of 320
Transistor 188
  bipolar 188
  complementary 383
  field-effect 188, 362
  junction 188
  monolithic 450
  nonjunction 462
  point-contact 188
  thin-film 421, 462
  unijunction 490
  unipolar 188, 362
Transistor as a control device 188
Transistor-biasing networks 308
Transistor equivalent-circuit models 241
  h-parameter 243
  hybrid-π 263
  simplified hybrid-π 270
  y-parameter 242
  z-parameter 241
Transistor notation 198
Transistor physical model 194
Transistor switch 332
Transistor test sets 279
Transition capacitance 94
Transition frequency 261
Transition region 68
Triac 530
Trigger circuit:
  avalanche-diode 517
  unijunction-transistor 516
Triggering bistable circuits 350
Triggering SCRs 514
Triode region, JFET 366
Triple-diffused BJT 192
Tuned amplifier 480
Tunnel diode 149
Tunnel interconnect, MOS IC 472
Turn-off base current 340
Turn-off current gain, GCS 527
Turn-off mechanism, SCR 513
Turn-off methods, GCS 527
Turn-off time:
  BJT 342
  SCR 513
Turn-on base current 340
Turn-on mechanism, SCR 512
Turn-on time 342
Turn-over error, electronic voltmeter 183
Two-port network representations 241
Type determination, semiconductor 10

UJT 188, 490
Unijunction transistor (UJT) 188, 490
  applications 498
  programmable 536
  structures 494
  trigger circuits 516
Unilateral trigger diode 517
Unipolar transistor 362

Valley point:
  tunnel diode 154
  UJT 492
Valence band 40
Valence bond 5
  disruption 5
  models of semiconductors 2
Valence energy level 37
Vapor-diffusion process 427
Voltage and current notation 198
Voltage-controlled resistor 368
Voltage gain:
  BJT stage 298
  common-gate stage 397
  common-source stage 393
  source follower 395
Voltage-reference diode 76
Voltage regulator 177
Voltage-transfer ratio 246
Voltmeter, a-c electronic 181

Wayne Kerr Bridge 281

y parameters 242

z parameters 241
Zener breakdown 73
Zener diode 76, 177
Zener-diode temperature coefficient 127
Zig-zag half-wave rectifier 167